THE SOFT X-RAY COSMOS
ROSAT SCIENCE SYMPOSIUM

AIP CONFERENCE PROCEEDINGS 313

THE SOFT X-RAY COSMOS

ROSAT SCIENCE SYMPOSIUM
COLLEGE PARK, MD NOVEMBER 1993

EDITORS: **ERIC M. SCHLEGEL**
NASA/GODDARD SPACE
FLIGHT CENTER
AND
UNIVERSITIES SPACE
RESEARCH ASSOCIATION

ROBERT PETRE
NASA/GODDARD SPACE
FLIGHT CENTER

American Institute of Physics   New York

Authorization to photocopy items for internal or personal use, beyond the free copying permitted under the 1978 U.S. Copyright Law (see statement below), is granted by the American Institute of Physics for users registered with the Copyright Clearance Center (CCC) Transactional Reporting Service, provided that the base fee of $2.00 per copy is paid directly to CCC, 27 Congress St., Salem, MA 01970. For those organizations that have been granted a photocopy license by CCC, a separate system of payment has been arranged. The fee code for users of the Transactional Reporting Service is: 0094-243X/87 $2.00.

© 1994 American Institute of Physics.

Individual readers of this volume and nonprofit libraries, acting for them, are permitted to make fair use of the material in it, such as copying an article for use in teaching or research. Permission is granted to quote from this volume in scientific work with the customary acknowledgment of the source. To reprint a figure, table, or other excerpt requires the consent of one of the original authors and notification to AIP. Republication or systematic or multiple reproduction of any material in this volume is permitted only under license from AIP. Address inquiries to Series Editor, AIP Conference Proceedings, AIP Press, American Institute of Physics, 500 Sunnyside Boulevard, Woodbury, NY 11797-2999.

L.C. Catalog Card No. 94-72499
ISBN 1-56396-327-2
DOE CONF-9311209

Printed in the United States of America.

CONTENTS*

Preface xiii
List of Color Plates xv
Color Plates xvii

ORAL PRESENTATIONS:
THE SOLAR NEIGHBORHOOD

ROSAT Survey Observations of the Monogem Ring 3c
    P. P. Plucinsky, S. L. Snowden, B. Aschenbach, R. Egger, R. J. Edgar,
    and D. McCammon

Stellar Clusters and Star-Forming Regions (I) 7c
    J.-P. Caillault

Low Latitude X-ray Shadows and the Soft X-ray Diffuse Background (I) 16c
    D. N. Burrows and J. A. Mendenhall

ROSAT Results on Hot and Cool Stars (I) 24
    J. H. M. M. Schmitt

The Corona of the K5 Giant γ Dra, and its Relation to the Hybrid-
Chromosphere Stars 36
    A. Brown, J. L. Linsky, and T. R. Ayres

Soft X-ray Observations of Isolated Radio Pulsars (I) 41
    J. P. Finley

Four New Radio Pulsars from High-Energy Selected Targets 51
    A. F. Zepka, J. M. Cordes, S. C. Lundgren, and I. M. Wasserman

ORAL PRESENTATIONS:
ACCRETION-POWERED SOURCES

Solving the Mystery of the Periodicity in the Seyfert Galaxy NGC 6814 59
    G. Madejski, C. Done, T. J. Turner, R. F. Mushotzky, P. Serlemitsos,
    F. Fiore, M. Sikora, and M. Begelman

ROSAT Results on Magnetic Cataclysmic Variables (I) 64
    M. G. Watson

Discovery of a Candidate Old, Isolated Neutron Star in the Field of a Galactic
Cirrus Cloud 75
    J. T. Stocke, Q. D. Wang, E. S. Perlman, M. Donahue, and J. Schachter

Evidence against Secular Evolution in the Orbital Period of the LMXB
EXO0748-676 80
    P. Hertz, Y. Ly, K. S. Wood, and L. R. Cominsky

*NOTE: A "c" after a page number indicates a color plate exists for that article. An "I" after a title indicates an invited talk.

ROSAT Observations of LINERs (I) 85
    G. A. Reichert, R. F. Mushotzky, and A. V. Filippenko
The Complex Soft X-ray Spectrum of Low Redshift Quasars (I) 95
    F. Fiore
IR to X-ray Spectral Energy Distributions of High Redshift Quasars....... 105
    J. Bechtold, R. Cutri, M. Rieke, M. Elvis, F. Fiore, B. Wilkes,
    J. McDowell, and A. Siemignowska
Separation of X-ray Emission Components in Radio Galaxies............. 110
    D. M. Worrall and M. Birkinshaw
Extended X-ray Emission in Seyfert Galaxies........................... 115
    A. S. Wilson
The Soft X-ray Properties of Quasars 121
    A. Laor, F. Fiore, M. Elvis, B. J. Wilkes, and J. C. McDowell

ORAL PRESENTATIONS:
DIFFUSE THERMAL EMISSION

Hot Gas and Iron Abundances in Galaxies (I) 129
    C. Jones and W. Forman
ROSAT Observations of Starburst Galaxies 139
    T. M. Heckman
ROSAT HRI Observations of Magellanic Cloud Supernova Remnants (I).... 144
    J. P. Hughes
Diffuse X-ray Emission from H II Complexes: Stellar Winds and Supernova
Remnants.. 154
    Y.-H. Chu
The Carina Nebula in X-rays.. 159
    M. F. Corcoran, J. Swank, G. Rawley, R. Petre, J. Schmitt, and C. Day
Halos, Starbursts, and Superbubbles in Spirals (I) 164
    J. N. Bregman
The Soft X-ray Halo of the Spiral Galaxy NGC4631.................... 173
    R. A. M. Walterbos, M. F. Steakley, Q. D. Wang, C. A. Norman,
    and R. Braun
Cosmological Implications of ROSAT Observations of Groups and Clusters
of Galaxies.. 178
    L. P. David, C. Jones, and W. Forman
Clumped X-ray Emission around Radio Galaxies in Clusters: New Tools for
Investigating Cluster Evolution (I) 183
    J. Burns, K. Roettiger, J. Pinkney, C. Loken, S. Doe, F. Owen, W. Voges,
    and R. White

ORAL PRESENTATIONS:
SERENDIPITY

ROSAT Observations of Supernovae (I) 195
    E. M. Schlegel
Detection of X-rays from SN 1987A with ROSAT 205
    P. Gorenstein, J. P. Hughes, and W. H. Tucker
ROSAT Observations of an Optical Quasar Survey Field.................. 211
    H. Brunner
Can AGN Alone Make the Cosmic X-ray Background? 216
    D. Leiter and E. Boldt

WORKSHOP PRESENTATIONS

Analysis Procedures for ROSAT PSPC Observations of Extended Targets ... 223
    J. A. Mendenhall, D. N. Burrows, S. L. Snowden, and D. McCammon
QPTOOLS: Tools for Creating and Manipulating IRAF/QPOE Data 233
    M. A. Conroy
Temporal Data Screening in PROS..................................... 236
    J. DePonte and M. A. Conroy
Evaluation of Source Counts and Upper Limits in Crowded ROSAT PSPC
Fields 239
    V. Kashyap, G. Micela, S. Sciortino, F. R. Harnden, Jr., and R. Rosner
Simulated PSPC Spectral Fits of Coronal X-ray Sources 244
    A. Maggio, S. Sciortino, and F. R. Harnden
Preparation of ROSAT PSPC Data for Diffuse Spatial Analysis 249
    J. A. Mendenhall and D. N. Burrows
Correcting for Aspect Solution Errors in ROSAT HRI Observations of Compact Sources 252
    J. A. Morse
Image Deconvolutions of ROSAT HRI Observations: Choosing a Point Spread Function.. 258
    J. A. Morse
Detecting X-ray Sources with the Wavelet Transform 260
    P. Rosati, R. Burg, and R. Giacconi
An IDL-Based ROSAT Data Analysis Package 263
    F. M. Walter

POSTERS

Stars

ROSAT HRI Observations of T Tauri Star Pairs in Taurus-Auriga 269
  F. Damiani, G. Micela, S. Sciortino, and F. R. Harnden, Jr.
An X-ray Spectral Study of Algol Binary Systems 272
  S. A. Drake, N. E. White, A. P. Smale, L. Angelini, F. E. Marshall, and S. H. Pravdo
Orion Trapezium—A View into the Astrophysics of Young Stars with the ROSAT PSPC 275
  S. Geier and H. J. Wendker
An Observational Test to Determine the Presence of Intrinsic X-ray Absorption 279
  W. L. Waldron
Star Formation in Orion (the Constellation) 282
  F. M. Walter

Cataclysmic Variables

Discovery of a New Cataclysmic Variable System 287
  F. M. Walter and S. Zoonematkermani
ROSAT Observations of 7 Cataclysmic Variables 291
  P. Szkody
Bayesian Timing Analysis of Three Cataclysmic Binary Systems 294
  P. E. Freeman, T. H. Metcalf, and D. Q. Lamb

Diffuse Gas

ROSAT PSPC Observations of the Eridanus Soft X-ray Enhancement 299c
  Z. Guo, D. N. Burrows, and W. T. Sanders
Exploring the Interstellar Medium with X-ray Shadows 301
  Q. D. Wang

Supernova Remnants

X-ray Emission from the Vela SNR Shock Region: Spectral Fitting with a Non-Equilibrium Ionization Model 309
  F. Bocchino, A. Maggio, and S. Sciortino
A High-Resolution X-ray Study of the SN1006 Supernova Remnant 312
  P. F. Winkler and K. S. Long

3C400.2—A SNR with a Centrally Condensed X-ray Morphology 315
    J. M. Saken, K. S. Long, W. P. Blair, P. F. Winkler,
    and B. M. DeChristopher
ROSAT Observations of IC 443....................................... 318c
    J.-H. Rho, R. Petre, and J. J. Hester
The Diffuse X-ray Emission of CTB 109 .... 320c
    J.-H. Rho and R. Petre
An X-ray and Optical Study of a Cloud-Blast Wave Interaction 322
    N. A. Levenson, J. R. Graham, J. J. Hester, J. C. Raymond,
    and R. Petre
ROSAT Observations of SNRs as Distance Indicators 325
    N. E. Kassim, P. Hertz, S. D. Van Dyk, and K. W. Weiler
X-ray and Radio Emission from W49B and Other Supernova Remnants 328
    J. R. Dickel, R. Murphy, Y.-H. Chu, and D. L. Goscha

Normal Galaxies

Star Formation Triggered by Galaxy Interactions—NGC 1792/NGC 1808 ... 333
    M. Dahlem
A ROSAT Observation of the Spiral Galaxy NGC 6946 336
    E. M. Schlegel
A Deep X-ray Image of M33 339
    K. S. Long, S. M. Gordon, W. P. Blair, and P. A. Charles
X-ray Emission from Giant H II Regions in M101 342
    R. Murphy and Y.-H. Chu
ROSAT HRI Observations of M33 345
    E. Schulman and J. N. Bregman

Galaxies and Hot Gas

Observations of the Antenna-like Interacting Galaxy Pair Arp270: From
X-rays to Radio Wavelengths... 349
    P. N. Appleton, C. Winrich, G. Fabbiano, and P. M. Marcum
Dark Matter in the Elliptical Galaxy NGC 1407 352
    R. E. White III, V. Andersen, and C. Williamson
PSPC Observations of the NGC 3607 Group of Galaxies 355
    M. Loewenstein and R. Petre
Hot Entrained Gas in the Jet of NGC 4258 (M 106) 358c
    G. Cecil, C. De Pree, and A. S. Wilson

Clusters of Galaxies

An X-ray Mosaic of the Perseus Cluster 363
    M. P. Kowalski
ROSAT PSPC Observations of the Shapley Supercluster 366
    J. O. Breen and S. Raychaudhury
A ROSAT PSPC Observation of the Lensing Cluster A1689................ 369
    S. Daines, C. Jones, W. Forman, and A. Tyson
ROSAT Observation of Abell 1795: Temperature and Mass Profile 372
    H. Brunner, S. Weimer, H. Westphal, and R. Staubert
ROSAT Observations of Distant Clusters of Galaxies 375
    M. Donahue, J. T. Stocke, S. L. Morris, and K. A. Arnaud
Complex Spatial Structures in Sunyaev-Zel'dovich Decrement Clusters
Abell 665 and CL0016+16 378
    J. P. Hughes and M. Birkinshaw
Evolution in Nearby Clusters of Galaxies and the Butcher-Oemler Effect.... 380
    J. A. Rose, A. Leonardi, and N. Caldwell
ROSAT Observations of Bright X-ray Clusters: Abell 2142 and Abell 2199 .. 383
    H. Siddiqui, G. Stewart, J. Butcher, A. Edge, and R. Johnstone
X-ray Luminosities of Distant Radio-Selected Clusters of Galaxies.......... 386
    J. L. Sokoloski, R. A. Daly, and S. J. Lilly
High Resolution X-ray and Optical Imaging of the Central Galaxy in the
Centaurus Cluster 389
    W. B. Sparks, R. I. Jedrzejewski, and F. Macchetto
The Evolution of the Intracluster Medium in Clusters with Extended Radio
Sources.. 392
    L. Wan, R. A. Daly, L. V. Jones, and S. J. Lilly

AGN

HI and the X-ray Spectrum of NGC 6251 397
    M. Birkinshaw and D. M. Worrall
ROSAT Observations of the Blazars PKS 1034-293 and PKS 1335-127 400
    L. Maraschi, A. Ciapi, G. Fossati, G. Tagliaferri, and A. Treves
X-ray Variability of the Quasar 4C 39.25 and Bent Relativistic Jets......... 403
    Y. F. Zhang and A. P. Marscher
Excess X-ray Absorption toward Giga-Hertz Peaked Radio Sources......... 406
    Y. F. Zhang and A. P. Marscher
X-ray Emission in Powerful Radio Galaxies and Quasars................... 409
    D. M. Worrall, C. R. Lawrence, T. J. Pearson, and A. C. S. Readhead
Optical/UV/Soft X-ray Quasar Spectra: Models vs. Observations 412
    A. Siemiginowska, F. Fiore, M. Elvis, B. J. Wilkes, J. C. McDowell,
    and S. Mathur

PSPC Observations of the 1 Jy BL Lacs **415**
  R. M. Sambruna, C. M. Urry, J. Stocke, E. Perlman, R. Kollgaard,
  E. Feigelson, D. Worrall, P. Padovani, L. Maraschi, and A. Treves
X-rays from the Lobes of Fornax A **418c**
  S. A. Laurent-Muehleisen, E. D. Feigelson, R. I. Kollgaard,
  and E. B. Fomalont
X-rays and Relativistic Beaming in Radio-Selected BL Lacertae Objects **420**
  R. I. Kollgaard, E. D. Feigelson, D. C. Gabuzda, R. M. Sambruna,
  and C. M. Urry
Multifrequency Studies of Optically Quiet Quasars **423**
  R. I. Kollgaard, S. A. Laurent-Muehleisen, E. D. Feigelson, H. Spinrad,
  and W. Brinkmann
Active and Passive Galaxies in Deep ROSAT Surveys **426**
  R. E. Griffiths, R. Della Ceca, B. J. Boyle, I. Georgantopoulos,
  G. C. Stewart, and T. Shanks
The ROSAT PSPC Spectrum of PKS2155-304 .... **429**
  D. M. Gilmore and C. M. Urry
ROSAT Observations of 3C390.3 .... **432**
  M. Fremont, J. H. Beall, B. N. Dorland, and W. A. Snyder
List of Participants **435**
Author Index **445**
Source Index **451**
Subject Index **461**

Preface

In the nearly four years since its launch, ROSAT, the Röntgen Satellit, has produced a startling array of discoveries across both the spectrum of objects encompassed by X-ray astronomy and the cosmic distance scale. Some of its achievements have brought to fruition long-standing searches, such as for shadows of interstellar clouds in the soft X-ray background, for diffuse ISM emission in nearby galaxies, and for X-rays from the Moon (which was the original justification for the sounding rocket experiment which in 1962 discovered Sco X-1 and the diffuse cosmic X-ray background, and thereby marked the foundation of the field of extrasolar X-ray astronomy). Others have resolved more recent mysteries, such as the nature of the mysterious γ-ray source Geminga, the apparent lack of X-ray pulsations from the Vela Pulsar, and the true source of the apparent periodicity in the nearby Seyfert galaxy NGC 6814. Still others have opened up entirely new vistas, such as the detection of X-rays from several supernovae, the discovery of massive halos around some of the humblest groups of galaxies, and the revelation in clusters of galaxies of the abundance of substructure, point sources, and relations to other classes of objects, such as radio galaxies. The 50,000–60,000 sources found in the ROSAT All-Sky Survey and the as yet undetermined number of serendipitous sources contained in the $\sim$5000 pointed observations is a treasure trove which will no doubt yield riches for years and possibly decades to come.

The ROSAT Science Symposium and Data Analysis Workshop was held in College Park, Maryland, on November 8–10, 1993, in order to showcase the results obtained thus far using ROSAT, and to provide a forum for discussion regarding the means for extracting the maximum amount of information from ROSAT data. The meeting attracted approximately 200 scientists from the U.S. and Europe, and covered a broad range of topics, ranging in size from pulsars to clusters of galaxies, and in distance from the solar neighborhood to the highest measured redshifts. This volume is the proceedings of that symposium. Like the symposium, it is not meant to be a compendium of all the great results ever to derive from ROSAT observations. Rather it is meant to offer a snapshot of work in progress: namely, what the astronomical community had learned through ROSAT observations as of November 1993. It succeeds in capturing the breadth of new results arising from ROSAT data, and the degree of excitement associated with the results.

The conference has turned out to be particularly timely. In the month immediately following it, two major developments occurred which have transformed the nature of the ROSAT mission. First, on December 20, the workhorse Position Sensitive Proportional Counter (PSPC) was removed from the focal plane, its official life terminated by the impending exhaustion of its gas supply. While it will see occasional use to complete high priority observations until the gas is fully spent, the emphasis of the mission was switched after three and a half years to usage of the High Resolution Imager (HRI), which had been used sparingly prior to that time. Secondly, in late November 1993, ROSAT lost the second of its complement of its four gyroscopes, bringing to an end an interval of 20 months of nearly flawless operations, and resulting in several months of reduced operations. As of the writing of this preface full operations have been restored, but it is now unclear whether that will be the case in the event of the loss of another Attitude Measurement and Control System failure. We in the project remain optimistic about the prospects for several additional years of operations, as none of the remaining components have displayed any signs of degradation.

Another, more fundamental, change is also affecting the conduct of the mission, at least in the United States. The substantial reduction of NASA funding for mission operations and data analysis, and particularly for grants to guest observers, comes at a time when ROSAT is reaching its peak scientific productivity. For a brief time the entire U.S. ROSAT program was threatened with extinction. How this new crisis will affect the rate at which fundamental results arise from ROSAT observations is unknown, but it is hoped that NASA will develop some means whereby support to ROSAT investigators can be continued.

The 1993 ROSAT Science Symposium is not intended to be a unique event. In fact, the U.S. ROSAT project hosted a one-day data analysis workshop in Boston, Massachusetts in November 1992, and the Max Planck Institut für Extreterrestrische Physik, the institution responsible for the development of ROSAT and the scientific conduct of the mission, has held two brief workshops. It is our intention to host symposia annually, at least through the life of the mission, or to ensure through some other means (such as the High Energy Astrophysics Division of the American Astronomical Society) that a forum exists for a thorough discussion of ROSAT results.

The editors owe a great debt to the members of the Scientific Organizing Committee and the Local Organizing committee, without whom the meeting would not have been nearly as exciting or run as smoothly. Special thanks are reserved for Ms. Paula Webber of Universities Space Research Association (USRA), who coordinated all of the conference activities, and is the individual most responsible for its success.

Rob Petre
April 1994

List of Color Plates

The color plates are listed in order with the corresponding author(s), paper title, and page numbers.

Plate Number	Title and Author(s)	Page
1	ROSAT Survey Observations of the Monogem Ring P. P. Plucinsky et al.	3
2, 3	Stellar Clusters and Star-Forming Regions (I) J.-P. Caillault	7
4, 5	Low Latitude X-ray Shadows and the Soft X-ray Diffuse Background D. N. Burrows and J. A. Mendenhall	16
6	ROSAT PSPC Observations of the Eridanus Soft X-ray Enhancement Z. Guo et al.	299
7	ROSAT Observations of IC 443 J.-H. Rho et al.	318
8	The Diffuse X-ray Emission of CTB 109 J.-H. Rho and R. Petre	320
9	Hot Entrained Gas in the Jet of NGC 4258 (M 106) G. Cecil et al.	358
10	X-rays from the Lobes of Fornax A S. Laurent-Muehliesen et al.	418

Color Plates

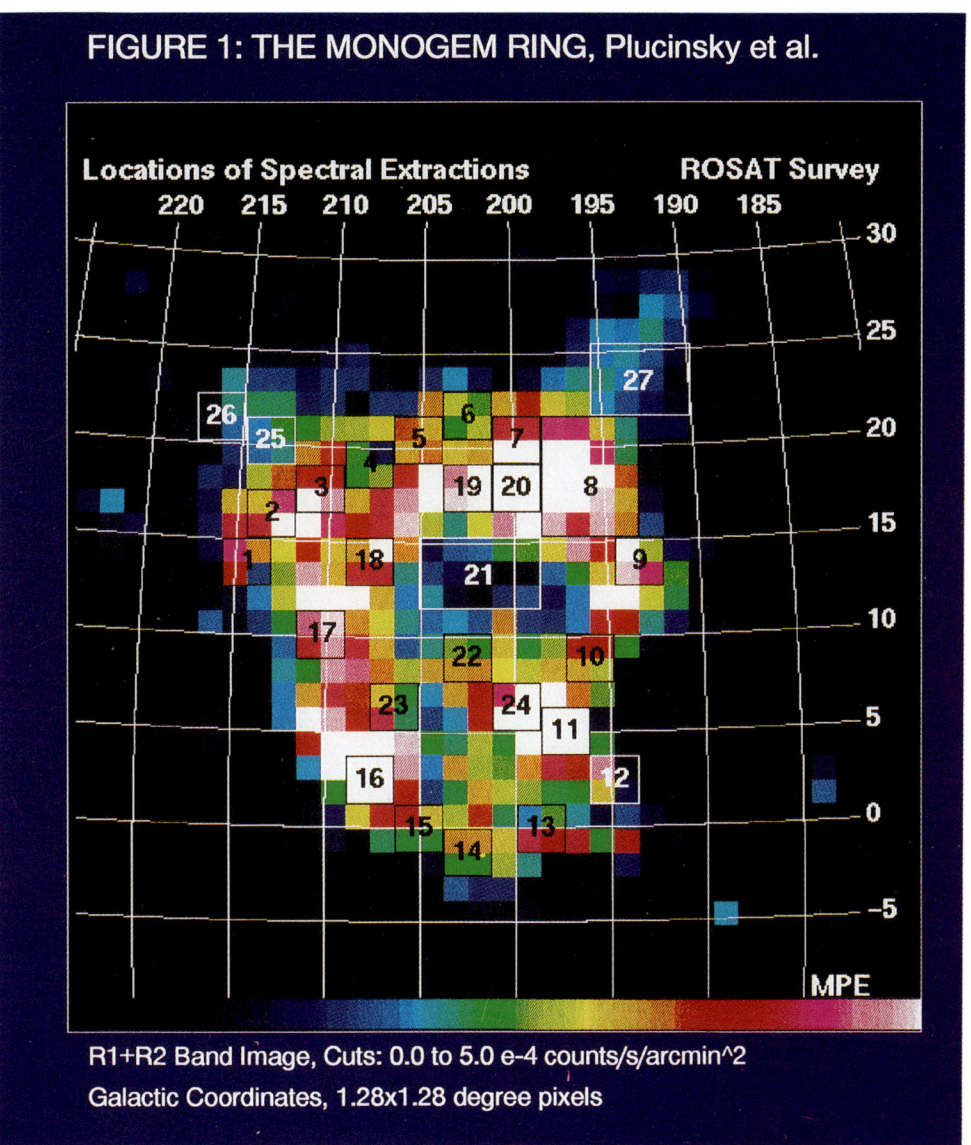

Plate 1.

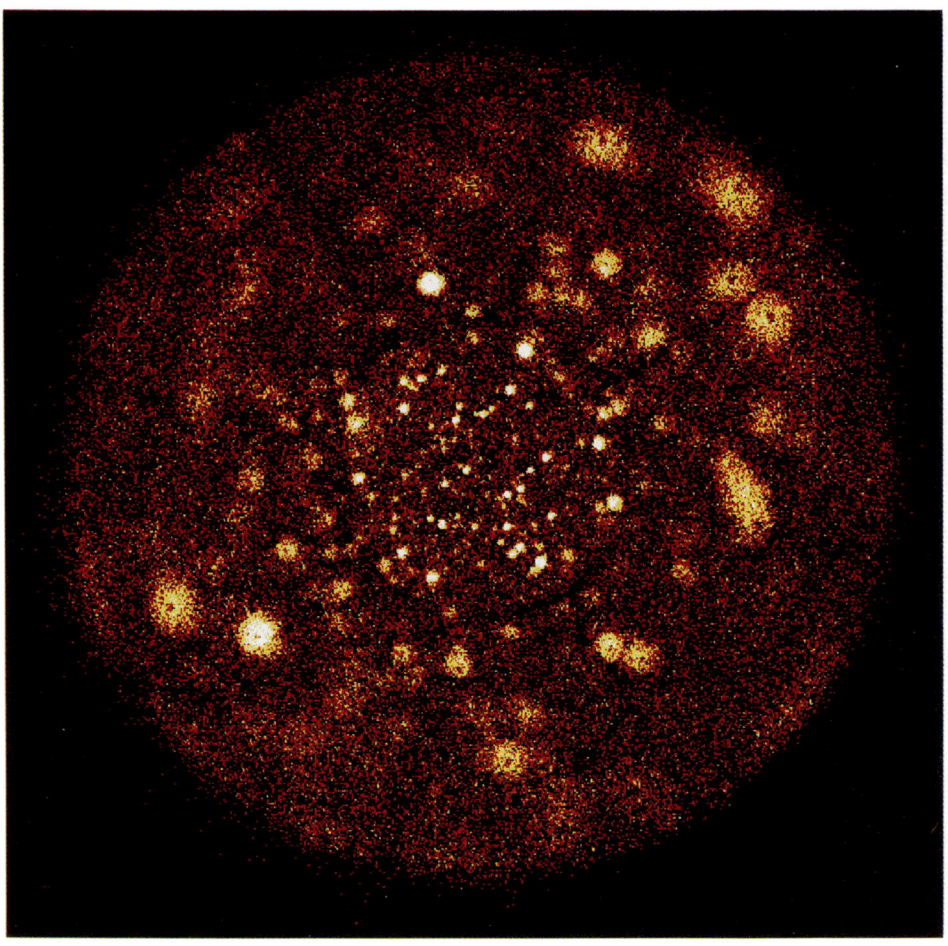

Figure 1: Hard band (0.5-2.0 keV), 2° diameter *ROSAT* PSPC image of the central portion of the Pleiades cluster; more than 140 stellar X-ray sources were detected in this ~ 27,000 sec exposure (Stauffer et al. 1994).

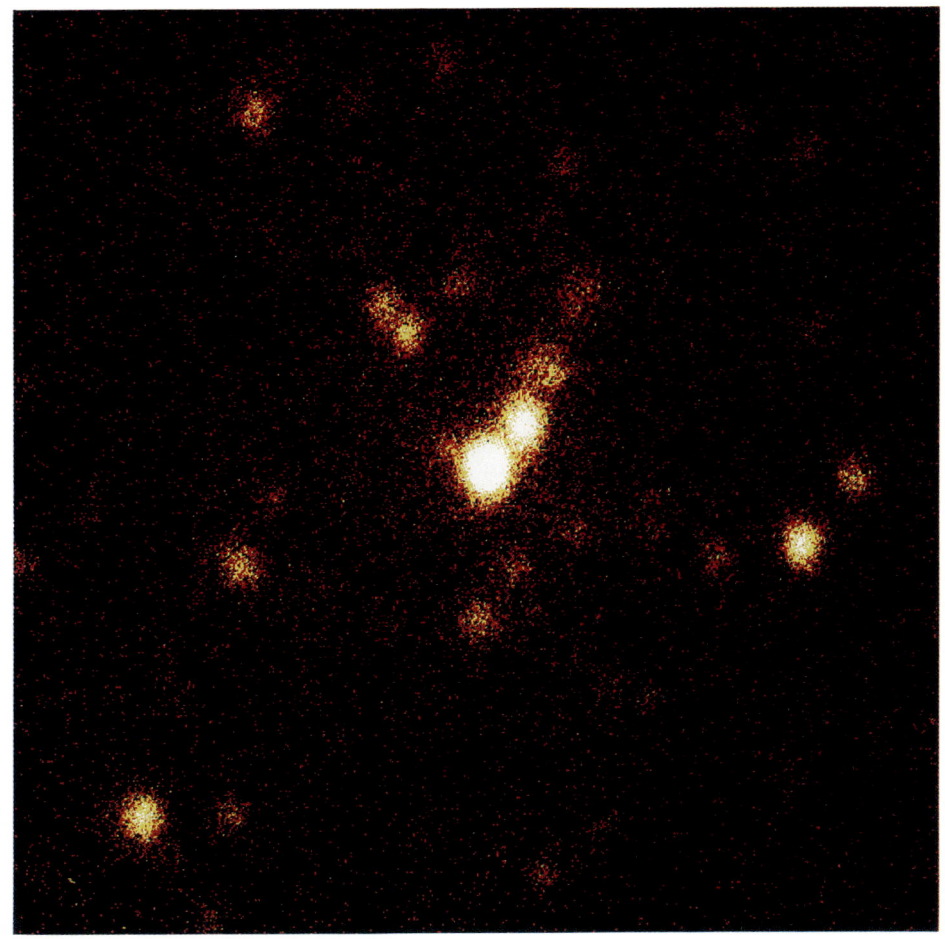

Figure 2: The inner 4.3′ × 4.3′ region of an ∼ 49,000 sec *ROSAT* HRI exposure of the Trapezium cluster. The central bright source is the main-sequence O7 star θ^1C Ori, the brightest component of the Trapezium; a total of ∼ 270 sources were detected in the full 40′ × 40′ image (Gagné et al. 1994a).

Plate 3.

a) $\frac{1}{4}$ keV images   Plate 4   b) $\frac{3}{4}$ keV images

(Burrows & Mendenhall: Low Latitude X-ray Shadows)

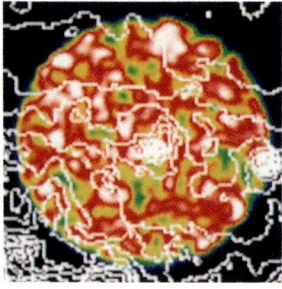
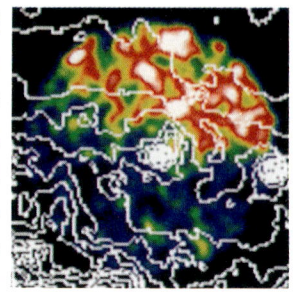

Fig. 1.— PSPC images of the Coalsack nebula. a) $\frac{1}{4}$ keV. b) $\frac{3}{4}$ keV. These images have been cleaned and smoothed as described in the text. The overlays are 100 μm contours from the IRAS Sky Survey Atlas.

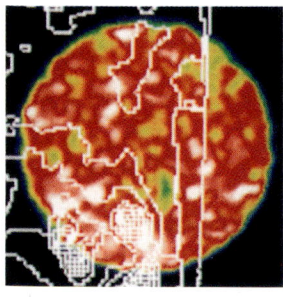
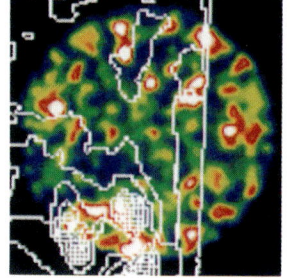

Fig. 2.— PSPC images of the Mon OB1 molecular cloud.

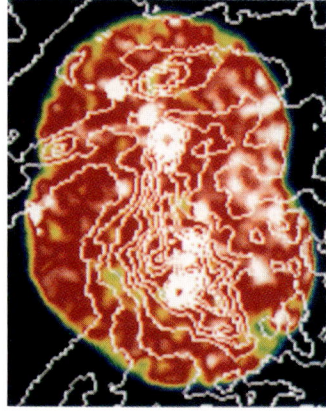
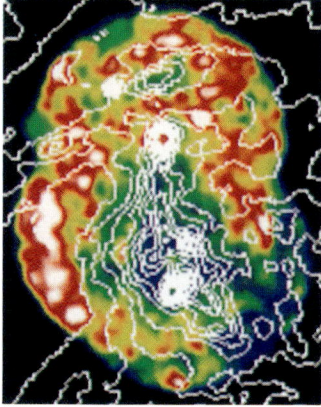

Fig. 3.— Images of the Chamæleon I molecular cloud.

a) $\frac{1}{4}$ keV images   Plate 5   b) $\frac{3}{4}$ keV images

(Burrows & Mendenhall: Low Latitude X-ray Shadows)

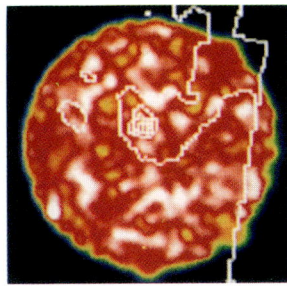 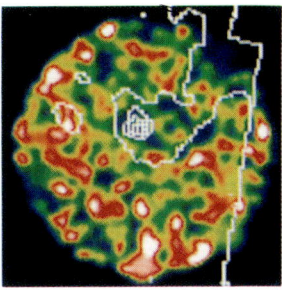

Fig. 4.— PSPC images of the L1551 cloud. a) $\frac{1}{4}$ keV. b) $\frac{3}{4}$ keV. These images have been cleaned and smoothed as described in the text. The overlays are 100μm contours from the IRAS Sky Survey Atlas.

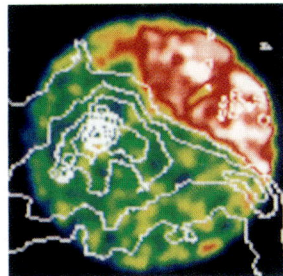 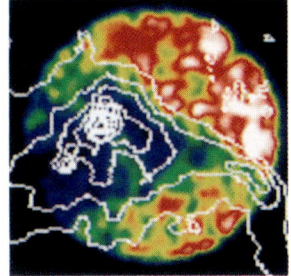

Fig. 5.— PSPC images of the R Cr A cloud.

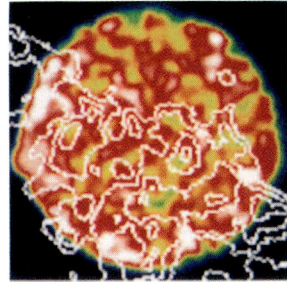 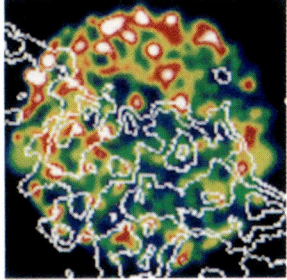

Fig. 6.— PSPC images of the Cepheus flare.

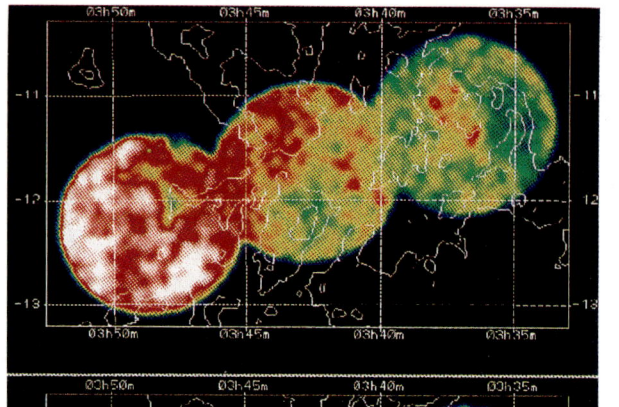
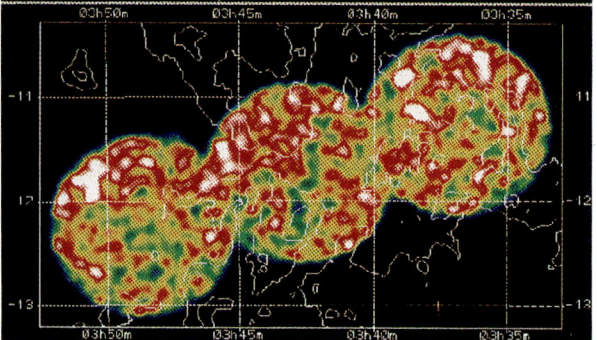

Plate 6a. Mosaic intensity maps of erid2 - erid4 (from right to left) with R1+R2 band on top and R4+R5 band at bottom. The overlaid contours are from ISSA coadded 100 micron map and start from 3.5 MJy/sr with interval 1.4 MJy/sr.
(Guo, Burrows and Sanders)

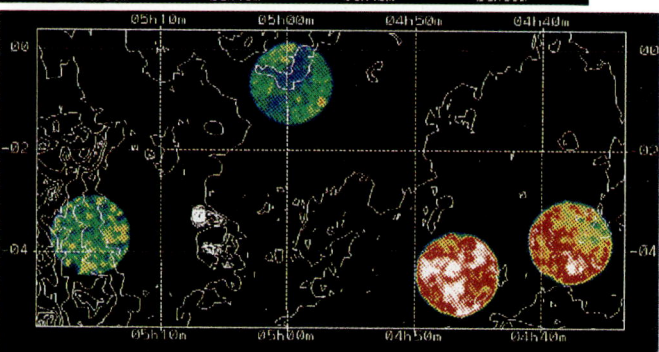
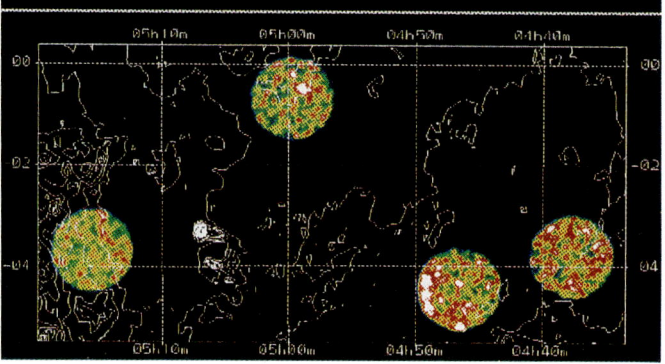

Plate 6b. Mosaic intensity maps of erid5 - erid8 (from right to left) with R1+R2 band on top and R4+R5 band at bottom. The overlaid contours are from ISSA coadded 100 micron map and start from 3.0 MJy/sr with interval 3.0 MJy/sr. Long term enhancement has been subtracted from the maps.
(Guo, Burrows and Sanders)

Plate 7.

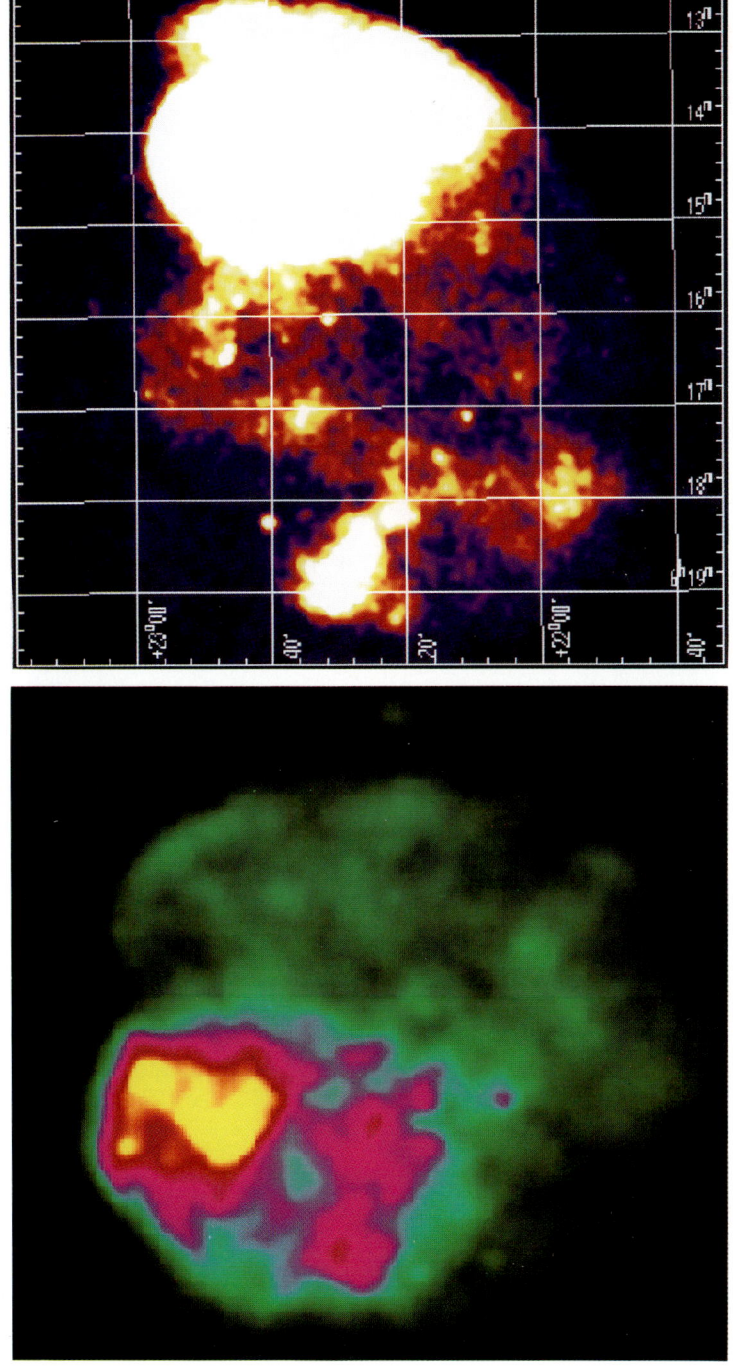

Fig. 1. The mosaiced PSPC image of IC 443

Fig. 2. The X-ray image of the IC 443 complex reveals IC 443 proper and a new quasi-shell structure to the southeast of IC 443

Figures from "ROSAT observations of IC 443" (Rho, Petre & Hester).

Fig. 1. The ROSAT PSPC image of CTB 109

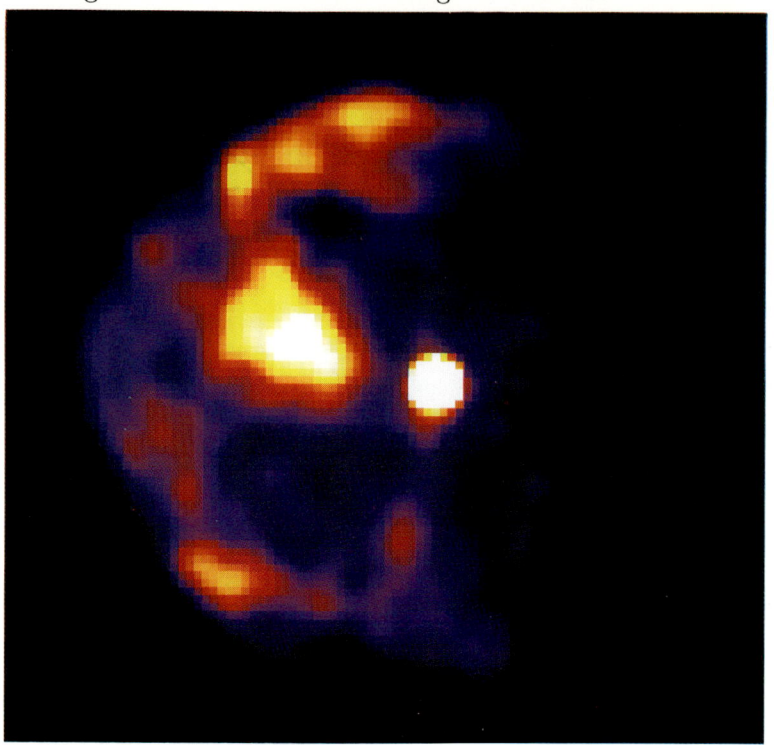

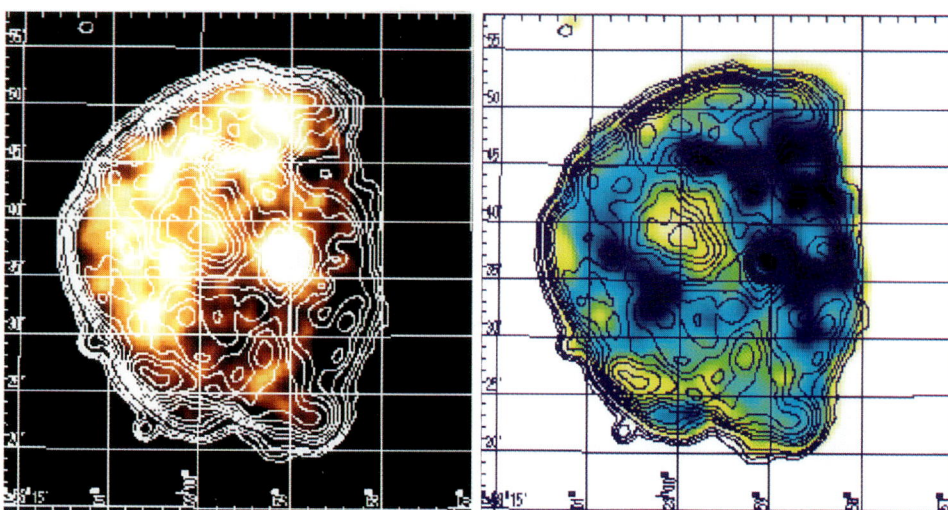

Fig. 2. The low E cut (0.7keV) hardness map and PSPC contours

Fig. 3. The high E cut (1.3keV) hardness map and PSPC contours

Figures from "The diffuse X-ray emission of CTB 109" (Rho and Petre).

Plate 8.

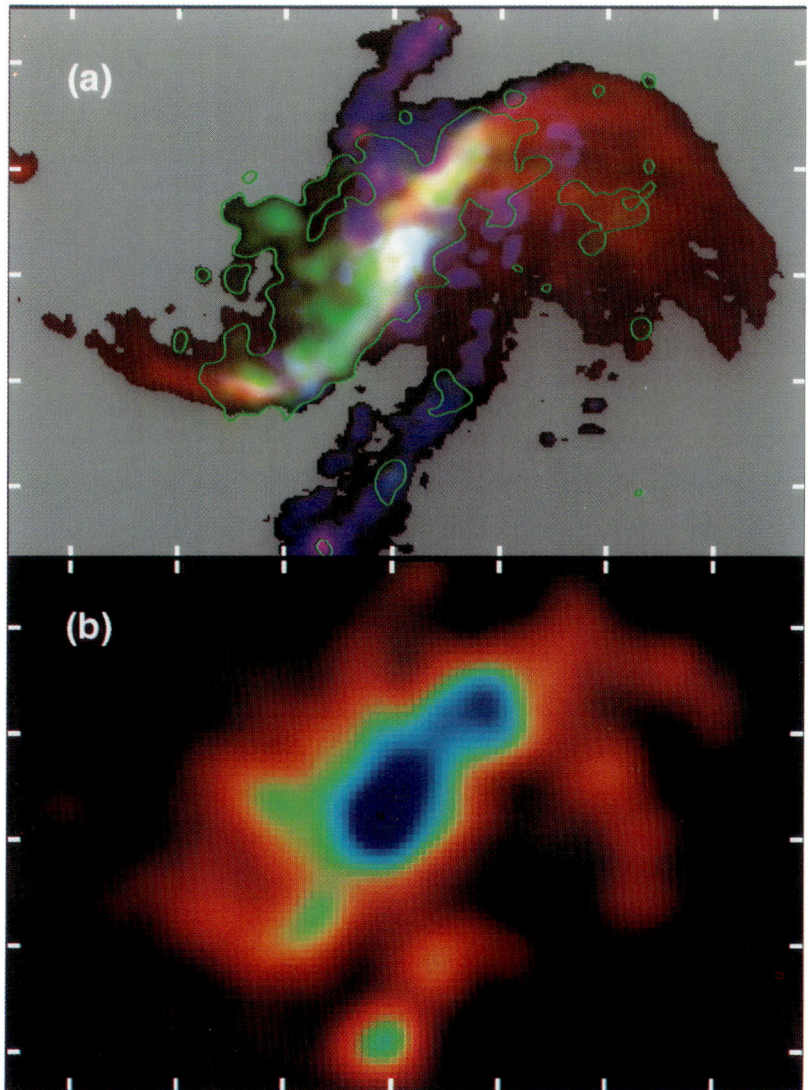

Color Plate 9: Cecil, DuPree and Wilson: Hot Entrained Gas

Fig. 1: (a) Multifrequency composite of NGC 4258, centered on the nucleus. Ticks at 2 kpc = 1' intervals. Red shows much of the jet in the λ6cm continuum, blue shows the Hα emission from the jet and normal spiral arms (top [N] and bottom) from fits to Fabry-Perot emission-line profiles (see CWT), and green shows the HRI x-ray map. The green contour shows the maximum extent of the brighter x-ray emission. (b) The PSPC colormap. Colors (red,green,blue) correspond to energy intervals (0.1:0.4,0.4:1,1:2.4) keV. The hardest photons are emitted at the nucleus and where the jet deflects to the NW. The NW jet branch has similar extent in the radio and x-ray bands, but the SE branch is less extensive in x-rays.

1.5 GHz image of Fornax A with X-ray contours of the 0.9-2.0 keV flux smoothed to 5 arcmin. X-ray contours are drawn at 65, 75, 85 and 95% of the peak flux. (See Laurent-Muehleisen et al.)

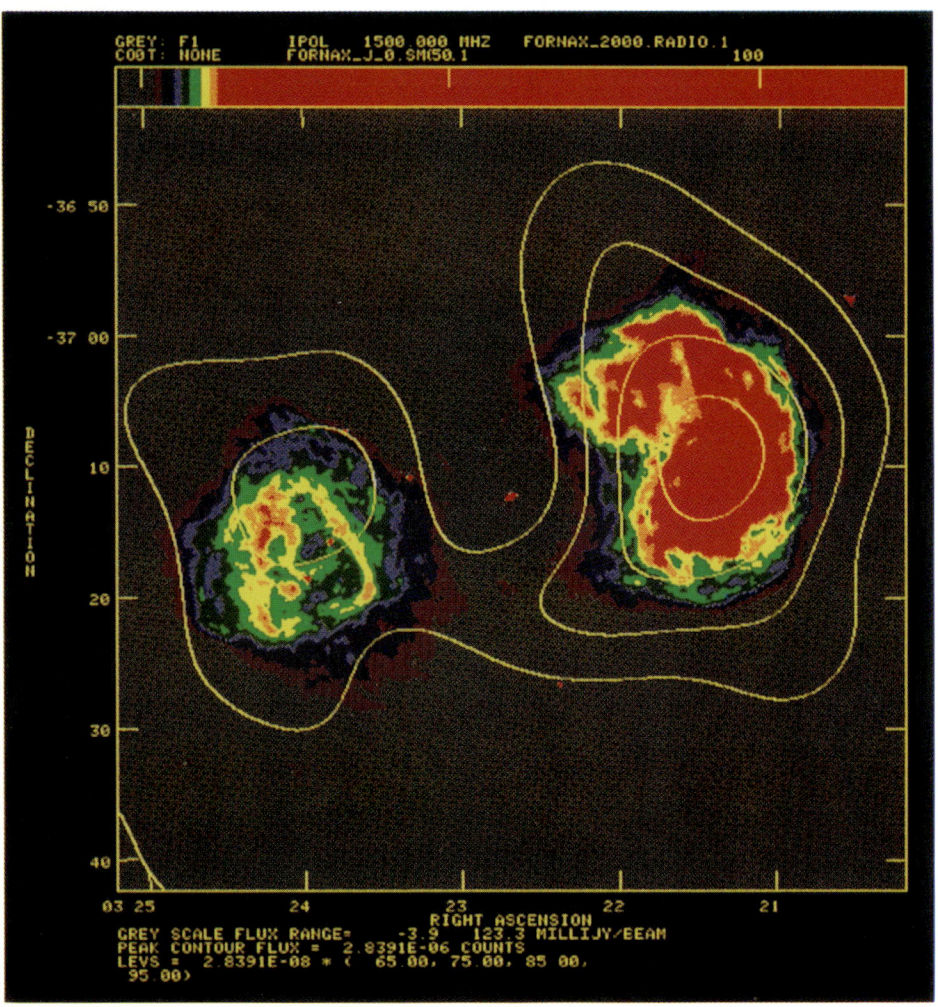

Plate 10.

ORAL PRESENTATIONS
THE SOLAR NEIGHBORHOOD

ROSAT SURVEY OBSERVATIONS OF THE MONOGEM RING

Paul P. Plucinsky
Harvard-Smithsonian Center for Astrophysics, 60 Garden St., Cambridge, MA 02138
Steven L. Snowden
NASA-Goddard Space Flight Center, Code 666, Greenbelt, MD 20771
Bernd Aschenbach, Roland Egger
Max Planck Institut für Extraterrestriche Physik, D85740 Garching, Germany
Richard J. Edgar, Dan McCammon
University of Wisconsin-Madison, 1150 University Avenue, Madison, WI 53706

ABSTRACT

Maps of the "Monogem Ring", also known as the Gemini-Monoceros X-ray enhancement, in the R1 (110–284 eV), R2 (140–284 eV), and R4+R5 (500–1100 eV) bands from the *ROSAT* All-Sky Survey have been generated. The Monogem Ring has a fragmented, shell-like structure in the R1 and R2 bands and extends across the galactic plane with little or no absorption; only part of the ring is visible in the R4+R5 band. There are intrinsic spectral variations within the ring, most notably from the low longitude side to the high longitude side, which would be consistent with a temperature increase from $\log(T/K) = 5.95$ to $\log(T/K) = 6.34$, assuming a Raymond & Smith (1977) equilibrium model with normal abundances and a constant neutral hydrogen column density (N_H) of 5.0×10^{19} cm^{-2}.

The Monogem Ring is modeled as a supernova remnant (SNR) in the adiabatic stage at various distances between 100 and 1300 pc. If the distance to the ring is 300 pc and the average temperature is $\log(T/K) = 6.15$, the initial ambient density is 5.2×10^{-3} cm^{-3}, the initial explosion energy is 1.9×10^{50} ergs, the luminosity in the 100–1000 eV band is 6.1×10^{34} ergs s^{-1}, and the age is $\sim 86,000$ yrs. The thermal pressure of the X-ray emitting gas is 4.5×10^4 K cm^{-3}, which is only 2–4 times higher than the pressure of the undisturbed interstellar medium (ISM).

1. INTRODUCTION

The Monogem Ring is a large region (diameter $\sim 25°$) of diffuse, soft X-ray emission centered at $l \sim 203°$ and $b \sim +12°$. This region has been known to be a source of enhanced $\frac{1}{4}$ keV emission since rocket flights in the 70's (see Bunner et al. (1973) and Long et al. (1977)). The first moderate spatial resolution (1.5°) data of this object were obtained with HEAO-1 in the 100–284 eV band. Nousek et al. (1981) determined that the object is truly ring-like in structure and interpreted it as a SNR.

The entire Monogem Ring region was observed during the *ROSAT* All-Sky Survey. A $39° \times 39°$ region was extracted in galactic coordinates, centered at $l = 202.5°, b = 11.5°$, requiring data from 51 days of the survey. There are a total of 1.9×10^6 counts in the R1, R2, and R4+R5 bands and the average exposure is ≈ 355 s. The non-cosmic background was modeled using the procedures described in Snowden et al. (1994) and the references therein. Point sources were identified and excluded from the analysis. The cosmic X-ray background was modeled empirically for each band by fitting a tilted and warped plane model to the regions in the field which are outside of the ring. For the R1 and R2 bands, $\sim 15\%$ of the counts were attributed to the non-cosmic background and $\sim 33\%$ of the counts were attributed to the cosmic background. For the R4+R5 band, $\sim 20\%$ of the counts were attributed to the non-cosmic background and $\sim 30\%$ to the cosmic background.

2. IMAGES AND SPECTRAL ANALYSIS

Figure 1 (color plate) displays the combined R1+R2 band image in galactic coordinates with $1.28° \times 1.28°$ pixels. The brightest parts of the ring have count rates greater than 7.5×10^{-4} counts s^{-1} arcmin^{-2} and the faintest parts have count rates of $\sim 2.0 \times 10^{-4}$ counts s^{-1} arcmin^{-2}. Typical errors for the bright and faint region count rates are 8% and 15% respectively. The ring-like structure is readily apparent; however, there are significant deviations from spherical symmetry, most notably the dark region centered at $l \sim 193°, b \sim 6°$, and the arc which abruptly ends at $l \sim 215°, b \sim 14°$. The R1 and R2 bands separately show the same ring-like structure, but with some subtle differences. The bright region located at $l \sim 197°, b \sim 17°$ is relatively brighter in the R1 band than in the R2 band and the bright region located at $l \sim 211°, b \sim 16°$ is relatively brighter in the R2 band than in the R1 band. The R4+R5 band image contains essentially no emission at $l \sim 197°, b \sim 17°$ but does contain a bright arc coincident with the region at $l \sim 211°, b \sim 16°$. The complete ring is not discernible in the R4+R5 band image. The absence or presence of emission in the R4+R5 band suggests differences in the physical conditions in the X-ray emitting gas such as temperature and density gradients.

In order to study these differences in more detail, spectra were extracted from 27 separate regions spanning the entire ring. These regions are indicated in Figure 1 as the boxes with their respective numbers and were required to have less than 25% errors in the count rates in the R1, R2 and R4+R5 bands. This requirement led to the large extraction areas for spectra #8, #21, and #27. The data were then fit with the Raymond & Smith (1977) spectral model for emission from an optically thin, hot plasma. Equilibrium conditions were assumed and the abundances of Allen (1976) were adopted. Table 1 lists the best-fit values of the temperature, N_H, and emission measure with the 2σ errors except where indicated. The best-fit temperatures span a range from $\log(T/K) = 5.95$ to $\log(T/K) = 6.34$. Region #8 has the lowest temperature which is consistent with the lack of R4+R5 band emission in that region, while regions #1–#6 have relatively high temperatures consistent with the bright arc of R4+R5 band emission. These data indicate a temperature gradient on a large scale in the ring but also on smaller scales of a degree or two. The temperatures for regions #11 & #12 and for #15 & #16 are different from each other at the 2σ confidence level. The best-fit N_H values span a range from 0 to 1.25×10^{20} cm^{-2}, consistent with N_H measurements made to various stars in the region. The most-likely value of the N_H for the 27 spectral fits combined is 5.0×10^{19} cm^{-2}. The emission measure is $\sim 1.0 \times 10^{-3}$ cm^{-6} pc for the faint regions and $\sim 1.0 \times 10^{-2}$ cm^{-6} pc for the bright regions.

3. INTERPRETATIONS

Three possible explanations were considered for the origin of this ring-like X-ray emission, an OB association bubble, an evaporating cloud shocked by the blast wave of the Local Bubble and a SNR. The stellar wind bubble explanation was rejected because of the lack of a candidate OB association at the right distance and position. The

Table 1: Spectral results derived from the Monogem Ring, T, N_H, and EM, and 2σ upper and lower limits.

region	log (T/K)	N_H (10^{20} cm^{-2})	EM (10^{-2} cm^{-6} pc)
1	$6.29^{6.37}_{6.22}$	$0.45^{1.50}_{0.0}$	$0.60^{0.98}_{0.42}$
2	$6.26^{6.32}_{6.18}$	$1.20^{2.93}_{0.44}$	$1.20^{2.41}_{0.83}$
3	$6.27^{6.33}_{6.22}$	$0.75^{1.54}_{0.14}$	$1.02^{1.43}_{0.73}$
4	$6.27^{6.38}_{6.20b}$	$1.20^{2.78b}_{0.0}$	$0.66^{1.22b}_{0.35}$
5	$6.28^{6.34}_{6.19}$	$0.15^{1.07}_{0.0}$	$0.46^{0.75}_{0.36}$
6	$6.32^{6.38}_{6.23}$	$0.10^{1.18}_{0.0}$	$0.39^{0.68}_{0.33}$
7	$6.17^{6.25}_{5.87}$	$0.00^{1.46}_{0.0}$	$0.42^{1.78}_{0.35}$
8	$5.95^{6.12}_{5.71a}$	$0.85^{2.45a}_{0.0}$	$0.82^{10.0a}_{0.37}$
9	$6.11^{6.21}_{5.92b}$	$0.00^{0.81b}_{0.0}$	$0.33^{0.75b}_{0.28}$
10	$6.28^{6.33}_{6.20}$	$0.00^{0.65}_{0.0}$	$0.41^{0.60}_{0.34}$
11	$6.17^{6.22}_{6.06}$	$0.25^{0.86}_{0.0}$	$0.88^{1.27}_{0.73}$
12	$6.34^{6.44}_{6.25}$	$0.60^{2.25}_{0.0}$	$0.57^{1.11}_{0.40}$
13	$6.14^{6.25}_{5.98b}$	$1.20^{3.79b}_{0.13}$	$0.67^{3.72b}_{0.34}$
14	$6.22^{6.30}_{6.17b}$	$0.95^{1.86b}_{0.01}$	$0.63^{0.96b}_{0.38}$
15	$6.32^{6.40}_{6.25}$	$0.40^{1.46}_{0.0}$	$0.56^{0.92}_{0.41}$
16	$6.14^{6.22}_{5.88}$	$0.35^{2.18}_{0.0}$	$0.77^{4.50}_{0.59}$
17	$6.25^{6.30}_{6.18}$	$0.40^{1.19}_{0.0}$	$0.70^{1.04}_{0.52}$
18	$6.16^{6.25}_{5.83}$	$0.60^{3.78}_{0.0}$	$0.61^{11.2}_{0.39}$
19	$6.19^{6.25}_{6.07}$	$0.35^{1.16}_{0.0}$	$0.67^{1.06}_{0.49}$
20	$6.14^{6.19}_{6.05b}$	$0.00^{0.36b}_{0.0}$	$0.61^{0.75b}_{0.53}$
21	$6.15^{6.27}_{5.93b}$	$0.40^{1.98b}_{0.0}$	$0.18^{0.69b}_{0.11}$
22	$6.27^{6.33}_{6.13}$	$0.00^{0.93}_{0.0}$	$0.34^{0.58}_{0.26}$
23	$6.29^{6.35}_{6.21}$	$0.15^{1.04}_{0.0}$	$0.42^{0.67}_{0.33}$
24	$6.21^{6.27}_{6.06}$	$0.00^{0.41}_{0.0}$	$0.49^{0.59}_{0.37}$
25	$6.23^{6.34}_{6.13b}$	$0.30^{1.10b}_{0.0}$	$0.29^{0.45b}_{0.16}$
26	$6.22^{6.39}_{5.74a}$	$1.25^{8.25a}_{0.0}$	$0.36^{10.0a}_{0.13}$
27	$6.06^{6.24}_{5.55a}$	$0.00^{2.85a}_{0.0}$	$0.12^{10.0a}_{0.42}$

^{a} limit determined from EM < 0.1 cm^{-6} pc
^{b} 1σ limit

evaporating cloud explanation was rejected because the temperature of the Monogem Ring is higher than the temperature of the Local Bubble and the temperature profile is flat as a function of radius. Therefore, the SNR hypothesis was adopted as the most likely, in agreement with Nousek et al.

The Monogem Ring was then modeled as a SNR in the adiabatic stage following the solution of Sedov (1959) and the formulation of Cox (1972). Clearly, the assumption of a uniform ambient medium is incorrect for this object; however, it is still useful to apply this simplified model to determine the parameter space which must be explored by more complicated models. The spectral fits for the entire region were redone with the N_H held fixed to 5.0×10^{19} cm^{-2} and an average temperature of log $(T/K) = 6.15$ was determined. It is possible to place an upper limit on the distance to the Monogem Ring of 1300 pc from N_H measurements. Therefore, models were calculated for distances between 100 and 1300 pc. Table 2 lists the values for the radius, initial ambient density, age, explosion energy, and thermal pressure. A reasonable solution for a single supernova explosion exists at distances between 300 and 600 pc; while distances greater than 600 pc require a multiple SN event. The ambient density and thermal pressure are low for all of the distances. The thermal pressures are roughly of the same order of magnitude as values which have been suggested for the pressure of the undisturbed medium. This raises the possibility that the pressure of the X-ray emitting gas will reach equilibrium with the surrounding medium before the remnant evolves into the radiative stage. For example, the solutions at 300 and 600 pc will have thermal pressures below 5000 K cm^{-3} when the remnant has expanded to the radius at which the radiative stage is expected to begin. Therefore, SNRs in relatively low-density media may produce long-lived bubbles of million degree gas.

Table 2: Radius, density, age, explosion energy, and thermal pressure as a function of distance for the Sedov model.

D (pc)	R (pc)	n_o (cm^{-3})	Age (yr)	ϵ_o (ergs)	P_{th}/k (K cm^{-3})
100	22.2	9.0×10^{-3}	2.9×10^4	0.01×10^{51}	7.9×10^4
300	66.5	5.2×10^{-3}	8.6×10^4	0.19×10^{51}	4.5×10^4
600	133.0	3.7×10^{-3}	17.1×10^4	1.10×10^{51}	3.2×10^4
1000	221.7	2.9×10^{-3}	28.5×10^4	3.93×10^{51}	2.5×10^4
1300	288.2	2.5×10^{-3}	37.1×10^4	7.58×10^{51}	2.2×10^4

4. REFERENCES

Allen, C.W. 1973, *Astrophysical Quantities* (3d ed; London:Athlone)
Bunner, A.N., Coleman, P.L., Kraushaar, W.L., McCammon, D. & Williamson, F.O. 1973, ApJ, **179**, 781
Cox, D. 1972, ApJ, **178**, 143
Long, K.S., Patterson, J.R., Moore, W.E., & Garmire, G.P. 1977, ApJ, **212**, 427
Nousek, J.A., Cowie, L.L., Hu, E., Linblad, C.J., & Garmire, G.P. 1981, ApJ, **248**, 152
Raymond, J.C., & Smith, B.W., 1977 ApJS, **35**, 419
Sedov, L.I., 1959 *Similiarity Solutions and Dimensional Methods in Mechanics*, (New York: Academic Press)
Snowden, S.L., McCammon, D., Burrows, D.N., & Mendenhall, J.A. 1994, ApJ, in press

STELLAR CLUSTERS AND STAR-FORMING REGIONS

Jean-Pierre Caillault
University of Georgia, Athens, GA 30602

Email ID
jpc@jove.physast.uga.edu

ABSTRACT

Stellar clusters and star-forming regions provide excellent laboratories for stellar astronomers since they consist of coeval populations of stars of varied mass but fixed distance, chemical composition and age. By comparing clusters of different ages, the evolution of a variety of stellar properties that can serve as diagnostics of the unseen physics controlling the interior evolution of stars can be determined.

ROSAT PSPC and HRI observations of more than a dozen star-forming regions and stellar clusters, ranging in age from $\sim 10^6$ yrs to $\sim 700 \times 10^6$ yrs, have provided an enormous amount of data with which questions left unanswered by *Einstein* observations can be addressed. The results from these *ROSAT* observations, as they pertain to the issues of stellar populations in star-forming regions, luminosity functions in stellar clusters, relations between coronal activity and stellar rotation, and temporal and spectral analyses, are presented.

1. INTRODUCTION

Clusters and star-forming regions are particularly interesting and useful targets for study with the instruments on *ROSAT*. Their constituent stars have essentially the same metallicity, are at the same distance, and are probably coeval, yet provide a richness of varied stellar masses. In addition, the clusters studied are very young (compared to field stars), hence likely to be very active; relatively nearby (most are within ~ 500 pc), thus possessing low absorption; and generally compact, allowing for the observation of many stars with only a few PSPC or HRI images. Also, the large spread in ages of the clusters (ranging from ~ 1 to $\sim 700 \times 10^6$ yr) allows for the study of the evolution of coronal properties, which should give clues about internal structure since X-ray activity in low-mass stars is believed to be primarily a function of a star's mass (i.e., convection zone depth) and rotation.

In Figures 1 and 2 (Plates 2 & 3) are shown two beautiful color examples of PSPC and HRI images. The former is a hard band (0.5-2.0 keV), 2° diameter *ROSAT* PSPC image of the central portion of the Pleiades cluster; more than 140 stellar X-ray sources were detected in this $\sim 27,000$ sec exposure (Stauffer et al. 1994). The inner $4.3' \times 4.3'$ region of an $\sim 49,000$ sec *ROSAT* HRI exposure of the Trapezium cluster is shown in Figure 2. The central bright source is the main-sequence O7 star θ^1C Ori, the brightest component of the Trapezium; a total of ~ 270 sources were detected in the full $40' \times 40'$ image (Gagné et al. 1994a).

The list of all of the *ROSAT* observations of star-forming regions or stellar clusters that have already been carried out or are scheduled is provided in Table 1. The Table includes the names, ages, and distances of all of the clusters, as well as the number of observations by the various *ROSAT* instruments and their corresponding total exposure times and, finally, the principal investigators. The clusters are listed in order of increasing age.

Table 1: Star-Forming Regions and Stellar Clusters Observed with ROSAT

Cluster	Age (10^6 yrs)	Distance (pc)	Observations[a] (#/ksecs)	Investigators
Trapezium	~1	440	H 1/50	Gagné et al.
			P 1/10	Geier et al.
Orion (other)	~1	440	P 17/140	Walter et al.
ρ Oph	~1	160	P 1/33	Casanova et al.
L1495E (Tau)	~0.5-10	160	P 1/26	Strom & Strom
Tau-Aur	~0.5-10	140	S	Neuhäuser et al.
			P 6/60	Kenyon et al.
Chamaeleon I	~10	140[b]	P 2/12	Feigelson et al.
NGC 2264	~20	750	H 2/30	Patten et al.
IC 2602	~30	160	R	Schmitt et al.
IC 2391	~30-50	160	P 1/23	Patten & Simon
α Per	~50	150	R	Schmitt et al.
			P 2/50	Stauffer et al.
Pleiades	~70	125	P 3/74	Stauffer et al.
			P 1/40	Micela et al.
			S	Schmitt et al.
			H 6/180	Harnden et al.
NGC 6475	~220	250	P 1/50	Prosser et al.
			P 1/29	Jeffries et al.
Hyades	~700	45	S	Schmitt & Stern
			P 11/200	Pye et al.
			P 1/40	Stern et al.
			P 3/50	Reid & Hawley
Praesepe	~700	150	R	Randich & Schmitt

[a] - H=HRI, P=PSPC, S=Survey, R=Raster scan
[b] - 215 pc according to Gauvin & Strom (1992)

Although the observations performed by the *Einstein* observatory essentially introduced an entire new discipline – stellar X-ray astronomy – many questions regarding stellar coronæ were left unanswered (see, e.g., the review by Rosner, Golub, & Vaiana 1985). As is clear throughout these proceedings, the *ROSAT* instruments provide improved sensitivity and spectral and spatial resolution over those aboard *Einstein*. As a result, *ROSAT* has already provided a rich stellar X-ray database from which answers to the questions left open by *Einstein* can be extracted (even though only a fraction of the observations listed in the Table have been fully analyzed). Hence, in the pages that follow, some of the more important (and controversial) issues concerning the coronal activity of stars in stellar clusters and star-forming regions are addressed.

2. LOW-MASS STELLAR POPULATIONS IN STAR-FORMING REGIONS

One of the key results stemming from the *Einstein* X-ray surveys of star-forming regions was the discovery that a new class of pre-main-sequence stars (the "naked" or weak T Tauri stars) could be readily selected via their strong X-ray emission (Walter 1986). As a result, an effectively new sub-discipline has burgeoned in the last 5-10 years; however, there are some outstanding questions concerning the X-ray data which remain from the *Einstein* analyses:

- What are the relative numbers of classical T Tauri stars (CTTS; $W_\lambda(H\alpha) > 10$ Å) and weak T Tauri stars (WTTS; $W_\lambda(H\alpha) < 10$ Å)?

- Are these types of stars co-spatial?

- How do their activity levels, variability, and spectral properties compare with those of main-sequence stars?

The *ROSAT* All-Sky Survey has been able to contribute greatly to the answers to these questions because, as a result of its greater areal coverage of the sky, it has detected so many more stars than *Einstein*. Neuhäuser et al. (1994) report that large fractions of previously known pre-main-sequence stars have been detected in the region of Taurus-Auriga (12% of CTTS and 54% of WTTS), as well as hundreds of new sources which they expect will ultimately be classified as WTTS. The pointed observations, too, have provided substantial improvements, since their greater sensitivity has allowed for the detection of many fainter X-ray sources in regions previously well-studied by *Einstein*. Gagné et al. (1994a), for example, have discovered hundreds of new PMS stars in the Orion Nebula star-forming region. Feigelson et al. (1993) have detected $\sim$ 20-40 new faint sources (probably WTTS) in Chamaeleon I; Strom & Strom's (1994) observations of L1495E have yielded $\sim$ 10 new faint PMS stars.

One interesting new aspect of this work is that there may be strong evidence that the WTTS are brighter X-ray sources than the CTTS. This can be seen from Figure 6b of Feigelson et al. (1993) and Figure 12 of Strom & Strom (1994). Neuhäuser et al. (1994) claim that the mean L_x of WTTS is $\sim 10^{30.3}$ ergs s^{-1}, while that of CTTS is only $\sim 10^{30.0}$ ergs s^{-1}. They also find that the mean L_x/L_{bol} ratios are greater for the WTTS: $10^{-3.36}$ vs. $10^{-4.05}$. Follow-up optical spectroscopy is needed to complete any census, though, since many of the new faint X-ray sources have not yet been spectrally classified.

Another interesting result of the Neuhäuser et al. (1994) survey is that the CTTS appear to have harder X-ray spectra than the WTTS. This result is based on hardness ratios, not actual spectral fits, but the evidence seems clear. The explanation is thought to be that the CTTS have a substantially greater amount of CS material surrounding them than the WTTS, hence, they display more extinction of soft X-rays.

As far as the relative numbers are concerned, Walter et al. (1988), based on *Einstein* data, have predicted that the ratio of WTTS to CTTS may be as high as 10:1. From their *ROSAT* survey of Chamaeleon I, Feigelson et al. (1993) determine that the ratio is $\sim$ 3:1. Neuhäuser et al.'s (1994) results, though, indicate variable ratios, depending on location within the star-forming region. For example, they find a ratio of $\sim$ 9:1 overall in Taurus, but only $\sim$ 4:1 in the central region and $\sim$ 1:1 in the darkest areas. This, of course, implies that the spatial distributions of the two classes of stars are different. Feigelson et al. (1993), though, assert that there is no evidence for spatial segregation in Chamaeleon I; instead they claim that there may be an age segregation: the younger stars, irrespective of whether they are CTTS or WTTS, seem to cluster closest to the darkest regions of Chamaeleon I, while the older stars are more dispersed. However, caution should be used before embracing this viewpoint: the age (and mass) determinations of these stars are very strongly dependent on the choice of pre-main-sequence evolutionary tracks, as can be clearly seen from Figure 9 of Strom & Strom (1994).

When the PMS stars are compared to MS stars, it is evident that the range of L_x values overlap: for example, the Chamaeleon I X-ray luminosity function extends over at least two orders of magnitude, from $10^{28.5}$ ergs s^{-1} to a peak of $\sim 10^{30.7}$ ergs s^{-1}, very similar to that of the stars in the ~ 70 Myr old Pleiades cluster (see Figure 14 of Stauffer et al. 1994). Also, Patten & Simon (1993) have compared the results of their *ROSAT* observation of the ~ 50 Myr old IC2391 cluster with those of Stern et al.'s (1992, 1994) observations of the ~ 700 Myr old Hyades cluster and find that the two clusters' L_x distributions overlap. Patten & Simon interpret this as indicative of clusters possessing stars with a wide range of coronal activity levels, independent of age. This problem is discussed further below.

3. MAIN-SEQUENCE LUMINOSITY FUNCTIONS

In principle, the comparison of the X-ray luminosity functions of these different aged clusters should provide enough diversity that reasonable conclusions might be drawn regarding the evolution of coronal activity. Most attempts at constructing stellar X-ray luminosity functions during and after the *Einstein* era necessarily invoked the use of survival analysis statistics (Isobe, Feigelson, & Nelson 1986) which allowed for the inclusion of the many X-ray upper limits obtained with the less sensitive IPC. An example of this application is shown in Figure 8 of Micela et al. (1988), where the stellar activity clearly seems to diminish from the Pleiades to the Hyades to the field. Hence, although the range of L_x values for different clusters might be the same (see previous section and below), the distributions for each cluster or star-forming region might be quite different.

Some of the pertinent questions which *ROSAT* can address are:

• Does the improved sensitivity remove the need for survival analysis?

• Can new cluster members be discovered via X-rays?

• Are late B, early A stars true sources of X-ray emission?

With the advent of the more sensitive *ROSAT* instruments, there is now no need, in many cases, to invoke survival analysis techniques to construct stellar X-ray luminosity functions. The Pleiades serves as an excellent example: Figures 10 & 11 in Stauffer et al. (1994) show that almost all (90%) of the stars in the cluster that were observed have been detected; the X-ray luminosity functions are shown in their Figure 13. In fact, there are probably many new cluster members being detected in the Pleiades and other clusters: there are numerous weak sources with uncatalogued optical counterparts which have been discovered. The optical work (spectroscopy, photometry, proper motions) must catch up to that of the X-ray – somewhat similar, although probably on a much smaller scale, to the situation concerning the newly discovered PMS stars described in the previous section.

An example of an unexpected result is the comparison by Randich & Schmitt (1994) of the L_x values of the stars in the Hyades to those in the similarly aged ($\sim$ 700 Myr) Praesepe cluster. They find that the Praesepe stars are much weaker X-ray sources than those in the Hyades. No obvious explanation springs to mind, although it is conceivable that one of the cluster's ages could be wrong or that one (or both) of the cluster's mean activity level is "unusual". Ultimately, comparisons of this type are essential for an understanding of the evolution of coronal activity.

One interesting side issue that seems to be clarified by the luminosity functions is that of the emission from late B, early A stars. Based on theoretical considerations, these stars are not expected to be X-ray emitters, since they lack the strong stellar winds necessary to provide the radiation-driven shock heating usually used to explain the emission from O and early B stars (Lucy & White 1980) and they lack the outer convection zones required for the magnetic dynamo models to work (Rosner et al. 1985). However, there exist many reports of detections of these stars (Walter et al. 1988; Caillault & Zoonematkermani 1989; Strom et al. 1990; Schmitt et al. 1993). Now, though, it seems as if *ROSAT*'s sensitivity and spatial resolution, which has reduced the possibility of candidate confusion, have led to a fairly clear demonstration, using the HRI observations of Orion, that the emission detected from these stars can be explained by the presence of unseen, low-mass companions (Caillault, Gagné, & Stauffer 1994). The bimodality of the luminosity function of the B/A stars in the Pleiades (Figure 13 of Stauffer et al. 1994) also indicates that the "detected" B/A Pleiads have L_x values identical to those of the low-mass stars in the cluster; the undetected B/A stars have substantially lower upper limits. However, Patten & Simon (1993) claim that the B/A stars that they detect in IC2391 could be true identifications and have peculiar L_x/L_{bol} ratios that are the result of variability in the wind geometry and/or mass loss rate.

4. ACTIVITY-ROTATION RELATION

Since the X-ray activity in a late-type star is expected to arise as the result of the magnetic dynamo, a relation between X-ray emission and rotation might be expected. Results from *Einstein* indicated that such relationships probably exist (e.g., Pallavicini et al. 1981; Bouvier 1990), but there were enough conflicting reports (e.g., Caillault & Helfand 1985) that a quantifiable relationship had not been achieved. The *ROSAT* results in this area have been spectacularly successful in demonstrating the verity of an activity-rotation relation.

The cleanest example of the dependence of coronal activity on rotation is shown in Figures 16 & 17 of Stauffer et al. (1994), which are plots of L_x/L_{bol} vs. $v \sin i$; the latter of these figures is reproduced here as Figure 3. It appears that the relation is mass dependent: it turns on at $(B-V)_o \sim 0.62$, which is equivalent to ~ 1 M$_\odot$. The activity increases quickly with $v \sin i$, from $L_x/L_{bol} \sim 10^{-4.5}$ up to $\sim 10^{-2.8}$ at $v \sin i \sim 15$ km s^{-1}, then seems to "saturate" for $v \sin i > 15$ km s^{-1}. Thus, since the range of X-ray luminosities of the Pleiades stars is almost two orders of magnitude, rotation is probably the dominant contributor to the dispersion seen in the luminosity function of the low-mass stars. Stauffer et al. (1994) also point out that the more massive stars do not display this relation between activity and rotation and that none of the high mass stars seem to reach the saturation level of $L_x/L_{bol} \sim 10^{-3}$.

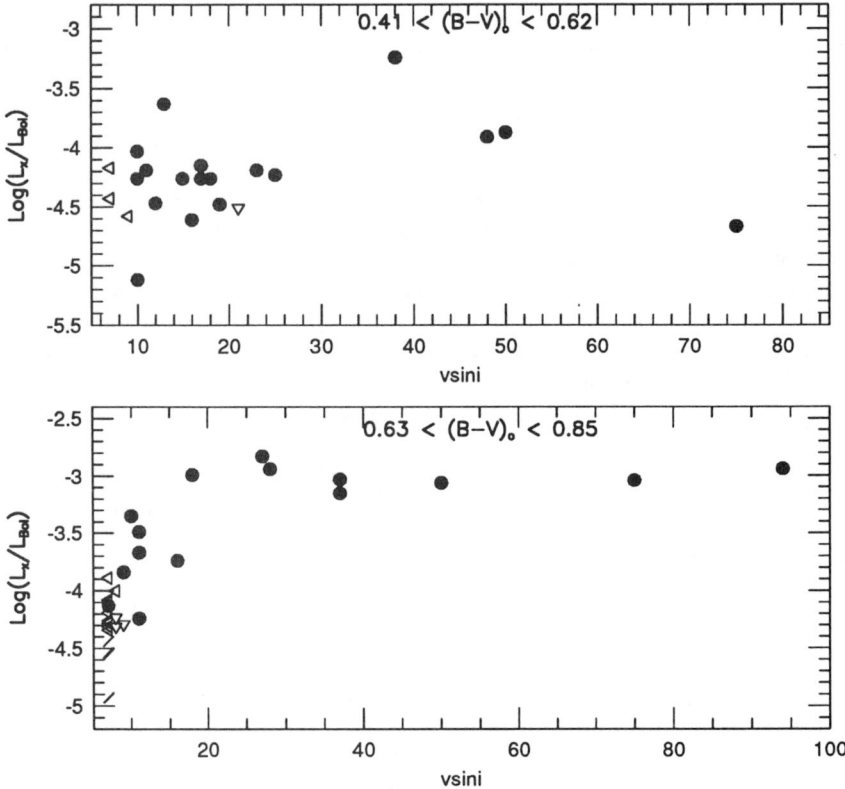

Figure 3: Plot of L_x/L_{bol} versus $v \sin i$ for solar-type Pleiades members (from Stauffer et al. 1994).

Other examples of the quest for an activity-rotation relation abound. Neuhäuser et al. (1994) have shown for a small sample of PMS stars that X-ray fluxes decrease with increasing rotational periods. Strom & Strom's (1994) Figure 14a shows a result very similar to that of Stauffer et al. (1994): the surface fluxes of the L1495E stars increase with increasing $v \sin i$. However, not everyone is in agreement: Feigelson et al. (1993) claim that there is no relation between activity and rotation for the stars in Chamaeleon I. Although they use L_x in their Figure 9, they assert that taking into account the radii of the stars (i.e., normalizing for surface area) does not affect their conclusion. However, their $v \sin i$ values only extend down to $\sim$ 15-20 km s^{-1}, above the point where Stauffer et al. (1994) notice the relation beginning to saturate, and their sample includes stars of masses above and below 1 $M_\odot$. Gagné et al. (1994a), in their survey of the stars in Orion, also find no dependence on rotation, but all of their stars have $v \sin i$ values $>$ 15 km s^{-1}, too, and are probably more massive than the Sun. They have the additional problem of having difficulty determining $L_{\rm bol}$ because of uncertainties in the effects of variability, unknown extinction, and presence or absence of disks. In the Hyades, all that is known is that the most rapidly rotating M dwarf ($v \sin i = 13$ km s^{-1}) is the second brightest star of its class, but that it lies below the saturation limit (Stern et al. 1994). More sensitive rotational velocity studies would be very useful for the Hyades stars and for the Chamaeleon and Orion star-forming regions, too.

As pointed out by Stauffer et al. (1994), one can understand the dispersion in X-ray luminosity by knowing the rotational velocity dispersion. By the Hyades age, G and K stars are rotating slowly with low dispersion, hence they have a small dispersion in their $L_{\rm x}$ values. The longer spindown timescales for M dwarfs are what give rise to their low $L_{\rm x}$ dispersion, though; i.e., they all have $v \sin i$ above the threshold of $\sim$ 15 km s^{-1} and, thus, are all in the saturation regime. Hence, it is expected that the M dwarfs in the Hyades should have the same dispersion in $L_{\rm x}$ as the Pleiades; unfortunately, the ROSAT observations of the Hyades M dwarfs have not yet been analyzed.

5. TEMPORAL AND SPECTRAL ANALYSES

a. Variability

Dozens of flares have been detected in the aforementioned clusters and star-forming regions, as expected. The likelihood of observing this short-term variability has been substantially enhanced with ROSAT largely because of the longer exposure times (20-50 ksecs per image in many cases) and, also, because of the greater sensitivity. However, analysis of variability on longer timescales is also possible, thanks to scheduling constraints which split many lengthy observations into two or more smaller pieces separated by a few months to more than a year. There is also, of course, the opportunity to compare the ROSAT results with those from Einstein – a timescale of about a decade. This has been done explicitly for the Orion Nebula and Pleiades stars (Gagné et al. 1994a & b).

In the Pleiades, comparisons were made on three different timescales: $\sim$ two weeks (between observations of two overlapping images), $\sim$ one year (between two segments of the central field observation), and $\sim$ 10 years (between ROSAT and Einstein). For each timescale Gagné et al. (1994b) compared, for each star, the ratio of count rates obtained during each time segment. Their analysis shows that the two-week timescale displays the tightest distribution of ratios, indicating a lower amount of significant ($\geq$ 50%) variability on that timescale. The one and 10 year distributions, similar to each other, are broader. They tentatively interpret these results as indicating that there is a very short-term component to the variability (flaring) and that there is also a long-term component which seems to be shorter than the $\sim$ 11 yr solar cycle. Gagné et al. (1994a) have performed a similar analysis on the stars in Orion; although no equivalent to the two-week timescale was available, the longer-term distributions were roughly the same as those in the Pleiades. This indicates that stellar coronal variability cycles may not be too different among main-sequence and pre-MS stars.

b. Temperatures and Emission Measures

Stern et al. (1994) find that only two-temperature Raymond & Smith (1977) thermal plasma models fit the spectra of the Hyades stars which they detected. They also show that the M dwarfs' fits yield hotter high-temperature components than those of G and K stars in the cluster.

Gagné et al. (1994b) performed a number of spectral fits to their Pleiades data. One aspect was modeling both the flaring and quiescent components of the emission from HII 2147, a late G star. Here, too, only two-temperature models produced acceptable fits. During the flare each temperature component increased in both temperature and volume emission measure. They also analyzed the spectra from two different spectral types of stars (G stars and K stars), further splitting the two groups into rapid and slow rotators (with a $v \sin i$ cutoff of ~ 15 km s^{-1}). They found that the more rapidly rotating stars have higher temperatures and volume emission measures than their slower counterparts.

6. THE FUTURE

ROSAT has obviously contributed greatly to the study of the coronæ of stars in stellar clusters and star-forming regions. Much of the future work necessary for improvement lies in the optical regime and not the X-ray. In particular, proper motions for determining cluster membership, spectroscopy and photometry to ascertain positions in the HR diagram, and rotational periods and velocities to elucidate further the activity-rotation relation, are required. In the X-ray regime, one expected result from the *ASCA*, *AXAF*, and *XMM* missions includes the discovery of many more low-mass PMS stars because of the decreased absorption effects at the higher energies to which those telescopes are sensitive, allowing for the potential to better calibrate the initial mass function. Also, as *EUVE* has shown in the ultraviolet regime, higher-resolution spectra will allow for better understanding of coronal temperatures, densities, and heating mechanisms. *In toto*, these data will contribute to substantial advancements in the theoretical models of coronal activity of stars in stellar clusters and star-forming regions.

7. ACKNOWLEDGEMENTS

The author is most grateful to Steve Snowden and the rest of the *ROSAT* Symposium SOC for the opportunity to present this review paper. The support of NASA Grant NAG5-1610 to the University of Georgia is also gratefully acknowledged.

8. REFERENCES

Bouvier, J. 1990, AJ, 99, 946
Caillault, J.-P. & Helfand, D. J. 1985, ApJ, 289, 279
Caillault, J.-P. & Zoonematkermani, S. 1989, ApJ, 338, L57
Caillault, J.-P., Gagné, M., & Stauffer, J. R. 1994, ApJ, submitted
Feigelson, E. D., Casanova, S., Montmerle, T., & Guibert, J. 1993, ApJ, 416, 623
Gagné, M., Caillault, J.-P., Sharkey, E., & Stauffer, J. R. 1994a, in *The Eighth*

Cambridge Workshop on Cool Stars, Stellar Systems, and The Sun,
    ed. J.-P. Caillault, ASP Conference Series, in press
Gagné, M., Caillault, J.-P., Hartmann, L. W., Prosser, C. F., & Stauffer, J. R. 1994b,
    in *The Eighth Cambridge Workshop on Cool Stars, Stellar Systems,
    and The Sun*, ed. J.-P. Caillault, ASP Conference Series, in press
Gauvin, L. S. & Strom, K. M. 1992, ApJ, 385, 217
Isobe, T., Feigelson, E. D., & Nelson, P. I. 1986, ApJ, 306, 490
Lucy, L. B. & White, R. L. 1980, ApJ, 241, 300
Micela, G., Sciortino, S., Vaiana, G. S., Schmitt, J. H. M. M., Stern, R. A.,
    Harnden, F. R. Jr., & Rosner, R. 1988, ApJ, 325, 798
Neuhäuser, R., Sterzik, M. F., Schmitt, J. H. M. M., & Morfill, G. E. 1994,
    in *The Eighth Cambridge Workshop on Cool Stars, Stellar Systems,
    and The Sun*, ed. J.-P. Caillault, ASP Conference Series, in press
Pallavicini, R., Golub, L., Rosner, R., Vaiana, G. S., Ayres, T., & Linsky, J. L.
    1981, ApJ, 248, 279
Patten, B. M. & Simon, T. 1993, ApJ, 415, L123
Randich, S. & Schmitt, J. H. M. M. 1994, in *The Eighth Cambridge Workshop on Cool
    Stars, Stellar Systems, and The Sun*, ed. J.-P. Caillault, ASP Conference Series,
    in press
Raymond, J. & Smith, B. 1977, ApJS, 35, 419
Rosner, R., Golub, L., & Vaiana, G. S. 1985, ARAA, 23, 413
Schmitt, J. H. M. M., Zinnecker, H., Cruddace, R., & Harnden, F. R., Jr. 1993,
    ApJ, 402, L13
Stauffer, J. R., Caillault, J.-P., Gagné, M., Prosser, C. F., & Hartmann, L. W. 1994,
    ApJS, in press
Stern, R. A., Schmitt, J. H. M. M., Rosso, C., Pye, J. P., Hodgkin, S. T., &
    Stauffer, J. R. 1992, ApJ, 399, L159
Stern, R. A., Schmitt, J. H. M. M., Pye, J. P., Hodgkin, S. T., Stauffer, J. R., &
    Simon, T. 1994, preprint
Strom, K. M., et al. 1990, ApJ, 362, 168
Strom, K. M. & Strom, S. E. 1994, preprint
Walter, F. M. 1986, ApJ, 306, 573
Walter, F. M., Brown, A., Mathieu, R. D., Myers, P. C., & Vrba, F. J. 1988,
    AJ, 96, 297

LOW LATITUDE X-RAY SHADOWS AND THE SOFT X-RAY DIFFUSE BACKGROUND

D. N. Burrows and J. A. Mendenhall
Department of Astronomy & Astrophysics, The Pennsylvania State University,
University Park, PA 16802
email: burrows@astro.psu.edu

Abstract

The ROSAT PSPC has proven to be an excellent instrument for detecting soft X-ray shadows cast by interstellar clouds. We have conducted a program of shadowing observations aimed at improving our understanding of the soft X-ray diffuse background and the hot phase of the ISM. In this program, dense interstellar clouds are used to place limits on the distance to the emission sources of the soft X-ray background by measuring the depths of the absorption shadows cast by these clouds. We will present some of the results obtained to date in this program, focussing particularly on shadows of low latitude clouds and what they tell us about the Local Bubble and the origin of $\frac{3}{4}$ keV galactic emission.

1. Introduction

The ROSAT PSPC instrument has made the study of soft X-ray shadows possible, opening up a powerful new probe into the interstellar medium. X-ray shadows, which are the analog of optical dark nebulae such as the Horsehead Nebula, are produced by absorption of diffuse X-rays from some background source by a foreground interstellar cloud (neutral or molecular). The shadows can be used to provide distance information on the X-ray emission, which has unambiguously proven the existence of million degree gas in the galactic halo (Burrows & Mendenhall 1991), and to study the absorbing clouds, which provides a new probe of the distribution of matter in interstellar clouds that is independent of previous techniques using $100 \mu m$ or radio observations (Wang & Walter 1994). In this paper, we present preliminary results of low latitude shadowing observations designed to determine the extent of emission regions contributing to the $\frac{1}{4}$ keV and $\frac{3}{4}$ keV diffuse backgrounds.

During the past decade, the predominant model of the $\frac{1}{4}$ keV diffuse background interpreted it as emission from million degree gas within a large cavity in the interstellar medium commonly known as the Local Bubble (Snowden et al. 1990). High latitude shadows clearly demonstrate that a substantial fraction of the $\frac{1}{4}$ keV flux in some directions comes from hot gas in the galactic halo (Burrows & Mendenhall 1993), and it is now clear that the Local Bubble emission, while important, cannot explain all of the observed flux. The low latitude observations discussed here show that we still have no evidence for widespread $\frac{1}{4}$ keV emission from beyond the Local Bubble at low latitudes.

The $\frac{3}{4}$ keV diffuse background has been a puzzle for many years (Nousek et al. 1982; Sanders et al. 1983; Snowden & Schmitt 1990). It appears that at least 50% of the high latitude flux in this band originates in the same extragalactic sources (primarily AGN) that produce the cosmic X-ray background seen above 2 keV. At low latitudes this extragalactic flux is absorbed by galactic gas, yet the count rate is nearly independent of galactic latitude (or 21 cm column density) over much of the sky, implying that some more local emission source "fills in" for the absorbed extragalactic photons. The papers cited above explored several explanations for this curious behavior including contributions from dwarf M stars, the Local Bubble, and distant SNRs, but a satisfactory solution to the puzzle still eludes us. Recently, Stanford & Caillault (1994) have modeled various components of the $\frac{3}{4}$ keV emission from the galaxy and fit these components to low latitude count rates derived from the *Einstein* IPC. For one of their lines of sight (along galactic longitude 24°) they find that the emissivity peaks at about 200 pc, drops slowly to about 2.6 kpc, and then drops rapidly to about 3.5 kpc. Such a model can be tested directly (for selected directions) with shadowing observations. Preliminary analysis of the observations discussed here appears to be consistent with this emissivity distribution for one line of sight, but inconsistent with it for two others which have no shadows cast by clouds at distances over 500 pc.

2. Data

The data presented here are derived from ROSAT PSPC pointed observations towards a variety of low latitude clouds selected on the basis of high column densities, well-defined boundaries, and a range of distance and galactic latitude. In the simplest analysis, we can assume that the $\frac{3}{4}$ keV diffuse background, which is fairly isotropic, arises from a uniform emission source. These observations then permit the global characteristics of this source (emissivity as a function of distance and latitude) to be sampled in a coarse way. For one region of the galactic plane in the constellation Monoceros, we have obtained observations of clouds at two distances to more directly sample the emissivity as a function of distance along this line of sight.

The ROSAT data were processed using the techniques described by Snowden et al. (1994), Mendenhall et al. (1994), and Mendenhall & Burrows (1994), which included cleaning the data of charged particles, scattered solar X-rays, afterpulses, and other known sources of contamination, and using energy-dependent exposure maps. Point sources were removed, and the cleaned images were smoothed with a 5' FWHM Gaussian for presentation to reduce statistical fluctuations and to provide a relatively uniform effective resolution over the PSPC field of view.

The clouds discussed here are presented in Table 1. With the exception of MBM 12, the shadow depths given in this table are calculated as $(1 - R_{ON}/R_{OFF})$, where R_{ON} and R_{OFF} are the observed count rates on and off the cloud, respectively. A horizontal dash indicates that no shadow was detected (with a typical upper limit of about 10% to the shadow depth). Because of the possibility of undetected diffuse contamination in these fields, the shadow depths must be treated as lower limits. For optically thick clouds, the shadow depths can be interpreted directly as the fraction of the observed flux that originates beyond the cloud, but for clouds with moderate optical depths (like L1495), a complete radiative transfer analysis must be performed to obtain this information.

The clouds discussed here fall into three groups on the basis of their galactic latitudes. The first cloud shown, MBM 12, is at a rather high latitude but is included because of its importance as the nearest known molecular cloud and because its deep $\frac{3}{4}$ keV shadow implies that the Local Bubble contributes little or none of the observed flux in this band. The four clouds in the next group are located in the galactic plane at a variety of longitudes and distances. The final group of five clouds is at intermediate latitudes ($15° < |b| < 20°$), again at a range of longitudes. Most of these clouds are relatively close, with distances between 150 and 200 pc, but we have one cloud in this latitude range at a larger distance of 450 pc.

Table 1: Low latitude X-ray shadows

Cloud Name	(ℓ, b)	d (pc)	τ_{max} ($\frac{1}{4}$ keV)	τ_{max} ($\frac{3}{4}$ keV)	Shadow Depth ($\frac{1}{4}$ keV)	Shadow Depth ($\frac{3}{4}$ keV)
MBM 12 [1]	(159°.6, −33°.6)	65	25	1.5	−	70%
Coalsack	(302°.8, −0°.1)	200	140	9	−	73%
B361 [2]	(89°.5, −0°.7)	500			−	−
Mon OB1	(201°.2, +1°.0)	800	20	2	−	−
Mon OB2	(207°.7, −0°.7)	1600	40	4	−	30%
Cham. I [3]	(297°.3, −15°.7)	140	160	10	−	46%
L1495	(168°.2, −16°.4)	150	9	0.6	−	17%
L1551	(178°.9, −20°.1)	160	150	9	−	30%
RCrA [4]	(0°.0, −17°.8)	< 200	400	27	46%	76%
Cepheus	(113°.4, 17°.5)	450	8	0.5	−	34%

[1] Snowden, McCammon, & Verter (1993).
[2] Steve Snowden, private communication.
[3] Data contributed by Eric Feigelson.
[4] Garmire (1992).

2.1. MBM 12

This cloud was discussed by Snowden, McCammon, & Verter (1993). The cloud has a well determined distance of 65 pc (Hobbs, Blitz, & Magnani 1986). Although it is not optically thick at $\frac{3}{4}$ keV, it casts a deep shadow in this band. Radiative transfer calculations for this cloud indicate that the on-cloud intensity is consistent with zero, with a 2σ upper limit of about 20% of the average $\frac{3}{4}$ keV rate, suggesting that little or none of the $\frac{3}{4}$ keV diffuse background originates in the Local Bubble.

By contrast, the $\frac{1}{4}$ keV data for this cloud indicate that virtually all of these photons come from the Local Bubble interior, with most, if not all, from the near side of the cloud.

2.2. Coalsack

The Coalsack is composed of at least two distinct clouds with distances of 188 pc and 243 pc (Seidensticker & Schmidt-Kaler 1989). For the purposes of this paper, we will treat these as a single cloud at a distance of about 200 pc. The X-ray images are shown in Figure 1 (Plate 4). Like the other images presented here, the $\frac{1}{4}$ keV image (Figure 1a) is on the left side of the page with the $\frac{3}{4}$ keV image of the same field (Figure 1b) presented to its right. Both X-ray images are overlaid with 100μm IRAS Sky Survey Atlas contours. The $\frac{1}{4}$ keV image has no shadow, consistent with emission from within the Local Bubble. In contrast, the $\frac{3}{4}$ keV image has a deep shadow, indicating that most of the $\frac{3}{4}$ keV flux comes from a larger distance in this direction.

2.3. B361

This observation (by Steve Snowden) has a deep shadow against the edge of the Cygnus Superbubble. Although the variable background intensity complicates the analysis somewhat, the intensity observed toward the cloud appears to be roughly consistent with the intensity at the edge of the field, away from the superbubble. If this result holds up under a more detailed analysis, then the observed $\frac{3}{4}$ keV flux in this direction comes from the near side of the cloud, which has a distance of about 500 pc. This implies either that the emission source is distributed between us and the cloud but not on the far side of the cloud, or that neutral material along this line of sight absorbs any $\frac{3}{4}$ keV emission from beyond the cloud.

2.4. Mon OB1

This observation is pointed towards the molecular cloud associated with the Mon OB1 association, which is located about 800 pc from the Sun. The X-ray images are shown in Figure 2 (Plate 4). Not surprisingly, there is no $\frac{1}{4}$ keV shadow. There is also no $\frac{3}{4}$ keV detected towards this cloud, indicating that the observed $\frac{3}{4}$ keV flux in this direction originates on the near side of the cloud.

2.5. Mon OB2

This observation is pointed at the edge of the molecular cloud associated with Mon OB2 and the Rosette Nebula. Analysis of this field is complicated by the fact that it is located at the southern edge of the Monogem ring (Plucinsky 1994), and much of the structure in the $\frac{1}{4}$ keV image is produced by the latter feature. However, the Monogem ring has no $\frac{3}{4}$ keV or 1.5 keV emission, and our data show shadows at both of these energies, so we believe that we are seeing a shadow of this cloud in the medium and high energy bands. Preliminary analysis suggests that the depth of the $\frac{3}{4}$ keV shadow is $\sim 30\%$, implying that nearly $\frac{1}{3}$ of the flux in this direction comes from beyond 1600 pc. This appears to be in conflict with the results obtained for B361 and Mon OB1 (but is at least qualitatively consistent with the emissivity distribution found by Stanford & Caillault [1994]), and indicates the possibility of substantial variations in the emissivity as a function of distance over relatively small angles (7° between Mon OB1 and Mon OB2).

2.6. Chamaeleon I

Our Chamaeleon images (Figure 3, Plate 4) are mosaics of three fields (two of which were contributed by Eric Feigelson). The lack of shadowing in the $\frac{1}{4}$ keV image is consistent with all of the flux in this band originating within the Local Bubble in this direction. The $\frac{3}{4}$ keV image shows a clear shadow cast by this cloud against background emission, with a shadowing depth of 46%, indicating that at least $\sim \frac{1}{2}$ of the $\frac{3}{4}$ keV flux in this direction comes from beyond 140 pc.

2.7. L1495

L1495 was observed by Eric Feigelson to search for T Tauri stars. We have used this field to check for X-ray shadows, with the result that we find no $\frac{1}{4}$ keV shadow, but find a weak $\frac{3}{4}$ keV shadow. The depth of the latter appears to be consistent with absorption of most of the $\frac{3}{4}$ keV diffuse background flux by the relatively low column density of the filament seen at 100μm, but a complete radiative transfer fit has not yet been performed for these data. Like the other mid-latitude fields, this cloud appears to be beyond the Local Bubble but absorbs a substantial fraction of the $\frac{3}{4}$ keV background in this direction.

2.8. L1551

Like L1495, L1551 has no $\frac{1}{4}$ keV shadow (Figure 4a, Plate 5) but has a small $\frac{3}{4}$ keV shadow (Figure 4b, Plate 5). The small angular size of this cloud limits our statistical accuracy, but the $\frac{3}{4}$ keV shadow depth appears to be about 30% for this cloud at 160 pc.

2.9. RCrA

This observation was discussed by Garmire (1992) and is shown in Figure 5 (Plate 5). This cloud is particularly interesting in that it is unique in having strong shadows in both the $\frac{1}{4}$ keV band and the $\frac{3}{4}$ keV band. However, it is silhouetted against a bright $\frac{1}{4}$ keV enhancement associated with either the interior of Loop I or with the galactic bulge, and is not representative of the general diffuse background found in the rest of the sky. The shadow depths found for this cloud indicate that both the $\frac{1}{4}$ keV and $\frac{3}{4}$ keV enhancements are located at distances greater than 200 pc.

2.10. Cepheus

The Cepheus flare (Lebrun 1986), located at a distance of 450 pc, has no $\frac{1}{4}$ keV shadow (Figure 6a, Plate 5) but a substantial $\frac{3}{4}$ keV shadow (Figure 6b, Plate 5), indicating that much of the latter emission comes from beyond this cloud.

3. Discussion

We would like to be able to summarize these observations in a simple, consistent model that could explain all of the data. This may eventually be possible, but at the moment the situation appears to be too complex for a single, simple explanation. However, we can provide a summary of the main points we can derive on the basis of the preliminary analysis performed so far.

3.1. 1/4 keV results

In the introduction, we alluded to the discovery of high latitude shadows in the $\frac{1}{4}$ keV diffuse background that demonstrate the existence of hot gas in the galactic halo. Models of the ISM predict that a large fraction of the volume of the galactic disk should also be filled with million degree gas, and one might hope to observe many low latitude shadows if the filling fraction of this hot gas in the disk were as high as 0.7 − 0.8, as suggested by McKee & Ostriker (1977) (although MO predict that the typical temperature of the hot ionized medium is only about 5×10^5 K, perhaps too low to produce observable shadows in the soft X-ray diffuse background). At any rate, with the exception of R Cr A, we find no low latitude $\frac{1}{4}$ keV shadows. Either the filling fraction of the hot medium is substantially lower than MO predicted, or we cannot see out of the Local Bubble to observe it (although we do see several other hot bubbles with temperatures of $1 − 3 \times 10^6$ K: the North Polar Spur/Loop I; the Eridion Bubble [Burrows et al. 1993]; the Monogem Bubble [Plucinsky 1993]; and the Cygnus Superbubble).

The exception to this rule is R Cr A, which has a substantial $\frac{1}{4}$ keV shadow cast by a region of enhanced emission in this band. It is not clear whether this background emission represents emission from the interior of Loop I, emission from a galactic wind emanating from the galactic bulge, or emission from the halo associated with the $\frac{1}{4}$ keV enhancement that peaks about 45° to the south in galactic coordinates. There are problems with all three interpretations. The first suffers from the fact that the cloud is too far away to clearly be located on the near side of the Loop I bubble (although this is not clearly ruled out either, since the distance to Loop I is not well known). The second and third possibilities must contend with the fact that the HI column density in this direction, 6.5×10^{20} cm^{-2}, represents many optical depths at this energy.

3.2. 3/4 keV results

We will first consider the group of clouds located in the galactic plane. For these clouds, the large background column density from the galactic plane absorbs any contribution from the Cosmic X-ray Background (CXB). Thus, these lines of sight provide the cleanest measurement of galactic contributions to this flux. Because of the large absorption length of these X-rays ($\sim 1 − 2$ kpc), we would expect that local variations in the structure of the ISM and in stellar populations would be relatively unimportant, and that a simple, average, large-scale model of the emission sources should adequately describe the observations. For three of the galactic plane observations, this might be achievable if a model can be constructed that provides most of the emission at distances between 150 pc and 500 pc. Such a model seems unlikely, though, since any emission source (such as dM stars) distributed fairly uniformly in the disk will be observable out to much larger distances than 500 pc unless the line of sight intersects nearby molecular clouds. On the other hand, the Mon OB2 observation, while more in keeping with the expected result that a significant fraction of the observed flux should originate at large distances, appears to be inconsistent with both B361 and Mon OB1. The Mon OB1 and Mon OB2 observations were made in an attempt to bracket the emission source by obtaining shadow depths for two clouds with small angular separations but very different distances; this objective is thwarted by the unexpected incompatibility of the shadow results, which may require large variations in emissivity distribution over angles of less than 10 degrees.

For the intermediate latitude clouds, the contributions from the CXB can be important and must be taken into account. These contributions are given in Table 2. This table gives the background column density from the Stark et al. (1992) 21 cm survey, except for the Chamaeleon cloud, where the background column density was taken from Cleary, Heiles, & Haslam (1979). In the $1 − 2$ keV band, 60% of the CXB has been resolved into point sources by ROSAT (Hasinger et al. 1993), and an upper limit of 25% has been placed on the truly diffuse

Table 2: Extragalactic contributions to $\frac{3}{4}$ keV flux

Cloud Name	Background N_H (10^{20} cm^{-2})	τ ($\frac{3}{4}$ keV)	F_{CXB} (50%)	F_{CXB} (75%)	Shadow Depth
Cham. I	9.8	0.49	0.30	0.46	46%
L1495	8.6	0.43	0.33	0.49	17%
L1551	20	1.0	0.18	0.27	30%
RCrA	6.5	0.32	0.36	0.54	76%
Cepheus	13.3	0.67	0.26	0.38	34%

component of the background. If a similar result applies in the $\frac{3}{4}$ keV band at high latitudes, then we can expect that between 50% and 75% of the high latitude background is extragalactic in origin. The column labeled τ gives the $\frac{3}{4}$ keV optical depth of the background gas toward the intermediate clouds, and the columns labeled F_{CXB} give the fraction of the observed count rate for each cloud due to the extragalactic CXB under the assumption that it contributes 50% or 75% of the high-latitude flux. Three of these clouds, Chamaeleon, L1495, and Cepheus, are only partially opaque at this energy (the Chamaeleon cloud has a large maximum optical depth, but this is concentrated in several small knots, and most of the cloud has optical depths closer to 1). This will reduce the contrast of the shadow and requires a careful radiative transfer calculation before definite conclusions can be drawn. However, the results for these clouds are not obviously in conflict with those obtained in the galactic plane.

We are grateful to Eric Feigelson, Gordon Garmire, and Steve Snowden for sharing their data on Chamaeleon, R Cr A, and B361, respectively. This work was supported by NASA grants NAG 5-1535 and NAG 5-1786.

4. References

Burrows, D. N., & Mendenhall, J. A. 1991, Nature, 351, 629

Burrows, D. N., & Mendenhall, J. A. 1993, in *Space Astronomy*, ed. J. Trümper, C. Cesarsky, G. G. C. Palumbo, & G. F. Bignami, Advances in Space Research, 13 (12), (12)83 – (12)92

Burrows, D. N., Singh, K. P., Good, J., Nousek, J. A., & Garmire, G. P. 1993, ApJ, 406, 97

Cleary, M. N., Heiles, C., & Haslam, C. G. T. 1979, ApJ, 230, L83

Garmire, G. P. 1992, BAAS, 24 (1), 689

Hasinger, G., Burg, R., Giacconi, R., Hartner, G., Schmidt, M., Trümper, J., & Zamorani, G. 1993, A&A, 271, 1

Hobbs, L. M., Blitz, L., & Magnani, L. 1986, ApJ, 306, L109

McKee, C. F., & Ostriker, J. P. 1977, ApJ, 218, 148

Mendenhall, J. A., & Burrows, D. N. 1994, this volume

Mendenhall, J. A., Burrows, D. N., Snowden, S. L., & McCammon, D. 1994, this volume

Nousek, J.A., Fried, P.M., Sanders, W.T., and Kraushaar, W.L. 1982, ApJ, 258, 83

Plucinsky, P. 1994, this volume

Sanders, W. T., Burrows, D. N., Kraushaar, W. L., & McCammon, D. 1983, in *Proceedings of IAU Symposium No. 101, Supernova Remnants and Their X-ray Emission*, ed. J. Danziger & P. Gorenstein, (Dordrecht:Reidel), 361

Seidensticker, K. J., & Schmidt-Kaler, Th. 1989, A&A, 192

Snowden, S. L., Cox, D. P., McCammon, D., & Sanders, W. T. 1990, ApJ, 354, 211

Snowden, S. L., McCammon, D., Burrows, D. N., & Mendenhall J. A. 1994, ApJ, in press

Snowden, S. L., McCammon, D., & Verter, F. 1993, ApJ, 409, L21

Snowden, S. L., & Schmitt, J. H. M. M. 1990, Ap&SS, 171, 207

Stanford, J. M., & Caillault, J.-P. 1994, ApJ, in press

Stark, A. A., Gammie, C. F., Wilson, R. W., Bally, J., Linke, R. A., Heiles, C., & Hurwitz, M. 1992, ApJS, 78, 77

Wang, Q. D., & Walter, F. M. 1994, ApJ, submitted

ROSAT RESULTS ON HOT AND COOL STARS

J.H.M.M. Schmitt
Max-Planck-Institut für Extraterrestrische Physik

Email ID
jhs@mpe.mpe-garching.mpg.de

ABSTRACT

I will present selected highlights from ROSAT observations of stellar X-ray emission. Specifically I will discuss the dearth of X-ray emission among white dwarfs and new evidence for the occurrence of X-ray emitting shocks in the winds of early-type stars. Lastly I will focus on ROSAT observations of the eclipsing cool star binary systems AR Lac, Algol, YY Gem and α CrB and the resulting implications for coronal structure on these stars.

1. Introduction

At the time of this ROSAT Science Symposium, the ROSAT satellite has been in operation for about 41 months. For most of this time science data were actually taken, the total satellite downtime (when no science data were output) so far amounts to less than five percent of the total operational time. Thus ROSAT has already now a longer operational lifetime than both the *Einstein* and EXOSAT Observatories, and more data is taken every day. Every astronomer, who has worked on a typical ROSAT pointing observation, knows the wealth of the information contained in the ROSAT images, and can appreciate their usefulness both in terms of the intended (i.e., "proposed") science as well archival research. Without being immodest, it is certainly in order to consider ROSAT as one of the more successful X-ray missions. By the same token it is however also appropriate to ask ourselves what has actually been learnt, what new information has been obtained and where have we "only" confirmed previous knowledge. This is the task I wish to tackle in this review for the field of X-ray observations of stars. As stars I consider in this context all "compact" X-ray sources whose X-ray emission does **not** directly result from accretion in whatever form. I purposely exclude all aspects of X-ray emission from young stars in open clusters such as Hyades, Pleiades, Praesepe, α Per (which have all been studied with ROSAT) as well as in various star forming regions such as Taurus-Auriga, Orion, Chamaeleon, Lupus, etc. since this topic is covered by J.-P. Caillault (1994) in this volume. My task is clearly impossible to solve in any satisfactory fashion at the moment; many of the ROSAT results have not been published yet, and more data are still being taken. I also admit to being biased towards my own research; many people have contributed towards the ROSAT pointing program, and any omission of any important published result is attributable only to my ignorance.

Stars exhibiting non accretion-driven X-ray emission can be grouped into three categories: White dwarfs, early type stars, and late type stars. In order to give a feeling for numbers, let me mention that among the approximately 60 000 sources in the ROSAT all-sky survey, about one third are stars (according to my definition from above). Among these 20 000 or so stellar sources, there are about 150 white dwarfs, about the same number of early type stars, the rest and obviously the vast majority are late type stars.

2. White Dwarfs

The (essentially) universally accepted view of the interpretation of the X-ray emission from white dwarfs holds that the X-ray emission is photospheric in the following sense: For sufficiently hot and sufficiently metal-deficient white dwarfs the atmospheric opacity is such that unity optical depth is reached at layers with temperatures greater than 10^5 K. X-ray astronomers therefore deduce a much higher "photospheric" temperature, and also measure considerably more flux (because of the strong temperature dependence of the emissivity) than optical astronomers. Prior to ROSAT, no sufficiently sensitive soft all-sky surveys were available, and it was therefore customary to estimate the number of X-ray emitting white dwarfs from optical surveys of faint blue objects. Fleming et al. (1993) juxtapose the number of expected white dwarf discoveries ($\approx$ 5500 with ROSAT PSPC) with those actually obtained (124); this difference between expectation and reality is significant even by astronomical standards and represents one of the major (unfortunately negative) discoveries obtained by ROSAT. The reason why many hot white dwarfs are X-ray faint is due to the fact that metal abundances (which are quite difficult to determine optically for hot white dwarfs) had been underestimated, and the addition of trace amounts of metals increases the atmospheric opacity such as to effectively block all X-ray radiation. Pure hydrogen white dwarfs like HZ 43 (which – despite being at a distance of 60 pc – produces almost 90 PSPC cts/s) must be quite rare, since such objects would have been detectable in the ROSAT all-sky survey over a sizable fraction of the Galaxy. As a consequence one can state that white dwarfs do not significantly contribute to the soft X-ray population at large.

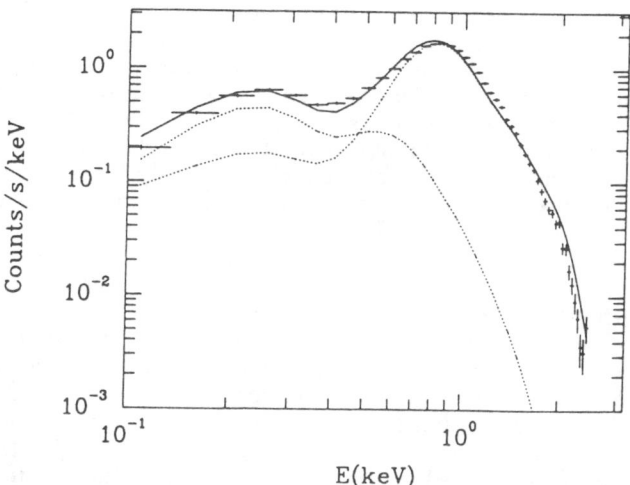

Fig.1: PSPC pulse height spectrum (from Hillier et al. 1993 for ζ Pup together with best fit two-component model (solid line), with the individual temperature components of log T = 6.66 and log T = 6.23 denoted by dots.

In the context of HZ 43 I wish to briefly remark on the calibration of the ROSAT PSPC since I have noticed (also at this meeting) a certain tendency among researchers to blame a failure of their favorite spectral model fits to a PSPC calibration error. The calibration of the ROSAT PSPC is certainly not perfect, just like any other X-ray detector or somebody's preferred spectral model. As far as HZ 43 is concerned, Napiwotzki et al. (1994) argue that the PSPC count rate of HZ 43 is too large (!) by a factor of 1.7 as compared to the WFC S1 and S2 rates which agree with their HZ 43 atmosphere model. Now, for the ROSAT PSPC HZ 43 is a difficult case because essentially all of the observed photon events are extremely soft, so that some of the response input parameters are not particularly well determined. On the other hand, to conclude from this that the calibration of the ROSAT WFC is correct but that of the ROSAT PSPC is not, appears to me somewhat premature, since HZ 43 has been used as a calibration source for the ROSAT WFC (but not for the ROSAT PSPC). All one can conclude is that the atmosphere model used by Napiwotzki et al. (1994) is consistent with that used for the ROSAT WFC calibration. As to higher energies, Briel and Henry (1993) compared the X-ray fluxes measured with the ROSAT PSPC and the *Einstein Observatory* IPC for the seven clusters of galaxies (which nobody would suspect to be variable) and find a mean flux ratio of $f_{PSPC}/f_{IPC} = 0.99 \pm 0.08$, again consistent with the advertised instrument calibrations.

3. Hot Stars

The discovery of X-ray emission from early type stars was one of the first discoveries made with the *Einstein Observatory* (Harnden et al. 1979). This discovery was – strictly speaking – not unexpected since Cassinelli and Olson (1979) actually predicted X-ray emission in order to explain the O star winds' superionisation (a prediction later to become obsolete by proper NLTE wind calculations; cf. Kudritzki 1985). Furthermore, the physical idea that gave rise to the prediction of X-ray emission in the first place turned out to be inconsistent with observed X-ray spectra of early type stars (which showed much less absorption than expected; cf. Cassinelli et al. 1981). In order to overcome these difficulties, Lucy and collaborators (Lucy and White 1980, Lucy 1982) proposed to appeal to the intrinsic instability of radiatively driven winds, and argued that the X-ray emission arises from shocks relatively far out in the wind, thus producing little observable X-ray absorption; this scenario has remained uncontested until the launch of ROSAT.

I am aware of two key observations of early type stars which are immediately relevant to this briefly sketched scenario of X-ray emission in early type stars, and in fact provide support for the validity of this picture. Hillier et al. (1993) present a well exposed ROSAT PSPC pulse height spectrum of the prototype O star ζ Pup (O4f). The remarkable property of ζ Pup is the rather low value of its line of sight interstellar colums density of $log N_H = 20.0 \pm 0.05$. As a consequence, X-ray absorption towards ζ Pup is comparatively low, unity optical being reached at $N_H = 3\ 10^{20} cm^{-2}$. Quite surprisingly, the PSPC pulse height spectrum (reproduced in Fig. 1, taken from Hillier et al. 1993) is strongly suppressed in the C-band below 0.28 keV. Since no reasonable model for the instrinsically emitted X-ray spectrum of a thermal plasma shows such a suppression, this can only be explained by X-ray absorption in the wind of ζ Pup itself. Hillier et al. (1993) model the PSPC pulse height spectrum by distributing "hot" X-ray emitting shocks uniformly in a "warm" wind which provides X-ray absorption as computed from a detailed non-LTE treatment. They further show that only models where He^{++} recombines into He^+ in the outer regions of the wind can account for the observed pulse height spectrum. Because of the wind's opacity, the soft X-ray emission below 0.28 keV must come from regions far out in the wind ($r > 100\ R_*$), while the harder X-ray emission comes from much further in. I emphasise that all of the observed absorption can be accounted for by the cold interstellar absorption and warm wind absorption, which have both been determined by other means; no "N_H-fitting" takes place.

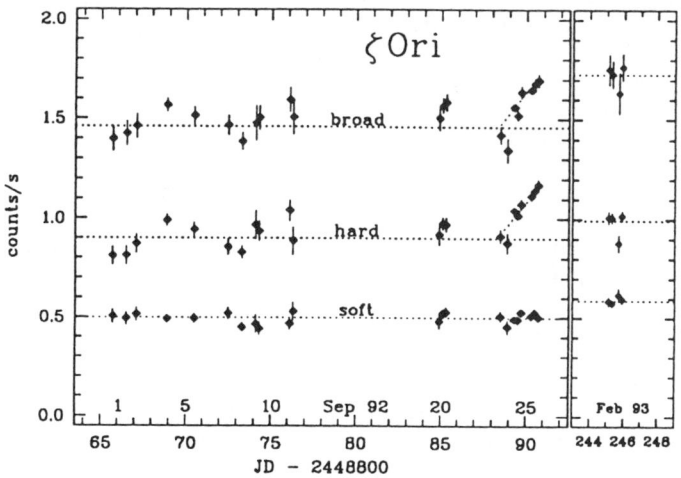

Fig. 2: X-ray light curve of ζ Ori (in the soft, hard and total) ROSAT band in September 1992 and February 1993 (taken from Berghöfer and Schmitt 1994).

A second important observation was obtained by Berghöfer and Schmitt (1994) in a long-term monitoring study of ζ Ori. While most of their program stars were constant, ζ Ori turned out to be constant for the first two years of monitoring, but in September 1992 the PSPC flux above 0.5 keV suddenly started in increase while the soft X-ray flux below 0.28 keV stayed constant (cf., Fig. 2). Half a year later, the hard flux was still enhanced as compared to the previous "constant" period, but had decreased compared to the end of the rise phase in September 1992; by that time, the soft X-ray flux had increased by 20 percent. Fitting the observed X-ray light curve to a simple model where all of the X-ray emission arises from a spherical shell propagating outward in the wind, it is possible to determine the velocity of this shell, which turns out to be $\approx$ 700 km/s in the star's frame. In the wind frame, which moves outwards - at the position of the shell - with a speed of 1200 km/s, the shell actually moves inwards, and can therefore be interpreted as a reverse shock. Instability theory of radiatively driven winds predicts that reverse shocks are much stronger than forward shocks, and in this sense this observation of ζ Ori probably represents the strongest evidence for the occurrence of such reverse shocks in radiatively driven winds of early-type stars.

4. Eclipse Studies of Cool Stars

One of the major discoveries made with the *Einstein Observatory* was the ubiquitous presence of coronae around late type stars (cf., Vaiana *et al.* 1981), a finding impressively confirmed by the vast numbers of coronal sources discovered in the ROSAT all-sky survey. A significant fraction of the ROSAT pointed observing program has been devoted to studying selected groups or specific examples of late type stars, and I just wish to mention in passing some of the highlights of these observations which include the detection of X-ray emission from hybrid stars (Reimers and Schmitt 1992, Brown 1994), the detection of X-ray emission from very low-mass stars down to almost the very bottom of the main sequence (Fleming *et al.* 1993) with a new candidate, i.e., LHS 3003, for the latest known X-ray emitter (Fleming 1993), as well as a complete sampling of detected X-ray emission of stars in the immediate solar vicinity with a construction of an X-ray luminosity distribution functions to levels below 10^{26} erg/s (Fleming and Schmitt 1994). Extensive studies of stellar flares have been undertaken, resulting in detections of flares from a hybrid star (α TrA, cf., Kashyap *et al.* 1994), a Be type star (λ Eri, cf., Smith *et al.* 1993), and flares on G and F type stars (cf., Schmitt 1994). The detection of microflares on UV Ceti was reported by Schmitt, Haisch and Barwig (1993), and Kürster (1994) reports a long-duration flare on the RS CVn system CF Tuc with a total duration of more than 10 days and a total released soft X-ray energy in excess of 10^{37} erg.

Space limitations prevent me from presenting a detailed discussion of the purposes and results of these studies, and therefore I will single out and concentrate upon only one item, i.e., eclipse studies of cool stars. Unlike the solar corona, stellar coronae can not be spatially resolved and imaged with presently available X-ray telescopes, and therefore rather fundamental physical parameters of stellar coronae such as their filling factors, volumes, pressures, and densities cannot be directly measured. If one attributes – as is customarily done – the X-ray emission from stellar coronae to thermal emission from magnetically confined plasma, at least some of these physical parameters **must** be different for the case of stellar coronae whose total soft X-ray luminosity can exceed the solar X-ray output by some orders of magnitude. Without measurements of coronal structure it is impossible to determine whether this is mainly due to enhanced coronal volumes (i.e., larger loops), enhanced filling factors (i.e., enhanced area covered by active regions), or enhanced densities (i.e., larger coronal pressures).

This was of course the background of the eclipse mapping studies carried out with the *Einstein Observtory* IPC, the EXOSAT LE and ME, and of course the ROSAT PSPC. Here I wish to discuss the results obtained on four such eclipsing systems, the eclipsing RS CV binary AR Lac, the Algol type binary β Per, the eclipsing dMe binary YY Gem, and the eclipsing solar-type star α CrB. The latter system an ideal one to study from the point of view of eclipse mapping since it consists of a smaller X-ray bright and a larger X-ray dark component. The α CrB system has not been studied before at X-ray wavelengths, while AR Lac has been observed both with the IPC (Walter, Gibson and Basri 1983) and EXOSAT (White *et al.* 1990), β Per (= Algol) with EXOSAT and GINGA, (White *et al.* 1986; Stern *et al.* 1992) and YY Gem with EXOSAT (Haisch *et al.* 1990). Nevertheless, in all cases the new ROSAT observations yielded new results which I will briefly present.

4.1 AR Lac

A system particularly well suited for eclipse studies is the brightest (as well as in the optical as in the X-ray wave band) and nearest ($d = 40 - 50\,pc$) eclipsing RS CVn AR Lac. The system consists of two components of spectral type G2 IV ($R_G = 1.54\,R_\odot$) and K0 IV ($R_K = 1.54\,R_\odot$) with an orbital period of 1.98 days. Because of its orbital inclination of $\sim 87°$, primary eclipse (i.e., G star eclipsed) is total and secondary eclipse (i.e., K star eclipsed) is annular.

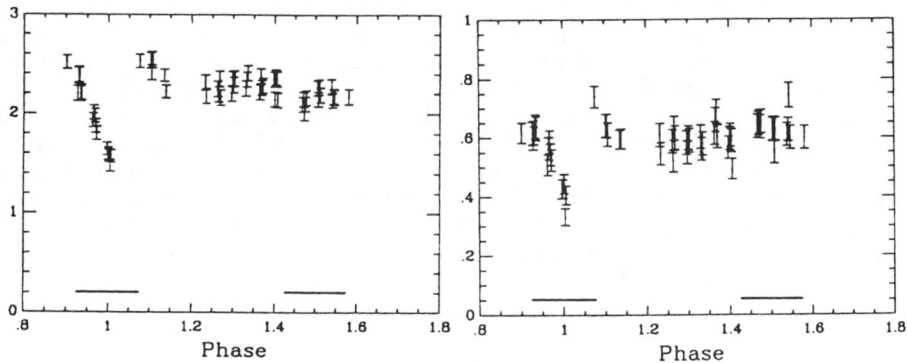

Fig.3 and 4: Low energy (0.1-0.4 keV; left panel) and high energy (1.1-2.4 keV; righ panel) light curve of AR Lac vs. binary phase (taken from Ottmann, Schmitt and Kürster 1993).

An X-ray eclipse at optical primary minimum was first discovered by Walter, Gibson, and Basri (1983), and later confirmed with the EXOSAT LE (White *et al.* 1990). Surprisingly, the simultaneously measured EXOSAT ME light curve (which recorded higher energy photons than the LE) did not show any eclipse from which White *et al.* (1990) concluded that the hot plasma (responsible for producing the higer energy photons recorded by the ME) pervaded the whole AR Lac binary system. In Figure 3 and 4 I show the ROSAT PSPC light curve of AR Lac at low energies (0.1 - 0.4 keV) and high energies (1.1 - 2.4 keV) as presented by Ottmann, Schmitt, and Kürster (1993). The light curve shows an obvious eclipse also at the higher photon energies, and no orbit-related spectral changes could be found. In addition, Ottmann, Schmitt, and Kürster (1993) show that the PSPC X-ray temperatures are consistent with the ones derived from the EXOSAT ME, from which it follows that the hot gas in AR Lac cannot pervade the whole binary system, but rather it appears to be confined to one of the binary components; this result has been confirmed by recent ASCA observations (White 1993).

4.2 Algol

The prototype Algol binary β Per consists of a B8V primary ($R_* = 2.9R_\odot$) and a K2IV secondary ($R_* = 3.5R_\odot$) with an orbital period of 2.87 d. Previous X-ray observation with EXOSAT covering half an orbit cycle centered on the optical secondary eclipse (cf., White et at 1986) as well as the GINGA observations presented by Stern et al. (1991) failed to detect any orbit-related variability. The ROSAT PSPC observations covered a total of two binary orbits (albeit with interruptions), the resultant PSPC light curve is shown in Figure 5 (taken from Ottmann 1994). The main characteristics of this light curve are, first, an enormous long-duration flare with a peak X-ray luminosity of 5 10^{31} erg/s, second, a smooth wave-like modulation with the orbital period of 2.7 days which is found before and after the giant X-ray flare, and third, a clear indication of an X-ray minimum at the time of secondary optical minimum, i.e., at the time when the (presumably X-ray dark) B star is front of the X-ray bright late type giant. From the X-ray minimum Ottmann (1994) infers a characteristic size scale of the corona of Algol of less than $0.8R_K$, a value significantly less than that deduced from the EXOSAT observations, while the wavelike variability pattern is interpreted as rotational modulation when X-ray bright regions on the K star are rotated behind the visible hemisphere. Other than a general hardening during the flare, no orbit-related spectral changes are observed, providing evidence that the coronal low-temperature and high-temperature regions in Algol must be cospatial.

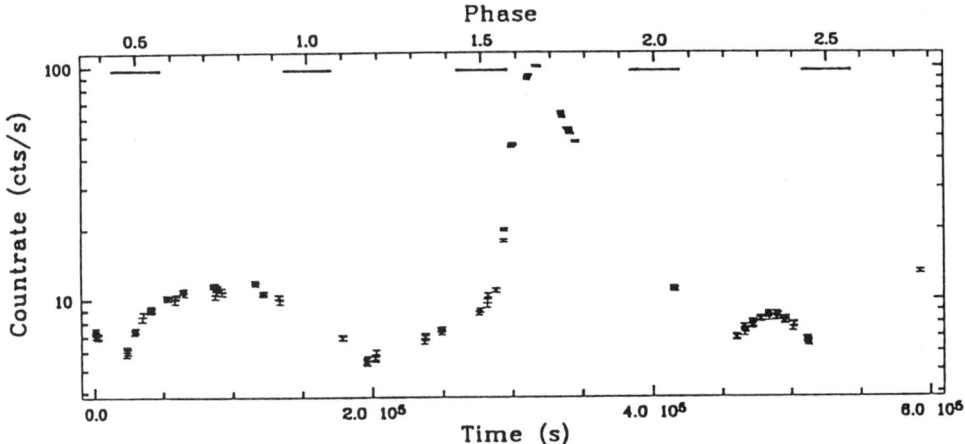

Fig.5: ROSAT PSPC light curve of Algol vs. time (lower tickmarks) and binary phase (upper tickmarks); taken from Ottmann 1994.

4.3 YY Gem

The system Castor C (= YY Gem) with a period of 0.814 days has been extensively studied in a variety of wavelength bands. Since YY Gem is a spectroscopic and eclipsing binary, stellar masses and radii could be determined optically; the system contains two almost identical M dwarfs with masses and radii of 0.57 $M_\odot$ and 0.62 $R_\odot$ respectively. From the point of view of magnetic activity, YY Gem is interesting because the tidally enforced rapid rotation is expected to lead to high levels of activity. Using the *Einstein Observatory*, Golub et al. (1983) detected X-ray emission from the Castor A+B+C system; from the positional coincidence of the X-ray source with YY Gem, it was argued that all of the observed X-ray emission should come from YY Gem.

Extensive studies of YY Gem with the EXOSAT Observatory showed, first, that X-ray emission does not only come from YY Gem, but also somewhat surprisingly from α Gem (Pallavicini et al. 1989), and second, that no obvious eclipses are seen. Specifically, Haisch et al. (1990) analysed all of the available EXOSAT YY Gem observations, and were only able to present a somewhat marginal case for an X-ray eclipse (cf., their Fig. 6).

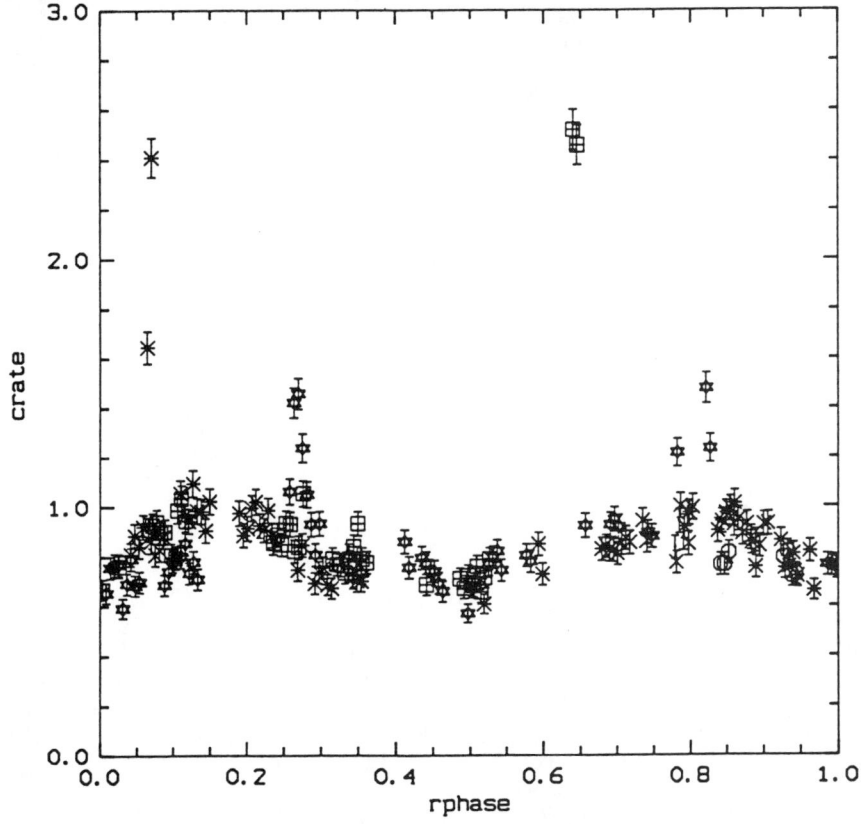

Fig.6: ROSAT PSPC light curve at high energies vs. binary phase for YY Gem; different symbols denote data points from different binary orbits.

A long observation of YY Gem was carried out with the ROSAT PSPC in the time between March, 29, 1991, 12 hours U.T. and April 3, 1991, 6 hours UT covering almost 6 binary orbits with somem interruptions. The PSPC image obtained by these observations is reproduced in Schmitt, Güdel and Predehl (1994); one has to keep in mind that because of the rather close angular separation of 72 arcsec between YY Gem and α Gem on the sky, the X-ray point response functions of the two sources overlap in the soft PSPC pulse height channels. A clear separation of the two sources is possible only at PSPC pulse height channels covering photon energies above 0.5 keV.

In Fig. 6 I show the recorded PSPC light curve (in pulse height channels above 50) vs. binary phase; the data were deadtime corrected (which is extremely important in this data set which suffered from enormous background variations) and background subtracted, the phase was computed from $T_{prim} = 2425698.3561\ JD$ and $P = 0.81428224\ d$ (cf., Haisch et al. (1989). The data from different binary orbits are denoted by different symbols; more detailed plots and a more detailed discussion and coronal map reconstruction are presented by Schmitt et al. (1994). From the light curve (cf., Fig. 6) we can draw the following conclusions: First, X-ray flares do frequently occur (after all, YY Gem is a flare star !), but usually such flares do not extend over more than $\approx$ 10 % of the binary orbit (i.e., about 2 hours); flares can be easily recognised once multiple phase coverage is available. Second, the X-ray light curve varies from orbit to orbit, but this orbit to orbit variability is at a level of about 10 % of the total signal; within a given orbit, the observed variations are much smoother. Third, clearly defined X-ray minima are found at the times of optical primary and secondary minimum; a detailed inspection of the recorded light curves shows that three subsequent primary and secondary minima have been detected. Fourth, there are definite out-of-eclipse variations, resembling to some extent the wave-like variability pattern observed on Algol by Ottmann (1994). Fifth, the depth of primary and secondary minimum is about the same. Sixth, primary minimum seems to be broader than secondary minimum, with the primary minimum being about twice as broad in binary phase than the the corresponding optical primary minimum (cf., Leung and Schneider 1978). And seventh, most of the X-ray emission seems to be associated with the secondary rather than the primary component, in accordance with the "photometric wave" in the optical (cf., Leung and Schneider 1978). For a detailed discussion of the YY Gem X-ray light curve the reader is referred to Schmitt (1994).

4.4 α CrB

The primary component of α CrB is of spectral type A0V, the seconday component of spectral type G5V; the two stars are in an eccentric orbit (e = 0.37) of 17.36 days, indicating that synchronisation has not (yet ?) taken place. The system shows photometric eclipses, with primary eclipse (i.e., G star in front of A star) being annular, and secondary eclipse (i.e., A star in front of G star) being total; further, the system is a double lined spectroscopic binary, and the radii of the binary components can be determined from light curve modelling as 3.0 and 0.9 $R_\odot$.

ROSAT PSPC X-ray observations of α CrB were obtained during secondary (optical) eclipse, on July 12, 1992, between 4:23 UT and 16:15 UT; the observation start and stop times span the time of the predicted secondary eclipse of July, 12, 1992 at UT 9:30. In Figure 7 the observed X-ray light curve of α CrB in terms of the observed PSPC count rate (binned into 200 second bins) vs. orbital phase is shown. The data have been corrected for all instrumental effects including dead time and are background subtracted, the phase refers to the stellar periastron and has been computed from the parameters given in Table 2 in Tomkin and Popper (1986). The light curve displayed in Figure 7 shows an obvious eclipse (at least) between phases 0.945 and 0.956. An inspection of the individual photons recorded between those times in comparison with the expected background count rates shows that the eclipse is consistent with being total (i.e., no source X-ray flux). Therefore the contribution of the A-type primary star to the total X-ray flux in the α CrB system must be negligible in accordance with expectations (Schmitt et al. 1985, Schmitt 1992).

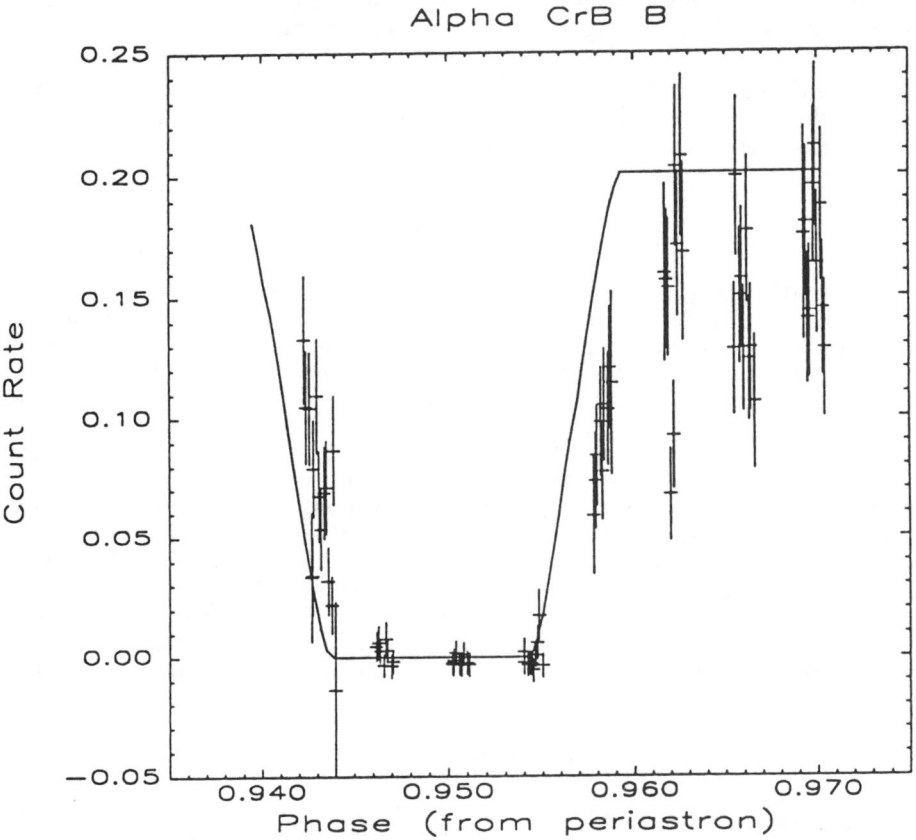

Fig.7: PSPC light curve (in counts/sec) vs. phase (with respect to stellar periastron) of the secondary minimum of α CrB on July 12, 1992. The solid line represents the light curve expected from a uniformly bright corona extending over the stellar disk.

From the out of eclipse count rate of α CrB of 0.17 cts/s, one estimates an X-ray luminosity of $\approx L_X \sim 6 \; 10^{28}$ erg/s, i.e., more than one order of magnitude larger than the typical soft X-ray luminosity of the Sun. If one had to account for the total X-ray output from α CrB with solar-like active regions, one would have to cover most of the star's surface with active regions and expect a coronal filling factor near unity. If one were then to model the corona as a uniformly emitting disk, a light curve drawn as a solid curve in Figure 7 would result. Clearly, the observed X-ray light curve is inconsistent with such a model. A detailed study of eclipse ingress and egress allows the construction of an eclipse map (cf., Schmitt and Kürster 1993), which shows that almost half of the total coronal flux comes from a rather small region which in fact may not even be resolved in our map. Assuming an upper limit to the characteristic spatial scale size of $L < 1.1 \; 10^{10} \; cm$, one can determine a characteristic coronal density. With a volume of $V < 1.5 \; 10^{30} \; cm^3$, one finds a density $n > 2.2 \; 10^{10} \; cm^{-3}$ using the the reconstructed emission measure. Therefore the main difference between active regions of the corona of α CrB B and solar active regions is the density.

5. Conclusions

I have attempted to review some of the highlights of ROSAT observations of stellar X-ray emission both from the ROSAT all-sky survey as well as the ROSAT pointing program. By necessity, this review had to be incomplete, but hopefully it can provide a glimpse of the many exciting ROSAT results on stellar X-ray sources. While the results on early type stars are "expected" in the sense that they provide support for the commonly accepted scenario of X-ray emission from shocks in the stars' winds, the results on hot white dwarfs and eclipsing cool star binary systems are "unexpected": Contrary to expectations, hot white dwarfs do not constitute a significant fraction of the soft X-ray sky, and therefore future high spectral resolution studies can be carried out only with a by comparison small number of sources. The ROSAT observations of eclipsing systems of cool stars reveal quite a few sources previously not known to show X-ray eclipses, hence casting doubt on the scenario derived from (eclipse-wise unsuccessful) EXOSAT observations that the X-ray emitting material in these systems pervades – in a fashion very similar to the microwave emitting material – the whole binary system.

Acknowledgments

I am grateful to the members of the MPE ROSAT stellar team, T. Berghöfer, M. Kürster, R. Neuhäuser, R. Ottmann, S. Randich, and C. Rosso.

References

Berghöfer, T.W. and Schmitt, J.H.M.M., 1994, Proceedings of Workshop on *Instability And Variability of Hot-Star Winds", Quebec, Kluwer.*
Briel, U.G. and Henry, J.P., 1993, *Astron. Ap.*, **278**, 379.
Brown, A., 1994, this volume
Caillault, J.-P., 1994, this volume.
Cassinelli, J.P., and Olson, G.L., 1979, *Ap. J.*, **229**, 304.
Cassinelli, J.P., Waldron, W.L., Sanders, W.T., Harnden F.R., Jr., Rosner, R., Vaiana, G.S., 1981, *Ap. J.*, **250**, 677

Fleming, T.A., Giampapa, M.S., Schmitt, J.H.M.M., and Bookbinder, J.A., 1993, *Ap. J.*, **410**, 387.
Fleming, T. *et al.*, 1993, in Proceedings of Eighth European Workshop on White Dwarfs, ed. M.A. Barstow, Kluwer Academic Press.
Fleming, T. 1993, personal communication
Fleming, T.A. and Schmitt, J.H.M.M., 1994, in *Proceedings of the Eighth Workshop on "The Sun, Cool Stars and Stellar Systems"*, Athens, Georgia, October 1993, ed. J.-P. Caillault, in press.
Golub, L., *et al.*, 1983, *Ap. J.*, **271**, 264.
Harnden, F.R., Jr., Branduardi, G., Elvis, M., Gorenstein, P., Grindlay, J., Pye, J.P., Rosner, R., Topka, K., Vaiana, G.S., 1979, *Ap. J. Lett.*, 234, L51
Haisch, B.M., Schmitt, J.H.M.M., Rodono, M., and Gibson, D.M., 1990, *Astron. Ap.*, **230**, 419.
Hillier, D.J., Kudritzki, R.-P., Pauldrach, A. W., Baade, D., Cassinelli, J. P., Puls, J., Schmitt, J.H.M.M., 1993, *Astron. Ap.*, **276, 117**.
Kashyap, V., *et al.*, 1994, *Ap. J. Lett.*, in press.
Kudritzki, R.-P., *et al.*, 1991, in **Extreme Ultraviolet Astronomy**, ed. R.F. Malina and S. Bowyer, Pergamon Press.
Kürster, M., 1994, in *Proceedings of the Eighth Workshop on "The Sun, Cool Stars and Stellar Systems"*, Athens, Georgia, October 1993, ed. J.-P. Caillault, in press.
Leung, K.-C. and Schneider, D.P., 1978, *Astron. J.*, **83**, 618.
Lucy, L.B., and White, R.L., 1980, *Ap. J.*, **241**, 300.
Lucy, L.B., 1982, *Ap. J.*, **255**, 286
Napiwotzki, R., Barstow, M.A., Fleming, T., Holweger, H., Jordan, S. and Werner, K., 1994, *Astron. Ap.*, in press.
Ottmann, R., Schmitt, J.H.M.M., and Kürster, M., 1993, *Ap. J.*, **413**, 710.
Ottmann, R., 1994, *Astron. Ap.*, submitted.
Pallavicini, R., Tagliaferri, G., Pollock, A.M.T., Schmitt, J.H.M.M., and Rosso, C., 1990, *Astron. Ap.*, **227**, 483.
Reimers, D., and Schmitt, J.H.M.M., 1992, *Ap. J. Lett.*, **392**, 55.
Schmitt, J.H.M.M., Golub, L., Harnden, F.R., Jr., Maxson, C.W., Rosner, R., and Vaiana, G.S., 1985, *Ap. J.*, **290**, 307.
Schmitt, J.H.M.M., 1992, in *Cool Stars, Stellar Systems and the Sun*, Seventh Cambridge Workshop, ed. M.S. Giampapa and J.A. Bookbinder, ASP Conference Series, **26**, 83.
Schmitt, J.H.M.M., Haisch, B.M., and Barwig,H.,1993, *Ap. J. Lett.*, **418**,L81.
Schmitt, J.H.M.M. and Kürster, M., 1993, *Science*, **262**, 215.
Schmitt, J.H.M.M., 1994, *Ap. J. Supp.*, , in press.
Schmitt J.H.M.M., Güdel, M., Predehl P., 1994, *Astron. Ap.*, in press.
Schmitt, J.H.M.M. *et al.*, 1994, to be submitted.
Smith, M.A., Grady, A.L., Peters, G.J., and Feigelson, E.D., *Ap. J. Lett.*, **409**, L49.
Stern, R.A., Uchida, Y., Tsuneta, S., and Nagase, F., 1992 *Ap. J.*, **499**, 321.
Tomkin, J., and Popper, D.M., 1986, *Astron. J.*, **91**, 1428.
Vaiana, G.S. *et al.* , 1981, *Ap. J.*, **245** , 163.
Walter, F.M., Gibson, D.M. and Basri, G.S., 1983, *Ap. J.*, **267**, 665.
White, N.E., Culhane, J.L., Parmar, A.N., Kellett, B.J., Kahn, S., van den Oord, G.H.J. and Kuijpers, J., 1986, *Ap. J.*, **301** 262.
White, N.E., Shafer, R.A., Horne, K., Parmar, A.N. and Culhane, J.L., 1990, *Ap. J.*, **350**, 776.
White, N.E., 1993, personal communication

THE CORONA OF THE K5 GIANT γ DRA, AND ITS RELATION TO THE HYBRID-CHROMOSPHERE STARS

Alexander Brown and Jeffrey L. Linsky[†]
Joint Institute for Laboratory Astrophysics,
University of Colorado & National Institute of Standards and Technology,
Boulder, CO 80309-0440, USA

Thomas R. Ayres
Center for Astrophysics and Space Astronomy,
University of Colorado, Boulder, CO 80309-0389, USA

Email IDs
ab@jila.colorado.edu; jlinsky@jila.colorado.edu; ayres@vulcan.colorado.edu

ABSTRACT

Gamma Draconis is the first, normal, single late K giant located on the red side of the coronal "dividing line" known to show conclusive evidence for hot ($\sim 10^5$ K) transition region (TR) and coronal plasma. We present ROSAT PSPC data and HST GHRS spectra of γ Dra and describe the coronal and TR properties of this K5 III star. The high temperature emissions of γ Dra are compared to those of a sample of hybrid-chromosphere bright giants and supergiants. New PSPC detections of the K3 giant α Hya and the G supergiant β Aqr are presented. Upper limits are found for the hybrid-chromosphere stars θ Her and α Aqr. These new measurements extend the X-ray to C IV flux-flux relations to significantly lower activity levels.

1. Why Is γ Dra Important?

The K giant γ Dra is particularly important in understanding coronal and transition region (TR) evolution in the region of the coronal dividing line (Ayres et al. 1981 ApJ 250 293 ; Schmitt 1992 in Cool Stars, Stellar Systems, and The Sun, (Eds. M.S. Giampapa & J.A. Bookbinder), PASP Conf. Ser. Vol. 26, 83) because:

i) γ Dra is a normal single K5 III giant star, either ascending the red giant branch or descending back down towards the giant clump after He ignition, and should show atmospheric structure typical of such stars.
ii) γ Dra shows extremely weak chromospheric emission, only a factor of 2 above the basal flux level (cf. Rutten et al. 1991 A&A 252 203), and thus offers the opportunity to investigate coronal structure at the lowest levels of magnetic activity currently observable.
iii) γ Dra lies on the "non-coronal" side of the coronal dividing line and deep X-ray observations can search for the presence of remnant coronal plasma in this region in which coronae are generally thought to be absent.
iv) γ Dra has been observed with the Goddard High Resolution Spectrograph (GHRS) on the Hubble Space Telescope (HST) and shows unusual chromospheric and TR properties that indicate the star might also have a corona (Linsky 1991 Mem. Soc. Ast. It. 63 577).

[†] Staff Member, Quantum Physics Division, NIST

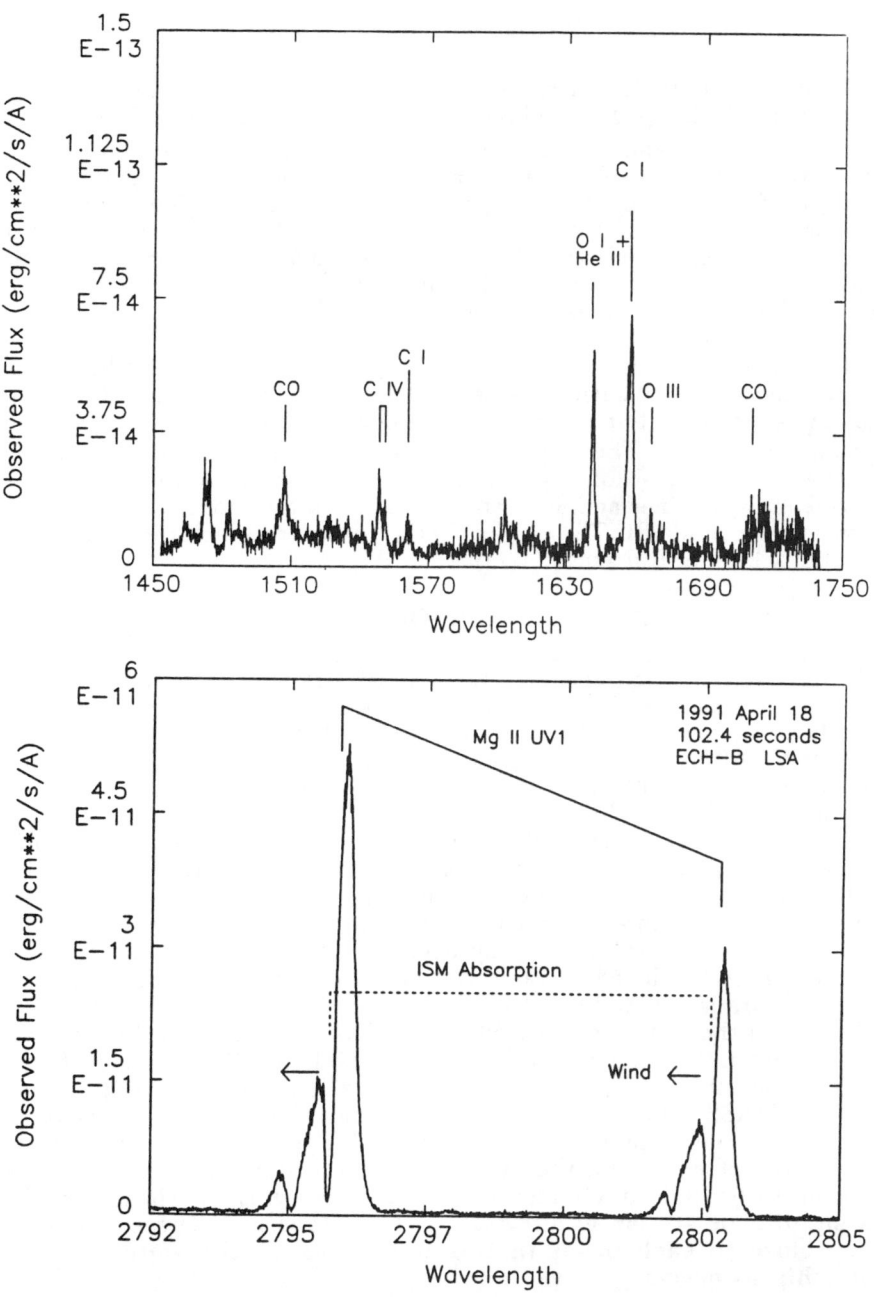

FIGURE I GHRS spectra of γ Dra. Upper panel: G140L low resolution spectrum showing C IV emission lines formed in the TR. Lower panel: Echelle spectrum of Mg II emission lines showing absorption due to a 70 km s^{-1} stellar wind.

2. GHRS Observations Of γ Dra

Ultraviolet spectra of γ Dra were obtained with the GHRS in April 1991. A set of low-resolution (G140L) observations of the complete far-UV spectrum from 1150 to 1850 Å, and high-resolution Mg II and O I emission line profiles show that the outer atmosphere of this giant has a hybrid-chromospheric structure, indicated by the simultaneous presence of transition region plasma **and** a high velocity cool wind. While there were some earlier indications that this star possessed a high velocity wind, the GHRS spectra clearly demonstrate the "hybrid-chromospheric" structure of its outer atmosphere. Figure I illustrates the low-resolution spectrum containing C IV and the Mg II echelle profiles. γ Dra is the first normal K giant known to have such a "hybrid' atmosphere. Since this star is an otherwise normal mid-K giant, it appears very likely that the outer atmospheres of such stars, which have come to be described as "non-coronal", possess small but non-negligible amounts of plasma at temperatures of up to several 10^5 K. These spectra represent the most sensitive ultraviolet observations yet obtained of any cool giant star. They allow the detection of many weak emission lines and accurate estimation of the radiative loss levels in the chromosphere and TR. The C IV radiative losses are a factor of 4 above the TR basal flux level of Rutten *et al.* (1991).

3. ROSAT PSPC Observations

Prior to our observations, only three hybrid-chromosphere stars had been detected as X-ray sources. Brown *et al.* (1991 ApJ 373 614) detected the K bright giant α TrA using the EXOSAT LE detectors and this star was also detected in the ROSAT All-Sky Survey (Haisch, Schmitt & Rosso 1992 ApJL 388 L61). Reimers & Schmitt (1992 ApJL 392 L55) detected the hybrid-chromosphere K bright giants γ Aql and β Ind.

Given a need for a better understanding of the coronal properties of hybrid-chromosphere stars, we observed five evolved stars with the ROSAT PSPC during AO 2 and 3: three known hybrid-chromosphere stars (β Aqr, α Aqr, θ Her) and two supected hybrid stars (γ Dra, α Hya). Subsequent analysis has shown α Hya to be a giant star without any detectable stellar wind. The results of these observations are listed in Table 1. The coronal volume emission measure (VEM) for γ Dra is 4-7 10^{50} cm^{-3} while the C IV VEM is 1.3 10^{51} cm^{-3}.

In Fig. II these stars are placed on the X-ray to C IV flux-flux diagram. The other stars in these figures are from the RIASS ROSAT/IUE sample (Ayres *et al.* 1994 ApJ submitted.) The dashed line in each panel is the 3/2 power law that fits the RIASS data for G and K main sequence stars. Except for the bright giants and supergiants (luminosity classes I and II), the stars all seem to fit this line relatively well except for the early F stars. The stars more luminous than giants seem to be uniformly underluminous in X-rays. **Even though γ Dra has a significant stellar wind while α Hya does not, these two stars lie very close to each other in this figure** *and* **on the same flux-flux relationship as active giants.**

4. Summary

- Coronae can exist on K giants beyond the "coronal dividing line" and the properties of γ Dra are likely to be typical for mid-K giants.

TABLE 1
ROSAT Observations of γ Dra and Other Evolved Stars

Star	Spectral Type	PSPC Count Rate (ct ksec^{-1})		Exposure Time (sec)	Date
γ Dra	K5 III	3.14 ± 1.02	3.1σ	18870	1992 Nov 21-29
α Hya	K3 III	7.06 ± 1.59	4.4σ	8528	1992 May 14-30
β Aqr	G0 Ib	10.48 ± 1.88	5.6σ	8032	1992 May 11-13
θ Her	K1 II	≤4.5	≤3.0σ	7434	1992 Mar 31
α Aqr	G2 Ib	≤17	≤3.0σ	5347	1992 May 12-13

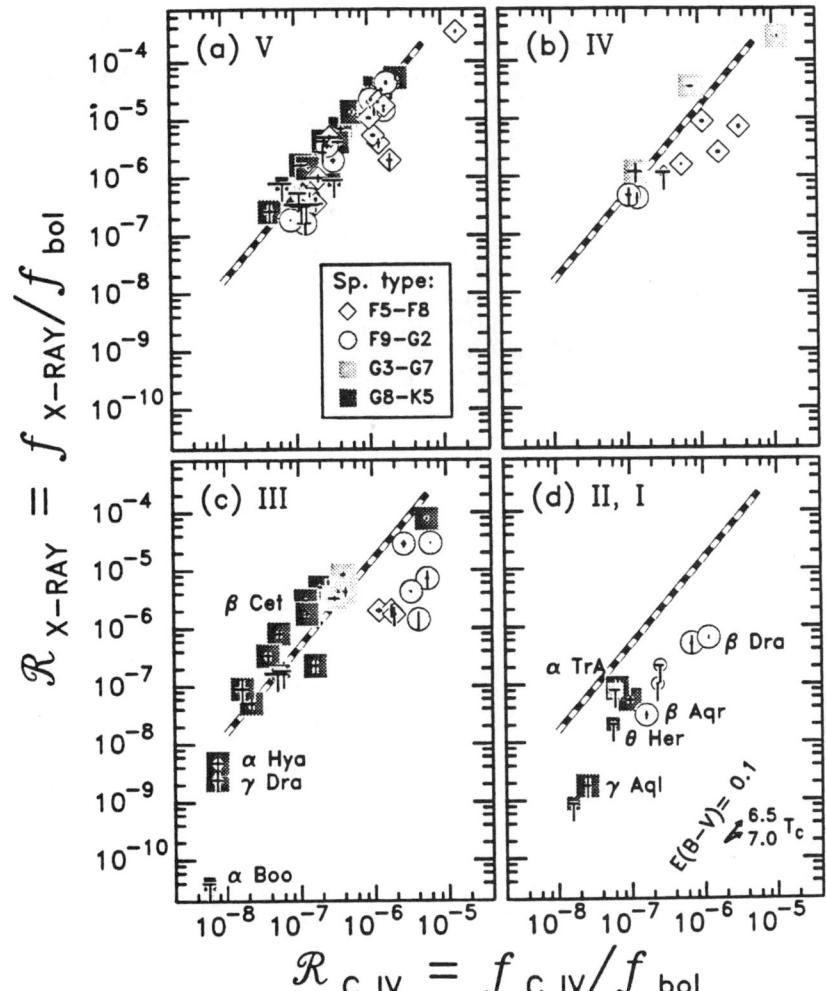

FIGURE II ROSAT PSPC X-ray to C IV flux-flux relations for stars of different luminosity class. γ Dra and α Hya extend this relation for giants by more than an order of magnitude.

- The coronae of stars like γ Dra are weak but not noticeably different from those of stars at the dividing line that **do not** have winds. Why should the coronae of γ Dra and α Hya be so similar if energy losses due to mass loss are important?
- The coronal and TR fluxes of γ Dra and α Hya extend and are totally consistent with the flux-flux relations for more active giants.
- G supergiant hybrid-chromosphere stars, e.g. β Aqr, also show X-rays and have coronae.
- Hybrid-chromosphere luminosity class I and II stars follow the same flux-flux relations as bright giants and supergiants with extremely active coronae.
- It is not clear whether these data confirm the picture that hot plasma on evolved stars is confined within ever decreasing TR and coronal filling factors as stellar winds increasingly dominate the outer atmospheric structure. Direct measurement of coronal densities for comparision with conditions in other atmospheric regions will be the best method to establish the degree of connectivity between the different temperature regimes.

We thank F. Walter for provision of IDL ROSAT data analysis software used for this work. This paper is based on data obtained with the ROSAT satellite and the NASA/ESA Hubble Space Telescope. The HST data were obtained at the Space Telescope Science Institute, which is operated by the Association of Universities for Research in Astronomy Inc. under NASA Contract NAS5-26555. This research was supported by NASA grants NAG5-1792 and NAGW-2904 to the University of Colorado and S-56460-D to the National Institute of Standards and Technology.

SOFT X-RAY OBSERVATIONS OF ISOLATED RADIO PULSARS

John P. Finley
Purdue University
Department of Physics
1396 Physics Bldg., West Lafayette, IN 47907

Email ID
finley@purds1.physics.purdue.edu

ABSTRACT

The soft X-ray response (below ~ 0.5 keV), spectral resolution, and imaging capabilities of *ROSAT* make it an ideal observatory for studying the high energy emission from isolated radio pulsars. Some of the results from the first three years of the *ROSAT* mission pertaining to this interesting class of neutron stars are summarized. The class is presented according to their dynamical age ($\tau \sim P/2\dot{P}$) which, by no accident, is also a presentation according to rotational energy loss rate. The implications for the thermal evolution of neutron stars will be highlighted.

1. Introduction

One of the first motivations of X-ray astronomy was the detection of thermal emission from the surface of a neutron star. Neutron stars, the ultra-compact result of the evolution of main sequence stars with masses greater than about 8 $M_\odot$, are born with internal temperatures on the order of 10^{11} K. Possessing no internal energy sources, within a few hundered years they rapidly cool through neutrino processes to internal temperatures of the order of 10^8 K and corresponding surface temperatures of a few 10^6 K. The neutrino processes are the dominant cooling mechanism for the first 10^5 to 10^6 years during which time the surface temperature slowly drops to around a few 10^5 K. Beyond the age of about 10^6 years surface photon emission becomes the dominant cooling mechanism and the surface temperature drops rapidly (for a review of "standard cooling" see Nomoto and Tsuruta 1987). Thus the hot surfaces of "young" neutron stars, dynamical ages ($\tau = P/2\dot{P}) < 10^6$ years, were expected to be detectable in the X-ray band. Unfortunately expectations and capabilities are not always in synch and detailed calculations of the thermal evolution of neutron stars (Chiu & Salpeter 1964) indicated that the X-ray detectors available in the early years lacked the necessary sensitivity to detect the surface emission. Hope does bloom eternal, however, and the problem was addressed with renewed vigor during the *Einstein* mission. The imaging capabilities of the mirrors and the spectral resolution of the IPC detector did progress the endeavor but, alas, a convincing case for the detection of surface emission from a neutron star still remained elusive. And so it remained.

The launch of *ROSAT* in June of 1990 once again offered the possibility of detecting the thermal surface emission of a neutron star. The excellent spatial and spectral resolution of the X-ray mirror/PSPC combination ($\sim 0.5'$ and $E/\triangle E \sim 4$ at 1 keV respectively) and the soft X-ray response (below ~ 0.5 keV) made the possibility, this time, a distinct one. The purpose of this paper will be to review some of the early results on isolated radio pulsars from the first few years of the *ROSAT* mission. By isolated radio pulsars I refer to "rotation powered" solitary neutron stars as distinct from the neutron stars found in "accretion powered" binary systems. The class will be discussed in order of their dynamical, or spindown, age as defined above (which, by no accident, is also an ordering according to rotational energy loss rate). *ROSAT*'s contributions to the present understanding of the thermal evolution of neutron stars will be an intended bias.

2. "Crab"-like Pulsars

This group of pulsars are young with dynamical ages in the range $\tau = P/2\dot{P}$ $\sim 10^3$ years. In addition to the Crab pulsar (PSR B0531+21), whose name I have borrowed for the group, we find PSR B1509-58, the pulsar associated with the supernova remnant (SNR) MSH15-52 (Seward and Harnden 1982), and the LMC pulsar PSR B0540-69 (Seward, Harden, and Helfand 1984) which is often referred to as the "twin of the Crab pulsar". The parameters of these objects are given below. All the pulsars

PSR B	Log τ (years)	Log $\dot{E}$ (erg/s)	Period (msec)
0531+21	3.09	38.65	33
1509-58	3.19	37.25	150
0540-69	3.22	38.17	50

in this group are associated with X-ray bright emission nebulae and this presents an insurmountable background which cloaks the thermal X-rays originating on the stellar surface. In addition, neutron stars this young and rapidly rotating are expected to possess a large non-thermal flux of X-ray emission originating in the magnetosphere. Two models have been extensively discussed in the literature; the outer gap model (Cheng, Ho, and Ruderman 1986a,b), and the polar cap model (Harding and Daugherty 1991). Both of these models are based upon a population of relativistic electrons created by the breakdown of voltage gaps near the boundaries present in the charge separated magnetosphere. The spindown power ($\dot{E} = I\omega\dot{\omega}$) is of order 10^{38} erg/s with a significant fraction emerging in the X-ray band (typically 10%). The most extensively studied of this group by *ROSAT* is PSR B0540-69 which will now be summarized.

2.1 The LMC Pulsar PSR B0540-69

The results of *ROSAT* PSPC observations (Finley *et al.* 1993) indicate a non-thermal spectrum for the pulsar plus SNR (the spatial resolution of the PSPC is insufficient to separate the two) best fit with a power law of photon index $\Gamma = -2.0$ observed through an equivalent interstellar hydrogen column density of 4×10^{21} cm^{-2}. This value for the column density implies that the pulsar plus SNR resides deep inside the gas clouds of the LMC. The X-ray flux of 5.7×10^{-11} erg/cm^2/s over the *ROSAT* bandpass of 0.1-2.4 keV corresponds to a luminosity of $\sim 2 \times 10^{37}$ erg/s for a distance to the LMC of 50 kpc, i.e. some 11% of the available spindown power. Timing results for the X-ray data best constrained by contemporaneous optical measurements (Gouiffes *et al.* 1992) indicate that the X-ray and optical pulses are in phase. Pulse phase resolved spectral modeling reveals a harder pulsed component (photon index $\Gamma = -1.3$) which agrees with previous optical to X-ray interpolations and is indicative of a synchrotron mechanism for the emission. The luminosity of the pulsed component is $\sim 1\%$ of the available spindown power. A weak upper limit to the surface temperature of $T_S \leq 5.3 \times 10^6$ K was derived from the data.

The good spatial resolution of the HRI ($\sim 5''$) allowed the study of the outer shell of SNR 0540-69.3 as well as the pulsar (Seward and Harnden 1993). Timing results are in agreement with the PSPC data and model dependent luminosity estimates indicate that approximately 20% of the X-ray flux originates in a diffuse outer shell of diameter $55''$. A blast-wave analysis yields an initial energy of the expanding supernova ejecta of $\sim 2 \times 10^{51}$ erg.

3. "Vela"-like Pulsars

This group, for lack of a better term, I refer to as juvenile with ages in the range of 10^4 years. Included in the group are the namebearer and canonical member the Vela pulsar (PSR B0833-45), PSR B1800-21, PSR B1706-44, PSR B1823-13, and PSR B1951+32 which is associated with the SNR CTB80. Parameters are given in the table below. All of these pulsars would be expected to be associated with a SNR given their

PSR B	Log τ (years)	Log $\dot{E}$ (erg/s)	Period (msec)
0833−45	4.05	36.84	89
1800−21	4.20	36.34	133
1706−44	4.25	36.53	102
1823−13	4.50	36.45	101
1951+32	5.02	36.58	40

youth (indeed the Vela pulsar and PSR B1951+32 are) but the reality is not so straightforward. In addition to the SNR a compact nebula is observed surrounding the pulsar in the 2 well studied members of this group, the Vela pulsar and PSR B1951+32. This compact nebula is believed to be the result of a relativistic wind of e^+e^-, accelerated by the low frequency dipole radiation from the pulsar, coming into pressure equilibrium with the surrounding interstellar medium. More than enough spindown energy is available, of order 10^{36} erg/s in this group, to fuel this nebula. The fraction of the spindown energy emerging in the X-ray band, however, has dropped to $\sim 1\%$ for this group.

3.1 The Vela Pulsar PSR B0833-45

The *ROSAT* All Sky Survey revealed that the Vela SNR is circular with a diameter of $\sim 5°$ (see the *ROSAT* calendar, May 1992). The pulsar resides at the geometric center of the SNR surrounded by a $2'$ compact nebula. A deep pointed observation with the PSPC (Ögelman, Finley, and Zimmerman 1993) detected X-ray pulsations for the first time, a long standing enigma which *ROSAT* has put to rest. The pulsed fraction is small, $\sim 4\%$, and resides in the soft channels (0.1 − 0.8 keV). The $0.5'$ spatial resolution allowed a spectral separation of the pulsar from the compact nebula. The pulsar is soft and can be described by a blackbody spectrum with kT = 0.15 keV implying a bolometric luminosity of $\sim 4\times10^{32}$ erg/s. The derived size of the emitting area of 3-4 km is smaller than the 11-18 km expected for the surface of a neutron star. The harder compact nebula can be described equally well by a hot thermal bremstrahlung plasma of kT = 1.6 keV or a power law distribution with photon index Γ = -2. Both descriptions yield luminosity estimates for the compact nebula about 5 times the bolometric luminosity of the pulsar. The low luminosity of this region compared to the available spindown power implies that the region must be highly transparent to the low frequency dipole radiation from the pulsar.

3.2 PSR B1800-21

This pulsar lies at the edge of a recently discovered radio SNR G8.7-0.1, part of the W30 complex, and a possible association was drawn due to coincidences on the sky, in space, and in age (Kassim and Weiler 1990). A 10,000 second PSPC exposure detected PSR B1800-21 at the 3σ level and is reported here for the first time. Bright diffuse emission was also observed north of the pulsar's position which corresponds with the northern extent of SNR G8.7-0.1. A preliminary analysis indicates that the emission is consistent with a SNR commensurate in age with PSR B1800-21 (Finley and Ögelman 1993a). Recent high resolution radio maps reveal a peculiar radio morphology for G8.7-0.1 (Frail, Kassim, and Weiler 1993). The SNR/PSR association can be salvaged if the initial supernova event occurred in a region of non-uniform density, a hypothesis which the X-ray data support (Finley and Ögelman 1993).

3.3 PSR B1706-44

This pulsar was detected by *ROSAT* (Becker et al. 1992) as well as *EGRET* aboard the Compton Gamma Ray Observatory (CGRO) above 100 MeV (CGRO EGRET Pulsar team 1992). The X-ray detection is consistent with a hard and heavily absorbed source. A limit to the pulsed fraction of $\leq 40\%$ was derived from the data.

3.4 PSR B1823-13

PSR B1823-13 was detected by the PSPC (Finley and Ögelman 1993b) during the AO3 period as part of the program which also detected PSR B1800-21. Both of these pulsars are peculiar in that no compact nebula is detected in the radio band (Braun, Goss, and Lyne 1989). The preliminary analysis indicates a hard and heavily absorbed spectrum and a limit to the pulsed fraction of $\leq 33\%$. If it is assumed that there is no compact nebula a surface temperature of $T_S \sim 0.15$ keV is consistent with the observed intensity for a neutron star of radius 12 km. A deep HRI exposure upcoming in AO4 should determine if the emission is from a point-like or extended object.

3.5 The CTB 80 Pulsar PSR B1951+32

The detection of pulsed X-rays at the radio period from PSR B1951+32 was another *ROSAT* first (Safi-Harb et al. 1993). The pulse profile displays a narrow peak with a small pulsed fraction in phase with the radio. The X-ray morphology is composed of several components. A compact core of 1' diameter contains the pulsar and a compact nebula each contributing ~ 40% and 60% of the X-ray luminosity from that region respectively. Seperation of the two components yields a surface temperature for the softer pulsar of $T_S \sim 0.17$ keV. The compact core is surrounded by a 5' diameter diffuse region consistent with a cooling SNR. No spectral change as a function of radius from the compact core is observed as had been previously reported. A weaker diffuse component extends to the E and NE of the position of the compact core.

4. Intermediate Age Pulsars

The intermediate aged pulsars are the group with $\tau \sim 10^5$ years. This group is composed of the Monogem pulsar PSR B0656+14 (but see below), Geminga (PSR B0630+18), and PSR B1055-52. The parameters are given in the table below. This

PSR B	Log τ (years)	Log $\dot{E}$ (erg/s)	Period (msec)
0656+14	5.04	34.58	385
0630+18	5.51	34.54	237
1055−52	5.73	34.48	197

group represents the best hope for detecting the thermal surface emission since at ages of $> 10^5$ years no complications due to bright SNR emission would be expected. In addition, HRI observations can detect or place limits on any compact nebular component. The spindown energy available is moderate, of order 10^{34}–10^{35} erg/s, and a very small fraction of typically $\leq 0.1\%$ emerges in the X-ray band.

4.1 PSR B0656+14

The Monogem pulsar PSR B0656+14, so called because of its proximity with the center of the Monoceres-Gemini soft X-ray enhancement (hereafter the Monogem ring), was found during the *Einstein* mission to be one of the brightest X-ray emitting isolated pulsars (Córdova *et al.* 1989). The association of PSR B0656+14 with the Monogem ring was bolstered by proper motion measurements which indicated that the pulsar was moving away from the geometric center of the ring (Thompson *et al.* 1991). However, a recent revised measurement show that the pulsar is actually moving toward the center of the Monogem ring rather than away (Thompson and Córdova 1993) and is therefore not associated. A 3,000 second PSPC observation acquired in March of 1991 revealed the 385 milli-second pulsations in the X-ray band (Finley, Ögelman, and Kızıloğlu 1992) confirming a previous marginal *Einstein* result (Córdova *et al.* 1989). The spectrum was very soft and well described by a blackbody of temperature 0.078 keV. The derived stellar radius was 10 km under the assumption of a 500 pc source distance. A weak high energy tail was also observed. The pulse shape was a broad sinusoid with a pulsed fraction of 14% and there was evidence that the spectrum was harder at the peak of the emission. A 10,000 second HRI pointing indicated that the source was point-like with less than 14% of the X-ray intensity originating in an extended component (Anderson *et al.* 1993) and lends further support to the cooling neutron star hypothesis. Since these early results an additional 13,000 seconds of PSPC data have been acquired and a presentation of this data can be found in the summary and conclusions.

4.2 The Geminga Pulsar PSR B0630+18

The fact that I can title this section "The Geminga Pulsar" is one of the truly outstanding accomplishments of the *ROSAT* mission. The Geminga pulsar is also the anomaly of the group of isolated radio pulsars presented here in that it is not a radio emitting pulsar. Oddly enough, however, it is γ-ray loud. First discovered in 1972 (Fichtel *et al.* 1975) during the SAS−2 mission its nature has been one of the long standing mysteries of high energy astrophysics although a similarity to the Vela pulsar had been suspected (Bignami *et al.* 1988, Halpern 1989). A 14,000 second PSPC exposure during March of 1991 put the mystery to rest with the detection of 237 millisecond pulsations from Geminga (Halpern and Holt 1992) and demonstrates its inclusion in the class of intermediate aged pulsars. The spectrum has two components (like PSR B0656+14) composed of a soft blackbody of temperature $kT = 0.045$ keV and a high energy tail which can be described by either a blackbody of temperature $kT = 0.26$ keV emitting from an area $\sim 10^{-5}$ the size of the soft component or a powerlaw of energy index ~ -1.5 (Halpern and Ruderman 1993). The pulse profile is rich in detail. In the soft band it is a broad sinusoid with a 33% pulsed fraction while the profile in the hard band has a similar pulsed fraction but is narrower than in the soft band. In addition, the hard component leads the soft component by $\sim 105°$.

4.3 PSR B1055-52

To tell the story of PSR B1055-52 is to rehash what I have just described for Geminga and PSR B0656+14. Another first time detection of pulsations in the X-ray band PSR B1055-52 also displays a soft component with a high energy tail (Ögelman and Finley 1993). The blackbody description yields a surface temperature of $kT = 0.065$ keV for the soft component while the hard component can be described by a powerlaw with photon index $\Gamma \sim -1.4$. The pulse profile is a broad sinusoid in the soft band with a very small pulsed fraction of $\sim 8\%$ while in the hard band it is narrower with a very large pulsed fraction of $\sim 50\%$ around 1 keV. The hard component like Geminga, leads the soft component. In all respects PSR B1055-52, PSR B0656+14, and Geminga are remarkably similar. In addition PSR B1055-52 was detected by EGRET aboard the CGRO above 300 MeV where it is seen to be 100% pulsed (Fierro *et al.* 1993). The γ-ray spectrum is a very hard powerlaw with a photon index of $\Gamma \sim -1.2$. I note that the powerlaw spectrum, if extrapolated to the *ROSAT* band, can describe the high energy tail.

5. Old Pulsars

The term old for this group with ages $\sim 10^6$ only refers to X-ray pulsars since the "average" radio pulsar is a few 10^6 years old. In this group I will discuss PSR B1929+10 the parameters of which are given below. The spindown power for pulsars in this age group are typically of order 10^{33} erg/s and, given the limited sample size, the fraction emerging in the X-ray band is a very small 0.01%. Since the initial heat of formation by this age is very small ($\sim 10^5$ K) and would be undetectable by *ROSAT* any X-ray emission must arise in the magnetosphere or, if originating on the surface, through a reheating mechanism.

PSR B	Log τ (years)	Log $\dot{E}$ (erg/s)	Period (msec)
1929+10	6.49	33.60	227

5.1 PSR B1929+10

One of the closest pulsars to Earth known is PSR B1929+10. Detected as an X-ray emitting source by *Einstein* (Helfand 1983) the modest counts collected did not allow any quantitative analysis. A spring 1992 observation with the PSPC yielded yet another first time pulsation detection (Yancopoulos *et al.* 1993). The pulse profile is a broad sinusoid with a pulsed fraction of $\sim$ 30% with a maximum that coincides with the radio pulse. The spectrum can be described by a blackbody with a temperature of kT = 0.34 keV but being emitted from a surface area of $\leq$ 50 m. The coincidence of the X-ray and radio pulse and the small emitting area of the X-rays may be indicative of reheating of the magnetic cap region. Lack of emission from the rest of the neutron star surface constrains the temperature to be < kT = 0.026 keV.

6. Summary and Conclusions

ROSAT with its superior spatial resolution, soft X-ray response, and large effective area compared to previous missions has had a large impact on the field of neutron star astrophysics even at this early stage. Prior to launch in 1990 the inventory of pulsed X-ray emitting neutron stars amounted to $\sim$ 5. At this point in time the list has almost tripled and results are continuing to come in. For an instrument not specifically designed to do timing the results in this area are significant. The richness of the data is also impressive. As an example of the richness I have compiled in Figure 1 the pulse profiles of the three intermediate aged pulsars discussed above; PSR B0656+14 (a-c), PSR B0630+18 (d-f), and PSR B1055-52 (g-i). The profiles are for the total *ROSAT* bandpass as well as in a soft and hard band (for the definition of the bands see the figure caption). The data for PSR B0656+14 is from a 13,000 second exposure not previously presented in the literature. The three pulsars display the same general behaviour, a broad sinusoidal pulse in the soft X-ray band and a narrower pulse in the hard band out of phase with the soft pulse. The pulsed fraction in the hard band is, in general, larger than is observed in the soft band. The soft pulse has been attributed to thermal emission from the surface of the cooling neutron star while the hard component has had various interpertations; non-thermal emission from the magnetosphere, emission from a reheated polar cap, etc... (see the literature cited in the text). The modulation of the soft component has been attributed to anisotropies on the stellar surface due to the intense surface magnetic field and opacity effects of the overlying stellar atmosphere (Ventura *et al.* 1993, Shibanov *et al.* 1993, Miller *et al.* 1993). The detail which the data in Figure 1 presents will force the emission models of neutron stars to a new level of sophistication which the previous data did not demand. I find the similarity of the data to be quite remarkble. Should that be surprising? Probably not, after all they are of the same feather and it is pleasing to actually see that each neutron star may not require its own model to explain the observations.

Another area in which significant progress has been made concerns the thermal evolution of neutron stars. As mentioned in the Introduction this was one of the first pursuits of X-ray astronomy and progress has been waiting some 30 years for definitive data. The data are now in hand and a strong case is being made that it is definitive. Here again, the intermediate aged pulsars are the cleanest group. However, the capabilities of *ROSAT* have provided essential data over the entire range of ages where the complications of measuring the surface temperature of the bare neutron star surface are, at least, tractable (10^4-10^6 years). To illustrate the early results some of the published surface temperatures as a function of the dynamical age are presented in Figure 2. The cooling curves are from the published literature for "standard" cooling with and without superfluid components in the stellar interior and, for comparison, the cooling curve if the direct URCA mechanism is operating (for a review of "standard" and non-standard cooling see Ögelman 1991). As the parameter space is populated (I have plotted only a small subset of the available data) the long standing goal of mapping the thermal history of a neutron star will finally progress beyond the simple academic exercise.

I have tried to give an overview of a small sample of the data from the isolated radio pulsars which has emerged from the early phase of the *ROSAT* mission. The treatment was by no means extensive nor exhaustive but I trust it will give a hint as to the type of detailed investigations which the data will allow to be carried out.

7. References

Anderson, S. B., Córdova, F. A., Pavlov, G. G., Robinson, C. R., & Thompson Jr., J. R. 1993, ApJ, 415, 423.
Becker, W., Predehl, P., Trümper, J., & Ögelman, H. 1992, IAUC 5554.
Bignami, G. F. et al. 1988, A&A, 202, L1.
Braun, R., Goss, W. M., & Lyne, A. G. 1989, ApJ, 340, 355.
CGRO EGRET Pulsar Team 1992, IAUC 5485.
Cheng, K. S., Ho, C., & Ruderman, M. 1986a, ApJ, 300, 500.
Cheng, K. S., Ho, C., & Ruderman, M. 1986b, ApJ, 300, 522.
Chiu, H-Y., & Salpeter, E. E. 1964, Phys Rev Letters, 12, 413.
Córdova, F. A., Hjellming, R. M., Mason, K. O., & Middleditch, J. 1989, ApJ, 345, 451.
Fichtel, C. E. et al. 1975, ApJ, 198, 163.
Fierro, J. M. et al. 1993, ApJ, 413, L27.
Finley, J. P., & Ögelman, H. 1993a, *in preparation*.
Finley, J. P., & Ögelman, H. 1993b, IAUC 5787.
Finley, J. P., Ögelman, H., & Kızıloğlu, Ü 1992, ApJ, 394, L21.
Finley, J. P., Ögelman, H., Hasinger, G., & Trümper, J. 1993, 410, 323.
Frail, D. A., Kassim, N. E., & Weiler, K. W. 1993, *preprint*.
Gouiffes, C., Finley, J. P., & Ögelman, H. 1992, ApJ, 394, 581.
Halpern, J. P. 1989, in *Proc. GRO Science Workshop* (ed. N. Johnson), 4-166.
Halpern, J. P., & Holt, S. S. 1992, Nature, 357, 222.
Halpern, J. P., & Ruderman, M. 1993, ApJ, 415, 286.
Harding, A. K., & Daugherty, J. 1991, ApJ, 374, 687.
Helfand, D. J. 1983, in *Supernova Remnants and Their X-ray Emission*, IAU Sympo-

sium 101 (ed. J. Danzinger & P. Gorenstein), 471.
Kassim, N. E., & Weiler, K. W. 1990, ApJ, 360, 184.
Miller, M. C. 1993, in *Isolated Pulsars* (ed. K. A. Van Riper, R. I. Epstein, & C. Ho), 255.
Nomoto, K., & Tsuruta, S. 1987, ApJ, 312, 711.
Ögelman, H. in *Neutron Stars: Theory and Observation*, NATO ASI Series, (ed. J. Ventura & D. Pines), 87.
Ögelman, H., & Finley, J. P. 1993, ApJ, 413, L31.
Ögelman, H., Finley, J. P., & Zimmerman, H. U. 1993, Nature, 361, 136.
Safi-Harb, S., *et al.* 1993, *in preparation*.
Seward, F. D., & Harnden, F. R. 1982, ApJ, 256, L45.
Seward, F. D., & Harnden, F. R. 1993, CfA Preprint No. 3684.
Seward, F. D., Harnden, F. R., & Helfand, D. J. 1984, ApJ, 287, L19.
Shibanov, Y. A. *et al.* 1993, in *Isolated Pulsars* (ed. K. A. Van Riper, R. I. Epstein, & C. Ho), 174.
Thompson, R. J., & Córdova, F. A. 1993, PSU Preprint.
Thompson, R. J., Córdova, F. A., Hjellming, R. M., & Fomalont, E. B. 1991, ApJ, 366, L83.
Ventura, J. *et al.* 1993, in *Isolated Pulsars* (ed. K. A. Van Riper, R. I. Epstein, & C. Ho), 168.
Yancopoulos, S., Hamilton, T. T., & Helfand, D. J. 1993, Columbia Univ. Preprint No. 525.

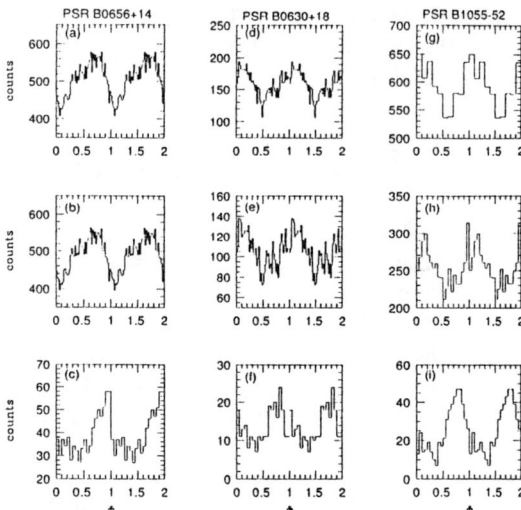

Figure 1. The pulse profiles of the three intermediate aged pulsars PSR B0656+14 (a-c), PSR B0630+18 (d-f), and PSR B1055-52 (g-i) in various energy bands. The bands for PSR B0656+14 are 0.1-2.4 keV (a), 0.1- 0.8 keV (b), and 0.8-2.4 keV (c). The corresponding bands for PSR B0630+18 are 0.07-1.5 keV (d), 0.07-0.28 (e), and 0.53-1.50 keV (f). The bands for PSR B1055-52 are 0.1-2.4 keV (g), 0.1-0.45 keV (h), and 0.55-2.4 keV (i).

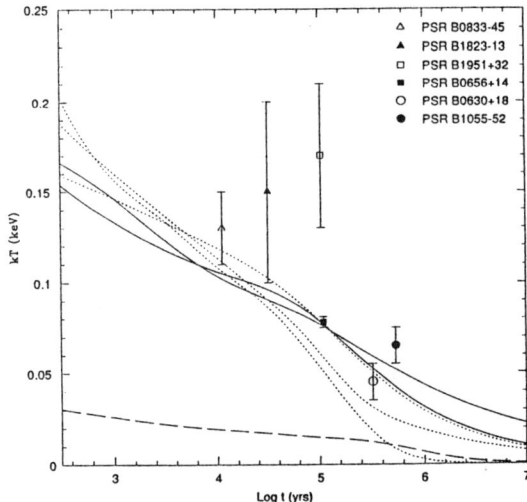

Figure 2. Some representative neutron star cooling curves as published in the literature: standard cooling without superfluid (solid), standard cooling with superfluid (dot), and direct URCA enhanced standard cooling (long dash). The data points are the measured *ROSAT* parameters as discussed in the text.

FOUR NEW RADIO PULSARS
FROM HIGH-ENERGY SELECTED TARGETS

Alex F. Zepka
James M. Cordes
Scott C. Lundgren
Ira M. Wasserman
Space Sciences Building, Cornell University, Ithaca, NY 14853

ABSTRACT

By applying a new method of source detection to the IPC data from the *Einstein* Satellite, we have obtained a list of 1900 X-ray targets at the declination range visible from the Arecibo Observatory. Over two-thirds of those have been observed in a search for radio pulsars that might be associated with X-ray emission. We report the discovery of three new radio pulsars near X-ray targets. One is a young anti-center pulsar (PSR J0631+10) with a shallow spectrum and spindown age of 43 kyr. The other two, PSR J1843+20 and J1908+04B, are slower pulsars with spin periods of 3.4 s and 0.84 s respectively. Both pulsars are at most 3' away from their X-ray counterparts, but the large positional uncertainties cannot yet rule out an association. PSR J1908+04B was found in the same 430 MHz beam of a previously known pulsar, B1905+04, which indicate that many pulsars may have gone undetected by previous searches. In a separate project, we have searched for radio pulsars inside the error boxes of unidentified γ-ray sources found by the EGRET detector aboard the GRO Observatory. Seven gamma-ray error boxes were covered with 430 MHz beam areas. The search has produced a new 3.5 millisecond pulsar towards the galactic anti-center. This pulsar is in a binary orbit with a period of only 6.3 hours. Preliminary results favor a low-mass companion. Investigation of the possible association of the pulsar and gamma-ray source as well as the existence of X-ray emission will have to await further measurements.

1. Introduction

The number of known pulsars today is approaching 600 (Taylor, Manchester and Lyne 1993). Still, only about a dozen of these are known to be associated with high energy sources. A detailed account of these is presented by Finley (these proceedings). Most are young pulsars (spindown age $\tau < 10^6$ years) for which the X-ray emission is either synchrotron radiation in the magnetosphere or thermal blackbody from a hot surface. X-rays have also been observed from a few nearby old neutron stars like the radio-quiet Geminga and the millisecond pulsar J0437-47. In addition, PSR B1929+10 (Wang, Li and Begelman 1993), and possibly B2224+65 (Cordes, Romani & Lundgren 1993), shows a trail of X-ray emission due to the interaction of its magnetosphere with the interstellar medium.

It is possible that some X-ray binaries experience periods of high-energy quiescence at which time the radio pulsar could be detected on its way to become a millisecond pulsar. Recently the binary pulsar B1259-63 (Cominsky, Roberts and Johnston 1993) was observed to emit X-rays in its highly elliptical orbit around a Be star (Johnston *et al.* 1992). This pulsar is believed to be a progenitor of a high-mass X-ray binary

(HMXRB). Also, the eclipsing pulsar B1957+20 emits X-rays as it ablates its gaseous companion (Fruchter et al. 1992).

These high-energy emitting objects have given us very important clues about the mechanisms of emission, neutron star cooling and the evolutionary model linking X-ray binaries and fast radio pulsars. The discovery of more such pulsars is obviously of great interest and, with that intent, we carried out a search of radio pulsars in a sample of X-ray targets obtained from the *Einstein* IPC data.

2. Selection of X-ray Targets

In the past, pulsar searches have been contained to the galactic plane. More recently it has been recognized that a population of nearby millisecond pulsars would be nearly isotropically distributed in the sky and untargetted searches were then carried out, blindly sweeping the sky at all galactic latitudes (Camilo 1994). Although these searches have been highly successful in the sense they discovered new fast and interesting pulsars, they usually have to cover large portions of sky to result in a discovery. In choosing a targetted search of high-energy sources, we expected to increase the efficiency in detecting new interesting pulsars.

Only a small fraction of the sources discovered by the *Einstein* X-ray Observatory were found to be associated with radio pulsars. Thus, in order to increase our chances of discovering new ones, we devised a new method of source detection which allowed us to enlarge the sample of targets for a radio search. This method (Zepka et al. 1994) compares the histogram of counts from the entire image with that expected for noise only. A test is applied to each histogram bin to determine the probability that the number of pixels are within the expected from statistical deviations of the background. If this probability is below a given preset threshold, all corresponding pixels in that bin are selected. Finally the selected pixels from all bins that pass this test are divided into groups of neighboring pixels, each group being a "detected source".

By applying the histogram method to a total of 2652 IPC images we obtained a list of 6783 X-ray sources. Of those, 3638 were in common with sources in the *Einstein* Observatory Catalog of X-ray sources (EOSCAT), and the 1451 for which an identification was available were excluded from our list. In addition, we cross-correlated our source coordinates with several other catalogs (of galaxies, white dwarfs, pulsars) to arrive at a final list of 4704 sources. We then selected 1900 in the range of observable declinations of the 305-meter Arecibo radio telescope for a search of radio pulsars.

3. Radio Search

To date, nearly 1200 of all selected targets have been observed. Each was observed for 3.1 minutes at 430 MHz. The data taking hardware includes a filterbank that divides the 8 MHz bandpass into 32 channels of 250 KHz bandwidth. The data was sampled every 180 μs, resulting in 32 time series (1 Mpoints each) digitized to 3 bits per sample. The off-line data processing consists of the following basic steps: de-dispersion, Fast Fourier Transform (FFT) and harmonic sum.

Since the 430 MHz radio beam at Arecibo has a radius of about 5', the total sky area coverage is about 26 deg^2. Previous untargetted searches at Arecibo using the same hardware and equivalent data processing have discovered about one new pulsar every 100 deg^2 at high galactic latitudes. Taking into account the smaller integration time used in these searches, we estimate a new pulsar should be detected serendipitously for each 30 deg^2 covered by our search. We have so far discovered 3 new pulsars and re-discovered two others previously known.

4. New Pulsars

J0631+10, is a high $\dot{P}$ pulsar ($\dot{P} = 104 \times 10^{-15}$ s s^{-1}), which combined with its 0.287 period, gives a spindown age $\tau = P/2\dot{P} \sim 43,000$ years. The pulsar at 1400 MHz shows a very sharp double peak, while at 430 MHz the pulse profile is much noisier, which is evidence of strong scattering. Perhaps this is due to the fact this pulsar is behind the Lynds 1605 dark cloud. It was easily detected at higher frequencies (1665 and 2380 MHz), corresponding to an unusually flat spectral index ($\alpha < 0.8$). The high DM (= 125 cm^{-3} pc) gives, according to the current model of the galactic electron density (Taylor and Cordes 1993), a distance of about 4 kpc from the solar system, but this is an upper limit since the Lynds cloud is likely to be responsible for some of the dispersion.

The IPC X-ray source is an unidentified EOSCAT source with a reported 5.6 signal-to-noise ratio. Subsequent 1400 MHz data allowed a better estimation of the radio pulsar with an accuracy of about 1.5'. With the distance between pulsar and X-ray source being at most 1.5' the association between the two is not only possible, but, given also its young age, likely. For the reported *Einstein* count rate, we estimate the X-ray luminosity for this source to be about $F_X \sim 2 \times 10^{31} d_{\rm kpc}^2$ erg s^{-1} (where $d_{\rm kpc}$ is the distance Earth-pulsar in kpc). The value of spindown luminosity $\dot{E} = I\Omega\dot{\Omega} \sim 5.4 \times 10^{34} I_{45}$ erg s^{-1} is then consistent with the observed X-ray emission.

In addition to searching for a connection with the X-ray source, we used available data from a few pointings of the CGRO EGRET near J0631+10 to check the possibility that this young pulsar may be a γ-ray emitter. We found some evidence of modulation at the radio pulsar period, although yet far from being conclusive.

Two other pulsars with longer periods were found from our search. PSR J1843+20 is a 3.4 s pulsar about 20° above the galactic plane, with DM = 80 cm^{-3} pc (which corresponds to a distance of about 2 kpc). The corresponding X-ray source was not detected by previous algorithms, probably due to the fact it is at the edge of the IPC image. A value of $\dot{P}$ will soon be available from current timing measurements, but using a typical value of $\dot{P} \sim 2 \times 10^{-15}$ s s^{-1}, we obtain a spindown luminosity $\dot{E} \sim 5 \times 10^{28}$ erg s^{-1} which is insufficient to power the X-ray source since $L_X \sim 8.4 \times 10^{30} d_{\rm kpc}^2$ erg s^{-1}. Observations at 1400 MHz put the distance between the center of the X-ray source and radio pulsar below 2', but since the radio position is uncertain to about the same amount and the X-ray error box is about 1.5', an association cannot be ruled out yet based on positional information only.

Another slow pulsar, J1908+04B ($P = 0.85$ s, DM = 375 cm^{-3} pc), was found at the edge of the supernova remnant W50. The X-ray source is compact and it appears as a 6 σ source in the EOSCAT catalog. Coincidentally, another pulsar was detected by the search algorithm in the same beam area that contains J1908+04B. This other pulsar, B1905+04, was found in a search by Fruchter (1989), but it was unknown to us at the time of our search. We re-processed data from B1905+04 to obtain the timing parameters and a better position for J1908+04B. Due to its large DM (and consequently large distance) and small value for $\dot{P}$, it is unlikely that J1908+04B can be the source of the X-rays since $\dot{E} \sim 4.2 \times 10^{31}$ erg s^{-1} and the X-ray source luminosity (assuming a distance of 5 kpc) is $L_X \sim 3.8 \times 10^{32}$ erg s^{-1}. The other pulsar in the beam, B1905+04, is about 4.5' from the X-ray source and is clearly not associated with it.

5. Search for γ-ray Emitting Radio Pulsars

In addition to X-ray targets, we searched for radio pulsars inside error boxes of unidentified sources detected by the EGRET detector aboard the CGRO γ-ray Observatory. Seven such sources with error boxes varying from 30' to 1° in radius were covered by 430 MHz radio beams placed in a hexagonal pattern with 8' of spacing between the beam centers. Such a pattern requires about 170 beams to cover an entire source of 1° radius with fairly uniform sensitivity. We used the same data taking system and processing as in the X-ray case. The total sky area covered by this search is less than 5 deg^2.

We discovered a new millisecond pulsar, J0751+18 (Lundgren, Zepka and Cordes 1993), about 20' from the center of error box of the γ-ray source J0752+17. The large and monotonic variations on the observed spin period were early evidence that this pulsar is in a tight binary orbit. The initial estimate for the orbital period was of 8.6 hours. Follow-up observations have shown this period to be actually 3/4 of that originally thought, or 6.3 hours, making it the fastest binary orbit known of a radio pulsar outside globular clusters. The reason for the discrepancy is due to zenith angle limitations of the Arecibo telescope (which allows this pulsar to be observed for a little more than 2 hours per day) and the fact that the orbit is close to a submultiple of 24 hours. The orbit was found to be very circular and the minimum companion mass is 0.14 $M_\odot$.

Recent untargetted searches for radio pulsars at Arecibo have found nearly one millisecond pulsar for every 250 deg^2 of sky. Although these searches have used the same hardware as ours, their integration time of about 40 seconds reduces the minimum detectable flux by a factor of two. Still, J0751+18 is strong enough that it would be easily detected by any of them. Since the amount of sky covered by our γ-ray search was of about 5 deg^2, the probability that this discovery was serendipitous is about 2%.

All new pulsars are currently being timed regularly at the Arecibo and we expect to obtain more precise coordinates and timing parameters over the next few months. In addition, we have scheduled optical observations at the Palomar observatory, and high resolution radio maps will be done at the VLA. More γ-ray data for J0631+10 and J0751+18 will be available soon. In the X-ray domain, ROSAT has been used to observe J0751+18 as a Target of Opportunity, and further observations of both pulsars with ASTRO-D are planned. We expect this future work to be more conclusive about the association between the high-energy sources and radio pulsars.

Acknowledgements

We acknowledge the work of F. Camilo (Princeton) and D. Nice (NRAO) in providing us with timing parameters for J1908+04B from the analysis of their B1905+04 data. J. Fierro (Stanford) kindly provided us with software to analyze EGRET data.

This work has been done under the NSF grant AST 9218075 and NASA GRO grants NAG 5-1452 and 5-2436. A. Zepka acknowledges support from the National Astronomy and Ionosphere Center.

References

Camilo, F. 1994, in *The Lives of Neutron Stars*, NATO ASI Ser., Doldrecht, in press.
Cominsky, L., Roberts, M., Johnston, S., 1993, *BAAS*, **24**, 912.
Cordes, J. M., Romani, R. and Lundgren, S. C., 1993, *Nature*, **362**, 133.
Fruchter, A. S., 1989, PhD thesis, Princeton University.
Fruchter, A. S., Bookbinder, J., Garcia, M. R., Bailyn, C. D., 1992, *Nature*, **359**, 303.
Johnston, S. et al. 1992, *Ap. J.*, **387**, L37.
Lundgren, S. C., Zepka, A. F, Cordes, J. M, 1993, *IAU Circ. No.* 5878.
Taylor, J. H. and Cordes, J. M., 1993, *Ap. J.*, **411**, 674.
Taylor, J. H., Manchester, R. N. and Lyne, A. G., 1993, *Ap. J. Supp. Ser.*, **88**, 529.
Wang, Q. D., Li, Z-Y., Begelman, M. 1993, preprint.
Zepka, A. F., Cordes, J. M., Wasserman, I. M, 1994, submitted to the *Ap. J.*.

ORAL PRESENTATIONS
ACCRETION-POWERED SOURCES

SOLVING THE MYSTERY OF THE PERIODICITY IN THE SEYFERT GALAXY NGC 6814

G. Madejski[1], C. Done[1,2], T. J. Turner[1], R. F. Mushotzky[1],

P. Serlemitsos[1], F. Fiore[3], M. Sikora[4,5], and M. Begelman[5]

[1]Code 666, NASA Goddard Space Flight Center, Greenbelt, MD 20771
[2]Currently at Physics Dept., Leicester University, Leicester, UK
[3]Center for Astrophysics, Cambridge, MA
[4]Copernicus Astron. Ctr., Warsaw, Poland
[5]JILA/University of Colorado, Boulder, CO

Email ID : madejski@heavax.gsfc.nasa.gov

ABSTRACT

The reports of periodic X–ray emission from the Seyfert galaxy NGC 6814 have motivated a number of exotic models for the active nucleus. We took advantage of the superior sensitivity and the wide field of view of the Position-Sensitive Proportional Counter on-board the ROSAT satellite to show that while the nucleus of NGC 6814 is indeed an X–ray emitter, the periodicity is due to another source, most likely a Galactic cataclysmic variable, $\sim$ 37 arc min away.

1. Introduction

NGC 6814, a low-luminosity ($\sim 10^{42}$ ergs/s) Seyfert galaxy has been a subject of much recent study due to its unique variability behavior. It was the only AGN so far in which a clear periodicity was detected in the X–ray flux. The existence of the $\sim$ 12,100 second period, first detected in the EXOSAT Medium Energy detector (ME; Mittaz and Branduardi-Raymont 1989; Fiore, Massaro, and Barone 1992), was subsequently confirmed in *Ginga* observations (Done *et al.* 1992). A number of interpretations were advanced for the periodicity, all based on orbital motion. These included, among others, gravitational lensing of X–ray emitting hot spots on an accretion disk by the central black hole (Abramowicz *et al.* 1991), or a captured star orbiting the black hole (Syer, Clarke, and Rees 1991; Sikora and Begelman 1992). The combination of very rapid ($\sim$ 50 sec) drops in the X–ray flux (Kunieda *et al.* 1990), short period, and large X–ray luminosity observed in NGC 6814 gave strong support for the black hole paradigm in AGN. Clearly, better understanding of this unusual object was needed, and to that end, we observed the field of NGC 6814 with the ROSAT X–ray satellite.

2. Observations and Results

An observation of NGC 6814 was conducted with the ROSAT X–ray telescope Position Sensitive Proportional Counter starting on March 31, 1993. It lasted for 180 ks of running time, and yielded $\sim$ 38 ks of useful data. Figure 1 shows the X–ray image over the 2 degree diameter field of view in the full 0.1–

2.4 keV bandpass of the instrument. The observation was centered within a few arc sec of the optical position of NGC 6814 nucleus, which is at RA(2000) = $19^h42^m40.4^s$, Dec(2000) = $-10°19'25''$. The positional coincidence of the strong X-ray source within a few arc sec from the the center of the image confirms that the nucleus of NGC 6814 is indeed the X-ray emitter.

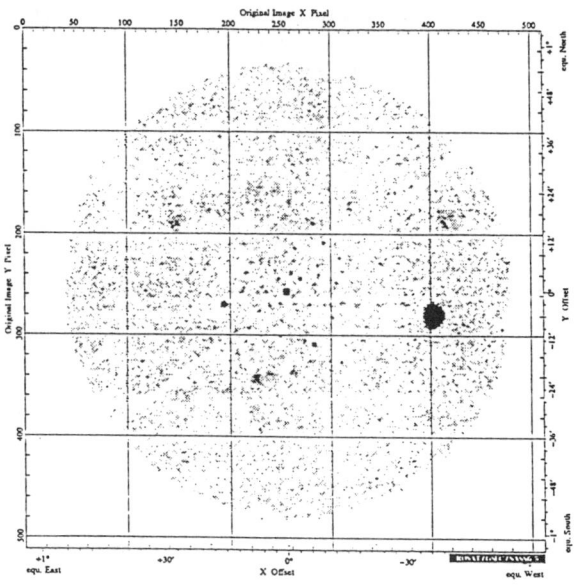

Figure 1. ROSAT PSPC image of the NGC 6814 field. The Seyfert galaxy NGC 6814 is the on-axis source, while the periodic X-ray emitter is 37' off-axis (the image of this second source is smeared by the degraded off-axis point spread function).

The ROSAT image reveals another bright point source ~ 37 arc min away, strongly distorted by the off-axis point spread function. The position derived from the ROSAT image is RA(2000) = $19^h40^m13^s$, Dec(2000) = $-10°25'30''$, where we estimate the position error to be ~ 30 arc sec, primarily because the off-axis image distortion as well as partial obscuration by the PSPC window supporting rib. We extracted light curves for both sources and tested them for periodicity in the 10,000 – 15,000 sec range by repeatedly folding the light curve, and measuring the significance of deviations from a constant by the L-statistic (related to χ^2 but following an F distribution; see Davies 1990; Done et al. 1992). Figures 2a and b show the L statistic for NGC 6814 and the second source; the ~ 12,000 second periodicity is clearly not associated with the Seyfert galaxy but with the off-axis source. In fact, the best fitting period of 12,142 s (see Fig. 2) is exactly that determined from the most recent *Ginga* observation of the NGC 6814 region (Done et al. 1992). We thus conclude that it is the second, off-axis source that is the origin of periodic X-ray emission.

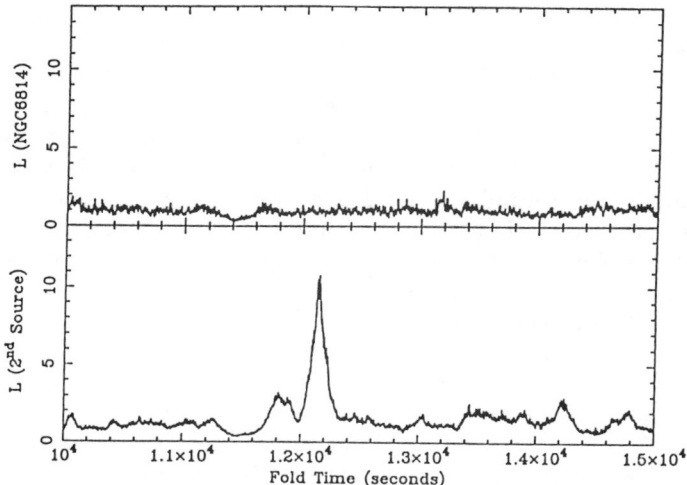

Figure 2. Value of L-statistic tests for periodicity over the range of 10,000 s to 15,000 s for NGC 6814 (a, top) and off-axis source (b, bottom). The 12,142 s period is clearly a property of the off-axis source.

The spectrum of this second source, derived from background-subtracted data, is well modeled as a black body plus a power law, absorbed by matter with Solar abundances. The fit yields a temperature of $kT_{bb} = 63 \pm 11$ eV and power law energy index of $\alpha = 0.1 \pm 0.25$, absorbed by an equivalent hydrogen column density of $N_H = 0.76 \pm 0.23 \times 10^{21}$ cm^{-2}. The black body contributes ~ 45 % of the 0.2 – 2 keV flux, with the power law contributing the rest. The 0.2 – 2 keV flux averaged over the entire observation is $\sim 6 \times 10^{-12}$ erg cm^{-2} s^{-1}, with $\sim 20\%$ uncertainty due to a partial obscuration by the PSPC support ribs. This flat power law index is in fact quite similar to that seen by *Ginga* for the spectrum from the NGC 6814 region (Kunieda et al. 1990; Turner et al. 1992). We estimate that the 2–10 keV flux, using the extrapolation of the ROSAT spectrum, is only $\sim 1.7 \times 10^{-11}$ ergs cm^{-2} s^{-1}, several times lower than the mean *Ginga* flux. However, this value is quite uncertain, since the overall source spectrum is likely to be more complex than we infer here. The nucleus of NGC 6814, on the other hand, shows a power law spectrum typical of other AGN, with a power law index $\alpha = 0.79^{+0.44}_{-0.33}$), absorbed by $N_H = 1.4^{+0.5}_{-0.7} \times 10^{21}$ cm^{-2}, consistent with the Galactic value. The mean 0.2 – 2 keV flux of NGC 6814 measured in our ROSAT observation is 7.1 $\times 10^{-13}$ erg cm^{-2} s^{-1}.

3. Discussion

Our observation shows that the X-ray emission with $\sim$ 12,100 sec periodicity, previously attributed to the Seyfert 1 galaxy NGC 6814, is due to an unrelated object that is located $\sim$ 37 arc minutes from the nucleus of NGC 6814. Why wasn't it discovered earlier? Of course the limited spatial resolution of collimated proportional counter instruments such as HEAO-A2, *Ginga* LAC, and EXOSAT ME – typically $\sim$ 1 degree – precluded the separation of the X-ray emission from the two sources. That was partially due to the fact that the off-axis source definitely is, and NGC 6814 is likely to be variable. As a result, a scanning technique, which relies on a transformation of data from temporal to spatial domain, could not be readily interpreted. As to the previous imaging X-ray instruments, the EXOSAT Low Energy telescope was not sensitive enough, as its off-axis point-spread function degraded the image such that it was not readily discernible from the instrumental background; the field is at low Galactic latitude, and thus the Galactic absorption severely reduced the soft (< 2 keV) X-ray flux of the source. In the case of the *Einstein* Observatory Imaging Proportional Counter, its field of view was simply too limited. A sensitive, large field-of-view imaging instrument such as ROSAT PSPC was necessary for the identification. The "discovery space" afforded by ROSAT + PSPC was thus an annulus, between $\sim 1/2$ and 1 degree.

This finding eliminates a lot of difficulties with modeling of X-ray emission from NGC 6814. The flat ($\alpha \sim 0.4$) *Ginga* X-ray spectrum (Kunieda et al. 1990; Turner et al. 1992) was quite unusual for an AGN, and $\alpha \sim 0.8$ is more typical. The inferred periodicity required fine-tuning of the source geometry, and the rapid drop of flux observed in the *Ginga* data (Kunieda et al. 1990; Done et al. 1992) implied an extremely compact source. Finally, the strength and rapid variability of the Fe K line flux (Kunieda et al. 1990) required a substantial amount of ionized material very close to the central source, and so a strong gravitational redshift was inferred to reconcile this with the measured line energy (Turner et al. 1992). These observational features will now have to be reconciled with another class of an X-ray source.

What is the other source? An optical follow-up (Rosen et al. 1993; see also Halpern 1993) indeed shows that there is a variable object at the ROSAT position, which has the expected $\sim 12,140$ sec period. The optical spectrum of it shows strong He II 4686 line emission, as well as possible cyclotron features, which are strongly suggestive that it is a magnetized CV system in our own Galaxy, consisting of a white dwarf fueled by accretion from a companion star. In fact, the ROSAT X-ray spectrum is typical of such systems (Cordova 1993).

We recognize that the observational data from what was thought to be NGC 6814 was considered a "rosetta stone" for the black hole paradigm for the structure of AGN, but we stress that the inferences from X-ray variability of Seyfert galaxies as a class are unaffected by this finding. These sources often exhibit non-periodic large amplitude variability on time scales of thousands of seconds (cf. McHardy 1989), and the removal of NGC 6814 does not substantially change the strong support that the rapid X-ray variability provides for the black hole hypothesis in AGN (see, e.g., Fabian 1992).

References

Abramowicz, M., Bao, G., Lanza, A., and Zhang, X.-H. 1991, *Astron. Astrophys.*, **245**, 454.
Cordova, F., 1993, in *X-ray Binaries*, ed. W. H. G. Lewin, J. van Paradijs, and E. P. J. van der Heuvel, in press.
Davies, S. R. 1990, *M.N.R.A.S.*, **244**, 93; see also *M.N.R.A.S.*, **251**, 64p.

Done, C., Madejski, G. M., Mushotzky, R. F., Turner, T. J., Koyama, K., and Kunieda, H. 1992, *Ap. J.*, **400**, 138.
Fabian, A. C. 1992, in *Frontiers of X-ray Astronomy*, ed. Y. Tanaka and K. Koyama (Tokyo: Universal Academy Press), p. 603.
Fiore, F., Massaro, E., and Barone, P. 1992, *Astron. Astrophys.*, **261**, 405.
Halpern, J. P. 1993, *Nature*, **365**, 607.
Kunieda, H., Turner, T. J., Awaki, H., Koyama, K., Mushotzky, R. F., and Tsusaka, Y. 1990, *Nature*, **345**, 786.
McHardy, I. 1989, in *Two Topics in X-ray Astronomy*, Proc. 23rd ESLAB Symposium, eds. N. White, J. Hunt, and B. Battrick (Paris: ESA), p. 1111.
Mittaz, J. P. D., and Branduardi-Raymont, G. 1989, *M.N.R.A.S.*, **238**, 1029.
Rosen, S., Done, C., Watson, M., and Madejski, G. M. 1993, *IAU Circ. No.* 5850.
Sikora, M., and Begelman, M. C. 1992, *Nature*, **356**, 224.
Syer, D., Clarke, C. J., and Rees, M. J. 1991, *M.N.R.A.S*, **250**, 505.
Turner, T., Done, C., Mushotzky, R., Madejski, G., and Kunieda, H. 1992, *Ap. J.*, **391**, 102.

ROSAT RESULTS ON MAGNETIC CATACLYSMIC VARIABLES

M G Watson
Department of Physics and Astronomy, University of Leicester
Leicester LE1 7RH, UK

Email ID
mgw@star.le.ac.uk

ABSTRACT

Recent ROSAT results on magnetic CV systems are reviewed with emphasis on the properties of the new sample of polars discovered from the ROSAT surveys.

1. Introduction

The importance of soft X-ray and EUV observations of cataclysmic variables (CVs) lies in the fact that a large fraction of their emitted radiation, at least in some systems, emerges at these wavelengths. It comes as no surprise, therefore, that ROSAT, with its good soft X-ray and EUV response, has already had a significant impact on this field. In this review I concentrate on magnetic CV systems, i.e. those systems where the magnetic field of the white dwarf is strong enough ($B_{wd} \gtrsim 1$ MGauss) to control the accretion flow from the secondary star over large distances within the binary. In these systems, the polars and intermediate polars, the infalling material shocks above the white dwarf surface (or in the white dwarf photosphere) and cools predominantly by producing bremsstrahlung hard X-rays (detected strongly in intermediate polars and some polars) and optical/infra-red cyclotron emission (responsible for the strongly polarised signal in polar systems). A very soft 'black-body' component is also present in polars (and may be present but unobservable in intermediate polars), due either to heating of the white dwarf surface by the hard X-rays, or to direct heating of the photosphere by the accretion flow. This component often contains a significant fraction of the available accretion luminosity.

2. The new sample of polars from the ROSAT all-sky surveys

Both of the all-sky surveys carried out by ROSAT: in the EUV by the UK Wide Field Camera (WFC; Pounds et al., 1993), and in the soft X-ray band by the German X-ray Telescope (XRT; e.g. Voges, 1994) have been used to discover new magnetic CV systems. Table 1 presents a combined list of the 17 new polar systems that have been discovered in the two largest programmes based on the WFC and XRT surveys, described below. (Six of these systems were identified independently in both programmes.) A number of new intermediate polars have also been discovered, but a discussion of these results is beyond the scope of this review.

2.1 ROSAT WFC Survey Results

The first catalogue of EUV sources from the ROSAT WFC survey (the 'Bright Source Catalogue' (BSC)), Pounds et al., 1993) contains 383 objects gleaned from an initial, rather conservative, analysis of the data. Seven new magnetic CVs, amounting to around 2.5% of the BSC, were discovered amongst the WFC survey sources as a result

Table 1: List of new polars from the ROSAT WFC and XRT surveys

Name (RXJ/RE)	PSPC[a] [c/s]	HR1[b]	WFC (S1)[c] [c/ksec]	Mag[d]	Type	Period [min.]	References
0132−65	0.6	−0.90		19	AM:	80:	[2]
0153−59	0.4	−0.91		15	AM:	80:	[2]
0203+29	0.5	−0.67		17	AM:	275	[2]
0453−42	1.7	−0.89	44	19	AM	95	[2,3]
0501−03	0.3	−0.75		15	AM:		[2]
0531−46	1.2	−0.70	16	16	AM:	135:	[1,2]
1002−19	0.6	−0.84		17	AM:	106	[2]
1007−20	2.1	−0.91		18	AM	208:	[2]
1015+09	1.3	−0.74		17	AM	80	[2]
1149+28	2.2	−0.95	138	17	AM	90	[1,2,4]
1307+53	2.5	−0.85	41	17−21	AM	80	[1,2,5]
1313−32	2.1	−0.79		16	AM	252	[2]
1844−74	1.4	−0.57	92	16	AM:	90	[1,2,6]
1938−46	8.9	−0.94	387	15	AM	140	[1,2,7]
1957−57	1.2	−0.89		17	AM:	99	[2]
2107−05	1.3	−0.91	19	15	AM	125	[1,2,8,9]
2316−05	1.5	−0.74		18	AM	209:	[2]

Notes
a: PSPC survey count rate (ref.2).
b: PSPC hardness ratio (ref.2).
c: WFC survey count rate in S1 filter (ref.1). Objects not detected have blank entries.
d: Typical V magnitude, taken from references quoted.

of optical follow-up observations carried out as part of the WFC Optical Identification Programme, a major campaign which aimed at identifying a substantial fraction, if not all, of the WFC survey sources with no obvious catalogued counterpart (Mason et al., 1991, 1994). Subsequent optical study of the seven new WFC CVs has lead to the classification and orbital period determinations of each object (e.g. Watson, 1993; see Table 1). Six of these systems turn out to be polar systems with periods in the range 80 minutes to 2.3 hours, whilst the seventh (but the first discovered) is an intermediate polar (see section 4.1). Further comments on some of the WFC discoveries appear in section 4.

2.2 ROSAT XRT Survey Results

The ROSAT XRT all-sky survey (RASS) in the soft X-ray band has the potential of discovering a substantial sample of new CV systems by virtue of its good sensitivity and all-sky coverage. The major problem posed by the RASS data is how to find the CVs amongst the > 50000 survey sources. The approach taken by Beuermann and colleagues (e.g. Beuermann and Thomas, 1993, Beuermann and Schwope, 1993) is to define a subsample of the RASS with the spectral parameters expected of a typical polar system. This reduces the number of objects very substantially, and this method, further restricted to sources at high galactic latitudes and relatively high count rates, has allowed Beuermann and colleagues to discover some 17 new polars to date (see Table 1). Other identification programmes based on the RASS are also underway and these have also contributed a few more magnetic CV identifications for RASS sources.

3. Properties of the new sample of polars

Before the launch of ROSAT, there were 17 or 18 catalogued polar systems (e.g. Ritter, 1990; Cropper, 1990). New results from ROSAT, primarily those from the surveys as discussed above, but also including some new discoveries from pointed observations, have increased the number of polars to 37, and new optically-selected polars increase the number to 42.[1] This dramatic increase in the number of known polars makes it interesting to re-examine their overall properties, and to compare the pre- and post-ROSAT objects.

3.1 Orbital period distribution

Orbital period distributions for CVs are one of the principal tools for testing evolutionary scenarios. As a result there has been considerable interest in the new sample of polars emerging from ROSAT studies. Figure 1 shows the period distributions for 3 different samples of CVs: non-magnetic CVs in general (taken from Ritter, 1990), polar systems pre-ROSAT (also from Ritter 1990), and the new, post- ROSAT sample of polars (based on Table 1 and Kolb & de Kool, 1993). Several features are clear in these plots. Both the non-magnetic and pre-ROSAT polar samples clearly demonstrate the period gap between ~ 2 and ~ 3 hours, and the sharp cut-off at a period ~ 80 minutes, features which now have a well-established interpretation in models of the secular evolution of CVs (see King, 1988 for a review) which involve angular momentum loss driven by magnetic braking at long periods, and by gravitational radiation at shorter periods. The period distribution for the pre-ROSAT sample of polars shows an additional feature, a prominent accumulation of systems around a period of 114 minutes. This 'period spike' has also received much attention in recent years. A possible explanation for this phenomenon, offered first by Hameury et al. (1988), is that systems spend rather longer at this period, and are rather brighter there, as it corresponds to the period at which they emerge from the period gap. However this interpretation places rather tight constraints on system parameters, and the white dwarf mass in particular must lie in the range $M_1 = 0.7$–$0.8\ M_\odot$.

The period distribution of the new, post-ROSAT sample of polars contains some surprises. The period spike is still present, but now looks somewhat less significant, and a new spike may be appearing at periods ~ 80 minutes (as is expected in most evolutionary models). The biggest surprise is that the period gap seems to be disappearing - there are now seven polars in the gap, five of which are well within it. Four of these are ROSAT discoveries, the other two resulting from recent optical work. One must be very cautious in interpreting these results because of the small numbers of objects involved,

---

[1] By the time this review is published this number is bound to be out of date!

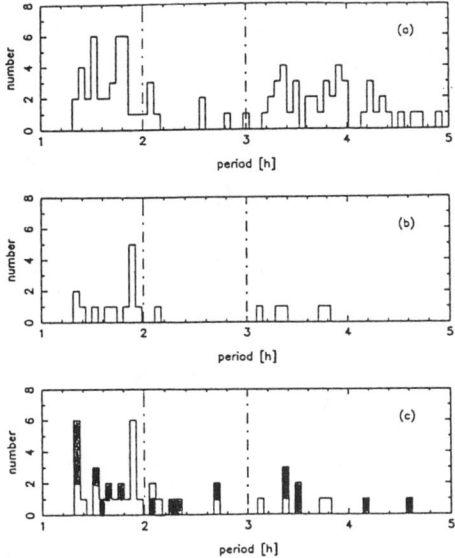

Figure 1: Orbital period distributions of CVs. (a) Non-magnetic CVs (from Ritter 1990); (b) pre-ROSAT sample of polars (from Ritter 1990); (c) post-ROSAT sample of polars (see text), shaded objects are ROSAT-discovered.

and because selection effects are very difficult to evaluate accurately, but at face value there does seem to be increasing evidence that the period gap may not as pronounced in polars as it is in non-magnetic systems.

Various reasons have been proposed to explain the lack of a gap in polars over the last year or so. The main contenders are: (a) suppression of magnetic braking entirely by the strong field of the white dwarf (Wu & Wickramasinghe, 1993; Wickramasinghe, 1993); (b) magnetic braking by the white dwarf (i.e. *enhanced* magnetic braking, King 1993); (c) deviations from the standard mass-radius relationship for the secondary (King 1993, Pylyser & Savonije, 1988). Each of these explanations has potential problems: (a) in particular has been shown by Kolb & de Kool to predict an overall period distribution for polars that is a poor fit to what is observed.

Even though the filling-in of the period gap is not entirely due to ROSAT discoveries, it is interesting to speculate why ROSAT might be particularly good at discovering polars in the period gap, i.e. whether the period distribution is strongly influenced by selection effects. Models (a) and (b) favour large magnetic fields, and such objects might be preferentially detected by ROSAT if strong fields boost their soft X-ray output. As discussed in section 3.2, a correlation between 'soft X-ray excess' and magnetic field strength has now been demonstrated, but it seems unlikely that this is enough to produce a strong selection effect since, at best, strong field systems would only have soft X-ray luminosities a factor two higher than weaker field polars. The last possibility, model (c), favours large white dwarf masses which might conceivably produce a strong enough selection effect to bias the ROSAT sample, as pointed out by King (1993).

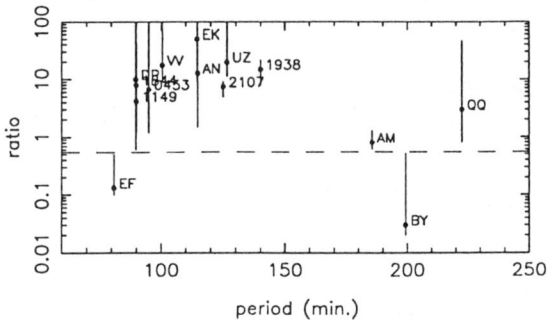

Figure 2: Black-body to bremsstrahlung (plus cyclotron) luminosity ratio for polar systems, plotted as function of orbital period (based on results in Ramsay et al., 1994). The dashed line for a ratio of 0.55 corresponds to the expected ratio for no 'soft X-ray excess'. Objects are identified by the first part of their name, or by the RA digits of their coordinate name for RE/RXJ sources.

3.2 Spectral properties of polars

The observed X-ray spectra of most polars have two components, a very soft 'black-body' component which dominates at EUV/soft X-ray wavelengths and a much harder 'bremsstrahlung' component dominating above ~ 0.5 keV. In the simplest models of accretion onto a magnetic white dwarf (e.g. King & Lasota 1979) the accretion flow is thermalised by a strong shock above the white dwarf surface, leading to hard X-ray production in this region, whilst the ultra-soft component is produced by X-ray heating of the white dwarf by that fraction of the hard component directed downwards. In this scenario, these two components are expected to be in approximate balance, more precisely (taking into account the hard X-ray albedo of the white dwarf) the expected ratio is $L_{bb}/L_{br} \approx 0.55$ if the cyclotron luminosity can be ignored (e.g. King & Watson, 1987).

Over the last decade it has become increasingly clear that the observed luminosity ratio in some polar systems does not match this expectation, in the sense that the ultra-soft luminosity is much larger than expected (e.g. Watson, 1986, Osborne, 1988). This discrepancy, often referred to as the 'soft X-ray problem', stimulated further theoretical work on accretion in magnetic CVs and led to the development of models in which some fraction of the accretion flow penetrates the white dwarf photosphere and deposits its energy there directly. Such 'blobby' accretion models, first suggested by Kuijpers & Pringle (1982) and developed by Frank et al. (1988), can solve the 'soft X-ray problem' trivially since the ratio of luminosities can be adjusted by altering the fraction of the flow in dense blobs (which then produce the soft component by heating the photosphere from inside) and in rarified blobs (which still shock above the surface yielding hard X-rays).

The relatively good spectral capabilities of ROSAT, in particular the fact that its band-pass extends to the EUV, provides the opportunity to have another look at the soft X-ray problem. Ramsay et al. (1994) have recently carried out such a study using

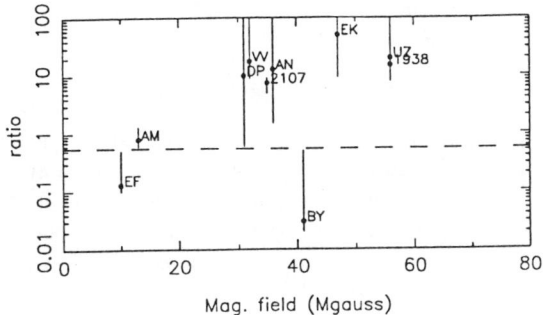

Figure 3: Black-body to bremsstrahlung (plus cyclotron) luminosity ratio for polar systems plotted as a function of white dwarf magnetic field (based on results in Ramsay et al., 1994). Objects are identified as in Fig.2

ROSAT pointed observations of a number of polar systems, both previously known and ROSAT-discovered objects. They find that the ROSAT PSPC spectra of most polars show the two spectral components as expected. Straight-forward fits to the PSPC spectra (combined with Ginga parameters for the hard component which is not well-constrained in most PSPC datasets) were used to determine the spectral parameters of each object and hence to derive the bolometric luminosities of the hard and soft components. The results, shown in Fig.2, demonstrate very clearly that the majority of polars, and the ROSAT-discovered polars in particular, have a large 'soft X-ray excess', i.e. $L_{bb}/L_{br} \gg 0.55$, amply confirming earlier observational studies. It is important to note that the PSPC spectra provide significantly better constraints on the parameters of the soft component than are available from most previous observations (with the exception of the few studies made with grating spectrometers).

For a subset of the polars studied by Ramsay et al. there are good estimates of the surface field of the white dwarf obtained from measurements of cyclotron and/or Zeeman features in their optical spectra. Fig.3 shows the luminosity ratio determined from the ROSAT measurements as a function of the surface field strength, with some indication of a positive correlation. In an independent study of a different subset of polars Beuermann and Schwope (1993) present a very similar correlation, adding weight to the supposition that this is a real effect. At least one system, BY Cam, does not fit in with the overall trend, but this may be related to the possibility that this is a slightly asynchronous system. The interpretation of this result is, at present, far from clear, although Beuermann & Schwope point out that this kind of correlation is consistent with the bombardment solutions developed by Woelk & Beuermann (1993). This discovery, if confirmed, seems guaranteed to stimulate more theoretical work on the details of the accretion process in polars.

4. Studies of individual objects

In this section I highlight the features of some of the new CVs discovered by ROSAT, and focus on one particular pointed observation to demonstrate the quality of the detailed timing studies that ROSAT can perform. My choice is necessarily very selective, and includes only work with which I have some personal involvement.

4.1 The intermediate polar RE0751+14

Initial follow-up studies of the WFC source RE0751+14 using IR, optical and hard X-ray observations established that it was an intermediate polar with a 13.9-minute white dwarf spin period and an orbital period $\sim$ 5 hours (Mason et al., 1992). This in itself is somewhat remarkable, since no other intermediate polar, with the possible exception of EX Hya, is known to have a detectable flux at EUV wavelengths. Further optical studies by Rosen et al. (1993) and Piirola et al. (1993) have detected circular polarisation modulated at the spin period, and Rosen et al. have also shown the presence of a second periodicity in the optical photometry at a period of 14.5 minutes. This second period is strongly wavelength-dependent and is very likely to be the beat period between the white dwarf spin and the orbit. More recently a ROSAT pointed observation has shown that this system has a bright, ultra-soft X-ray component, confirming that the WFC survey detection relates to a true EUV component (Duck et al., 1994; see Fig.4).

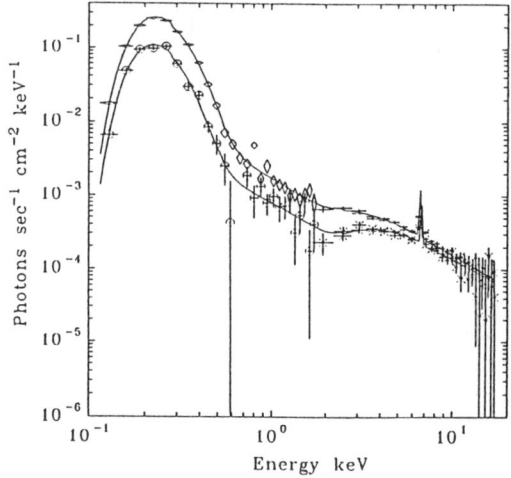

Figure 4: The combined ROSAT PSPC and Ginga spectra of RE0751+14 sampling the bright state (upper curve) and the dip (lower curve). The model fits consist of a a soft blackbody component ($kT \sim 54$ eV) and a hard bremsstrahlung component ($kT \sim 45$ keV) with an Fe emission line at 6.65 keV, subject to partial covering by a two-component absorber with column densities $N_H \sim 1 - 2 \times 10^{20}$ cm^{-2} and $N_H \sim 5 - 10 \times 10^{22}$ cm^{-2}. During the dip the covering fraction of the denser absorber increases. Taken from Duck et al. (1994).

RE0751+14 thus appears to have all the properties normally associated with **polar** systems, although it is clearly an synchronous system. This makes it in some sense the first real 'intermediate' polar, and an object of some importance for understanding what determines the detailed differences between the two classes of magnetic CV system. The detection of modulated circular polarisation, and the presence of a significant cyclotron component inferred from the colour dependence of the photometric light curves, argues for a white dwarf field $\gtrsim 10$ MGauss (e.g. Piirola et al., 1993). RE0751+14 might thus be one of the long-sought asynchronous progenitor systems for polars at lower orbital periods. The fact that only one such system is presently known does not fit well the fact that several dozen polars are now catalogued at lower orbital periods, but it may be that there are strong selection effects that prevent most objects like RE0751+14 being discovered. At present we don't know what these might be.

4.2 The halo CV RE1307+53

The optical counterpart of RE1307+53 was one of the faintest discovered in the WFC identification programme, but subsequent observations revealed it to be around 3 magnitudes fainter at V≈ 20–21 mag. Optical photometry nevertheless revealed a clear modulation at the orbital period of 79.7 minutes, and the same periodicity is clearly present in the WFC survey light curve (Osborne et al., 1994). The fact the optical modulation is similar to other polar systems strongly indicates that accretion still takes place in this low optical state. The optical faintness of the counterpart already suggested a distant object. This was amply confirmed by near infra-red observations with IRCAM on UKIRT which gave K=19.3 ± 0.2 mag., yielding a robust lower limit to the distance $d > 1160$ pc assuming a Roche-lobe filling secondary, since the K magnitude is insensitive to either the temperature or surface gravity (Bailey, 1981). (The likely dilution by an accretion-powered component emission merely increases the distance estimate.) This distance clearly implies RE1307+53 is a halo object, since at $b = 63°$ it has a scale height $z > 1$ kpc above the galactic plane. The discovery of a polar system in the galactic halo has important evolutionary implications, since the halo location implies a very old system (age $> 10^{10}$ y) yet the evolutionary timescales of most CVs are much shorter than this ($\tau \sim 5 \times 10^9$ y). Osborne et al. suggest that RE1307+53 must be a system with an extended evolutionary timescale, which might arise if it had a very low mass secondary and long orbital period when it emerged from the common envelope phase.

4.3 The eclipsing polar UZ For

The eclipsing polar system UZ For was the subject of a long ROSAT PSPC observation made in August 1991, together with fast optical photometry obtained at the AAT (Hakala et al., 1994). The resultant X-ray and optical light curves, shown in Fig.5, provide an impressive dataset with which to study this object, and demonstrate the power of ROSAT for making detailed timing studies of CVs. The optical eclipse profile, although complex, can be decomposed into transitions corresponding to the four important sources of optical emission: two emitting poles on the white dwarf surface, the body of the white dwarf and the accretion stream, as is indicated in Fig.6. Modelling of the optical profile, following the approach of Bailey & Cropper (1992), allows strong constraints to be placed on the binary system parameters, and the location and extent of the optically-emitting poles.

The X-ray eclipse (Fig.6), in contrast, is much simpler, consisting of a sharp

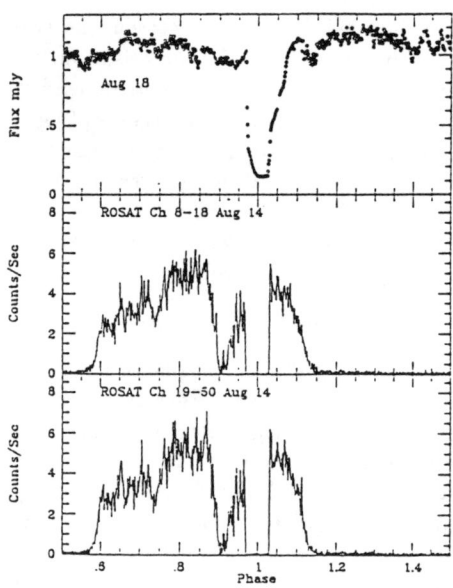

Figure 5: X-ray and optical light curves of UZ For folded at the orbital period (from Hakala et al., 1994). The ROSAT data are shown in two pulse height channel ranges: channels 8-18 correspond to photon energies $\sim 0.1 - 0.2$ keV and channels 19-50 to energies $\sim 0.2 - 0.5$ keV.

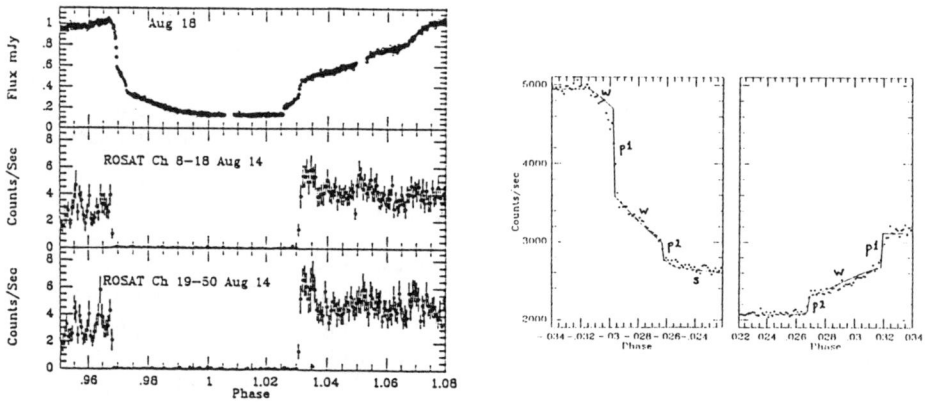

Figure 6: Optical and X-ray eclipse profiles for UZ For. The right-hand panel shows the detailed optical ingress and egress (from Hakala et al., 1994). Transitions in the optical eclipse are identified as follows: **w**: white dwarf; **p1**: main optical pole; **p2**: second optical pole; **s**: accretion stream (the long stream egress can be seen more clearly in Fig.5).

ingress and egress which are essentially simultaneous with the fastest parts of the optical profile which correspond to the eclipse of the dominant optical pole. This means that the X-ray and the optical emission from the dominant pole lie very close together on the white dwarf surface. This is expected in most models of accretion in polar systems, but this is the first time, to my knowledge, that this has been demonstrated observationally. The ingress and egress are marginally resolved in both the X-ray and optical data, implying a projected size of the emission regions has a radius $\lesssim 5°$.

As well as the prominent eclipse, the X-ray light curve also shows two other features: a deep, erratic minimum preceding the eclipse by ~ 0.2 in phase, and a low intensity state lasting ~ 0.45 of the binary cycle. The deep minimum is very likely to be an absorption dip, caused by the accretion stream crossing the line of sight to the white dwarf, a phenomenon seen in other polars. Spectral variations occurring at the edges of this feature support this interpretation. The X-ray low-state is also a common feature of the light curves of several polars, being due to the emitting pole rotating behind the body of the white dwarf. The duration of the low-state, and the shape of the transitions between the low- and high-states, provide additional information about both the location and extent of the X-ray emission region. Detailed modelling of the intensity transitions (Hakala et al., 1994) show that the 'horizontal' extent of the X-ray emission region is larger at the lowest photon energies, implying some temperature structure for the region. (Fig.5 shows some indication of this effect, the intensity transitions are clearly steeper in the higher energy ROSAT light curve.) The modelling also shows that the shape of the intensity transitions is completely inconsistent with an emission region that is confined to the surface of the white dwarf, since the transitions are too fast. (Similar conclusions are reached by Heise et al. (1985) in their discusion of the EXOSAT observations of AM Her). The ROSAT observations of UZ For require an emission region that has finite vertical extent, amounting to $\sim 0.003 - 0.005 R_{wd}$, i.e. a vertical height of ~ 20 km. This is a surprising result, since it corresponds to several thermal scale-heights. Assuming the interpretation is correct, it means that the structure we see in soft X-rays is a splash, i.e. matter thrown up above its natural scale-height, which falls back on dynamical timescale. There must be constant splashes to explain the observations. This interpretation is, however, broadly consistent with the Hameury & King (1988) model of anomalous state light curves.

References

Bailey, J., 1981. *MNRAS*, **197**, 31.
Bailey, J. and Cropper M. 1991. *MNRAS*, **253**, p.27.
Beuermann, K. & Schwope, A.D., 1993. In *Proc. 105th Meeting of the ASP*, in press. [ref.2]
Beuermann, K. & Thomas, H-C., 1993. *Advances in Space Research*, **13**, No.12, p??.
Buckley, D.A.H., O'Donoghue, D., Hassall, B.J.M., Kellett, B.J., Mason, K.O., Sekiguchi, K., Watson, M.G., Wheatley, P.J. & Chen, A., 1993. *MNRAS*, **262**, 93. [ref.7]
Cropper, M. 1990. *Space Science Rev.* **54**, 195.
Duck, S.R., Rosen, S.R., Ponman, T.J., Norton, A.J., Watson, M.G. & Mason, K.O., 1994. Submitted to *MNRAS*.
Frank, J., King, A.R., & Lasota, J-P., 1988. *Astron.Astrophys.*, **193**, 113.
Hakala, P.J., Watson, M.G., Vilhu, O., Hassall, B.J.M., Kellett, B.J., Mason, K.O. &

Piirola, V., 1993. *MNRAS*, **263**, 61. [ref.8]
Hakala, P.J., Watson, M.G. & Bailey, J., 1994. In preparation.
Hameury J-M.& King, A.R., 1988. *MNRAS*, **235**, 433.
Heise, J., Brinkmann, A.C., Gronenschild, E., Watson, M.G., King, A.R., Stella, L., & Kieboom, K., 1985. *Astr.Astrophys.*, **148**, L14.
King, A.R., 1988. *QJRAS*, **29**, 1.
King, A.R., 1993. In *Proc. 105th Meeting of the ASP*, in press.
King, A.R., & Lasota, J-P., 1979. *MNRAS*, **188**, 653.
King, A.R. & Watson, M.G., 1987. *MNRAS*, **227**, 205.
Kuijpers, J. & Pringle, J.E. (1982). *Astron.Astrophys.*, **114**, L4.
Kolb, U. & de Kool, M., 1993. *Astron.Astrophys.*, in press.
Mason, K.O., Branduardi-Raymont, G., Bromage, G.E., Buckley, D., Charles, P.A., Hassall, B.J.M., Hawkins, M.R.S., Hodgkin, S., Pike, C.D., Jomaron, C.M., Jones, D.H.P.,McHardy, I., Naylor, T.T., Ponman, T.J. & Watson, M.G., 1991. *Vistas in Astronomy*, **34**, 343.
Mason, K.O., Watson, M.G., Ponman, T.J., Charles, P.J., Duck, S.R., Hassall, B.J.M., Howell, S.B., Ishida, M., Jones D.H.P. & Mittaz, J.P.D., 1992. *MNRAS*, **258**, 749.
Mason, K.O.,et al., 1994. In preparation.
Mittaz, J.P.D., Rosen, S.R., Mason, K.O. & Howell, S.B., 1992. *MNRAS*, **258**, 277. [ref.4]
O'Donoghue, D., Mason, K.O., Chen, A., Hassall, B.J.M. & Watson, M.G., 1994. *MNRAS*, in press. [ref.6]
Osborne, J.O., 1988. *Mem. S. A. It.*, **59**, 117.
Osborne, J.O., Beardmore, A.P., Wheatley, P.J., Hakala, P., Watson, M.G., Mason, K.O. & Hassall, B.J.M., 1994. Submitted to *MNRAS*. [ref.5]
Piirola, V., Hakala, P. & Coyne, G.V., 1993. *Ap.J.*, **410**, L107.
Pounds, K.A., et al., 1993. *MNRAS*, **260**, 77.
Pylyser, E. & Savonije, G.J., 1988. *Astron.Astrophys.*, **191**, 57.
Ramsay, G., Mason, K.O., Cropper, M., Watson, M.G. & Clayton, K., 1994. Submitted to *MNRAS*. [ref.3]
Ritter, H., 1990. *Astr.Astrophys.Supp.Ser.*, **85**, 1179.
Rosen, S.R., Mittaz, J.P.D. & Hakala, P.J., 1993. *MNRAS*, **264**, 171.
Schwope, A.D., Thomas, H-C. & Beuermann, K., 1993. *Astron.Astrophys.*, **271**, L25. [ref.9]
Voges, W.H., 1994. This volume.
Watson, M.G., 1986. In *The Physics of Accretion onto Compact Objects*, Proceedings Tenerife Workshop, p.97, eds. K.O.Mason, M.G.Watson & N.E.White. [Springer-Verlag]
Watson, M.G., 1993. *Advances in Space Research*, **13**, No.12, 125. [ref.1]
Wickramasinghe, D.T., 1993. In *Cataclysmic Variables and Related Physics*, O. Regev & G.Shaviv eds, *Annals of the Israel Physical Society*, **Vol.10**, p.208.
Woelk, U. & Beuermann, K., 1993. *Astron.Astrophys.*, in press.
Wu, K. & Wickramasinghe, D.T., 1993. In *Cataclysmic Variables and Related Physics*, O. Regev & G.Shaviv eds, *Annals of the Israel Physical Society*, **Vol.10**, p.336.

DISCOVERY OF A CANDIDATE OLD, ISOLATED NEUTRON STAR IN THE FIELD OF A GALACTIC CIRRUS CLOUD

John T. Stocke, Q. Daniel Wang, and Eric S. Perlman
CASA, University of Colorado
stocke@hyades.colorado.edu, (wqd,perlman)@casa.colorado.edu

Megan Donahue
STScI
donahue@stsci.edu

Jonathan Schachter
Harvard/Smithsonian Center for Astrophysics
shaks@cfa237.harvard.edu

Abstract

Using the example of a still unidentifed "serendiptious" X-ray source in the field of the nearby galaxy NGC 1313, we describe a procedure for determining if unidentified X-ray sources are likely compact objects. The very high X-ray to optical flux ratio and other unusual properties of the X-ray source MS 0317.7-6647 eliminate all the usual classes of optical counterparts except for either a very massive X-ray binary (and Black Hole candidate) in NGC 1313 or a very nearby ($\sim$ 100 pc) isolated neutron star slowly accreting interstellar matter. The presence of an IR cirrus cloud which shadows the 1/4 keV X-ray background in this field supports the latter possibility.

1. The Method as Applied to MS 0317.7-6647

In every **ROSAT** PSPC image there are numerous "serendipitous", unidentifed X-ray sources. Based upon an extensive survey made by identifying $>$ 800 such sources discovered with the **Einstein** IPC (the **Einstein** Extended Medium Sensitivity Survey: EMSS; Gioia et al. 1990; Stocke et al. 1991; Maccacaro et al. 1994), these sources are: 51% Active Galactic Nuclei (AGN), 25% Galactic stars, 13% clusters of galaxies, 4% BL Lac Objects, 2% normal galaxies, 4% still unidentified and 1% compact objects (cataclysmic variables, AM Her binaries, X-ray binaries). The EMSS established criteria for determining which of the several potential optical counterparts within the X-ray positional error circles was the X-ray emitter (see Stocke et al. 1991 for details). Chief among these criteria is that each class of optical counterpart possesses a limited range of X-ray to optical flux ratios; e.g. all stars have $f_x/f_v < 1$, AGN exhibit a Gaussian distribution from $f_x/f_v = 0.1$ to 20 and clusters and BL Lacs have f_x/f_v ratios that extend as high as 60. Thus most classes of optical counterparts for faint X- ray sources can be eliminated if $f_x/f_v > 20$. In addition, all BL Lac Objects are radio emitters (Stocke et al, 1990) and all clusters of galaxies have relatively hard soft X-ray spectra, are non-variable sources and are extended at HRI resolution. Thus, additional data can eliminate all of the other classes of optical counterparts except compact objects like X-ray binaries which can have extremely large f_x/f_v ratios. This process of elimination approach may be most fruitfully applied in the fields of nearby galaxies since some of the excess numbers of sources found in X-ray images surrounding nearby galaxies (Fabbiano, 1989) could be extremely luminous (and therefore extremely massive) X-ray binaries in these other galaxies (i.e. Black Hole candidates; see Stocke, Wurtz and Kuhr 1991).

With this potential in mind a **ROSAT** HRI image of a field containing the nearby (4.5 Mpc) low luminosity SBc galaxy NGC 1313 was obtained in the Spring of 1992 to study a still unidentified EMSS source 6 arcmin south of NGC 1313 with an apparently very large f_x/f_v ratio. A PSPC exposure of this field also detects this source with sufficient counts to provide a spectral analysis and rudimentary variability study. The source is extremely soft; e.g. if fit by a power law, the best fit photon spectral index is -4 with a substantial total column of $n(H) = 4 \times 10^{21}$ cm^{-2}. The available X-ray data shows that the source is variable on a timescale of years and a portion of the **ROSAT** PSPC data also show a factor of three variability on a timescale of hundreds of minutes. MS 0317.7-6647 is point-like in the HRI image at J2000 position: 03 18 22.33 -66 36 07.6 (corrected for a 6 arcsec aspect error by using the optical and radio positions of SN 1978K in NGC 1313; Ryder et al. 1993).

Optical images of this field were obtained in B, R and Hα passbands using the 1.1m telescope on Las Campanas and a radio image at 6 cm was obtained using the Australia Telescope array. The optical R band image is shown in Figure 1. Near the HRI position and at the edge of the plausible HRI positional error circle is a faint optical point source with $R = 20.3 \pm 0.25$, $B = 21.4 \pm 0.5$ and no detectable Hα emission. However, the available Hα images are not deep and can only eliminate the possibility that the faint optical point source is not pure Hα. At any rate, the brightest that the optical counterpart to MS 0317.7-6647 could be is set by this source, $V = 20.8$. If the faint optical point source is not the optical counterpart, the true optical counterpart is fainter than 23rd magnitude. The AT failed to detect radio emission inside the HRI error circle to 0.3 mJy.

Using the brightest X-ray flux observed for this source and assuming minimal reddening, $f_x/f_v > 64$, too large to be consistent with any optical counterpart except an extreme BL Lac Object, cluster of galaxies or compact object. A BL Lac is ruled out by our failure to detect radio emission and the cluster ID is eliminated by the intrinsic source variability, very soft observed spectrum and point-like HRI detection. By using a reddening value implied by the best-fit PSPC hydrogen column (Shull and van Steenberg, 1985), $f_x/f_v > 15$. While this value is marginally consistent with an AGN identification, the coincidence of such an extreme X-ray to optical flux ratio and an extremely soft power-law spectrum (see other papers in this conference by Elvis and collaborators) is most unlikely. Therefore, the only possibility for this source appears to be a compact object, an X-ray binary in our Galaxy or in NGC 1313 or an isolated neutron star either still slowly rotating or only accreting interstellar matter.

2. The "MACHO" Option

The proximity of MS 0317.7-6647 to NGC 1313 (10 kpc on the sky from the nucleus), suggests that it could be an X-ray binary source in that galaxy. That the X-ray spectra of all three sources in this field have the same total absorbing columns (Petre, private communication) is supportive of this possibility. If MS 0317.7-6647 is in NGC 1313 (4.5 Mpc; de Vaucouleurs, 1963), the X-ray luminosity for this binary source is extremely high, $L(x; 0.2 - 2.4 \text{ keV}) =$

7.5×10^{39} ergs s^{-1}. Because this is ~ 50 times the zero-metallicity Eddington limit, a **minimum** primary mass of $50 M_\odot$ is implied; i.e. this source would be a Black Hole candidate. Thus, the acronym MACHO, "massive compact halo object" would apply to this source.

Although the MACHO option cannot be ruled out with the data at present, the very large minimum mass for a stellar remnant and the somewhat large galactocentric distance argue against this option. Even stronger evidence against the MACHO option is the extremely soft PSPC spectrum, which is unprecedented for an X-ray binary. Although it has been noted that Black Hole binaries, as a class, seem to possess "ultra-soft" X-ray components (White and Marshall, 1984), these components have derived temperatures of a few keV, not 0.2 keV as seen for this source.

If this source is in NGC 1313 and an extinction derived from the X-ray determined HI column is assumed, even an early O star secondary for this X-ray binary system would need to be ≥ 2 mag fainter than the observed optical point source in Figure 1. Thus, in the MACHO model, the optical detection is present in the HRI error circle by chance.

3. The "WIMP" Option

The high Galactic latitude makes this source an unlikely X-ray binary in our own Galaxy. For example, the X-ray luminosity of this source could be $\sim 10^{33}$ ergs s^{-1} if this source were 2.5 kpc away (and so 1.7 kpc out of the plane). In this case the observed optical object in Figure 1 would have approximately the correct absolute magnitude to be a late M star secondary (although it is much too blue for this to be the case). Placing it further from us would make it an extreme halo object, which, while not impossible, would be unprecedented. Placing it closer would lower the X-ray luminosity and the distance that the source would be below the Galactic plane but it would create a secondary too faint to be stellar. Thus, a Galactic X-ray binary identification seems unlikely.

Another possibility is that this source is really quite close to us and has a very low X-ray luminosity as might be expected for a companion-less old pulsar (e.g. Geminga) or an isolated neutron star slowly accreting interstellar gas (Lepp and McCray, 1983; Blaes and Madau, 1993). Several authors have suggested that isolated neutron stars accreting gas from interstellar or molecular clouds should be observable in sizeable numbers by X-ray satellites. We term this the "WIMP" option for MS 0317.7-6647, which stands for "Weak Infall onto the Magnetic Pole of a neutron star". The WIMP option is supported by two observations of this source:

The observed **ROSAT** PSPC spectrum of this source is very well-fit by an 0.2 keV Black Body (see Figure 2) which yields an unobscured X-ray luminosity of 1.7×10^{30} ergs s^{-1}(d/100pc) where d is the unknown distance to this source. This temperature and luminosity is in the expected range for a WIMP (see e.g. Colpi et al. 1993) and requires an emitting region of size 250 m (d/100 pc) at the neutron star polar cap.

Based upon IRAS 100 micron scans, Wang and Yu (1994) have discovered an IR cirrus cloud in this region that is spatially coincident with MS 0317.7-6647. Based upon the X-ray shadow that this cloud casts from the 1/4 keV X-ray background, the estimated distance to this cirrus cloud is ≤ 100 pc. An isolated neutron star imbedded in this cloud would yield X-ray parameters similar to those observed for this source.

The optical point source we detect cannot be the continuum emission of a nearby neutron star since the expected optical magnitude of such emission would be fainter than 30th magnitude. So, in the WIMP model the optical object is in the HRI error circle by chance.

4. Conclusion

Based primarily upon its extreme f_x/f_v ratio, the "serendipitous" X-ray source MS 0317.7-6647 is either a massive X-ray binary and Black Hole candidate in the nearby (4.5 Mpc) spiral galaxy NGC 1313 or a very nearby (100 pc) isolated neutron star accreting interstellar matter out of a cirrus cloud which is present at this location in the sky. The **ROSAT** PSPC spectrum of this source is very well-fit by a 0.2 keV Black Body with a luminosity as predicted for an isolated neutron star if it is imbedded in the observed cirrus cloud. New observations which will help determine the nature of this unusual source include deeper Hα images to search for a region ionized by the neutron star X-ray continuum, optical spectroscopy of the 21st magnitude point source within the HRI error circle (although this source may be present just by chance), an ASCA spectrum of this source which could detect non-thermal cyclotron emission or a high energy power-law tail and a radio search for pulses since the radio upper limit may not definitively eliminate a pulsar with small duty cycle.

References

Blaes, O. and Madau, P. 1993 ApJ 403, 690.
Colpi, M., Campana, S. and Treves, A. 1993 A&A, in press.
de Vaucouleurs, G. 1963 ApJ 137, 720.
Fabbiano, G. 1989 Ann Rev Astr Ap 27, 87.
Gioia, I. et al. 1990 ApJS 72, 567.
Lepp, S. and McCray, R. 1983 ApJ 269, 560.
Maccacaro, T. et al. 1994, in prep.
Ryder, S. et al. 1993 ApJ 416, 167.
Shull, J.M. and van Steenberg, D. 1985 ApJ 294, 599.
Stocke, J.T. et al. 1990 ApJ 348, 141.
Stocke, J.T. et al. 1991a ApJS 76, 813.
Stocke, J.T., Wurtz, R. and Kuhr, H. 1991b AJ 102, 1724.
Wang, Q. and Yu, K. 1994, in prep.
White, N.E. and Marshall, F.E. 1984 ApJ 281, 354.

Figure Captions

FIGURE 1: A portion of the R-band CCD image of this field. The field-of-view of this picture is 1.6 arcmin square. The **ROSAT** HRI position is marked with a cross and the object ESE from the cross is the possible optical counterpart to MS 0317.7-6647.

FIGURE 2: The **ROSAT** PSPC spectrum of MS 0317.7-6647 is well-fit by a Black Body (solid-line histogram) with parameters: $kT = 0.17(\pm 0.04)$ keV and $n(H) = 4.3(\pm 2.2) \times 10^{21}$ cm^{-2}. $\chi^2 = 22.5$ for 25 degrees of freedom.

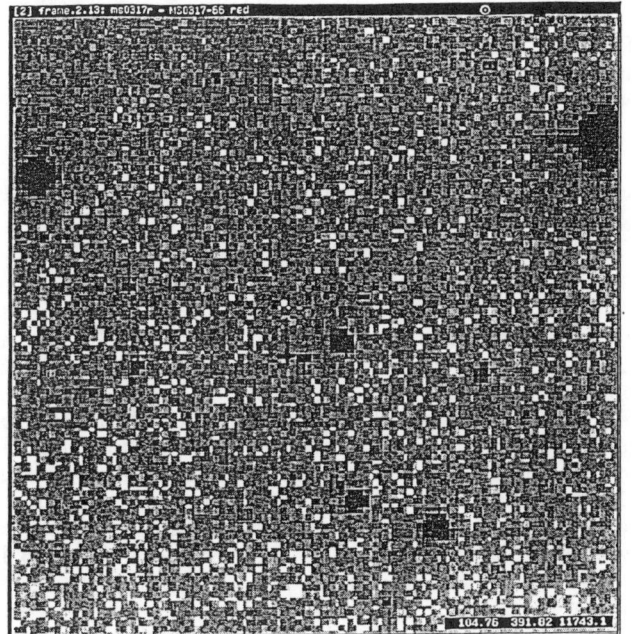

Figure 1.

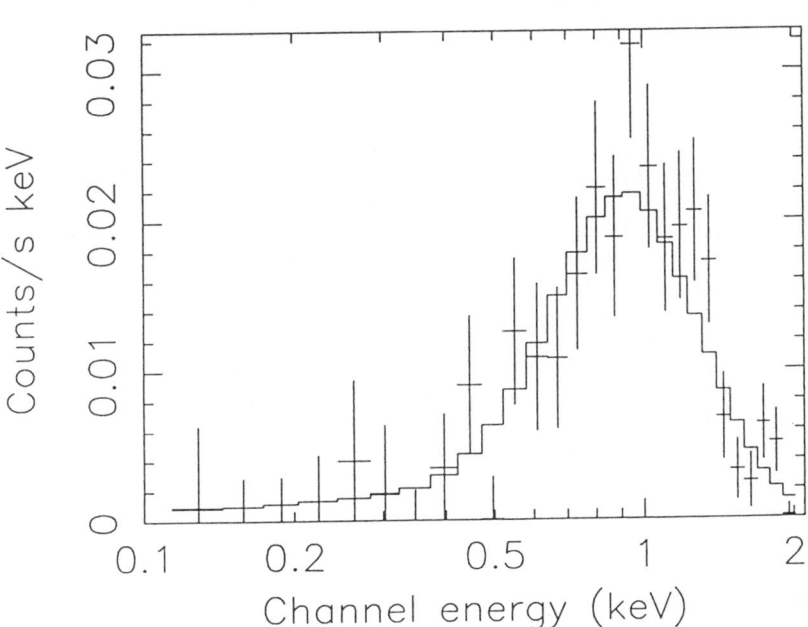

Figure 2.

EVIDENCE AGAINST SECULAR EVOLUTION IN THE ORBITAL PERIOD OF THE LMXB EXO0748-676

P. Hertz[1], Y. Ly[2], and K. S. Wood[1]
Code 7621, Naval Research Laboratory, Washington, DC 20375
hertz@xip.nrl.navy.mil, yly@beauty.tjhsst.edu, wood@ssd0.nrl.navy.mil

L. R. Cominsky[1]
Phys. and Astron. Dept., Sonoma State University, Rohnert Park, CA 94928
lynnc@charmian.sonoma.edu

ABSTRACT

We have observed with ROSAT a single eclipse egress for the low mass X-ray binary EXO0748-676. We have combined EXOSAT, Ginga, and ROSAT timings to fit the ephemeris of EXO0748-676 and measure the evolution of its orbital period. Assuming that the observed eclipse times are determined solely by a deterministic period plus measurement error, we confirm Asai et al. — a constant period and a constant period derivative are both ruled out. A constant period acceleration or 12 yr sinusoidal variability provide equally good fits.

The sinusoidal variation in orbital period is reminiscent of several cataclysmic variables. However no CV has been observed for more than 1.5 long cycles, and many show orbital period timings that deviate significantly from the predictions of the sinusoidal ephemerides. We suggest that there may be intrinsic, stochastic jitter in the time of individual eclipses. If so, the timing residuals plotted in an O-C diagram represent a random walk in orbital phase and can usually be fitted by a sinusoid with a period 1-2 times the duration of the data set. We present statistical tests in support of this interpretation.

1. INTRODUCTION

Low mass X-ray binaries (LMXBs) are the brightest galactic X-ray sources. They are powered by the accretion of matter from a Roche-lobe filling, low mass companion onto a neutron star primary. The accretion may be driven by the loss of angular momentum through gravitational radiation or magnetic braking, the nuclear evolution of the companion star, or perhaps a radiation driven stellar wind. Each model makes specific predictions on the evolution of the LMXB.

The sign and magnitude of the orbital period derivative are perhaps the most critical diagnostics of LMXB evolution. However only 4 LMXBs have measured orbital period derivatives, and in each of these there are problems rectifying the measured $P_{orb}/\dot{P}_{orb}$ with theory (White et al. 1994). One of these systems is the eclipsing X-ray binary EXO0748-676, whose sharp edged eclipse transitions provide a good fiducial marker for timing the orbital period P_{orb}.

---

1   ROSAT Observatory Guest Investigator.
2   also Thomas Jefferson High School, Alexandria, VA.

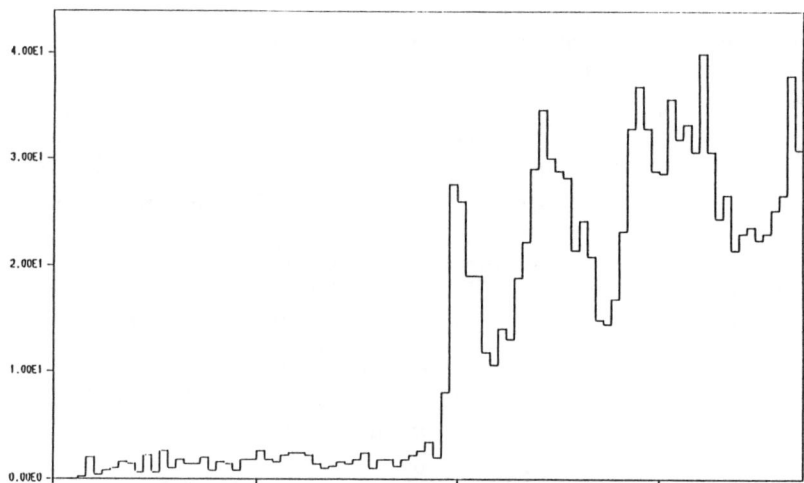

Figure 1: The ROSAT light curve in 5 s bins. EXO0748-676 is in eclipse at the beginning of the observation. The approximately 80 s quasi-periodicity due to occultation of the source by support wires in the PSPC window as ROSAT wobbles is clearly visible.

2. ROSAT DATA

EXO0748-676 was observed three times during ROSAT AO1 and AO2. The purpose was to time eclipses and extend the EXO0748-676 data base. Although 23 OBIs of data were obtained, only one eclipse egress was observed (Figure 1). The source was observed on axis, so the ROSAT wobble is clearly visible in the data.

We have begun a program to measure the response of the PSPC as a function of detector coordinate. We then propose to remove the effect of the wobble by correcting each photon for the effective response of the PSPC at that photon's location in the detector. This should fill in the dips in the light curve caused by the occultation of the source by the coarse support grid for the PSPC window. The response map is being created as the quotient of an exposure map (the convolution of the detector blur function with the histogram of source exposure in detector coordinates) and a detector map (histogram of detected photons in detector coordinates).

Until the wobble is corrected for, we adopt an uncertainty of 5 s in the time of eclipse egress, and we have corrected the time to the solar system barycenter. With ROSAT we measured the egress of eclipse 15440 at barycentric TJD 8571.752298 ± 0.000058.

3. THE O-C DIAGRAM

The eclipse duration for EXO0748-676 has been shown to be variable (Parmar et al. 1991) so we have timed eclipse egress rather than correct the ROSAT measurement to mid-eclipse using an unknown eclipse duration. In Figure 2 we plot the residuals from a constant peiod model of all observed EXOSAT (Parmar et al. 1991), Ginga (Asai et al. 1992), and ROSAT eclipse egress times in barycentric TJD as a function of TJD. This is the standard O-C diagram. Deviations from zero indicate a changing orbital period.

We have fitted linear (constant period), quadratic (constant period derivative), cubic (constant period acceleration), and sinusoidal (long periodw) models to the O-C data. The results are also shown in Figure 2, and the best fitted coefficients are given in Table 1. The cubic and sinusoidal models give fits of comparable quality. We predict that ASCA observations during the 1993 PV phase will be sufficient to distinguish between the cubic and sinusoidal models.

TABLE 1: Models Fitted to O-C Data

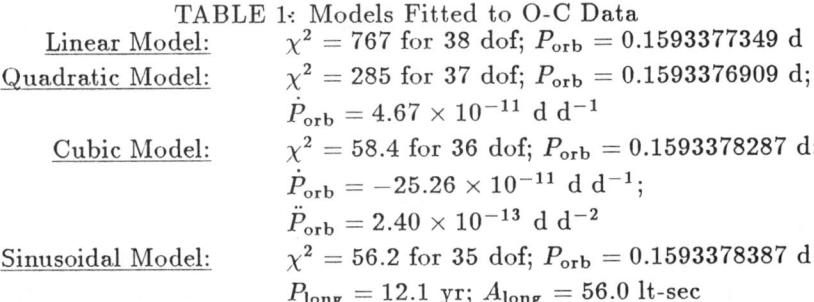

Linear Model: $\chi^2 = 767$ for 38 dof; $P_{\rm orb} = 0.1593377349$ d

Quadratic Model: $\chi^2 = 285$ for 37 dof; $P_{\rm orb} = 0.1593376909$ d; $\dot{P}_{\rm orb} = 4.67 \times 10^{-11}$ d d^{-1}

Cubic Model: $\chi^2 = 58.4$ for 36 dof; $P_{\rm orb} = 0.1593378287$ d; $\dot{P}_{\rm orb} = -25.26 \times 10^{-11}$ d d^{-1}; $\ddot{P}_{\rm orb} = 2.40 \times 10^{-13}$ d d^{-2}

Sinusoidal Model: $\chi^2 = 56.2$ for 35 dof; $P_{\rm orb} = 0.1593378387$ d; $P_{\rm long} = 12.1$ yr; $A_{\rm long} = 56.0$ lt-sec

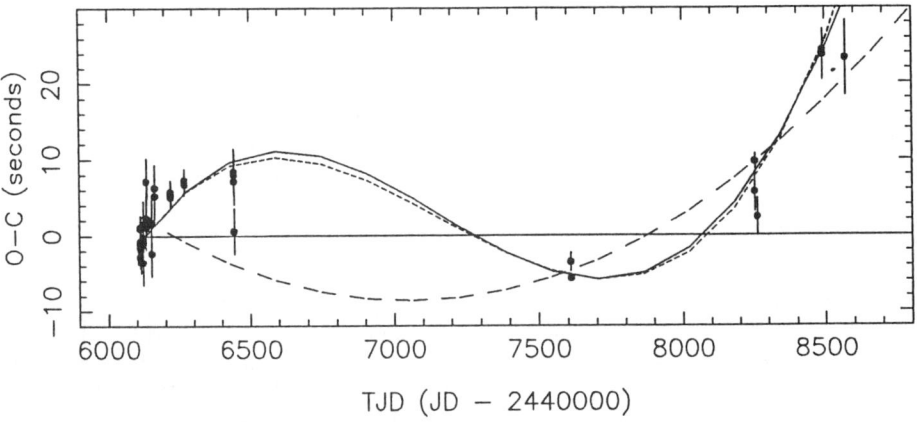

Figure 2: The O-C Diagram. The residuals of measured eclipse egresses from a constant period ephemeris are plotted as a function of time. The best fitted linear (light solid), quadratic (dashed), cubic (dotted), and sinusoidal (heavy solid) models are shown.

4. DISCUSSION

It is usually assumed that the time of eclipse is determined solely by the deterministic ephemeris plus measurement error. Under this assumption, the orbital period, as determined by the time between eclipses, has varied over the last 6.5 years by more than 20 seconds. A constant period derivative is inconsistent with the data. Asai et al. (1991) have suggested that the period is varying sinusoidally with a period of approximately 12 years, almost twice the time that the source has been observed. We confirm their fit, and obtain a best fitted long period of 12.1 years and an amplitude of 56 lt-sec.

A number of cataclysmic variables have been observed with variations in orbital period that are also interpreted as sinusoidal variations, e.g., U Gem (18 yr, 1 cycle, Eason et al. 1983, Beuermann & Pakull 1984), IP Peg (5 yr, 1.5 cycles, Wolf et al. 1993), RW Tri (8 or 14 yr, 0.5 or 1 cycle, Africano et al. 1978), DQ Her (14 yr, 1.7 cycles, Patterson et al. 1978), UX UMa (29 yr, 1.5 cycles, Mandel 1965, Quigley & Africano 1978), EX Hyd (20 yr, 1.3 cycles, Bond & Freeth 1988). None of these systems have been observed for more than 1.7 cycles. The sinusoidal variations in cataclysmic variables are variously interpreted as due to the presence of a third body, loss of angular momentum from the system, exchange of orbital and spin angular momentum, apsidal motion, and motion of the "hot spot" relative to the two stars. Only the last can be safely eliminated for EXO0748-676.

5. STATISTICS OF THE O-C DIAGRAM

An examination of the closely spaced EXOSAT eclipse timings shows that there is an rms residual of ~ 3 seconds about any of the model ephemerides. This indicates that there may be intrinsic variability in the eclipse timings. The variable eclipse durations give further evidence of this. If there is some "phase jitter" in the eclipse timings, then the O-C residuals contain serially correlated errors and the statistics used to show that P_{orb} is changing are invalid. The following comments are based on Lombard (1993) and Sterne (1934).

Let T_n be the time of eclipse egress for cycle number n. If the only uncertainty in the measurement of T_n is measurement error, then $T_n = F(n) + e_n$, where $F(n)$ is the ephemeris and e_n are independent identically distributed measurement errors. If this assumption is true, then the analysis in §3–4 is correct.

If there is intrinsic jitter in every cycle, then the instantaneous period during cycle n is $D_n = T_n - T_{n-1} = F(n) - F(n-1) + \varepsilon_n$, where ε_n is the jitter in cycle n. Now we have $T_n = T_0 + \sum_{i=1}^{n} D_n + e_n = F(n) + \sum_{i=1}^{n} \varepsilon_i + e_n$. The T_n are no longer independent, and they are no longer distributed about $F(n)$. They execute a random walk in O-C space, with the random walk controlled by the stochastic process $\{\varepsilon_i\}$. The random walk will move away from a constant period ephemeris and then back. If you fit an ephemeris to an O-C diagram when the source is random walking, you will fit a sinusoid with a period approximately the length of the data base.

Is this reasonable? It is statistically. Lombard (1993) has worked out the cumulative statistics for the case when every cycle is observed and shown that several variable stars (RR Sco, T Cen, DY Peg) exhibit this behavior. The complete statistical description of the case where cycles are observed randomly, as for catalysmic variables and EXO0748-676, has not been worked out in full (Lombard & Koen 1993). The fact that the number of cycles observed is always 1-1.5, as well as the fact there are always rms residuals of 5-10 s over any model, support the intrinsic jitter hypothesis.

For EXO0748-676, we show an indicative plot. If the eclipse timing yields T_n at cycle n and T_m at cycle m, $m > n$, then the period residual is $R_m = (T_m - T_n) - (m - n)\overline{P}$. Here $\overline{P}$ is the mean period over the entire data base. In Figure 3 we plot R_m vs. m. Note that R_m is larger when $m - n$ is larger (long gaps between timings), and that the largest values of R_m increase as m

increases. This behavior is expected for a random walk process. If there truly is a period derivative, then a regression of $(T_m - T_n)/(m-n) - \overline{P}$ vs. $(m+n+1)/2$ should have non-zero slope even in the presence of random walk noise. However the errors on the data points are highly correlated, and a standard least squares analysis can not be performed. The data are not good enough (yet) to detect a period derivative in the presence of intrinsic jitter.

REFERENCES

Africano, J. L., et al. 1978, PASP, 90, 568
Asai, K., et al. 1992, PASJ, 44, 633
Beuermann, K., & Pakull, M. W. 1984, A&A, 136, 250
Bond, I. A., & Freeth, R. V. 1988, MNRAS, 232, 753
Eason, E. L. E., et al. 1983, PASP, 95, 58
Lombard, F. 1993, presented at Applications of Time Series Analysis in Astronomy and Meteorology, Padua, Italy, 6-10 September 1993
Lombard, F., & Koen, C. 1993, in preparation
Mandel, O. 1965, Perem. Zvesdy, 15, 474
Parmar, A. N., et al. 1991, ApJ, 366, 253
Patterson, J., Robinson, E. L., & Nather, R. E. 1978, ApJ, 224, 570
Quigley, R., & Africano, J. 1978, PASP, 90, 445
Sterne, T. E. 1934, Harvard College Observatory Circular, 366, 1
White, N.E., Nagase, F., & Parmar, A. N. 1994, in X-ray Binaries, eds. W. H. G. Lewin, J. van Paradijs, & E. P. J. van den Heuvel (Cambridge U. Press).
Wolf, S., et al. 1993, A&A, 273, 160

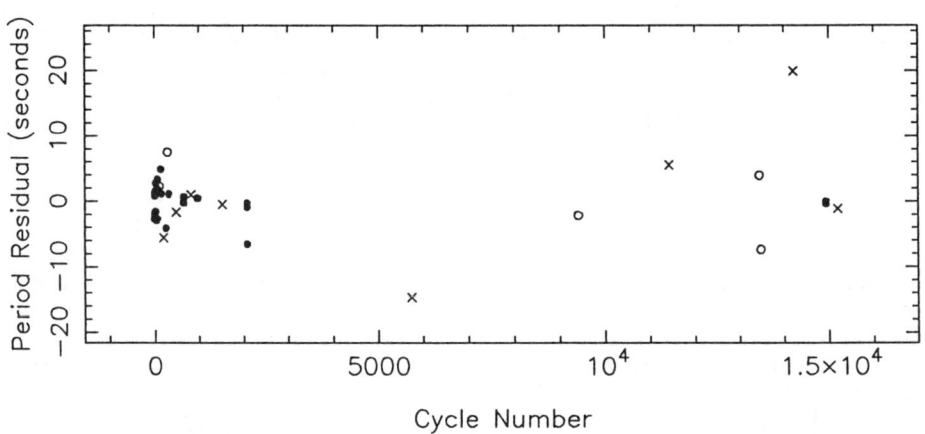

Figure 3: Period residuals vs. cycle number. Single cycle gaps ($m - n = 1$) are plotted with filled circles, short gaps ($m - n < 100$) are plotted with open circles, and long gaps ($m - n > 100$) are plotted with crosses.

ROSAT OBSERVATIONS OF LINERS

Gail A. Reichert
MC 668, NASA Goddard Space Flight Center, Greenbelt, MD 20771
reichert@rosgip.gsfc.nasa.gov

Richard F. Mushotzky
MC 666, NASA Goddard Space Flight Center, Greenbelt, MD 20771
mushotzky@lheavx.gsfc.nasa.gov

Alexei V. Filippenko
Astronomy Dept., University of California, Berkeley, CA 94720
alex@bkyast.berkeley.edu

ABSTRACT

Results of recent ROSAT observations of five LINERs – NGC 3998, NGC 4579, NGC 3079, NGC 2639, and Pictor A – are presented. Except for NGC 3079, the x-ray sources are unresolved. The emission from NGC 3079 is dominated by a point source, but also shows extended emission, including (at energies $\lesssim 0.5$ keV) a "patch" of emission likely associated with the galaxy superwind. No significant short timescale variability was observed. Results of spectral fitting were inconclusive. Pure power law models appear to be "disfavored," but can only be rejected for NGC 3079 and NGC 3998. Spectral indices for the pure power law models are steeper than the average $\Gamma \sim 1.7$ for AGN at higher energies, but adding, e.g., a component of coronal plasma emission allows the power law component to flatten. For the most part, thermal models also give acceptable fits. Independent of spectral model, it is clear that high intrinsic absorbing columns and strong soft x-ray excesses are not observed.

1. Introduction

Low Ionization Nuclear Emission-line Regions, or "LINERs" (Heckman 1980), are found in the majority of early-type galaxies (type Sb or earlier), some late-type and peculiar galaxies, and many radio galaxies. Together with active galactic nuclei (AGN) and giant HII region galaxies, LINERs constitute one of the three major classes of activity within the nuclei of galaxies. The distinction is based purely on optical line intensity ratios. In LINERs, lines from low ionization species that tend to be weak in AGN, such as [OII] $\lambda 3727$, are relatively strong, while lines from high ionization species that tend to be strong in AGN, such as [OIII] $\lambda 5007$, are relatively weak[1]. Conversely, LINERs differ from HII region galaxies (and are similar to AGN) in that lines from warm, predominately neutral or singly ionized gases, such as [OI] $\lambda 6300$, [NII] $\lambda 6583$, and [SII] $\lambda\lambda 6716$, 6731, are relatively strong.

The origins of the emission line activity in AGN and in HII region galaxies are relatively well understood. It is generally accepted that the line emitting gas is photoionized, in AGN by a central, compact source of nonstellar radiation

---

[1] Heckman (1980) defined LINERs to have [OII] $\lambda 3727$/[OIII] $\lambda 5007 > 1$ and [OI] $\lambda 6300$/[OIII] $\lambda 5007 > 1/3$.

which extends to very high frequencies (e.g., Kwan & Krolik 1981), and in HII region galaxies by the quasi-blackbody emission from OB stars. It is not clear what causes the activity in LINERs. Several models have been proposed, including shock heating, cooling flows, photoionization by a diluted power law continuum as might be produced by a "dwarf" AGN, and photoionization by very hot young stars (see Heckman 1987, Filippenko 1989 for reviews).

The idea that LINERs may be "dwarf" AGN is an attractive one. A number of LINERs are known to resemble AGN in several aspects, having broad components of Hα, compact, flat spectrum, nuclear radio sources, forbidden lines whose widths increase with increasing critical density, and in some cases broad UV emission lines. If a large fraction of LINERs are "dwarf" AGN, then by sheer number they would dominate the low luminosity end of the AGN luminosity function, and would have important bearing on such questions as the fraction(s) of galaxies which contain AGN, the evolution of AGN within galaxies, and the evolution of normal galaxies. LINERs may also make a substantial contribution to the diffuse x-ray background at energies $\lesssim$ 2 keV (Elvis, Soltan, & Keel 1984), if they emit as much x-radiation relative to their Hα fluxes as do so-called "classical" AGN (i.e., Seyfert galaxies, QSOs).

The primary stumbling block to the "dwarf" AGN hypothesis for LINERs is the fact that, with few exceptions (e.g., Halpern & Filippenko 1984; Filippenko 1985), no clearly nonstellar continuum sufficient to produce the observed emission line fluxes has yet been detected. Detection of such a continuum has proven to be very difficult in the optical and UV bands due to the overwhelming predominance of the background stellar component. The problem becomes much easier at higher frequencies, as the background stellar emission is much diminished. The x-ray spectra of AGN are also very different from the spectra that would be expected for the other models. Clearly it is important to study the characteristics of the x-ray emission from LINERs.

2. Results of Previous X-ray Observations

Like most galaxies, LINERs tend to be faint x-ray sources. Until recently little has been known concerning their x-ray properties. A few LINERs were detected in *Einstein* observations (Fabbiano, Kim, & Trinchieri 1992, and references therein), but the low statistical quality of the data, coupled with the relatively coarse spatial and spectral resolutions of the *Einstein* instruments, meant for the most part that only total fluxes could be obtained.

Prior to our ROSAT observations, limited x-ray spatial and/or spectral information had been obtained for 3 x-ray bright LINERs:

M81: At a distance of only 3.6 Mpc, M81 is close enough that the *Einstein* IPC could begin to resolve x-ray sources within the galaxy. *Einstein* observations (Fabbiano 1988) showed that the emission from M81 is dominated by a central point source with a very steep spectrum, requiring a power law photon spectral index Γ of about 4. Exosat observations when the nucleus was in a brighter state showed the x-ray emission to vary by amplitudes of $\sim$ 100% on timescales of hours (Barr, Mushotzky, & Giommi 1993), comparable to the fastest variability timescales observed for low luminosity AGN. *Ginga* and BBXRT spectra are well fit at energies above 2 keV by a single power law with $\Gamma \sim 2.2$ (Ohashi & Tsuru 1992; Petre et al. 1993), steeper than the average of $\sim$ 1.7 observed for AGN (Mushotzky 1993), but much flatter than that required by the *Einstein* IPC. The BBXRT spectrum also requires an additional component of coronal emission from plasma at kT~ 0.4 keV, which Petre et al. (1993) attribute to diffuse emission from the underlying galaxy, and an absorbing column $N_H \sim 4 \times 10^{21}$ cm^{-2}. The upper limit to any Fe Kα emission is 90 eV (Petre et al. 1993). Comparison of the fluxes from the *Einstein*, Exosat, *Ginga*, ROSAT (Boller et al. 1992) and BBXRT fluxes shows that the total emission has varied by more than a factor of 10 over the past 14 years (Petre et al. 1993).

NGC 3998: *Einstein* IPC observations of NGC 3998 showed an unresolved central point source whose x-ray spectrum could be well fit by either power law or thermal models (Dressel & Wilson 1985). The *Ginga* spectrum was well fit by a single power law model with $\Gamma = 2$; thermal models gave significantly poorer fits and could be rejected (Awaki et al. 1991). As discussed by Awaki et al., comparison of the IPC and *Ginga* fluxes showed that the brightness had varied by a factor of 3.

Pictor A: Pictor A was detected as an x-ray source by *HEAO-1* A2, *Einstein*, and Exosat (Marshall et al. 1979; Kriss 1985; Singh, Rao, & Vahia 1990). Singh, Rao, & Vahia (1990) reported that the Exosat spectrum can be adequately fit by a single power law model with $\Gamma \sim 1.8$ and a column density consistent with the Galactic column along our line of sight. However, the x-ray data are not restrictive and other spectral models (e.g., thermal, "broken" power law) cannot be rejected. In contrast with M81 and NGC 3998, the *HEAO-1* A2, *Einstein*, and Exosat fluxes are in good agreement (Singh, Rao, & Vahia 1990).

3. ROSAT Observations

The motivation for our ROSAT observations was to study LINERs which, like M81, closely resemble AGN in their properties at other wavelengths. In particular, we wished to search for unambiguous evidence for the presence of an active nucleus which could explain the optical line activity. Five LINERs were observed: NGC 3998, NGC 4579, NGC 3079, NGC 2639, and Pictor A. Like M81, most or all of these LINERs resemble "classical" AGN in the following ways: 1) the Hα emission lines show broad components, 2) the nuclei contain compact, flat spectrum radio sources, and 3) the widths of the emission lines increase with increasing critical density. Like AGN, the ultraviolet spectra of NGC 3998, NGC 4579, and Pictor A also show broad UV emission lines (Reichert et al. 1993; Keel & Windhorst 1991). NGC 3998 was observed in reduced pointing mode for ~50 ksec; Pictor A, NGC 4579, NGC 2639, and NGC 3079 were observed for ~5 ksec, 9 ksec, 9 ksec, and 19 ksec, respectively.

The results of our ROSAT observations can be summarized as follows.

3.1 Spatial Characteristics

Except for NGC 3079, all of the x-ray sources are unresolved. Conservative upper limits to the extents of the emission are ~5" for NGC 3998, NGC 4579, and Pictor A, and ~12" for NGC 2639. The radially averaged profile for NGC 3998 is shown in Figure 1.

The emission from NGC 3079 is more complex than for the other objects. As shown in Figure 2, the profile is dominated by a point source, but also shows emission extending up to about 2.5' from the center. The profile also depends on energy (Figures 2, 3). At energies above ~0.5 keV, the extended emission is slightly elongated along the major axis of the galaxy as determined from optical, 6 cm radio, HI 21 cm, and CO measurements (Figure 4). Below ~0.5 keV, there

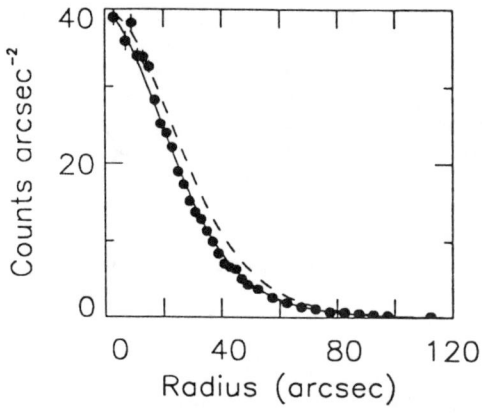

Fig. 1—Radially averaged counts profile (0.2 – 2 keV) for NGC 3998. Solid line shows the point spread function (PSF) expected for the observed spectrum. Dashed line shows the PSF convolved with a Gaussian with $\sigma = 5''$.

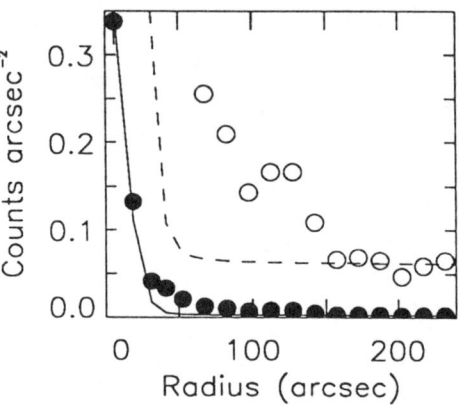

Fig. 2—Radially averaged counts profile for NGC 3998, for pulse height channels 52 to 199 (~0.5 – 2 keV). Profile was corrected for variations in mean exposure. Solid circles show the oberved profile. Solid line shows the observed background plus the PSF whose norm best matches the observed profile. Open circles and dashed line show the profile and background plus best fitting PSF scaled by a factor of 20.

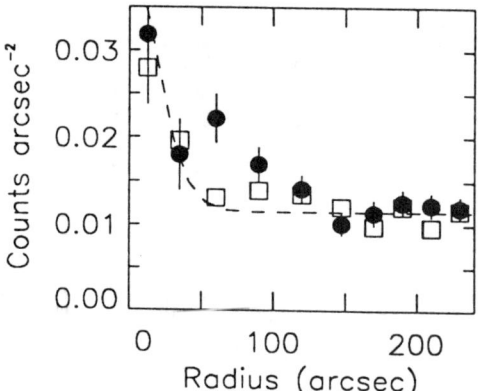

Fig. 3—Radially averaged counts profiles for NGC 3079, for pulse height channels 8 to 51 (~0.1 – 0.5 keV). Profiles were corrected for variations in mean exposure. Solid circles show the oberved profile averaged over the northeast quadrant; open squares the profile averaged over the remaining 3 quadrants. Dashed line shows the observed background plus the PSF whose norm best matches the observed profiles.

is an additional soft "patch" of emission to the northeast about 1 to 2′ from the center.

NGC 3079 is an especially interesting case as it exhibits an unusual range of nuclear activity. (For a more extensive discussion of its properties, see, e.g., Filippenko & Sargent 1992.) A powerful far-infrared galaxy, it shows well-defined,

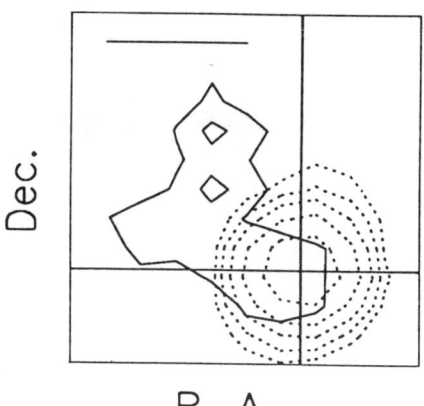

Fig. 4—Contours of the low energy ($\lesssim 0.5$ keV; solid lines) versus higher energy (0.9 – 2.1 keV; dashed lines) x-ray emission from NGC 3079. Contours were smoothed using a boxcar 45' on a side. The cross marks the center of the higher energy emission, which coincides with the radio nucleus. A 1' bar is also shown.

kiloparsec-scale radio lobes of considerable complexity, vigorous, ongoing star formation within the nucleus, and spatially and kinematically complex optical line emission. Maps of the Hα+[NII] emission show a loop which extends approximately along the galaxy minor axis in the same direction as the radio lobes (Ford et al. 1986), as well as long filaments or plumes which appear to originate from the surface of the disk (Heckman, Armus, & Miley 1990; hereafter HAM). The optical emission lines indicate that gas is outflowing from the nucleus in an energetic, bipolar outflow or "galactic superwind" (HAM), which may be driven by the central starburst (Armus, Heckman, and Miley 1990), by an active nucleus (Irwin & Sofue 1992), or by both (Filippenko & Sargent 1992).

It is of interest to note that the soft x-ray patch extends in roughly the same direction as the Hα+[NII] loop, radio jet, and radio lobes. The patch is detected at just 3σ confidence, and so a great deal of weight should not be given to the exact shape of the contours shown in Figure 4. However, according to our current best guess, the x-ray patch appears to fill a gap between two of the filaments in the deep Hα+[NII] map (HAM). The patch is therefore likely to be associated with the superwind and may consist of a filled bubble or a sheath-like structure. For reasonable spectral assumptions, the patch contributes a luminosity (0.2 – 0.5 keV) in the range $(0.2 - 4) \times 10^{39}$ erg s^{-1}, which is not unreasonable for this picture.

3.2 Flux Variability

Pronounced flux variability over long and short timescales is often a characteristic of low luminosity AGN. As discussed in section 2, previous observations have shown that the x-ray fluxes of LINERs can vary by as much as a factor of 10 over long timescales. Comparison of the ROSAT fluxes for NGC 3998, NGC 4579, and NGC 3079 with the fluxes from previous observations confirm the long timescale variability at low energies. Interestingly, the ROSAT flux for Pictor A is consistent with the fluxes from all previous observations over the past 14 years.

In contrast to the long timescale variability, no short timescale variability was observed during any of our ROSAT observations. Upper limits to any short term variability are of order 10 to 30% (at 90% confidence). Our best coverage was obtained for NGC 3998, which (except for Earth blockage, SAA passage,

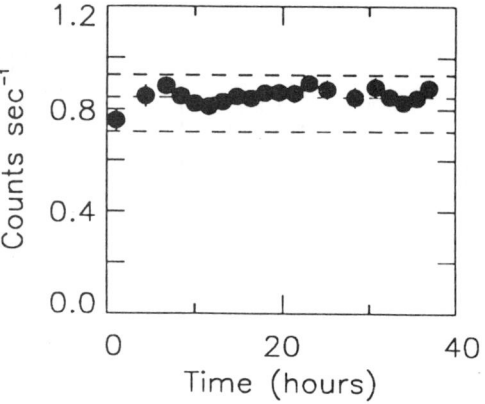

Fig. 5—X-ray light curve for NGC 3998, plotted in counts sec^{-1} versus time in hours. Fluxes were not corrected for variations in mean exposure caused by the wire mesh. Dashed lines show the average count rate and the 90% confidence limits to variability (+11%, −15%). The reduced χ^2 for the hypothesis of constant flux is 1.06.

etc.) was observed continuously for almost 40 hours. Figure 5 shows the light curve for the NGC 3998 observation; the fluxes have not been corrected for differences in mean exposure due to coverage by the wire mesh. It is clear that this light curve is markedly different from those obtained in Exosat observations of, e.g., the low luminosity AGN NGC 5506 and MCG-6-30-15.

3.3 Spectral Fits

The results of spectral fits to the ROSAT data are much less clear. At this point, the strongest general statement that we can make based on the ROSAT data is that pure power law models appear to be "disfavored." For example, a pure power law model can be ruled out for both NGC 3079 and NGC 3998. For the other sources, pure power law models do not give the best fits, but the fits are acceptable and the model cannot be rejected.

The ROSAT spectral data for the five objects are not restrictive, and allow a multitude of models, especially when multiple components are allowed. The results for NGC 3998, NGC 3079, NGC 4579, and Pictor A are as follows:

Pure power law: Based on goodness-of-fit, can be rejected at 95% and 99% confidence levels for NGC 3998 and NGC 3079, respectively. Gives acceptable fits to NGC 4579 and Pictor A spectra, with photon spectral index $\Gamma \sim 2$ in each case.

Pure Thermal Bremsstrahlung: Can be ruled out for NGC 3998 by the *Ginga* data, and for NGC 3079 at 95% confidence based on goodness-of-fit. Gives acceptable fits to NGC 4579 and Pictor A spectra, with temperature $kT \sim 1.6$ keV in each case.

"Broken" Power Law: Gives acceptable fits for all four objects. Best fitting model for NGC 3998 and Pictor A. "Break" energy ~ 0.8 keV, lower energy spectral index Γ_1 flatter than higher energy spectral index Γ_2 in all cases. Best fit lower energy photon indices are $\Gamma_1 \sim 1.3$ (NGC 3998), ~ 0.1 (NGC 3079), ~ 1.5 (NGC 4579), ~ 0 (Pictor A). Best fit higher energy indices are $\Gamma_2 \sim 2.1$ (NGC 3998), ~ 3.4 (NGC 3079), ~ 2.1 (NGC 4579), ~ 1.9 (Pictor A).

Power Law plus Coronal Plasma (Raymond–Smith): Gives good fits for all four objects. Best fitting model for NGC 3079 and NGC 4579. Best fitting photon spectral indices are flatter than for the pure power law model; best fit values are $\Gamma \sim 1.9$ (NGC 3998), ~ 1.6 (NGC 3079), ~ 1.9 (NGC 4579), ~ 1.6 (Pictor A). Best fit plasma temperatures are $kT \sim 0.8$ keV (NGC 3998), ~ 0.4 keV

(NGC 3079), ~ 0.5 keV (NGC 4579, Pictor A). The plasma contributes $\sim 4\%$, $\sim 40\%$, $\sim 5\%$, and $\sim 12\%$ of the total ROSAT flux for NGC 3998, NGC 3079, NGC 4579, and Pictor A, respectively.

Bremsstrahlung plus Coronal Plasma: Gives good fits to ROSAT spectra for all four objects. Can be ruled out for NGC 3998 by the *Ginga* data. Best fit bremsstrahlung temperatures are hotter than for the pure bremsstrahlung model; best fit values are $kT_b \sim 3.0$ keV (NGC 3079), ~ 2.2 keV (NGC 4579), ~ 1.9 keV (Pictor A). Best fit plasma temperatures are ~ 0.5 keV (NGC 3079, NGC 4579), ~ 0.3 keV (Pictor A). The plasma contributes $\sim 40\%$, $\sim 7\%$, and $\sim 12\%$ of the total ROSAT flux for NGC 3079, NGC 4579, and Pictor A, respectively.

The ROSAT spectrum of NGC 2639 appears to be qualitatively different from the spectra of the other four objects. There appears to be a strong emission feature at ~ 0.98 keV, which may contribute a substantial fraction (30%!) of the total ROSAT flux. However, the x-ray source is quite faint ($\sim 1/30$ as bright as NGC 3998) and the data are of low statistical quality; only 150 source counts were obtained for the spectrum. With so few counts, it is not possible to rule out any of the models discussed above on statistical grounds alone.

Given the ambiguity in the spectral fits, what can we say about the x-ray spectral characteristics of these LINERs? One model independent statement is that, unlike many low luminosity AGN, the LINERs do not appear to be significantly absorbed beyond that expected from the line of sight Galactic column. The table below compares the Galactic column with the ranges allowed by the ROSAT data, giving the extreme values of the 90% confidence ranges for *all* of the fitted models:

	ROSAT N_H^a	Galactic N_H^a
NGC 3998	0 – 4.1	1.2
NGC 3079	0 – 4.2	0.84
NGC 4579	0 – 4.5	2.5
Pictor A	0 – 5.7	1.6 – 2.1
NGC 2639[b]	1 – 3	~ 7

[a] In units of 10^{20} cm^{-2}.
[b] Range of best fit values for the various spectral models. Ranges for other objects give the extrema of the 90% confidence ranges for the various fits.

The BBXRT column for M81 is somewhat higher than these values, but is still consistent with the line of sight column from our Galaxy combined with that expected for the host galaxy of M81 (Petre et al. 1993). Clearly these LINERs do not show the high intrinsic columns ($\gtrsim 10^{22}$ cm^{-2}) observed for low luminosity AGN.

AGN spectra also commonly show curvature in their x-ray spectra, such that the spectrum at higher energies is flatter than that at lower energies (Turner, George, & Mushotzky 1993, and references therein). Independent of spectral model, it is clear that similar excesses in the x-ray spectra of these LINERs must be far less pronounced. The curvature may even be in the opposite sense, with the spectrum steepening at higher energies rather than flattening.

4. Discussion

Interpretation of these results is hampered by our inability to restrict the

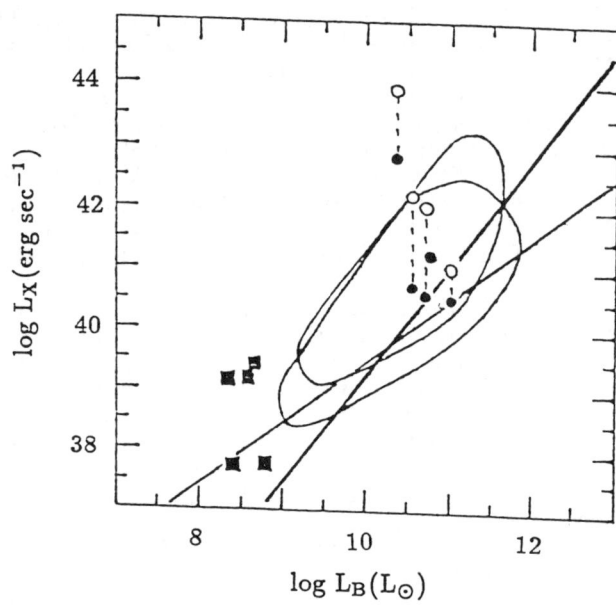

Fig. 6—X-ray luminosity L_X (0.2–2 keV) plotted against blue luminosity L_B (in solar luminosities). Open and solid circles show the total luminosities and the luminosities of the plasma components (for the power law plus coronal plasma models), respectively. Solid curves delineate the loci of points for normal elliptical and spiral galaxies from the compilation of Fabbiano, Kim, & Trinchieri (1992) (with a few outliers shown by solid squares); solid straight lines their best fit linear regressions of L_X against L_B.

range of acceptable spectral models. Some simple consequences of the various models are discussed below.

Pure power law: If the spectra of these LINERs are best described as pure power laws, then their spectra are steeper than the $\Gamma \sim 1.7$ typical of AGN over the range 2 – 10 keV. ROSAT spectra of AGN also tend to be steeper than their higher energy spectra (e.g., Brinkmann 1992). However, for NGC 3998 and M81, the steepness of the spectrum is known to extend to $\gtrsim 10$ keV. Based on the spectral steepness, Petre et al. (1993) argue that the x-ray production mechanism in M81 is qualitatively different from that in higher luminosity AGN.

"Broken" power law: Although the best fit spectral indices (above the break energy) are still steep, the allowed confidence ranges are larger and include the average (2–10 keV) value. The spectral curvature is qualitatively different from that observed for higher luminosity AGN, however.

Power law plus coronal plasma (Raymond–Smith): Such a model is attractive by reason of consistency; these LINERs are among the best candidates to contain "dwarf" AGN, and we would also expect to detect some emission from the host galaxy. This type of model is already required to explain the spectrum of M81 (Petre et al. 1993). As shown in Figure 6, except for Pictor A, the luminosities in the plasma components agree quite well with the luminosities expected given their blue magnitudes. Similarly, the luminosities in the power law components are roughly consistent with the relation between x-ray and optical luminosity derived for higher luminosity AGN (see Figure 7; Mushotzky 1993). The spectral fits with this model allow flatter spectral indices, and the confidence ranges include the average (2–10 keV) value.

Pure bremsstrahlung, or bremsstrahlung plus coronal plasma: Although the x-ray sources are unresolved by the ROSAT PSPC, the upper limits to the

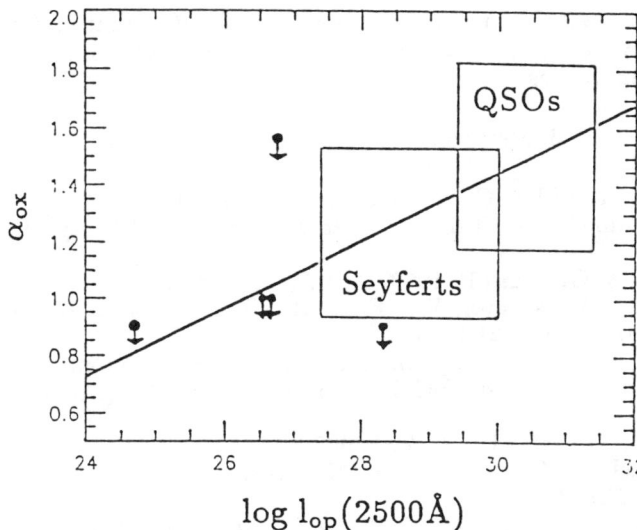

Fig. 7—Effective logarithmic slope α_{ox} (from 2500Å to 2 keV) plotted against monochromatic optical luminosity l_{op} (2500Å). Optical luminosities are derived from the fluxes at 2500Å obtained from *IUE* observations, and hence are upper limits to the luminosities in any AGN-like continuum. The straight line shows the relation derived for higher luminosity AGN, $l_x \sim l_{op}^{-0.7}$. Also shown are the ranges in α_{ox} observed for higher luminosity AGN. Adapted from Mushotzky (1993).

extents of the emission are not particularly restrictive. At the distance of our closest target (16 Mpc), 5″ corresponds to ∼370 pc. NGC 3998, NGC 4579, and Pictor A have been observed with *IUE*, and their UV profiles also show unresolved components of at most a few arcseconds in size (Reichert et al. 1992; Keel & Windhorst 1991). For NGC 3998 and NGC 4579, at least, the UV continua appear to be dominated by starlight (Reichert et al. 1992). Hence, it seems not unreasonable that the x-ray emission from at least some of these LINERs might arise in stellar or stellar-related sources, such as x-ray binaries and/or supernova remnants. NGC 3079 in particular is known to contain vigorous, ongoing star formation in the nuclear region, which adds weight to this idea. On the other hand, x-ray variability is an argument against the thermal hypothesis. We note that if some of these LINERs are thermal x-ray sources, then the ratios of x-ray to blue luminosity L_x/L_B are high, although (excepting Pictor A) they still fall within the loci of points defined by *Einstein* observations of normal galaxies (Fabbiano, Kim, & Trinchieri 1992).

5. Conclusions

Is the activity in LINERs caused by "dwarf" AGN? For M81, at least, the answer appears to be yes. Unfortunately, current x-ray data do not allow us to answer the question for other LINERs. However, if LINERs are generally caused by faint active nuclei, then they differ from their more luminous counterparts in the following ways: short timescale variability, high intrinsic absorption, and strong soft x-ray excesses are (so far) not observed, and the shape of the x-ray spectrum may be qualitatively different. For M81 and NGC 3998, the relative spectral steepness is known to extend to ∼10 keV. In M81, any Fe K emission must have equivalent width $\lesssim 90$ eV, at the lower bound of the distribution for Seyfert galaxies (Mushotzky 1993). Any theories of LINERs as AGN must be able to account for these differences.

6. References

Armus, L., Heckman, T. M., & Miley, G. K. 1990, *Ap. J.*, **364**, 471.
Awaki, H., Koyama, K., Kunieda, H., Takano, S., & Tawara, Y. 1991, *Ap. J.*, **366**, 88.
Barr, P., Mushotzky, R. F., & Giommi, P. 1993, *Ap. J.*, submitted.
Boller, T., Meurs, E. J. A., Brinkmann, W., Fink, H., Zimmermann, U., & Adorf, H.-M. 1992, *Astr. & Ap.*, **261**, 57.
Brinkmann, W. 1992, in *X-Ray Emission from Active Galactic Nuclei and the Cosmic X-ray Background*, ed. W. Brinkmann & J. Trumper (Munich: MPE Report 235), 1.
Dressel, L. L., & Wilson, A. S. 1985, *Ap. J.*, **291**, 668.
Elvis, M., Soltan, A., & Keel, W. C. 1984, *Ap. J.*, **283**, 479.
Fabbiano, G. 1988, *Ap. J.*, **325**, 544.
Fabbiano, G., Kim, D.-W., & Trinchieri, G. 1992, *Ap. J. Suppl.*, **80**, 531.
Filippenko, A. V. 1985, *Ap. J.*, **289**, 475.
Filippenko, A. V. 1989, in *IAU Symposium 134: Active Galactic Nuclei*, ed. D. E. Osterbrock and J. S. Miller (Dordrecht:Kluwer), 495.
Filippenko, A. V., & Sargent, W. L. W. 1992, *Astron. J.*, **103**, 28.
Ford, H. C., Dahari, O., Jacoby, G. H., Crane, P. C., & Ciardullo, R. 1986, *Ap. J. (Letters)*, **311**, L7.
Halpern, J. P., & Filippenko, A. V. 1984, *Ap. J.*, **285**, 475.
Heckman, T. M. 1980, *Astr. Ap.*, **87**, 152.
Heckman, T. M. 1987, *IAU Symposium 121: Observational Evidence of Activity in Galaxies*, ed. E. Ye. Khachikian, K. J. Fricke, and J. Melnick (Dordrecht: Reidel), 421.
Heckman, T. M., Armus, L., & Miley, G. K. 1990, *Ap. J. Suppl.*, **74**, 833.
Irwin, J. A., & Sofue, Y. 1992, it Ap. J. (Letters), **396**, L75.
Kriss, G. A. 1985, *Astron. J.*, **90**, 1.
Kwan, J., & Krolik, J. H. 1981, *Ap. J.*, **250**, 478.
Marshall, F. E., Boldt, E. A., Holt, S. S., Mushotzky, R. F., Pravdo, S. H., Rothschild, R. E., & Serlemitsos, P. J. 1979, *Ap. J. Suppl.*, **40**, 657.
Mushotzky, R. F. 1993, in *The Nearest Active Galaxies*, ed. J. E. Beckman, L. Colina, & H. Netzer.
Ohashi, T. & Tsuru, T. 1992, in *Frontiers of X-Ray Astronomy*, ed. Y. Tanaka & K. Koyama (Tokyo: Universal Aacdemy Press), 435.
Petre, R., Mushotzky, R. F., Serlemitsos, P., Jahoda, K., & Marshall, F. E. 1993, *Ap. J.*, **418**, 644.
Reichert, G. A., Puchnarewicz, E. M., Filippenko, A. V., Mason, K. O., Branduardi-Raymont, G., & Wu, C.-C. 1992, in *Relationships Between Active Galactic Nuclei and Starburst Galaxies*, ed. A. V. Filippenko (San Francisco: Astronomical Society of the Pacific), 277.
Reichert, G. A., Puchnarewicz, E. M., Filippenko, A. V., Mason, K. O., Branduardi-Raymont, G., and Wu, C.-C. 1993, in *The Nearest Active Galaxies*, ed. J. E. Beckman, L. Colina, & H. Netzer, 85.
Singh, K. P., Rao, A. R., & Vahia, M. N. 1990, *M.N.R.A.S.*, **246**, 711.
Turner, T. J., George, I. M., and Mushotzky, R. F. 1993, *Ap. J.*, **412**, 72.

THE COMPLEX SOFT X-RAY SPECTRUM OF LOW REDSHIFT QUASARS

Fabrizio Fiore
Harvard-Smithsonian Center for Astrophysics, 60 Garden st. Cambridge, MA 02138

Email ID
fiore@cfa.harvard.edu

ABSTRACT

Several ROSAT PSPC observations of low-redshift AGN have been analyzed to study their soft X-ray spectrum. Most spectra can, at least roughly, be described in the 0.1-2.5 keV band by simple power laws with energy indices in the range 1.0-2.5, reduced at low energies by Galactic absorption. The energy indices are systematically steeper than those found in the same sources at higher energies (by $\Delta\alpha_E = 0.5 - 1$ with respect to *Ginga* or EXOSAT - 2-10 keV - measurements, and, surprisingly, by $\Delta\alpha_E$ up to 0.7 with respect to IPC - 0.2-3.5 keV - measurements). These differences are larger than the evaluated systematic error on PSPC spectral indices and therefore they suggest a break between the soft and hard components in the "keV" region. In fact, fits to high signal-to-noise spectra reveal a significant excess above ~ 1 keV with respect to the simple power law model. No evidence for strong emission lines is found in any of the objects. The implications of these findings for AGN models where the soft excess is interpreted in terms of free-free emission from optically thin plasma or of optically thick thermal emission from the innermost region of an accretion disk, are discussed.

1. Introduction

It is well known that the single power law model with Galactic or intrinsic absorption (PL model) is an inadequate description of the 0.2-10 keV spectrum in $\sim 50\%$ of luminous quasars (Wilkes & Elvis 1987, Masnou et al 1992) and of Seyfert galaxies (Turner & Pounds 1989). The spectra of these AGN show in the 0.2-2 keV band some sort of "soft excess", above the extrapolation of the 2-10 keV power law. This low energy X-ray turn-up, is interpreted in many cases as the high energy tail of the blue bump, although such a connection is not compelling given the pattern of the data: both components increase into the unobservable extreme UV region; no direct evidence, such as simultaneous/correlated variations in both the UV and soft X-ray emission, yet exists. Other types of "soft excess" include emission associated with an extended component around the active nucleus (e.g. NGC4151, Elvis et al 1990, Perola et al 1986; NGC1068, Wilson et al 1992); emission line features in the 0.7-1 keV band (Turner et al 1991); the recovery of the spectrum below the oxygen edge in "warm absorber" sources (e.g. MCG-6-30-15, Nandra & Pounds 1992; 3C351, Fiore et al 1993).

The poor energy resolution of the instruments used prior to ROSAT in the 0.1-2 keV band resulted in little/no information about the spectral form of the "soft excess". The softer energy band and improved resolution of the ROSAT (Trümper 1983) PSPC (Pfeffermann et al 1987) allow study of its spectrum and thus constraints upon and discrimination between different models for the blue bump-soft X-ray components. These models include: optically thick emission from the inner edge of an accretion disk (e.g. MKN 335, Turner & Pounds 1988; PG1211+143, Elvis et al 1991) and optically thin free-free emission from gas with temperature the order of 1 million K (Barvainis 1993).

2. The X-ray Data

We observed a sample of low-redshift quasars, selected to be the brightest with known soft excesses, to acquire high signal-to-noise ($S/N \gtrsim 100$) spectra with the PSPC. Results from the analysis of 6 radio-quiet quasars are presented in Fiore et al (1994a, Paper I). Figure 1a,b,c shows the residuals, as a fraction of the counts in each channel, after subtracting the best fitting PL model from the spectra of the Seyfert galaxy MKN110, the radio-loud quasar 3C273, and the composite spectrum obtained adding together 7 PSPC spectra of 5 low redshift radio-quiet quasars (Paper I). Information on the objects and on the fits are given in Table 1.

TABLE 1: AGN observations; power law fits

Name	redshift	Counts^a (S/N)	$N^b_{H Gal}$	α_E	N^b_H	χ^2 (dof)^c
MKN110	0.036	59 (170)	1.5^d	1.38 ± 0.03	1.36 ± 0.06	57.4 (29)
3C273	0.158	38 (190)	1.68^e	0.92 ± 0.03	1.61 ± 0.08	74.1 (29)
Soft-7^f	0.048–0.155	49 (150)	1.91–2.99^g	1.70 ± 0.02	2.59 ± 0.06	59.6 (29)
PG1114+445	0.144	0.78 (26)	1.8^d	$1.13^{+0.24}_{-0.31}$	1.04 ± 0.62	59.7 (25)

^{a} in units of 10^3, PI=11-245; ^{b} in units of 10^{20} cm^{-2}; ^{c} fit on SASS channels 3-34; ^{d} Stark et al 1992; ^{e} Savage et al 1993; ^{f} composite spectrum obtained adding together seven PSPC spectra of 5 quasars: NAB0205+025, PG1211+143, MKN205, TON1542, and PG1244+026; ^{g} Elvis et al 1989.

All the observations were performed in the low gain setting (i.e. after October 1991) and the spectra were fitted using the January 1993 resolution matrix. The simple power law model is a rather good description of these high quality spectra, and the deviations are smaller than 10–20 % over the whole 0.1-2.5 keV energy band. The largest features (5 to 10 % of the counts per channel) are an excess of counts between 0.2 and 0.4 keV, a deficit between 0.6 and 1.5 keV and an excess above 1.5 keV.

This contrasts strongly with the results found in the so called "warm absorber" sources (e.g. Nandra & Pounds 1992, Fiore et al 1993, Turner et al 1993). In these AGN strong absorption edges were found at $\sim 0.7-0.8$ keV (rest frame) and were interpreted as due to highly ionized oxygen (OVII, OVIII) in an ionized "warm" absorber of column density $10^{22} - 10^{23}$ cm^{-2} covering the central X-ray sources. An example is given on Figure 1d (see also Table 1), where we show the residuals after the subtraction of the best fit PL model from the PSPC spectrum of PG1114+445. The deviations from the model are very large ($\gtrsim 50\%$ of the counts per channel) and the edge structure is clear even with a signal to noise in the whole spectrum of only ~ 26.

The N_H found by the PSPC is close to the Galactic column along the line of sight in most cases (we found however in a few sources N_H significantly less than the Galactic one). This is shown in Figure 2a,b where we plot the best fit N_H as a function of the Galactic one for a sample of 26 AGN with more than 1000 counts each. The fits were performed with both the two "official" response matrices, i.e. those released on March 1992 and on January 1993. Column densities obtained with the March 92 matrix agree with the Galactic column generally better than those obtained with the Jan 93 matrix. There is a systematic difference of $\sim 5 \times 10^{19}$ cm^{-2} between the best fit N_H obtained with the two matrices. This is clear from Figure 3a, were the best fit N_H obtained with both matrices are plotted. Figure 3b gives the best fit α_E. The systematic difference in best fit spectral index is ~ 0.2. The two matrices were calibrated using observations performed at the beginning of 1991 in the high gain setting (March 1992 matrix), and in May 1992 in the low gain setting (Jan 1993 matrix). Thus, the above differences could be due to time variations of the PSPC gain not included in correction process currently applied to data in SASS (see also Paper I and Turner, these proceedings). The two matrices should be appropriate, at least with regard to the best fit parameters, to observations performed around the above two periods. Observations performed in between could have gains slightly different from those in the observations used to calibrate the matrices, and therefore fitting them with one of the two matrices would produce a small systematic error in the best fit parameters. We evaluate the maximum amplitude of this systematic error as the difference in the best fit parameters obtained using both matrices, i.e. $\sim 5 \times 10^{19}$ cm^{-2} on N_H and ~ 0.2 on α_E.

The PSPC spectral indices of low redshift AGN span a broad range ($1.0 < \alpha_E < 2.5$). They are systematically steeper by $\Delta\alpha_E \sim 0.5$ than the 2-10 keV spectral indices found by EXOSAT and *Ginga* (Turner & Pounds 1989, Lawson et al 1992, Nandra 1991, Williams et al 1992). This difference is much larger than the evaluated maximum systematic error on PSPC slopes. Figure 4a shows the best fit α_E for a single power law model for a sample of 16 low-z AGN with both PSPC and 2-10 keV measurements and with no evidence of intrinsic neutral or ionized absorption. This difference immediately suggests a) that the PSPC spectra are dominated by the soft excess, and b) that the break between the soft and hard components is in the "keV" region. In fact, the residuals after subtracting the single power law model in MKN110, 3C273 and in the composite spectrum show a significant excess above 1-1.5 keV. In the cases of MKN110 and the composite spectrum (radio-quiet objects), a good fit can be obtained using a broken power law model with high energy slopes in the range observed by EXOSAT and *Ginga* for radio-quiet sources (Table 2). The case of the radio-loud quasar 3C273 is more complex. Fitting a broken power law model fixing the high energy slope to the value observed by Williams et al (1992) improves the χ^2 but does not yield an acceptable fit. The soft X-ray spectrum of this source is probably more complicated than the parameterization adopted.

TABLE 2: broken power law fits

Name	$\alpha_E 1$	E^a_{break}	$\alpha_E 2$	χ^2 (dof)
MKN110	1.5 ± 0.1	$1.2^{+0.3}_{-0.5}$	$1.0^{+0.3}_{-0.4}$	27.5 (27)
3C273	$1.0^{+0.2}_{-0.1}$	$1.3^{+0.6}_{-0.2}$	$0.6 - 0.2$	60.8 (27)
Soft-7	1.8 ± 0.1	1.3 ± 0.3	$1.1^{+0.3}_{-0.6}$	25.9 (27)

[a] in keV, rest frame

Fitting the broken power law model to the individual spectra making up the composite spectrum improves significantly the χ^2, with respect to the PL model, in three cases out of five (Paper I). The break energy is between 0.7 and 1.5 keV (rest frame energies) in all cases. This break energy is higher than that of about 0.3-0.6 keV inferred from IPC studies. The discrepancy between PSPC and IPC measurements is also clear from figure 4b where we plot the best fit single power law α_E for 18 quasars with IPC (Elvis et al 1986, Wilkes & Elvis 1987, and Shastri et al 1993) and PSPC (Paper I, Laor et al 1994, Walter & Fink 1993) observations. The IPC slopes are flatter than the PSPC slopes by $\Delta\alpha_E \sim 0.5$ (again significantly more than the systematic error on PSPC slopes). This is surprising since the energy band of the two instruments is similar (the largest difference being the IPC response between 2 keV and 3.5 keV, where the PSPC is insensitive). In fact, simulating a two component model or a broken power law model (with high energy slopes similar to those observed in the 2-10 keV range) in the two instruments, and fitting the spectra with the PL model, produces similar α_E (within $\Delta\alpha_E = 0.2$). If the discrepancy is not due to calibration errors fitting the spectra simultaneously in the two instruments should provide stronger constraints on the models. Unfortunately, one cannot be sure that this is the case (Paper I), so to make full use of the high signal to noise IPC and PSPC spectra a major effort to verify and possibly revise the relative IPC-PSPC calibration is unavoidable.

The PSPC alone can statistically discriminate between a number of different curved or multicomponent models that can be reasonably associated with the soft X-ray spectrum of quasars. A black body plus power law model or a line plus power law model does not give a good fit of the data fixing the power law slope at the typical value measured above 2 keV; a Raymond-Smith plus power law model gives a good fit either for very small metal abundances ($A < 0.1$ in all cases analyzed), or two temperatures. The PSPC data of PG1211+143 and PG1244+026 are well fitted by a single power law model and no improvement in χ^2 is obtained by fitting more complex models. (This is likely to be due to the fact that these two sources have the worst signal to noise ratio in the present sample, and to the relatively small contribution of a hard component in the PSPC band.) A sharp break between soft and hard components is suggested in PG1426+015 and NAB0205+024. A shallower break is suggested in TON1542 (see Paper I for details). Examples of the diversity of quasar soft X-ray spectra are shown in Figure 5 where we plot the ratio of the PSPC counts of NAB0205+024 to that of MKN205, and to that of TON1542 (the quasars have similar redshift and Galactic N_H). The ratios show that the soft excess in NAB0205+024 is much stronger than in MKN205 and that the break between the soft and hard components in the former source is probably steeper than in TON1542.

3. Comparison With Models

A longstanding question in AGN research is whether the soft X-ray emission and the blue bump are one and the same component. Walter & Fink (1993) reported a strong correlation between the soft X-ray (PSPC) slope and the ratio between the flux at 1350Å to that at 2 keV in a sample of about 50 AGN. These authors interpreted the correlation as evidence for a link between soft X-ray and OUV emission. Their sample, however, is soft X-ray selected, since it includes the brightest quasars in the all sky survey, and thus it is strongly biased against flat spectrum, soft X-ray weak sources. The correlation may therefore be due to selection effects. Laor et al. (1994 and these proceedings) report a significant correlation between the PSPC slope and the optical (2500Å) to X-ray (2 keV) spectral index α_{OX} in a sample of 10 optically selected quasars with z< 0.4, high optical luminosity ($M_B < -23$), and very low Galactic N_H, a result similar to the Walter & Fink correlation. Laor et al. (1994) interpret their result as due to a decrease in the importance of the hard X-ray component, as measured by the 2 keV flux, rather than an increase in the strength of the soft component as the blue bump increases. This results applies to high optical luminosity quasars only. Low optical luminosity AGN tend to have flatter α_{OX} (by ~ 0.2, or a factor ~ 3.5 in the ratio of optical to X-ray luminosities, for optical luminosities differing by a factor 300, Worrall et al 1987) but similar PSPC slope (the mean α_E in the Laor et al sample is 1.5, equal to the value found by Walter & Fink for a sample consisting of 24 objects classified as quasars and 34 lower luminosity AGN). As an example, MKN110 is a low redshift low optical luminosity AGN in the PG survey with very low Galactic column. It has $\alpha_{OX} \sim 1.0$, much flatter than the mean of 1.5 in Laor et al (or a factor 20 in the ratio of optical to X-ray luminosities), but $\alpha_E \sim 1.4$, only slightly flatter than the mean of 1.5 in the Laor et al sample. It therefore falls quite far from the correlation found by Laor et al. (1994). Clearly a larger, unbiased sample of AGN with a large range of luminosities is required to reach a final conclusion.

Under the assumption that the soft X-ray component dominating the PSPC quasar spectra is related to the optical/UV blue bump, the following section discusses the ability of optically thin plasma models to reproduce both X-ray and optical/UV data. The alternative case of optically thick emission from an accretion disk is presented by Siemiginowska et al (these proceedings).

3.1 Free-free Models

Thermal bremsstrahlung emitted by an optically thin cloud can contribute to the optical/UV/soft-X-ray band of the spectrum (Barvainis 1993). In the soft X-ray band the free-free emission from the ionized plasma represents only part of the total emission, since recombination, 2-photon, and line emission from the same plasma can also be important, depending on the temperature (Raymond & Smith 1977). The main emission lines for temperatures $10^6 < T < 5 \times 10^6$ K are the iron L complex at about 0.9 keV and the O VII and O VIII K$-\alpha$ lines at 0.57 and 0.65 keV. We find strong limits on the amount of line emission at those energies (see Figure 1 and Paper I). Thus the contribution from an optically thin ionized gas to the soft X-ray excess in these sources is important only if the metal abundance is very small, if the soft X-ray emission lines are optically thick, or if there is a distribution in temperatures (see Paper I sections 3.5 and 3.6). In all these cases the soft X-ray spectra are indistinguishable (for the PSPC) from a single temperature thermal bremsstrahlung.

In Figure 6 we plot for a sample of six, low-z, radio-quiet quasars (NAB0205+024, PG1211+143, MKN205, TON1542, PG1244+026 and PG1426+015) the soft X-ray color ($1keV L_{1keV}/0.4keV L_{0.4keV}$, soft component only), against the optical-UV color ($1325Å L_{1325Å}/5500Å L_{5500Å}$, proportional to the optical-UV slope). The soft X-ray color is calculated using the X-ray monochromatic luminosities obtained by fitting to the PSPC data: a two power law model (solid line error bars); and a power law plus thermal bremsstrahlung model (dashed line error bars). The optical and UV data are from Elvis et al (1994) and Fiore et al (1994b).

The dotted line and open squares show the predictions of a pure free-free models with temperature 10^6 K – 10^7 K (see Fiore et al 1994b for details). The contribution of the recombination and 2-photon continua increases the soft X-ray color by $\sim 300\%$ for $T = 10^6$ K and by $\sim 10\%$ for $T = 10^7$ K (from figure 2 in Raymond & Smith 1977). The model reproduces the spread in the soft X-ray color for $10^6 \lesssim T \lesssim 10^7$ K. The model also predicts an optical-UV slope within the observed range. However the optical-UV slope is insensitive to temperature in this range and thus the model cannot simultaneously give the observed spread in optical-UV slopes, if this is intrinsic to the source emission spectrum, and a soft X-ray excess.

In Figure 7 we plot the observed and predicted (by the best fit power law plus thermal bremsstrahlung model) 1325 Å luminosities for the sample of six low-z radio-quiet quasars. We assume a 100 % upper error on the predicted luminosity. When more than one IUE observation is available we used the smallest observed 1325 Å luminosity. It is clear that this model systematically underpredicts the 1325 Å luminosity by a factor 3 to 10. Fitting the data with this model and requiring that the predicted 1325Å luminosity matches the observed one, yields χ^2 much worse than in the previous cases (by $\Delta\chi^2 = 20$ to 100). This argues against single temperature free-free models to explain both the optical-UV and soft X-ray excess emission, a conclusion similar to that already reached based upon the absence of strong emission lines in the PSPC band.

4. Conclusions

We have presented high signal to noise soft X-ray (0.2-2 keV) spectra of a number of low-redshift quasars. We summarize in the following our main results.

1) We compared the two "official" PSPC resolution matrices. We found a systematic difference in best fit parameters fitting a single power law model to the data. We evaluate the maximum amplitude of the systematic error in PSPC spectral measurements equal to these differences ($\sim 5 \times 10^{19}$ cm^{-2} in N_H and ~ 0.2 in α_E).

2) The PSPC spectra span a broad range of slopes ($1.0 < \alpha_E < 2.5$) and are steeper than at higher energies. They are dominated by the so called "soft X-ray excess" (above the extrapolation of the higher energy power law).

3) The deviations from a single power law model are small ($< 10\%$ of the counts per channel) but still significant. The "break point" between the hard and soft components is typically in the "keV" region. The PSPC spectra can exclude "narrow" models, such as a line or a black body, for the soft X-ray component.

4) The strength of any line emission feature in the soft X-ray band is small ($< 10 - 20\%$) unless the continuum is very complicated. This is puzzling, since many different models predict the existence of such lines: "warm absorbers", Netzer 1993 but see also Fiore et al 1993, reprocessing from a disk, Ross & Fabian 1993, optically thin ionized plasmas with temperatures between 0.1 keV and 1 keV. In the latter case the absence of strong line features argues against emission from an isothermal ionized plasma as the main contributor to the soft X-ray component. More generally this suggests peculiar physical conditions for the soft X-ray emission region and the nearby regions, that is, high temperatures and ionization states (high $\frac{L}{L_{Edd}}$ in the models of Ross & Fabian 1993), or, alternatively, a non-thermal origin for the soft X-ray emission.

5) Assuming a physical link between the blue bump and the soft X-ray component, emission from optically thin isothermal plasma can explain the observed soft X-ray and mean optical-UV colors, but not, simultaneously, the spread in optical-UV

color. Furthermore, these models underpredict the 1325Å luminosity.

6) Pure disk models, even in a Kerr geometry, do not seem to have the necessary flexibility to account for the observed spread in optical-UV and soft X-ray slopes and luminosities (Siemiginowska et al, these proceedings, Fiore et al 1994b). The PSPC soft X-ray component slope requires high inclinations and high accretion rates, which in turn overestimate the soft X-ray luminosity, when producing the correct UV luminosity.

7) Reflection from an ionized accretion disk is probably an important component in explaining the observed soft X-ray spectra and luminosities. From the absence of strong line features in the PSPC spectra, we speculate that high accretion rates are needed in this model to reproduce the data. The photoionized gas can imprint sharp "jumps" on the emitted spectrum, since the gas opacity can suddenly change in the proximity of the many strong absorption edges present in the soft X-ray band. This is intriguing, since at least in some cases our fits to the PSPC spectra prefer a sharp break between the soft and hard components at energies close to that of the OVIII edge (0.87 keV), the distinctive feature in the reflection models by Życki et al (1994).

ACKNOWLEDGEMENTS: Martin Elvis, Ari Laor, Andy Lawrence, Smita Mathur, Jonathan McDowell, Aneta Siemiginowska, and Belinda Wilkes have all contributed to the present investigation. This work was supported by NASA grants NAGW-2201 (LTSARP), NAG5-1872, NAG5-1883 and NAG5-1536 (ROSAT), and NASA contracts NAS5-30934 (RSDC), NAS5-30751 (HEAO-2) and NAS8-39073 (ASC).

References

Barvainis, R. 1993, ApJ, 412, 513

Elvis, M., Green, R.F., Bechtold, J., Schmidt, M., Neugebauer, B.T., Soifer, B.T., Matthews, K., & Fabbiano, G. 1986, ApJ, 310, 291

Elvis, M., Lockman, F.J. and Wilkes, B.J. 1989, ApJ, 97, 777

Elvis, M., Fassnacht, C., Wilson, A.S., & Briel, U. 1990, ApJ, 361 459

Elvis, M., Giommi, P., Wilkes, B.J., & McDowell, J.C. 1991, ApJ, 378, 537

Elvis, M., Wilkes, B.J., McDowell, J.C., Green, R.F., Bechtold, J., Willner, S.P., Polomski, E. & Cutri, R. 1994, ApJ Supp., submitted

Fiore, F., Elvis, E., Mathur, S, Wilkes, B.J., & McDowell, J.C. 1993, ApJ, 415, 129

Fiore, F., Elvis, M., Siemiginowska, A., Wilkes, B. J. and McDowell, J. C. 1994a, ApJ, submitted (Paper I)

Fiore, F., Elvis, E., Siemiginowska, A., Wilkes, B.J., McDowell, J.C., & Mathur, S. 1994b, ApJ, submitted

Laor, A., Fiore, F. Elvis, M., Wilkes, B.J., & McDowell, J.C. 1994, ApJ, submitted

Lawson, A.J., Turner, M.J.L., Williams, O.R., Stewart, G.C., & Saxton, R.D. 1992, MNRAS, 259, 743

Masnou, J.-L., Wilkes, B.J., Elvis, M., McDowell, J.C., & Arnaud K.A. 1992, A&A, 253, 35

Nandra, K. 1991, Ph.D. Thesis, University of Leicester UK

Nandra, K., & Pounds, K.A. 1992, Nature, 356, 215

Netzer, H. 1993, ApJ, 411, 594

Perola, G.C. et al 1986, ApJ, 306, 508

Pfefferman, E., et al. 1987, Proc SPIE, 733, 519

Raymond, J.C., & Smith, B.W. 1977, ApJS, 35. 419

Ross, R.R. & Fabian, A.C. 1993, MNRAS, 261, 74
Savage, B.D. et al 1993 ApJ, 413, 116
Shastri, P., Wilkes, B.J., Elvis, M., & McDowell, J.C. 1993, ApJ, 410, 29
Stark, A.A., Gammie, C.F., Wilson, R.W., Bally, J., Linke, R., Heiles, C., & Hurwitz, M. 1992, ApJ Supp., 79, 77
Trümper, J. 1983, Adv. Space Res., 2, No. 4, 241
Turner, T. J., & Pounds, K. A. 1988, MNRAS, 232, 463
Turner, T. J., & Pounds, K. A. 1989, MNRAS, 240, 833
Turner, T.J., Weaver, K.A., Mushotzky, R.F., Holt, S.S., & Madejsky, G.M. 1991, ApJ, 381, 85
Turner, T.J., Nandra, K, George, I.A., Fabian, A.C., & Pounds, K.A. 1993, ApJ, in press
Walter, R., & Fink, H.H. 1993, A&A, 274, 105
Wilkes, B.J., & Elvis. M. 1987, ApJ, 323, 243
Williams, O.R., et al. 1992, ApJ, 389, 157
Wilson, A.S., Elvis, M., Lawrence, A., & Bland-Hawthorn, J. 1992, ApJ, L75
Worrall, D.M., Giommi, P., Tananbaum, H., & Zamorani, G. 1987, ApJ, 313, 596
Życki, P.T., Krolik, J.H., Zdziarski, A.A., & Kallman, T.R. 1994, in preparation

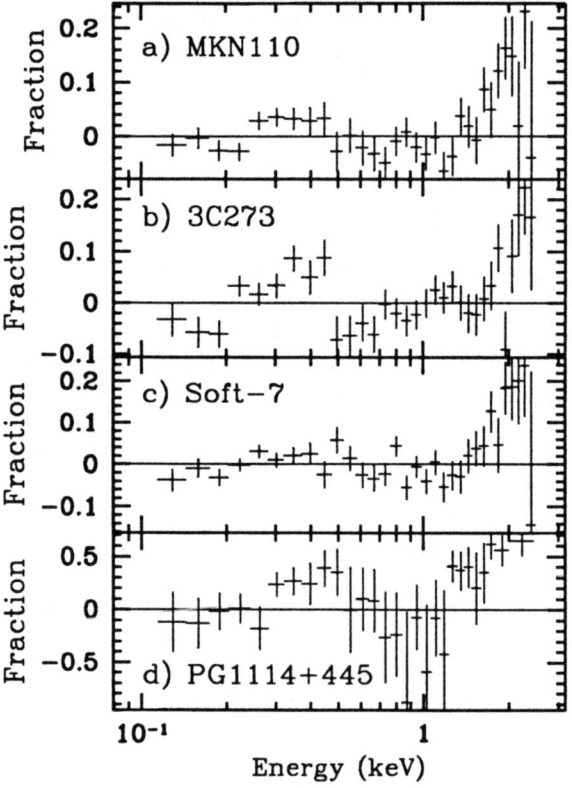

Figure 1. Residuals after the subtraction of the best fitting single power law model from the PSPC spectra of MKN110 a), 3C273 b), PG1114+445 d), and the composite spectrum obtained adding together 7 spectra of the 5 quasars NAB0205+024, PG1211+143, MKN205, TON1542 and PG1244+026 (c). Energies are in the observed frame.

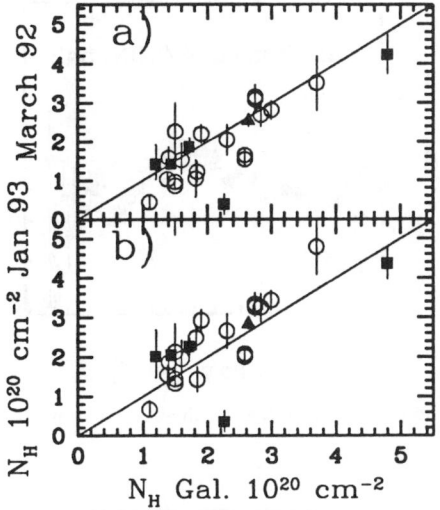

Figure 2. The N_H obtained for a sample of 26 AGN with more than 1000 counts each with the March 1992 (a) and the January 1993 (b) resolution matrices, plotted as a function of the Galactic N_H. Circles identify PSPC-B observations performed after October 1991; squares identify PSPC-B observations performed before October 91; The triangle symbol identifies one PSPC-C observation.

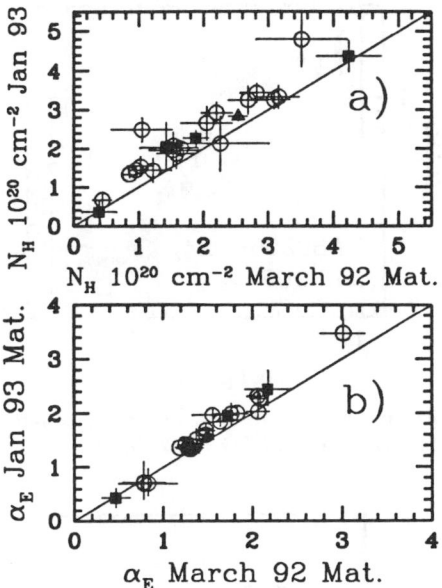

Figure 3. The N_H (a) and the α_E (b) obtained for the same sample as Fig. 2 with both matrices. Symbols are the same as in Fig. 2

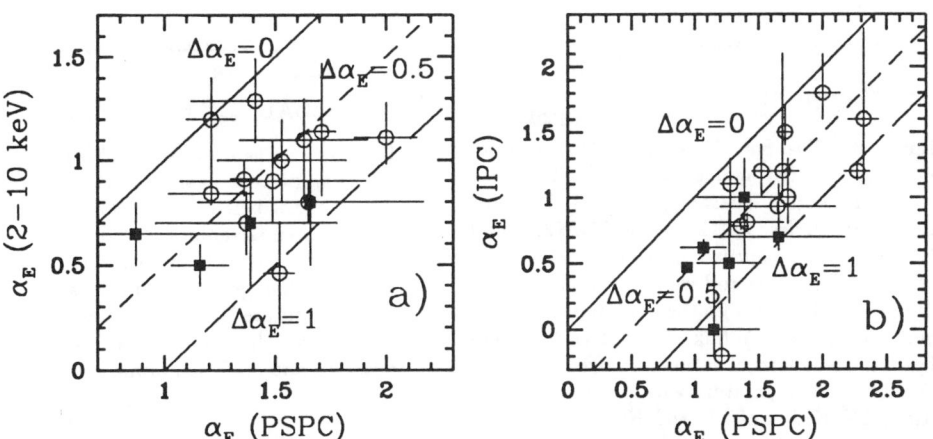

Figure 4. The 0.1-2.5 PSPC α_E plotted against the 2-10 keV EXOSAT or *Ginga* α_E for a sample of 16 AGN (a). The PSPC α_E plotted against the IPC α_E for a sample of 18 AGN (b). Filled squares identify the radio-loud quasars, and open circles radio-quiet AGN.

104   X-ray Spectrum of Low Redshift Quasars

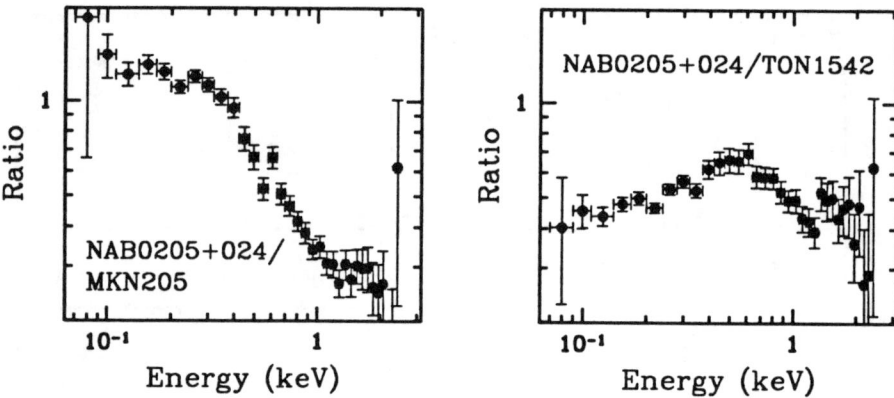

Figure 5. The ratio of the PSPC counts in the spectrum of NAB0205+024 to those of MKN205 (a), and TON1542 (b).

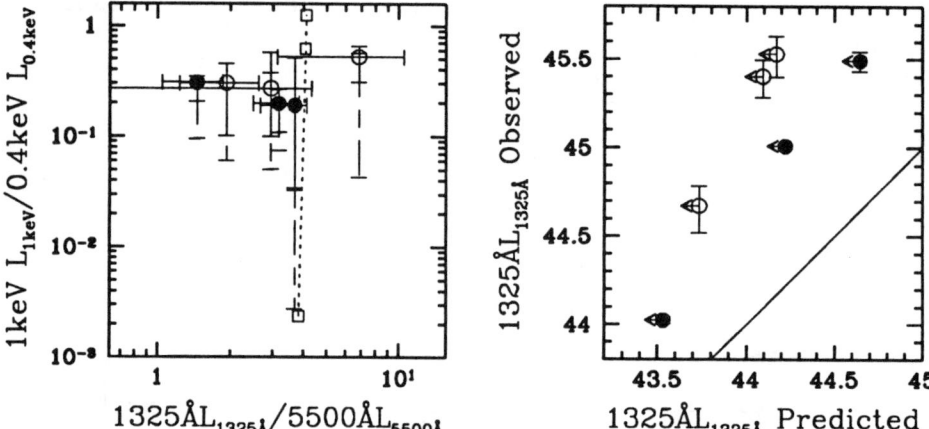

Figure 6. The Soft X-ray color (soft component only) plotted against the optical-UV color for six radio-quiet quasars (see text); In three cases (open circles, PG1211+143, MKN205, and PG1426+015) the quasars have been observed by IUE more than once and we plot the optical-UV color obtained using both the maximum and minimum optical and UV flux. The dotted line identifies the prediction of a pure free-free model as a function of temperature with open boxes indicating: 10^6 K, 5×10^6 K and 10^7 K (soft X-ray color increases with temperature).

Figure 7. The observed 1325Å luminosity plotted against that predicted by the best fit power law plus thermal bremsstrahlung model for the X-ray spectrum of the six quasars in Figure 6. The predicted luminosity upper limits are 100 % of the best fit luminosity. The line indicates equal observed and predicted luminosities.

IR TO X-RAY SPECTRAL ENERGY DISTRIBUTIONS OF HIGH REDSHIFT QUASARS

Jill Bechtold, Roc Cutri, Marcia Rieke,
Steward Observatory, University of Arizona

Martin Elvis, Fabrizio Fiore, Belinda Wilkes,
Jonathan McDowell, Aneta Siemignowska
Harvard-Smithsonian Center for Astrophysics

ABSTRACT

We have observed 13 quasars having z > 2.8 with the ROSAT PSPC, and detected 12 of them, including the z=4.11 quasar 0000-263. For the radio quiet quasars with z > 2.5, the mean $< \alpha_{ox} > \sim 1.8$. Thus, the high redshift quasars are relatively more X-ray quiet than quasars with z < 2.5. However given their high optical luminosities, they are consistent with the extrapolation of the dependence of $< \alpha_{ox} >$ on l_{opt} seen at low redshift. For the radio-loud quasars, $< \alpha_{ox} > \sim 1.4$, independent of redshift. This is smaller than the expected value for the optically luminous, high redshift objects, if they are comparable to steep-spectrum, compact radio sources at low redshift.

For 6 of the quasars, there are sufficient counts detected to provide meaningful constraints to the X-ray spectrum. For the others, a PSPC hardness ratio is used to constrain the X-ray spectral properties. The observations imply that at $z \approx 3$, the X-ray spectra of radio-loud and radio-quiet quasars are different. Implications for the interpretation of the evolution of the luminosity function of quasars are discussed. Models where quasars are numerous and short-lived are favored.

1. Introduction

We have observed a sample of bright, high redshift quasars with the PSPC. Most of these quasars were discovered since the demise of the Einstein Observatory and EXOSAT, and so had not been previously studied in the X-rays. The objects in the sample were chosen because they are very bright at optical wavelengths; 6 are radio-loud and 7 (including the one not detected) are radio-quiet. While this sample is clearly heterogeneous, there seemed to be no other way to get a comparable data set to study the X-ray properties of high redshift quasars: the ROSAT all-sky survey is generally too shallow to detect many z>3 quasars, and the deep pointed PSPC surveys cover too little of the sky to expect many of these rare objects. Also, these quasars are bright at all wavelengths, and so it was relatively easy to construct the spectral energy distributions for these quasars from the radio to X-rays.

For the objects in this sample we obtained near-IR spectrophotometry with the GESPEC at the Multiple Mirror Telescope, optical spectrophotometry with the B&C Spectrograph at the Steward Observatory 2.3 meter, and IR photometry at the MMT. This work is discussed in greater detail in Bechtold et al. 1994a,b, and in Fiore (1994, these proceedings). The X-ray spectral properties of 4 of the quasars were discussed in Elvis et al. (1994). The infrared data is also discussed by Kuhn (1994).

2. Optical to X-ray Spectral Index

The mean spectral index between the optical and X-ray bands, α_{ox}, has been studied extensively for low redshift quasars, based on observations with the Einstein IPC (Zamorani et al 1981; Avni & Tananbaum 1982, 1986; Kriss & Canizares 1985; Worrall et al. 1987). For optically selected, and presumably mostly radio quiet quasars, the ratio of X-ray to optical flux decreases with increasing optical luminosity, and is only weakly dependent on redshift. Radio loud quasars were found to have larger X-ray to optical flux ratios compared to radio quiet objects of the same optical luminosity, implying that an extra component to the X-ray emission is present that is associated with the radio emission. No significant dependence on redshift is seen for this effect. Since our sample extends the redshift range observed, we re-examined these trends.

We computed α_{ox} for the quasars in our sample, where

$$\alpha_{ox} = \frac{-log(l_x/l_{opt})}{log(\nu_x/\nu_{opt})}.$$

Here $log\nu_x$ is the frequency corresponding to 2 keV and $log\nu_{opt}$ corresponds to 2500 Å; all quantities are in the rest frame of the quasar. Note that we have measured the flux at 2500 Å directly from the GESPEC data at an observed wavelength of ~ 1 μm. We assume that the X-ray spectra are power laws with $\alpha_E = 0.5$ and absorption by the known Galactic N(HI) only. For a comparison low redshift sample, we used the results from the Einstein Quasar and Seyfert 1 Galaxy Database by Wilkes et al. (1994) as obtained from the Einline on-line service. To investigate the radio properties of the sample, we cross-correlated this list with the QCAT database, which is based on the Veron-Cetty & Veron (1989) and Hewitt & Burbidge (1987) catalogs. In addition we have included several serendipitously detected quasars at $z \approx 1.5$ from the PSPC image of 0130-403, and 2 other high redshift quasars discussed by Henry et al. (1994) and Fink and Briel (1993).

In Figure 1, we plot α_{ox} as a function of redshift for the radio quiet and radio loud quasars. The high-redshift radio quiet objects appear to be significantly more X-ray quiet (i.e. larger α_{ox}) than their low redshift counterparts. The 7 objects with z > 2.5 have a mean $<\alpha_{ox}> = 1.80 \pm 0.01$, compared to $<\alpha_{ox}> = 1.38 \pm 0.05$ for the 111 quasars detected in the IPC with z < 2.5. The radio loud objects, instead, appear to have nearly constant α_{ox} with redshift, with a mean of ~ 1.4. If we extrapolate the dependence of α_{ox} on l_{opt} for the radio quiet quasars given by Worrall et al. (1987), then for log $l_{opt} = 33$ ergs sec^{-1} Hz^{-1} (typical of the quasars in our sample), the expected α_{ox} is 1.78, which is consistent with the observed mean $<\alpha_{ox}> \sim 1.8$. At the same l_{opt}, the radio loud, steep spectrum quasars are expected to have $\alpha_{ox} = 1.66$; the radio-loud, flat spectrum sources, $\alpha_{ox} = 1.45$; and the radio-loud, steep-spectrum, but compact sources, $\alpha_{ox} \approx 1.6$. Thus, given the dependence of α_{ox} on l_{opt}, it appears that the radio quiet quasars at z > 2.5 show no evolution of α_{ox} with z. On the other hand, the radio loud quasars at high redshift may be more X-ray loud than expected, if they are similar in nature to the steep spectrum compact radio sources at low z.

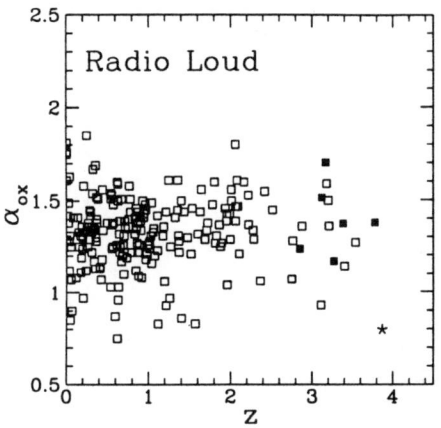 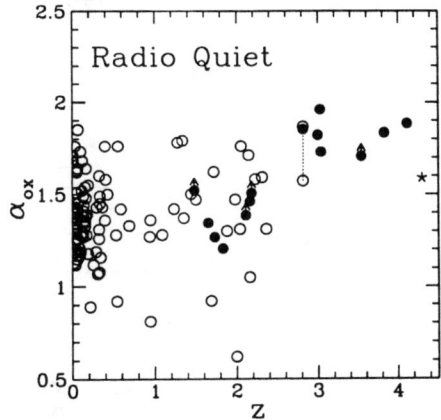

Figure 1: α_{ox} as a function of redshift for radio quiet and radio loud quasars. Open symbols are IPC results and closed symbols are ROSAT PSPC results presented here. The IPC and PSPC values for 0207-389 are connected by a dashed line. The z=4.3 quasar of Henry et al. (1993) and the z=3.87 quasar of Fink and Briel (1993) are shown as 5 pointed stars. Upper limits for the IPC sample are not shown for clarity, but were included in the analysis described in the text.

3. PSPC X-ray Spectra of High Redshift Quasars

Elvis et al. (1994) presented the PSPC spectra of 4 radio-loud quasars, and found that two showed evidence for soft X-ray absorption in excess of the Galactic HI column density. This is different from the situation at low redshift, where few high luminosity quasars show absorption. Since the spectral resolution is not high enough to measure the redshift of the absorbing material, it's location could be almost anywhere along the line of sight. Elvis et al. (1994) give a discussion of the possibilities: it could be related the radio source, in the host galaxy, or originate in an intervening galaxy along the line of sight. To begin to disentangle these possibilities, we have obtained spectra of two radio quiet quasars, 1946+76 at z=3.03 and 0000-263 at z=4.11. Neither show absorption in excess of the Galactic value, with 3σ upper limits of 1.3×10^{22} cm^{-2} and 5.6×10^{22} cm^{-2} respectively, assuming the absorption is at the emission line redshift. However, these upper limits only marginally rule out the absorption seen by Elvis et al. (1994) in 2126-158 (z=3.27, $N_H=1.4 \times 10^{22}$ cm^{-2}) and are larger than the absorption seen in 0438-436 (z=2.85, $N_H=0.9 \times 10^{22}$ cm^{-2}). Interestingly, 1946+76 and 0000-263 have damped Lyα absorbers in their optical spectra, as does the radio loud 0420-388 which also has no absorption (z=3.12, $N_H < 1 \times 10^{22}$ cm^{-2} 3σ). Therefore, since no soft X-ray absorption is seen in 3 quasars that are known to have intervening galaxies, it is more likely that the absorption is intrinsic to the quasars and may well be related to the radio source.

To study the X-ray spectral properties of the quasars in our sample with less than 100 photons detected, we have computed an X-ray color, or hardness ratio, R =

IR to X-ray Spectral Energy Distributions

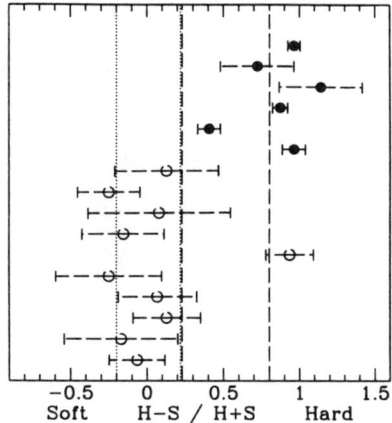

Figure 2: The X-ray hardness ratio R for high redshift radio quiet quasars (open symbols) and radio loud quasars (filled symbols). The dashed and dotted lines show the expected R for radio-loud and radio quiet quasars respectively, using the mean spectral index found at low redshift by GINGA (Williams et al. 1992) and the minimum and maximum Galactic N_H values in our samples.

(H-S)/(H+S), where H = the number of photons between 0.44 and 2.48 keV and S = the number of soft photons between 0.11 and 0.44 keV (c.f. Maccacaro et al. 1988). This hardness ratio is shown in Figure 2. Clearly the radio loud quasars are harder than the radio quiet ones. The mean $<R> = -0.018 \pm 0.110$ for the radio quiet quasars, whereas $<R> = 0.740 \pm 0.030$ for the radio loud quasars, a 6.6σ difference.

If the radio quiet quasars have no soft absorption in excess of the Galactic column, then their power law indices are $<\alpha_E> \approx 1.0$, similar to the low redshift quasars observed in the same rest energy range by GINGA (Williams et al. 1992). If they do have absorption, then their hardness ratios imply $<\alpha_E> \gg 1$, in which case the intrinsic power law spectrum has evolved significantly with redshift.

On the other hand, the hardness ratios of the radio loud quasars imply that if they have no soft absorption in excess of the Galactic value, then $\alpha_E < 0$, much flatter than the $<\alpha_E> = 0.7$ average observed in the GINGA sample. In this case, their intrinsic spectra have evolved with redshift. Of course, in 2 cases there is clear evidence from the X-ray spectra for absorption, and the hardness ratios are consistent with absorption being present in all the high redshift radio loud objects, with an unabsorbed $<\alpha_E> = 0.7$.

4. Discussion

One would like to relate the changes in the spectral properties of individual quasars with redshift to the changes in the ensemble properties of the quasars, that is, the evolution of the luminosity function of quasars at high z. The X-rays in particular probably originate close to the central engine in the quasars and so may give direct clues to the nature of the evolution of the black hole and accretion. If the soft X-ray absorption seen by Elvis et al. (1994) originates outside the central engine, then the X-ray colors imply that the intrinsic X-ray spectra of radio loud quasars does not evolve with z. On the other hand, the radio quiet objects have steeper X-ray power laws at high z than at low z.

This may be interpreted in the context of recent models for the X-ray emission of quasars by Haardt and Maraschi (1993). In their models, the 2-20 keV X-rays are inverse compton emission from a hot corona above a colder accretion disk. More massive black holes would be expected to have steeper X-ray spectra than less massive ones. If quasars are long-lived and rare, so that the evolution of the luminosity function is the result of the slow increase of black hole mass with decreasing redshift, then one expects flatter X-ray spectra at high z. This is inconsistent with our results for radio quiet objects. On the other hand, if quasars are numerous and short-lived, they may be better understood in the hierarchical collapse of CDM halos described by Haehnelt and Rees (1993). More massive fluctuations collapse at high z than at low z, so one expects quasars at high z to have more massive black holes. In this case, the X-ray spectra would be steeper at high z, as is seen for the radio quiet quasars. The radio loud quasars on the other hand, may have a significant fraction of their X-ray emission from some other source.

Acknowledgements

Data was obtained with the Multiple Mirror Telescope, a joint facility of the University of Arizona and the Harvard-Smithsonian Astrophysical Observatory. This work was supported by NASA grants NAGW-2201 (LTSARP), NAG5-1872, NAG5-1536 and NAG5-1680 (ROSAT), and NASA contract NAS8-39073 (ASC). This work was also supported by NSF grants RII-8800660, INT-9010583 and AST-9058510, and a gift from Sun Microsystems.

REFERENCES

Avni, Y. & Tananbaum, H. 1982 Ap.J.(Letters), 262, L17
Avni, Y. & Tananbaum, H. 1986 Ap.J., 305, 83
Bechtold, J., et al. 1994a Ap.J., submitted
Bechtold, J. et al. 1994b Ap.J. (Letters), submitted
Elvis, M., Fiore, F., Wilkes, B., McDowell, J. & Bechtold, J. 1994, Ap.J., 422, in press
Fink, H. H. & Briel, U. G. 1993 A.A., 274, L45
Haardt, F. & Maraschi, L. 1993 Ap.J., 413, 507
Haehnelt, M.G. and Rees, M.J. 1993 MNRAS, 263, 168
Henry, J. P. et al. 1994, in preparation
Hewitt, A. & Burbidge, G. 1987 Ap.J.Supp., 63, 1
Hewitt, A. & Burbidge, G. 1993 Ap.J.Supp., 87, 451
Kriss, G.A. & Canizares, C. R. 1985 Ap.J., 297, 177
Kuhn, O. 1994, Ph.D. thesis
Maccacaro, T. et al. 1988 Ap.J., 326, 680
Veron-Cetty, M.-P. & Veron,P. 1989, ESO Scientific Report, No. 7
Wilkes, B.J. et al. 1994 Ap.J. Supp., in press
Williams, O.R. et al. 1992 Ap.J., 389, 157
Worrall, D.M., Giommi, P. Tananbaum, H. & Zamorani, G. 1987 Ap.J., 313, 596
Zamorani, G. et al. 1981 Ap.J., 245, 357

SEPARATION OF X-RAY EMISSION COMPONENTS IN RADIO GALAXIES

D.M. Worrall and M. Birkinshaw
Harvard-Smithsonian Center for Astrophysics, Cambridge, MA 02138-1596

Email ID
dmw@cfa.harvard.edu, mb1@cfa.harvard.edu

ABSTRACT

One of ROSAT's major achievements has been its ability to separate X-ray emission components in many radio galaxies, where this was previously possible only for a very few well-known sources, e.g., M 87 and Cen A. The dominant X-ray emission mechanism in radio galaxies as a class was unclear, with correlations between the X-ray and radio emissions used on one hand to argue for a nuclear origin for the X-rays, and on the other hand for a thermal origin. Now, with ROSAT we find the presence of both resolved (thermal) and unresolved emission to be typical.

Our results are illustrated with PSPC data from the first six radio galaxies in our study. Spectral and spatial measurements independently support the presence of multiple emission components. The resolved emission can be modeled as thermal radiation from gas of galaxy, group, or cluster dimension depending on object. The unresolved emission may be thermal or non-thermal. Evidence is presented to support a non-thermal origin for most of the unresolved emission: for NGC 6251, where this component is dominant, gas-confinement properties argue against a thermal origin; for the sample as a whole, a proportionality between the unresolved X-ray and the radio-core luminosity densities supports the existence of non-thermal X-ray radiation from the inner regions of a parsec-scale radio jet. Our results have implications for the unification of BL Lac objects with low-power radio galaxies; part of the unresolved X-ray emission in the radio galaxies is probably unbeamed.

1. X-ray Observations and Data Reduction

The radio galaxies listed in Table 1 are the first sources which were observed during two of our programs: to study a subsample of radio galaxies which Ulrich (1989) proposed are drawn from the parent population of BL Lac objects, and to study the emission in and around low-power radio galaxies with prominent radio jets.

For data reduction we used the Post Reduction Off-line Software (PROS; Worrall et al. 1992) and additional software which we developed to convolve the energy-dependent PSPC point response function (PRF; Hasinger et al. 1992) with radially-symmetric spatial models. Only counts within the energy band for which the PRF is well modeled (0.2 – 1.9 keV) are used. NGC 326 is the only source to show obvious *asymmetric* extended emission; the analysis presented here is for a pie-slice region where the source is least extended, and it is presented only to provide a qualitative comparison with the other sources.

For each radial profile, best fits were determined to models consisting of an unresolved component, a β model which is appropriate for gas in hydrostatic equilibrium (e.g., Sarazin 1986), and a combination of an unresolved component and a β model. The combined fits (Fig. 1) give a significantly smaller value for

Table 1: PSPC Observations

Object	Name B1950	z	Date	Exposure[a](s)
NGC 315	0055+300	0.0165	1992 Jan 17–Feb 1	10,343
			1992 Jul 19–21	17,869
NGC 326	0055+265	0.047	1992 Jul 24–29	20,969
4C 35.03	0206+355	0.0375	1992 Jul 24–27	14,843
NGC 2484	0755+379	0.0413	1991 Oct 30–Nov 1	15,172
NGC 4261	1216+061	0.0073	1991 Dec 24–30	22,042
NGC 6251	1637+826	0.024	1991 Mar 13–16	14,830

a. Time used in analysis, excluding some intervals of high background.

χ^2 than a fit to either model alone in all cases except for NGC 315, for which the β-model component is barely resolved. To check that systematic errors in the aspect determination have a negligible effect on our results, we applied our analysis to two BL Lac objects and two quasars spanning similar exposure times and net counts to the radio galaxies; all fitted an unresolved component and excluded the presence of additional resolved X-ray emission of similar strength to that detected in the radio galaxies.

The same net counts which give the radial profiles of Figure 1 are used for our spectral fitting. The presence of a resolved X-ray component suggests at least some thermal emission, but for all sources the fits improve if either a power-law model or a second thermal model is added. In either case, the percentage of counts in the unresolved component agrees well with that in one of the spectral components, indicating self-consistency between our spectral and spatial analyses; in the case of a power-law plus thermal spectral model, we are required to identify the power law with the unresolved emission (as is physically most reasonable) to get agreement (Fig. 2). The spectral fits to a power-law plus thermal model give spectral energy indices of $\alpha \sim 0.8$, but with large errors, and the temperatures are tightly distributed about $kT \sim 0.7$ keV (except for the anomalous case of NGC 326). The power-law indices, although not well constrained, are generally consistent with values for BL Lac objects (Worrall & Wilkes 1990), the sources proposed as low-power radio galaxies with their relativistic jets pointed towards the observer. The fits to two thermal components give one with $kT < 1$ keV and the other with $kT > 1$ keV.

2. The Nature of the Unresolved X-ray Emission

The X-ray data alone do not determine the nature of the unresolved component, but the integrated emission of discrete X-ray sources similar to those in spiral galaxies is unimportant; the X-ray luminosities of all the separate spectral components for the radio galaxies lie at least an order of magnitude above the X-ray/optical correlation for spiral galaxies of Fabbiano, Gioia & Trinchieri (1989), and are comparable with or exceed 10^{34} W (the limiting X-ray luminosity of spiral galaxies). The resolved emission is therefore certainly hot gas; the measured scale-size associates it primarily with the galaxy for NGC 4261 (but see indication for a larger scale-size excess in Fig. 1, and the paper by Mushotzky in this volume) and NGC 315, with cluster gas for NGC 326, and with group gas for the other sources.

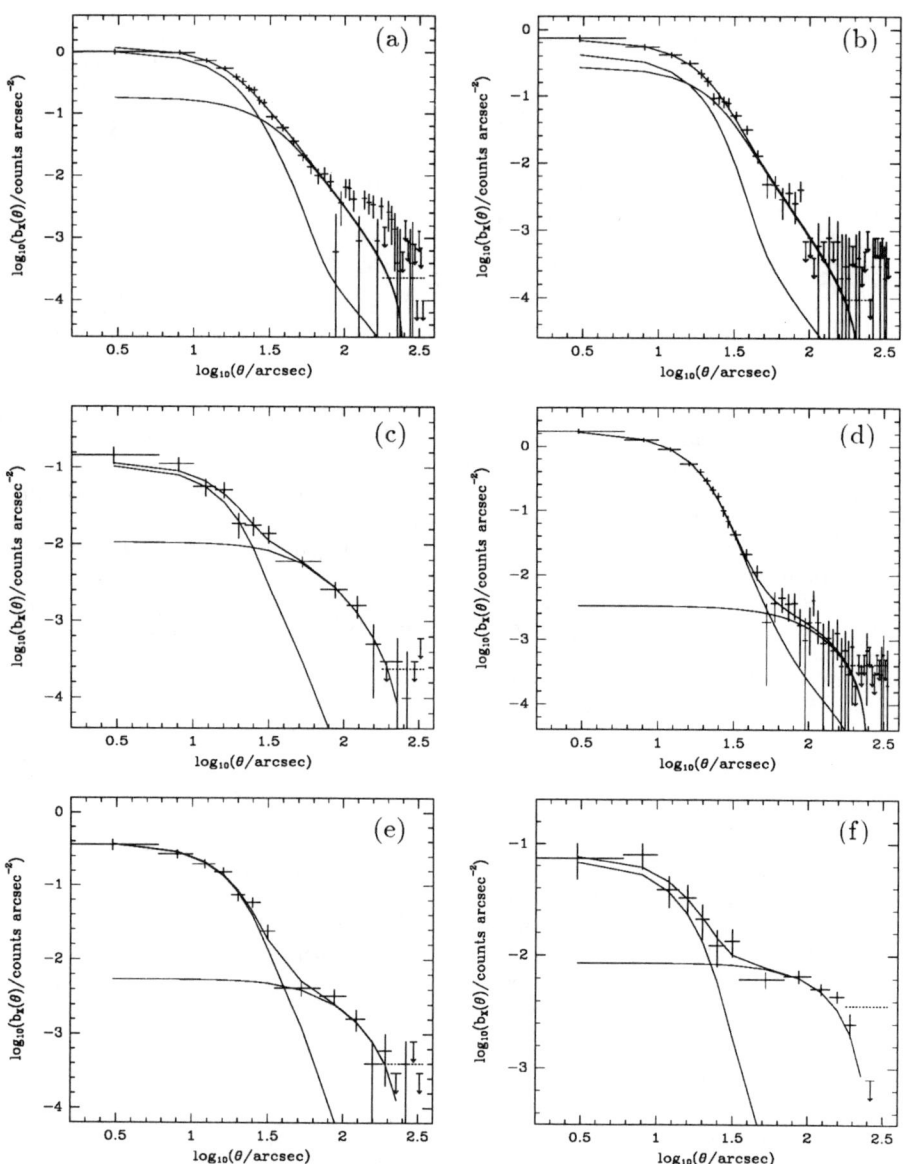

Figure 1: Background-subtracted X-ray radial profiles for the radio galaxies in order of increasing intrinsic core-radius of the β model. (a) NGC 4261, (b) NGC 315, second exposure, (c) 4C 35.03, (d) NGC 6251, (e) NGC 2484, and (f) NGC 326. Data are fit to a combination of an unresolved component (narrow curve) and a β model (broad curve) convolved with the PSPC PRF. The dotted curve shows the contribution, taken into account in the fitting, of the model to the background annulus. (Figure taken from Worrall & Birkinshaw 1994.)

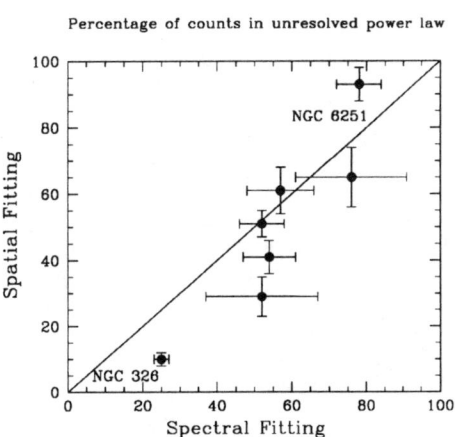

Figure 2: The fraction of counts in the unresolved component from spatial fitting agrees well with that in the power law from the spectral fitting, especially considering that these errors do not include uncertainties in the model parameters and the spatial analysis did not take into account the different spectral distributions of the β model and the unresolved component. The discrepancy for NGC 326 is not surprising given the inadequacies of a radially-symmetrical analysis for its complex emission region. The discrepancy for NGC 6251 is discussed in the text.

The unresolved X-ray component is allowed either a thermal or non-thermal origin from our X-ray analysis alone. However, for NGC 6251, which has the brightest unresolved emission, a detailed analysis has shown that this inner emission requires a small ($\sim 20\%$) contribution of cool ($kT \sim 0.5$ keV) gas added to either a power-law or a hot ($kT \sim 5$ keV) thermal component (Birkinshaw & Worrall 1993; but see also Birkinshaw & Worrall, this volume). The difficulty in interpreting the dominant spectral contribution to the unresolved emission as thermal is the high temperature; neither the stars in NGC 6251 nor the surrounding galaxies display a sufficiently large velocity dispersion for ~ 5 keV gas to be hydrostatically confined. This points to the interpretation of the unresolved X-ray component of NGC 6251 being $\sim 80\%$ non-thermal and $\sim 20\%$ 0.5 keV gas (lowering its position slightly in Fig. 2).

General support for the non-thermal nature of most of the unresolved X-ray emission is shown in Figure 3 where the X-ray spectral luminosity of the fitted power-law component is plotted against the spectral luminosity of the radio core. Although NGC 326 is shown in the figure, its X-ray spectral luminosity is highly uncertain and is likely to be overestimated (see caption). The other sources shown are consistent with proportionality between their X-ray and radio emissions, supporting a model where these X-rays are non-thermal emission from the inner regions of a parsec-scale radio jet.

3. Implications for Unified Schemes

The measurement of multiple X-ray emission components has implications for Unified Schemes in which BL Lac objects are low-power radio galaxies with their parsec-scale relativistic jets pointed towards the observer. Padovani & Urry (1990) used *Einstein* Observatory data to construct X-ray luminosity functions for BL Lac objects and radio galaxies assuming that the intrinsic luminosity of the jet is some fixed fraction of an unbeamed luminosity. Birkinshaw & Worrall (1993) argued on the basis of results for NGC 6251 that thermal emission may be the source of the required unbeamed X-rays. In NGC 6251 the emission interpreted as non-thermal outshines the thermal emission by a factor of ~ 4, placing the source as 'transitional' between radio galaxies and BL Lac ob-

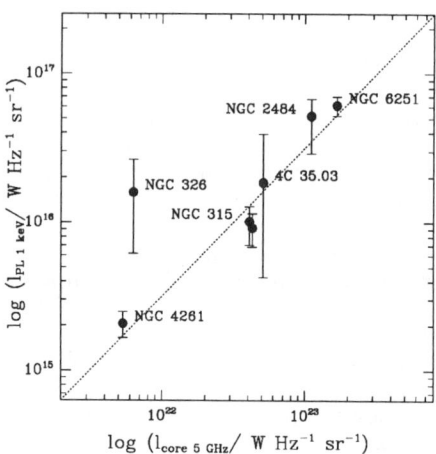

Figure 3: 1 keV X-ray spectral luminosity for the power-law component plotted against 5 GHz radio-core spectral luminosity. X-ray errors are 1σ for one interesting parameter, allowing the other parameters to vary. An additional large systematic error should be applied to the X-ray point for NGC 326; this source is dominated by non-radially symmetric extended emission and the unresolved component may contain substantial thermal emission. The dotted line is of slope unity. (Figure taken from Worrall & Birkinshaw 1994.)

jects. The model parameters of Padovani & Urry predict that for > 90% of the radio-galaxy population the beamed emission should be ≲ 30% of the unbeamed emission, and for > 70% of the population the beamed percentage drops to ≲ 1%. Since we find that all the sources under study here (with the exception of NGC 326) have a significant fraction of their emission unresolved on PSPC scales, consistency appears to require much of the unresolved X-ray emission to be unbeamed, perhaps through a contribution from hot gas or because of a velocity or collimation gradient in an X-ray-emitting jet (e.g., Ghisellini & Maraschi 1989; Maraschi, Celotti & Ghisellini 1992). We await approved observations of four of these galaxies with the ROSAT HRI to search for resolved components on spatial scales as small as $\sim 5''$ to test the fraction of the PSPC unresolved emission which might arise from thermal gas.

Further details of this work can be found in Worrall & Birkinshaw (1994). The work was funded by NASA grants NAG5-1882 and NAG5-2312, and NASA contract NAS8-39073.

References

Birkinshaw, M. & Worrall, D.M. 1993, ApJ, 412, 568
Fabbiano, G., Gioia, I.M. & Trinchieri, G. 1989, ApJ, 347, 127
Hasinger, G., Turner, T.J., George, I.M. & Boese, G. 1992, NASA/GSFC/OGIP. Calibration Memo CAL/ROS/92-001
Ghisellini, G. & Maraschi, L. 1989, ApJ, 340, 181
Maraschi, L., Celotti, A. & Ghisellini, G. 1992, in Physics of Active Nuclei. eds. W.J.Duschl & S.J.Wagner (Berlin: Springer-Verlag), 605
Padovani, P. & Urry, C.M. 1990, ApJ, 356, 75
Sarazin, C.L. 1986, Rev.Mod.Phys., 58, 1
Ulrich, M.-H. 1989, in BL Lac Objects, eds. L. Maraschi, T. Maccacaro & M.-H. Ulrich (Berlin: Springer-Verlag), 45
Worrall, D.M. & Wilkes B.J. 1990, ApJ, 360, 396
Worrall, D.M. & Birkinshaw, M. 1994, ApJ, in press
Worrall, D.M. et al. 1992, in Data Analysis in Astronomy IV, eds. V. Di Gesu et al. (New York: Plenum Press), 145

EXTENDED X-RAY EMISSION IN SEYFERT GALAXIES

Andrew S. Wilson
Space Telescope Science Institute, 3700 San Martin Drive,
Baltimore, MD 21218
and
Astronomy Department, University of Maryland,
College Park, MD 20742

Email ID
awilson@stsci.edu

ABSTRACT

I report on a program of imaging nearby Seyfert galaxies with the ROSAT HRI. Of the six galaxies observed to date, at least four show spatially extended X-rays. In all cases, the direction of the X-ray extent coincides with that of the nuclear radio source and/or the extended narrow line region. Further, the spatial scale ($\approx$ kpc) of the resolved emission is similar to that of the extended narrow line region. In NGC 2110, the extended component contributes the "soft excess" found near 1 keV in spectroscopic observations. I argue that the resolved X-rays result from thermal emission of a 10^{6-7} K gas. This hot gas has a pressure similar to that of both the narrow optical line-emitting gas and the synchrotron plasma responsible for the kpc-scale radio sources. The hot gas may represent a wind driven by the compact nuclear X-ray source or gas shocked and entrained by the outflowing radio jets and lobes.

1. INTRODUCTION

The properties of spatially-extended X-ray emission in active galaxies are of interest for several reasons. It has often been argued that the release of large amounts of energy by an AGN should drive a hot wind into the surrounding medium (e.g. Wolfe 1974). This wind should be detectable as an X-ray source if its density is high enough. The clouds in the narrow line region (NLR) may condense out of the wind (e.g. Krolik & Vrtilek 1984). Alternatively, ambient gas on the kpc scale may be heated to X-ray emitting temperatures by shocks driven by radio jets or lobes (e.g. Wilson & Willis 1980). A "halo" of extended X-ray emission may result from electron scattering of X-ray photons from the compact nuclear source (e.g. Fabian 1977). Lastly, the presence of spatially extended X-rays will "dilute" any variability of the compact nuclear source and confuse interpretation of its spectrum. For all of these reasons, it is valuable to map the X-ray structure of nearby active galaxies at the highest possible spatial resolution.

The largest NLR's and radio sources associated with Seyfert galaxies are of order tens of arc secs (few kpc) in extent. Observations with the ROSAT HRI, with resolution $\simeq$ 5 arc secs, thus hold out the prospect of resolving these structures. So far, we have imaged half a dozen galaxies and this paper summarises the current state of the program. Section 2 describes the results on a galaxy by galaxy basis, while Section 3 summarizes trends in the sample and possible interpretations.

2. INDIVIDUAL GALAXIES

2.1 NGC 1068

As the results on this galaxy have already been published (Halpern 1992; Wilson et al. 1992), I shall just give a brief summary of the results. The image reveals a compact source, coincident with the optical nucleus to within the errors and containing $\simeq 55\%$ of the total X-ray flux, plus emission extending 10 – 15 arc secs (1 arc sec = 73 pc for $H_0 = 75$ km s^{-1} Mpc^{-1}) from the nucleus and containing $\simeq 23\%$ of the total flux. The resolved emission extends asymmetrically towards the N and NE, which is the preferred axis of the Seyfert activity as seen in other wavebands. We preferred a thermal interpretation of the circumnuclear extended emission: a plausible model for a gas in collisional equilibrium would have a temperature of 10^{6-7}K, a density of 1-2 cm^{-3}, a gas mass of $(9-4) \times 10^7 M_\odot$, and a pressure of $(3-12) \times 10^{-10}$ ergs cm^{-3}. This pressure is very similar to that of the lower density NLR (Shields & Oke 1975), consistent with models in which the optical line-emitting filaments condense out of, or are entrained by, a hot outflowing wind.

There is also large–scale (radius ≈ 60 arc secs) X-ray emission with similar morphology to the starburst disk, although an extension of the nuclear wind to large radii is another possible interpretation. This emission contains the remaining 22% of the flux. An origin in the starburst disk is favored by the ratios of the large-scale X-ray to B band and X-ray to 60 μm fluxes, both of which ratios are identical to those seen in starburst galaxies.

A study of the HRI spectral "hardness ratio" suggests an increasing hardness with increasing distance from the nucleus, a result which we hope to check with a longer HRI integration. Such changes in "hardness ratio" can imply changes in either the continuum slope or the equivalent width of Fe L emission.

2.2 NGC 2110

The results on this galaxy are taken from a joint ROSAT HRI – BBXRT study by Weaver (1993) and Weaver et al. (1994). NGC 2110 has an optical emission-line nuclear spectrum with properties intermediate between a Seyfert 2 and a LINER. There is also a high excitation, emission-line nebulosity and an inverted 'S'-shaped radio jet, which extend 10 arc secs and 2.5 arc secs from the nucleus, respectively (1 arc sec = 150 pc). Both the inner part of the emission line nebulosity and the radio jet align approximately N–S. The galaxy is a luminous ($\simeq 10^{43}$ ergs s^{-1}), variable source of hard X-rays, which are photoelectrically absorbed below a few keV by a large column ($N_H \simeq 3 \times 10^{22}$ cm^{-2}) of cold gas. The X-ray spectrum of NGC 2110 is known to show soft X-ray (below 1 keV) emission above that expected from a model of a hard X-ray power law fully covered by the absorbing column (Reichert et al. 1985; Turner & Pounds 1989).

The ROSAT HRI image shows an extension to the N; we do not believe this extent is a result of aspect errors, because other sources in the field show a slight E–W elongation. We used a star about 6 arc min from the galaxy as a PSF. By scaling the peak flux of the star to that of NGC 2110 and subtracting it from NGC 2110, a map of the extended X-ray emission could be obtained. As shown in Fig. 1, this emission is centered roughly 4 arc sec N of the nuclear point source, close to where the optical emission-line gas is of highest excitation and along the extension of the radio axis. It should be emphasised that details of the X-ray distribution in Fig. 1 are very uncertain and only the existence of asymmetrically extended X-ray emission seems secure. The extended emission

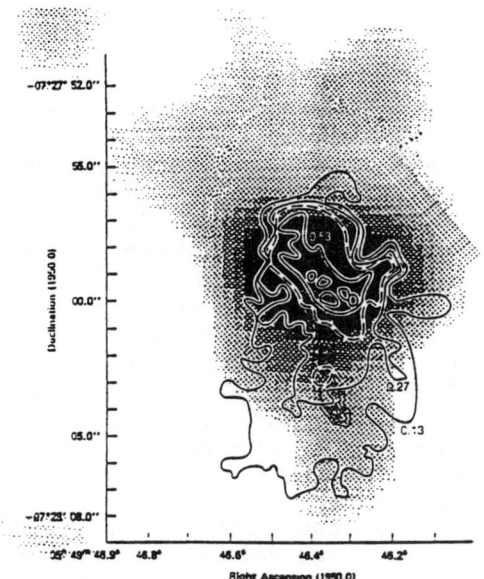

Figure 1 (from Weaver 1993) - Grey scale map of the result of subtracting the PSF from the ROSAT HRI image of NGC 2110. The peak of the original map is indicated by the cross, which is assumed to coincide with the optical and radio nucleus. Thin contours are the 6 cm VLA map (Ulvestad & Wilson 1983). Thick contours represent the gaseous excitation ratio R = [OIII]λ5007/Hα+[NII]$\lambda\lambda$6548, 6584 (from data in Haniff, Ward & Wilson 1991). Contours of R are 0.13, 0.27, 0.40 (dotted), 0.53 and 0.67. Note the correspondence between the extended X-rays and the highest excitation optical emission-line gas. The contours of R near the edge of the optical nebulosity, where the lines are weak, are extremely uncertain.

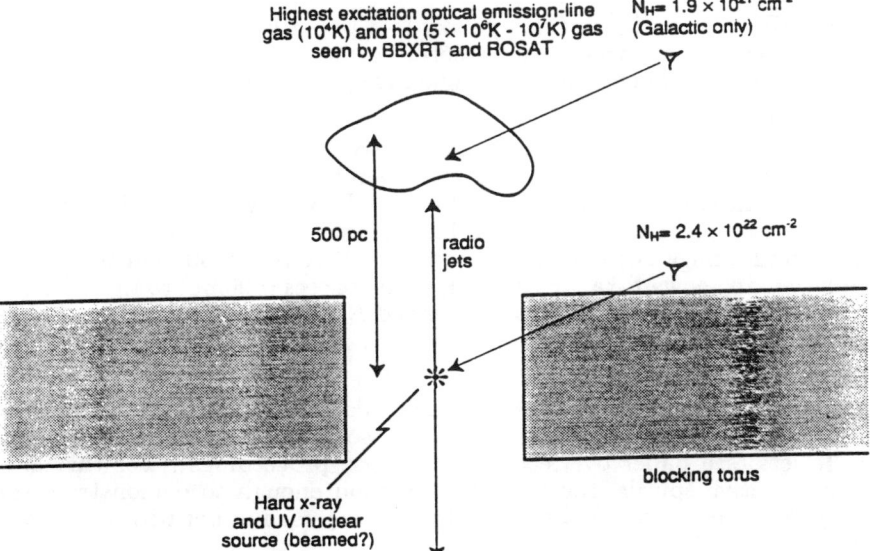

Figure 2 - Schematic diagram of our preferred model for NGC 2110.

comprises 11±3% of the total HRI flux. For the best thermal bremsstrahlung (T = 5×10^6K) or Raymond-Smith plasma (T = 1×10^7K) spectral fits to the BBXRT soft excess (see below), the flux and luminosity of the extended component are $(3.5-10)\times10^{-13}$ ergs cm^{-2} s^{-1} and $(3.7-11))\times10^{40}$ ergs s^{-1}, respectively.

The BBXRT spectrum requires three components for an acceptable fit – (i) a power law of photon index Γ = 1.41±0.15 photoelectrically absorbed by cold gas of column density $N_H \simeq (2.4\pm0.3)\times10^{22}$ cm^{-2}, (ii) an Fe Kα emission feature at 6.32±0.08 keV, and (iii) a soft component detected between 0.6 and 1.2 keV. The spectral form of the soft component is not well determined and various models – partial covering of the directly seen power-law source, electron-scattered nuclear emission, and a separate thermal bremsstrahlung continuum or line source – adequately describe the data. For a Gaussian line (which would plausibly represent a blend of Fe L and Ne K emission) description, the soft excess has a flux of 3.4×10^{-13} erg cm^{-2} s^{-1} and comprises 14±6% of the total flux between 0.1 and 2.4 keV. This flux and percentage are in excellent agreement with the corresponding numbers for the extended component seen with the ROSAT HRI. Further, the effective area curve of the ROSAT telescope plus HRI peaks near 1 keV, and is very well matched to the energy range in which BBXRT sees the soft excess. Thus, *the extended emission and the soft excess are very likely one and the same component.* This result strongly supports the hypothesis of Turner & Pounds (1989) that the soft excess in NGC 2110 arises in a separate component from the hard power-law source.

Fig. 2 is a cartoon of the nuclear region of NGC 2110. The nuclear hard X-ray source is seen through a large column (a "blocking torus"?), while the soft excess originates from the kpc scale, and suffers attenuation by only the Galactic column density. The soft excess, the high excitation optical emission lines, and the radio jets are all powered by the nuclear source, the output of which is channelled along the axis of the torus. Recent HST observations (Mulchaey et al. 1994) confirm that the nucleus of NGC 2110 is heavily obscured.

2.3 NGC 3516

Our HRI observations of this galaxy (see paper by J. Morse in these proceedings) provide some evidence for resolved emission along the same direction as the extended optical emission-line region (Miyaji et al. 1992).

2.4 NGC 4151

Elvis et al. (1983, 1990) found that the soft X-ray emission of this galaxy is extended towards the SW and that the extended component contains about 15% of the total Einstein HRI flux. Fig. 3 shows a deconvolution (using a Lucy-Richardson algorithm, with a BL Lac object in the field taken as the PSF) of our ROSAT image superposed on contours of [OIII] λ5007 emission. This preliminary map shows X-rays extending $\approx$ 10 arc secs (600 pc) to the SW, close to the direction of elongation of the extended NLR and apparently confirming the Einstein results.

2.5 NGC 4258

The ROSAT HRI image of this SABbc galaxy with a LINER-type nuclear spectrum shows spectacular X-ray emission associated with the famous optical and radio jets (see paper by Cecil et al in these proceedings). For this single case among active spirals, the resolution is good enough to demonstrate that the X-rays are predominantly associated with the jets, and not with a hot wind.

2.6 Mkn 3

Although the HRI observation of this galaxy has not yet been analysed in depth, preliminary evaluation suggests the nuclear source is unresolved. We confirm the discovery by Turner et al. (1993) of a "companion" X-ray source some 100 arc sec (1 arc sec = 260 pc at Mkn 3) to the W of the nucleus. The direction to this "companion" (p.a. 265°) coincides accurately with the p.a. of the jet-like nuclear radio source (p.a. 264°, e.g. Kukula et al. 1993) observed on the 2 arc sec scale.

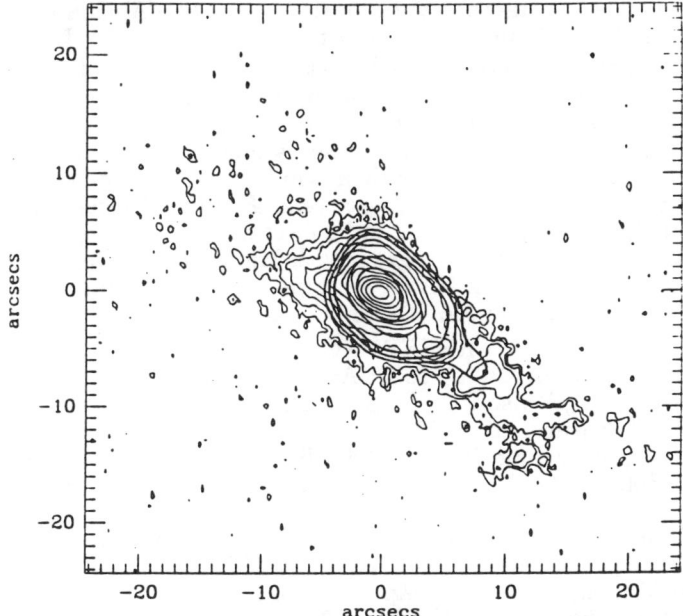

Figure 3 - ROSAT HRI and optical emission-line images of NGC 4151. The thick contours (plotted at 1, 2, 5, 25, and 50% of the peak) are the deconvolved ROSAT HRI image, and the thin contours the [OIII]λ5007 distribution (from Pérez-Fournon & Wilson 1990). Figure prepared by Jon Morse.

3. CONCLUDING REMARKS

The chief results of this work are: (i) Kpc scale X-ray emission is common among Seyfert galaxies, (ii) The soft X-ray luminosity of the extended emission is typically in the range 10^{40-42} erg s^{-1}, and (iii) The extended X-rays align with the radio and/or NLR structures, and have a spatial extent comparable to that of the extended emission-line gas.

The extended X-rays are unlikely to result from synchrotron radiation or inverse Compton scattering (the standard arguments, deployed by e.g. Wilson et al. [1992] for NGC 1068, have general applicability to these radio-weak AGN). If the extended X-rays are electron-scattered nuclear radiation, a sufficient column density (N) of extended gas must be present to produce the scattered fraction ($f_{scatt} = N\sigma_T \Delta\Omega / 4\pi$, where $\Delta\Omega$ is the solid angle subtended by the scattering material at the central source). If the extended X-rays make up the soft excesses seen in Seyfert 2's (Mulchaey et al. 1993), then $f_{scatt} \approx 10^{-2}$, $\Delta\Omega/4\pi$ = 0.16 for a bi-cone of opening angle 65°, and N $\approx 10^{23}$ cm^{-2}. This gas must be sufficiently highly ionized that it does not absorb the soft X-rays. If this ionization is achieved thermally, a temperature above 3×10^6K is needed and direct bremsstrahlung dominates scattered radiation (Elvis et al. 1990). If the ionization is achieved radiatively, the ionization parameter of the extended gas must be so high that the luminosity of the central source along the radio/NLR axis becomes much larger than is observed. While such a situation cannot be ruled out, the magnitude of the required luminosity ($> 2\times10^{46}$ erg s^{-1} for NGC 1068) renders a scattering interpretation of the extended emission implausible.

We consider thermal emission from a hot gas the most likely mechanism for the extended X-ray emission. This hot gas has a pressure of $10^{-10} - 10^{-9}$ erg cm^{-3}, which is similar to that of both the optical line-emitting gas in the NLR and the synchrotron emitting plasma responsible for the kpc-scale radio jets and lobes. Higher resolution X-ray observations are needed to determine whether this hot gas is associated with a wind (i.e. a wide-angle outflow), perhaps driven by the nuclear X-ray source (e.g. Krolik & Begelman 1986), or with gas entrained and shocked by the radio ejecta.

This work is a collaborative program with Martin Elvis, Gerald Cecil, Jon Morse and Kim Weaver. I thank NASA for support under grants NAGW-2689, NAGW-3268, NAG5-1532, and NAG8-793.

REFERENCES

Elvis, M., Briel, U., & Henry, J. P. 1983, ApJ, 268, 105.
Elvis, M., Fassnacht, C., Wilson, A. S., & Briel, U. 1990, ApJ, 361, 459.
Fabian, A. C. 1977, Nature, 269, 672.
Halpern, J. P. 1992, In Testing the AGN Paradigm, AIP Conference Proceedings 254, Eds S. S. Holt, S. G. Neff, & C. M. Urry, p524 (American Institute of Physics).
Haniff, C. A., Ward, M. J. & Wilson, A. S. 1991, ApJ, 368, 167.
Krolik, J. H., & Begelman, M. C. 1986, ApJ, 308, L55.
Krolik, J. H., & Vrtilek, J. M. 1984, ApJ, 279, 521.
Kukula, M. J. et al. 1993, MNRAS, 264, 893.
Miyaji, T., Wilson, A. S., & Pérez-Fournon, I. 1992, ApJ, 385, 137.
Mulchaey, J. S., Colbert, E., Wilson, A. S., Mushotzky, R. F. & Weaver, K. A. ApJ, 414, 144.
Mulchaey, J. S., et al. 1994, In Preparation.
Pérez-Fournon & Wilson, A. S. 1990, ApJ, 356, 456.
Reichert, G. A., Mushotzky, R. F., Petre, R., & Holt, S. S. 1985, ApJ, 296, 69.
Shields, G. A., & Oke, J. B. 1975, ApJ, 197, 5.
Turner, T. J., & Pounds, K. A. 1989, MNRAS, 240, 833.
Turner, T. J., Urry, C. M., & Mushotzky, R. F. 1993, ApJ, 418, 653.
Ulvestad, J. S. & Wilson, A. S. 1983, ApJ, 264, L7.
Weaver, K. A. 1993, Ph. D. thesis, University of Maryland.
Weaver, K. A., Mushotzky, R. F., Serlemitsos, P. J., Wilson, A. S., Elvis, M., & Briel, U. 1994, In Preparation.
Wilson, A. S., & Willis, A. G. 1980, ApJ, 240, 429.
Wilson, A. S., Elvis, M., Lawrence, A., & Bland-Hawthorn, J. 1992, ApJ, 391, L75.
Wolfe, A. M. 1974, ApJ, 188, 243.

THE SOFT X-RAY PROPERTIES OF QUASARS

Ari Laor
Institute for Advanced Study, School of Natural Sciences Olden Lane,
Princeton, NJ 08540

Email ID
laor@guinness.ias.edu

Fabrizio Fiore, Martin Elvis, Belinda J. Wilkes, and Jonathan C. McDowell
Harvard-Smithsonian Center for Astrophysics, 60 Garden Street,
Cambridge, MA 02138

Email ID
fiore@garth.harvard.edu, elvis@cfa222.harvard.edu, belinda@cfa222.harvard.edu,
mcdowell@urania.harvard.edu

ABSTRACT

We present first results of our *ROSAT* program to observe a complete sample of optically selected quasars. The sample selection criteria, combined with the PSPC capabilities, allow us to determine the soft ($\sim$ 0.2-2 keV) X-ray spectra of quasars with about an order of magnitude higher precision compared with earlier soft X-ray observations.

The complete sample includes 23 quasars, of which 10 were already analyzed. Most spectra are well characterized by a simple power-law, with $\langle \alpha_x \rangle = -1.50 \pm 0.40$. This average is significantly steeper than suggested by earlier soft X-ray observations of quasars. The 0.3 keV flux is well correlated with the 1.69 μm flux, which implies that the X-ray variability power spectrum of quasars flattens out between $f \sim 10^{-5}$ and $f \sim 10^{-8}$ Hz.

Strong correlations are also present between α_x and the following emission line parameters: Hβ FWHM, $L_{\text{[O III]}}$, and Fe II/Hβ. These correlations suggest that: 1. The quasars' environment is optically thin down to $\sim$ 0.2 keV, 2. α_x is not significantly variable, 3. α_x might be a useful absolute luminosity indicator, and 4. The Galactic He I and H I column densities are well correlated.

Continuum-continuum correlations in our sample argue against either thin or thick accretion disks as the origin of the soft X-ray emission.

1. Introduction

The X-ray properties of quasars have been studied extensively over the last decade using *HEAO-1, EINSTEIN, EXOSAT,* and *GINGA* (e.g. Mushotzky 1984; Wilkes & Elvis 1987; Comastri et al. 1992; Lawson et al. 1992; Williams et al. 1992). These observations indicate that the X-ray emission above 1-2 keV is well described by a power law with a spectral slope $\alpha_x = d\ln f_\nu/d\ln \nu$ of about -0.5 for radio loud quasars and about -1.0 for radio quiet quasars.

X-ray observations below 1 keV indicate a spectral steepening, or equivalently an excess emission, relative to the flux predicted by an extrapolation of the hard X-ray power-law (e.g. Arnaud et al. 1985; Wilkes & Elvis 1987; Turner & Pounds 1989; Masnou et al. 1992; Comastri et al. 1992). In some objects the excess can be described as a very steep and soft component, which is consistent with the Wien tail of a hot thermal component dominating the UV emission. However, these studies were of limited quality because of the low signal to noise ratio (S/N) and low energy resolution of the *EINSTEIN* IPC, and the *EXOSAT* LE detectors, in particular in the crucial energy range below 0.5 keV. Furthermore, the objects studied do not form a complete sample, and these results are likely to be biased by various selection effects which were not well defined a priori. In particular, most studied objects are nearby, intrinsically X-ray bright, AGNs.

We use the PSPC detector to make an accurate determination of the soft X-ray properties of a well defined, complete, and otherwise well explored sample of quasars. This survey will allow us to address the following questions: 1. What are the soft X-ray spectral properties of normal, optically selected, low redshift quasars? 2. Are simple thin accretion disk models able to fit the observed optical/UV/soft X-ray continuum? are other modifying mechanisms, such as a hot corona required? are models invoking optically thin free-free emission possible? 3. Do the observed soft X-ray properties display any significant correlations with other properties of these quasars? Are these correlations compatible with various models for the continuum and line emission mechanisms?

2. The Sample

We found the BQS sample, a subset of the PG survey defined by Schmidt & Green (1983), to be particularly suitable for our purpose for the following reasons: 1. These objects are selected only by their optical properties, thus they are not biased in any direct way in terms of their X-ray properties. 2. This sample has already been studied extensively, and uniformly, in other parts of the spectrum. These studies provide us with the most complete and coherent picture of the emission properties of bright AGNs, and allow us to make a detailed study of possible correlations between the soft X-ray properties and various other emission properties. 3. This sample includes only bright quasars, thus high S/N spectra could be obtained within a reasonable amount of spacecraft time.

The complete PG sample includes 114 AGNs, of which 92 are quasars (i.e. $M_B < -23$). These objects were selected by a color criterion which according to Schmidt and Green is satisfied by $\sim 90\%$ of all known quasars. We select a subsample of the PG quasars which is optimally suitable for soft X-ray observations. The following two selection criteria were used: 1. $z \leq 0.400$. This prevents the rest-frame 0.2 keV from being redshifted beyond the observable range. 2. $N_{H\,I}^{Gal} < 1.9 \times 10^{20}$ cm^{-2}, where $N_{H\,I}^{Gal}$ is the H I Galactic column density as measured in 21 cm. This low $N_{H\,I}^{Gal}$ cutoff is critical for minimizing the effects of Galactic absorption. This cutoff implies an upper limit on the Galactic optical depth in our sample of $\tau_{0.2keV} < 1.6$. These criteria limit our sample to 23 quasars. The 10 quasars reported here are: PG 1114+445, PG 1115+407, PG 1216+069, PG 1226+023, PG 1309+355, PG 1322+659, PG 1352+183, PG 1415+451, PG 1512+370, PG 1543+489. These quasars were selected on the basis of their availability, and we do not expect a systematic bias in their intrinsic properties relative to the complete sample.

3. The Observations and Analysis of the Spectra

All the quasars were readily detected with a minimum of ~ 600 net counts, and typically ~ 2000, which allows an accurate determination of the spectral slope.

We fit each spectrum with a single power-law of the form $f_E = e^{-N_H \sigma_E} f_0 E^{\alpha_x}$, where σ_E is the absorption cross section per H atom, f_0 is the flux density at 1 keV, and E is in units of keV. We make three different fits for each object with: 1. N_H a free parameter, 2. $N_H = N_{H\,I}^{Gal}$. 3. $N_H = N_{H\,I}^{Gal}$, and $0.47 \leq E \leq 2.5$ keV. A comparison of these fits allows us to identify an intrinsic absorption or emission excess relative to a single power-law fit, to determine whether the 21 cm measurement of $N_{H\,I}^{Gal}$ is a reliable measure of the Galactic soft X-ray opacity, and to look for a dependence of α_x on energy.

In order to look for curvature in the spectrum we fitted a power-law model which includes an additional curvature term of the form $f_E = e^{-N_H \sigma_E} f_0 E^{\alpha(E)}$, where $\alpha(E) = \alpha_0 + \beta \log E$. These fittings were made using software developed by one of us (F.F.).

4. Results and Discussion

We find an average spectral index $\langle \alpha_x \rangle = -1.50 \pm 0.40$ for our 10 quasars, which is consistent with the RASS observations of AGNs analyzed by Walter & Fink (1993). The typical statistical error in α_x is only $2-4\%$, which is about an order of magnitude smaller than the typical error available for quasars in earlier observations. The *ROSAT* spectra of quasars clearly indicate that the soft X-ray slope is steeper than the hard X-ray slope. However, they suggest a steepening by $\Delta\alpha \sim 0.5$ below ~ 1 keV, rather than $\Delta\alpha \gtrsim 1$, and in some case a significantly larger steepening, below ~ 0.5 keV suggested by earlier X-ray observations (e.g. Turner & Pounds 1989; Masnou et al. 1992; Comastri et al. 1992).

The X-ray spectra of 9 of the 10 quasars described here (excluding PG 1114+445) are consistent with a single power-law shape at 0.15-2 keV. Deviations from a single power-law are typically 30% or less. Statistically significant deviations at a level of less than 30% are present in the spectra of 3C 273 and of PG 1512+370. PG 1114+445 is the only quasar where the deviations are significantly larger than 30%, and its spectrum is well described by a power-law + an absorption edge from highly ionized O.

The curved power-law model provides a significantly better fit in PG 1114+445 and 3C 273, but in both cases the new χ^2 is still high and more complicated models are suggested. None of the remaining 8 objects yielded a significant improvement of χ^2 at better than the 1% level. The best fit value for the curvature term β is within 2 σ of zero in these 8 objects.

We find a rather small scatter in the near IR to X-ray flux ratios. In particular, $f_{0.3\,keV}$ in our sample can be predicted to within a factor of two, once $f_{1.69\,\mu m}$ is given. This result, together with the general lack of significant near IR variability in AGNs (Neugebauer et al. 1989), implies that the soft X-ray flux does not vary by typically more than a factor of two from the mean over timescales shorter than a few years. This indicates that the variability power spectrum in quasars does not continue to increase with increasing time scales, as observed at $f = 10^{-5} - 10^{-3}$ Hz, but flattens out somewhere between $f \sim 10^{-5}$ and $f \sim 10^{-8}$ Hz.

Both thin and thick accretion disk models predict the soft X-ray flux to be strongly dependent on inclination. These models therefore suggest a large range in the optical to soft X-ray flux ratio for accretion disks seen at a range of inclinations. These predictions contrast with the small range of optical to soft X-ray flux observed here.

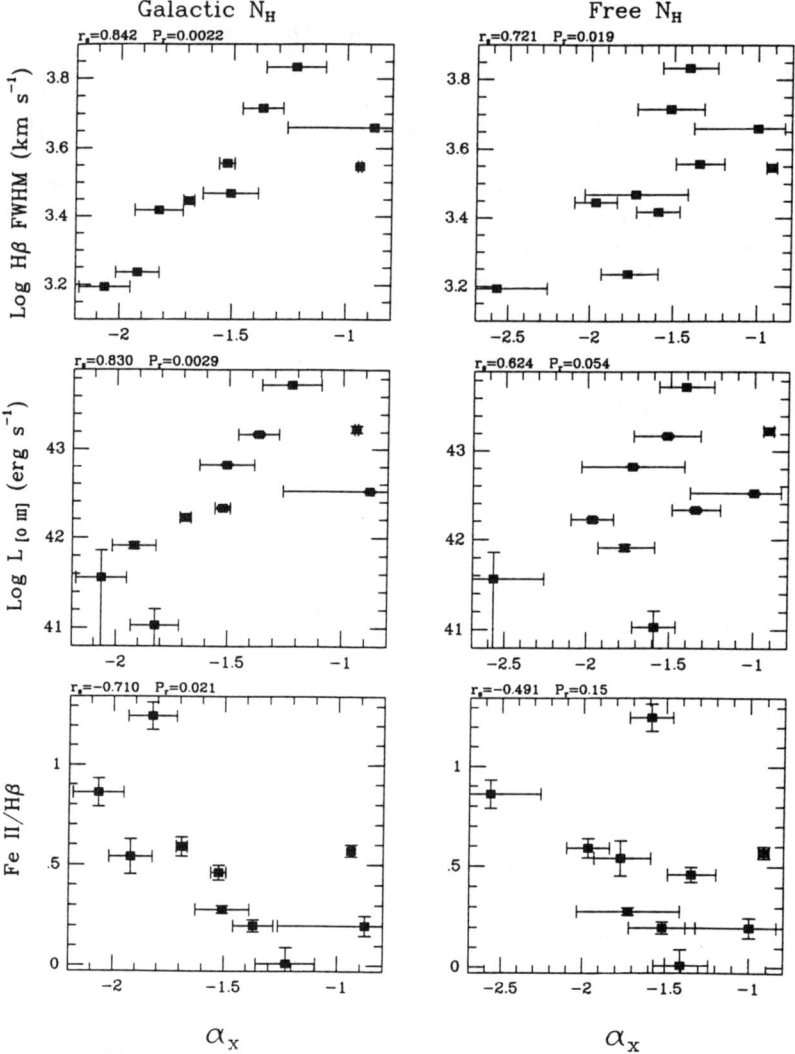

Figure 1. The emission properties which most strongly correlate with α_x. The Spearman rank-order correlation coefficients (r_S) and the significance levels (P_r) are indicated. Left panels, best fit α_x with $N_H = N_{H\,I}^{Gal}$. Right panels, best fit α_x with N_H as a free parameter. All correlations with a fixed N_H are significantly stronger, which indicates that $N_{H\,I}^{Gal}$ is a good measure of the soft X-ray opacity. The strong correlation with $L_{[O\,III]}$ indicates that α_x does not vary significantly in most objects.

Remarkably strong correlation are suggested in our sample between α_x and the following emission line parameters: the FWHM of Hβ, the absolute luminosity of the narrow [O III] line $L_{\text{[O III]}}$, and the Fe II/Hβ flux ratio (see Figure 1). These three emission line parameters are already known to be correlated with each other (Boroson & Green 1992). These correlations become significantly weaker when a free N_H rather than $N_H = N_{H\,I}^{\text{Gal}}$ is used. This indicates that the H I column in the Galaxy, as measured in 21 cm, is a good indicator of the Galactic soft X-ray opacity. The typical $N_{H\,I}^{\text{Gal}}$ in our sample is $\sim 1.5 \times 10^{20}$ cm^{-2}, and at this column the Galaxy becomes optically thick only below ~ 0.2 keV. The main opacity source at 0.2 keV is He I, which therefore implies that the H I column at high Galactic latitudes is very well correlated with the He I column. The presence of the strong correlations for $N_H = N_{H\,I}^{\text{Gal}}$ also suggests the lack of a significant column of He I at the quasar's rest frame.

The strong correlation of α_x with $L_{\text{[O III]}}$, together with the fact that $L_{\text{[O III]}}$ is very unlikely to vary in quasars on time scales shorter than a few years, indicates that α_x also does not vary significantly, i.e. by more than $\sim 10\%$ (as deduced from the scatter in Fig.1). This correlations connects the continuum shape of a quasar (α_x) with its absolute luminosity ($L_{\text{[O III]}}$). This suggests that α_x might be a useful absolute luminosity indicator for quasars, which would allow quasars to be used as probes for the value of the cosmological density parameter Ω_0.

It is not clear what physical mechanisms are responsible for the strong correlations suggested here between α_x and the emission line parameters mentioned above. Further theoretical studies on the effects of the X-ray emission on the broad and narrow line emission might clarify the physical source of these correlations.

A complete description of the work described here is given in Laor et al. (1994). This work was supported in part by NASA grants NAG 5-2087, NAG 5-1618, NAG 5-30934, NAGW 2201 (LTSARP), and NASA contract NAS 8-30751. A. L. acknowledges support by NSF grant PHY 92-45317.

5. References

Arnaud, K. A. et al., 1985, MNRAS, 217, 105

Boroson, T. A. & Green, R. F., 1992, ApJS, 80, 109

Comastri, A., Setti, G., Zamorani, G., Elvis, M., Giommi, P., Wilkes, B. J., & McDowell, J. C. 1992, ApJ, 384, 62

Laor, A., Fiore, F., Elvis, M., Wilkes, B. J. & McDowell, J. C. 1994. ApJ, submitted

Lawson, A. J., Turner, M. J. L., Williams, O. R., Stewart, G. C., & Saxton, R. D. 1992, MNRAS, 259, 743

Masnou, J-L, Wilkes, B. J., Elvis, M., Arnaud, K. A. & McDowell, J. C. 1992, A&A, 253, 35.

Mushotzky, R. F. 1984, Adv. Space Res., 3, No. 10, 157

Neugebauer, G., Soifer, B. T., Mathews, K. & Elias, J. H. 1989, AJ, 97, 957

Schmidt, M. & Green, R. F. 1983, ApJ, 269, 352

Turner, T. J. & Pounds, K. A. 1989, MNRAS, 240, 833

Walter, R. & Fink, H. H. 1993, A&A, submitted

Wilkes, B. J., Elvis, M. & McHardy, 1987, ApJ, 321, L23

Williams, O. R. et al., 1992, ApJ, 389, 157

ORAL PRESENTATIONS
DIFFUSE THERMAL EMISSION

HOT GAS AND IRON ABUNDANCES IN GALAXIES

C. Jones and W. Forman
Harvard-Smithsonian Center for Astrophysics

Email ID
cjf@cfa.harvard.edu

ABSTRACT

ROSAT PSPC observations of the hot gas in galaxies and clusters can be used to measure the gas temperature and density distibutions from which the gas mass and total mass of the system can be derived. These observations show that outside central cooling regions, galaxies (and the cores of clusters) have isothermal temperature distributions. The masses measured for bright early type galaxies require the presence of dark halos. PSPC spectral observations also are used to determine the heavy element abundances in hot coronae. For luminous galaxies, the abundance is approximately solar which excludes models with high supernova rates in the past. Low iron abundances are found in cool, less luminous galaxies.

1. Introduction

The presence of large amounts of hot gas in hydrostatic equilibrium in clusters, groups, and galaxies allows one to measure the structure and the total mass of the system, and, since this gas is the repository of enriched material, the history of stellar evolution. In this review, we discuss first the observations of hot gas as a new component of the interstellar medium in spiral galaxies, second the ROSAT observations of the hot coronae around NGC4472 and NGC1399 to measure their dark halo masses, and third the ROSAT measurements of the iron abundance in galactic coronae.

2. A Hot Component of the ISM in Spiral Galaxies

Einstein and ROSAT images showed a hot component of the interstellar medium in the Large Magellanic Cloud (Wang et al. 1990 and Trumper et al. 1991). From the ROSAT PSPC observations, Trumper et al. reported a gas temperature of several 10^6 K. Deep ROSAT HRI observations of the bulge of M31 show diffuse x-ray emission with a luminosity of 6×10^{38} ergs sec^{-1}, which cannot be explained by an extrapolation of the known x-ray sources to lower count rates or by the contribution of other stellar populations (Primini et al. 1993). The small amount of cool material in the M31 bulge suggests that galactic winds sweep the region (Soifer et al. 1986, Ciardullo et al. 1988). The faint x-ray emission may be due to such a galactic wind. In the disk of NGC 5194 (M51), the "Whirlpool Galaxy," the ROSAT PSPC image shows both emission from individual bright sources and emission from a hot interstellar medium (Marston et al. 1994). This hot gas has an x-ray luminosity of a few 10^{40} ergs sec^{-1} and a temperature of about four million degrees. As in the hot ISM in the LMC, the gas temperature in M51 is hotter than expected in the McKee-Ostriker Hot Ionized Medium.

3. Mass Measurements in Early-Type Galaxies

The discovery of x-ray luminous hot coronae around elliptical and S0 galaxies (Forman et al. 1979; Nulsen, Stewart, & Fabian 1984; Forman, Jones, & Tucker 1985) dramatically changed our view of these galaxies. Prior to the x-ray observations, it was believed that the gas shed by evolving stars in early type galaxies was expelled by galactic winds (e.g., Mathews and Baker 1971 and Faber and Gallagher 1976). However, the x-ray observations showed that early type galaxies were characterized by luminosities up to 10^{42} ergs sec^{-1}, temperatures $\sim$ 1 keV (10^7K), and gas masses up to 10^{10} M$_\odot$. The presence of heavy halos was required to suppress galactic winds to permit the buildup of gas to the values observed (Forman et al. 1985; Fabian et al. 1986; Mathews and Loewenstein 1986; Sarazin and White 1988; David, Forman, & Jones 1991; Loewenstein 1992). ROSAT observations now provide precise measurements of the gas temperature distribution and definitively demonstrate the presence of high mass-to-light ratios around early type galaxies.

In the sections which follow, we illustrate mass measurements through a discussion of the Virgo galaxy NGC4472 and the Fornax galaxy NGC1399. NGC4472 is representative of early type galaxies and, although relatively bright in both x-rays and optical light, it lies almost exactly on the best-fitting correlation of optical vs. x-ray luminosity (e.g., Donnelly, Faber & O'Connell 1990). In a study of the mass distribution around NGC4472 using globular cluster velocities, Mould et al. (1990) remarked "If we wish to learn about the mass distribution of an individual elliptical galaxy... this is the key galaxy." Thus, although NGC4472 lies near a loose concentration of galaxies in the southern portion of the Virgo cluster (Binggeli et al. 1987; Huchra 1990), its x-ray emission and its corona, appear representative of an individual galaxy.

Determining Gravitational Masses Around NGC4472 and NGC1399

The foundation for deriving total (gravitating) masses from x-ray observations of hot gas in early type galaxies is the assumption of hydrostatic equilibrium. The hydrostatic assumption relies on several properties of the x-ray emitting gas. First, gas particle mean free paths are short which assures that particles have an isotropic velocity dispersion. Second, the gas relaxation time is short (comparable to the sound crossing time) which assures that hydrostatic equilibrium is a reasonable assumption.

The x-ray observations have advantages over optical observations for measuring the mass of a galaxy or of a cluster. With spatially resolved x-ray spectra, it is straightforward to derive the gravitating mass as a function of radius. The precision of the x-ray determinations are not limited by lack of "particles" (galaxies/globular clusters) to be observed, since longer x-ray observations readily yield more precise measurements.

The gravitating mass $M < r$, interior to a radius can be written as:

$$M_{grav}(<r) = \frac{-kT(r)}{G\mu m_p}\left(\frac{d\ln\rho}{d\ln r} + \frac{d\ln T}{d\ln r}\right)r \quad (1)$$

where μ = mean molecular weight, m_p = hydrogen mass, k = Boltzmann's constant, ρ = gas density, and $T(r)$ is the gas temperature at a radius r. Thus, by measuring the gas density and temperature profiles from the x-ray observations, one can determine the total gravitating mass. This approach has become quite standard and was first demonstrated by Bahcall and Sarazin (1977) and Mathews (1978). Fabricant and

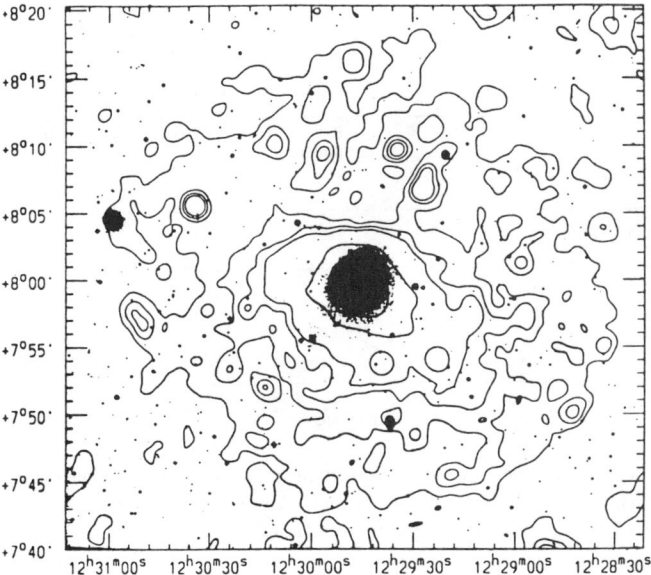

Figure 1 shows the x-ray contour map derived from a smoothed ROSAT PSPC observation of NGC4472 superposed on an optical photograph.

collaborators (Fabricant et al. 1980; Fabricant and Gorenstein 1983) provided the first detailed analysis of M87 and found a mass-to-light ratio increasing with radius up to ~ 180 in solar units.

Forman et al. (1993) analyzed the ROSAT PSPC image of NGC4472. The galaxy exhibits an extensive corona in x-rays as the contour map in Figure 1 shows, with emission visible to beyond $10'$ (~ 50 kpc). To derive the x-ray parameters needed for the mass determination, we exclude point sources, generate a radial profile to determine the surface brightness distribution, and extract spectra in annuli. For each annulus, the observed source spectrum is compared to models of an optically thin thermal spectrum. The free parameters include the gas temperature, the hydrogen column density, the elemental abundance (we vary the abundance of all elements heavier than helium, but keep their relative abundances with respect to each other fixed at the solar value), and the normalization of the model. The hydrogen column density showed no evidence of variability over the galaxy and the values measured in each annulus were consistent with the galactic value (1.64×10^{20} cm^{-2}; Stark et al. 1992). We performed a similar analysis for the hot gas around NCG1399. Figure 2 shows the results for the gas temperature. The temperatures in the inner annuli are less than in the outer parts of the galaxies. Hence, significant radiative cooling of the gas could be occurring in these inner regions, as suggested by Thomas et al. (1986) and single temperature models are probably inappropriate since the gas within this region is most likely inhomogeneous with gas at many different temperatures (see Fabian 1993).

The second parameter needed to determine the gravitating mass distribution is

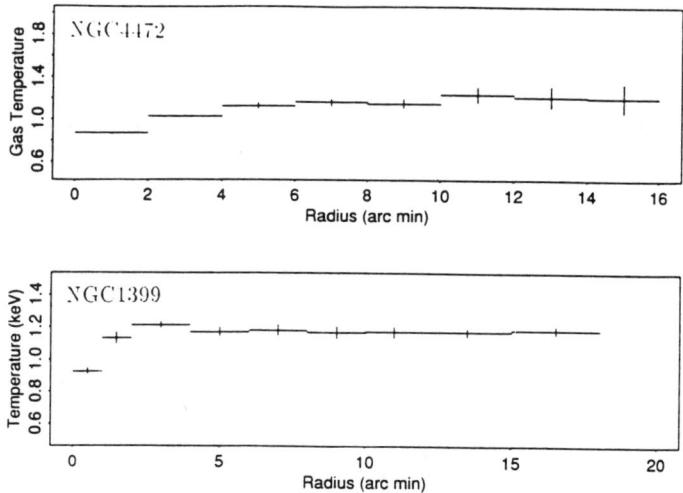

Figure 2 shows the best fitting temperature as a function of radius for the ROSAT observation of NGC4472 (top) and NGC1399 (bottom). There is cooling in the inner annuli. In the outer regions of both galaxies, the gas is essentially isothermal.

the gas density. As shown in Figure 2, the gas temperature outside the core is essentially isothermal. This allows a direct determination of the gas density distribution from the observed surface brightness distribution. For isothermal gas, a surface brightness distribution characterized by a power law, $S(r) \propto r^{-3\beta+1/2}$, directly yields a gas density distribution given by $n_{gas} \propto r^{-3\beta/2}$. We analyzed the emission around NGC4472 in a variety of ways to determine the surface brightness distribution. As Figure 1 shows, the emission around NGC4472 does not exhibit perfect azimuthal symmetry. Depending on details of the choice of azimuth for determining the radial profile, we find a range in slope of surface brightness corresponding to $\beta = 0.50 - 0.59$. Using the temperature distribution and this range in β characterizing the gas density distribution, we find

$$M(r) = 6.5 \pm 0.5 \times 10^{12} T_7 \; r_{100 \text{kpc}} \; M_\odot$$

where the error includes the uncertainty on the slope of the surface brightness distribution. Taking the optical luminosity for NGC4472 as $L_B = 7.4 \times 10^{10} L_\odot$ for a distance of 16 Mpc, we constrain the mass-to-light ratio (in solar units) to lie in the range 78-93 for $r = 100$ kpc and in the range 54-65 for $r = 70$ kpc (where the temperature is accurately measured). For NGC4472, Lauer (1985) gives a core mass-to-light ratio of 14. Hence, the ROSAT PSPC observations definitively demonstrate the existence of a massive dark halo surrounding NGC4472 as first suggested by Forman et al. (1985).

We also analyzed the ROSAT PSPC observation of NGC1399, an E1 galaxy located near the center of the Fornax cluster. We detect a hot galactic corona around the galaxy with a gas temperature of 1.21 ± 0.01 keV. Spatially resolved spectral data provide

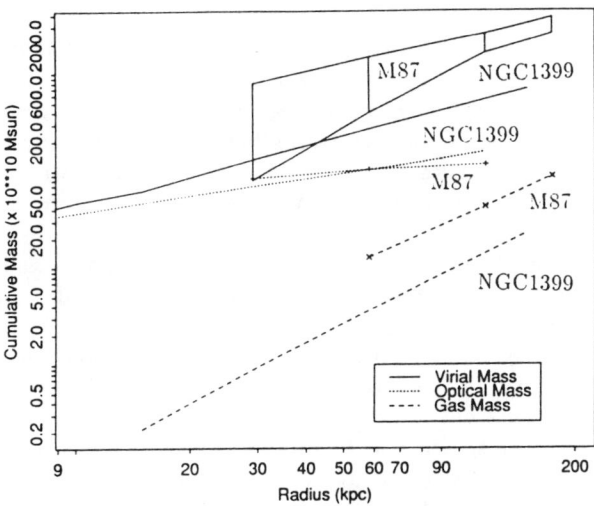

Figure 3 shows the optical (galaxy), X-ray (gas), and total mass distributions as a function of radius for NGC1399 and M87.

both gas temperature and gas abundance profiles extending to 18 arcminutes (125 kpc) from the galaxy. The temperature distribution is shown in Figure 2. It is apparent that beyond the central two arcminute region, the gas temperature is consistent with being isothermal since it varies by less than 0.05 keV over the full radial range from 2 to 18 arcminutes. The temperature distribution, combined with the x-ray surface brightness profile, yields a determination of the gravitating mass within 125 kpc which ranges from $4.3 - 8.1 \times 10^{12}$ $M_\odot$ (95% confidence range including systematic uncertainties). If we include the extended optical halo around NGC1399, the mass-to-light ratio increases with radius from 28-51 $M_\odot/L_\odot$ at 4'-6' to 38-75 $M_\odot/L_\odot$ at 16'-18'.

In Figure 3 we show the the optical, x-ray, and total mass distributions measured around NGC1399. Within 125 kpc, the gas mass is a relatively insignificant contributor to the total mass (less than 5% of the total). The optically luminous matter contributes 80% of the total mass at 10 kpc and remains significant but decreases to 20% of the gravitating mass at 125 kpc. The large contribution of the optically luminous matter to the total gravitating matter arises from a large extended envelope (Schombert 1986).

To better understand NGC1399 and the Fornax system, we compared the mass distributions described above to those measured in the M87/Virgo system (see Figure 3). For M87 we used the masses given in Table 5 of Fabricant and Gorenstein (1983) but scaled these values to an assumed distance for M87 of 20 Mpc.

Although, as Figure 3 shows, the optically luminous matter in M87 and NGC1399 is quite similar, the total mass and mass in gas in the two systems differ significantly with M87 having roughly 30 times more mass in both gas and gravitational mass than NGC1399, at about 100 kpc. The large ratio of gas masses accounts for the very large ratio in x-ray luminosity between these two systems.

Although the gas and gravitating masses of M87 and NGC1399 differ considerably, the ratio of gas to gravitating mass is comparable. Both M87 and NGC1399 have a ratio of gas to gravitating mass of about 4% at 100 kpc. This is rather remarkable since most of the luminous baryonic matter is found in the optically luminous stellar component and the ratio of gas to stellar mass varies so widely between the two systems (30% of the luminous baryons lie in gas in M87 while 1% of the luminous baryons are found in gas in NGC1399).

The similarity of the optical masses for NGC1399 and M87 and the differences between their total masses and the gas masses in these systems lead us to speculate that the galaxies may have formed through similar mechanisms at an early stage in the formation of the surrounding cluster, when the central potentials were similar. Over time, the greater underlying density perturbation in Virgo led to a deeper potential around M87 containing both more gas and more total mass than in the shallower potential around NGC1399. In both systems, as the potential deepened, the gas, now hotter, no longer contributed significantly to the growth of the galaxies, so that the optical masses of M87 and NGC1399 remain at their early values. Thus the optical mass of the central galaxy in a cluster is a poor estimator of the total mass of the system.

In summary, the results of the spectral analysis demonstrate that the ROSAT PSPC observations provide accurate determinations of the gas temperature distribution for the hot gas around early type galaxies. For the galaxies that have been studied, the gas outside the central cooling region is nearly isothermal. The derived gravitating mass implies a mass-to-light ratio well in excess of that measured in galaxy cores, requiring the presence of dark matter in the halo.

We note that the asymmetries in the x-ray emission around NGC1399 and NGC4472 suggest that the central region of the galaxies may not be at rest, but may be moving slowly in a larger potential. Similar asymmetries also have been found in the x-ray emission around NGC5044 (David et al. 1993). In each, the maximum in the x-ray emission lies at the optical galaxy center, but is offset with respect to the centroid of the x-ray emission at large radii.

4. The Heavy Element Abundances in the Hot Coronae of NGC4472 and NGC1399

Spatially resolved spectra, obtained with the ROSAT PSPC, provide a measurement of the heavy element abundance in the hot gas. The hot corona, as the repository of the stellar mass loss (that not expelled from the galaxy or recycled into stars), contains a fossil record of past supernova activity in the galaxy. Since the radial flow time of the gas is long, the hot gas contains a history of the metals injected into it by evolving stars. As Loewenstein and Mathews (1991) emphasized, measuring radial abundance gradients determines both the history of the past supernova rate, as well as its present value. Abundance measurements also distinguish among different models for the gas dynamics in the corona. In particular, different evolutionary models of the hot coronae predict different abundances. For example, predicted abundances in long-lived wind models (D'Ercole et al. 1989) exceed several times the solar value and can readily be distinguished from those with nearly solar abundances expected in quasi-hydrostatic or inflow models (Loewenstein and Mathews 1987; David et al. 1991).

The iron abundances in the hot corona around NGC4472 and NGC1399 are

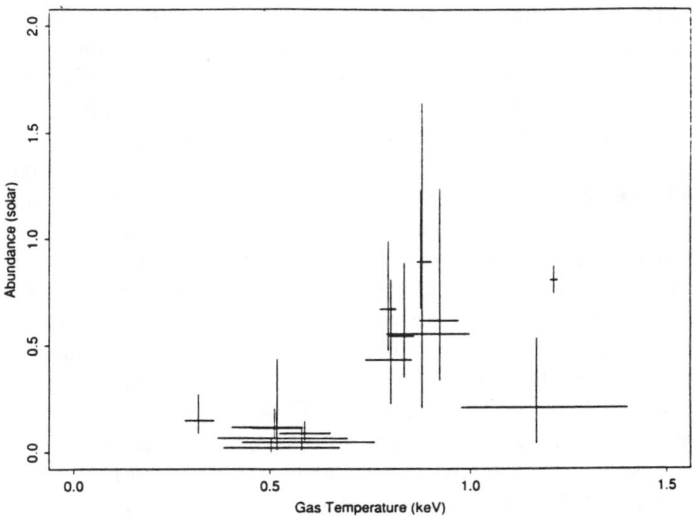

Figure 5 shows the iron abundance measured in PSPC observations of early type galaxies plotted as a function of the temperature of the corona. The hotter galaxies (1 keV) have solar abundances of iron, while the cooler galaxies have lower abundances.

supernova rate in early type galaxies remains uncertain, the stellar component should provide a lower limit to the iron abundance. Thus, our expectation, would be that the iron abundance should be roughly solar for the integrated spectrum and that there should be a decreasing gradient with increasing radius from the galaxy center (see section 4. and Forman et al. 1993). We find that while some galaxies do seem to confirm these general expectations, others do not. In this section, we summarize the available x-ray measurements of the iron abundance in the hot coronae of early type galaxies and discuss their implications.

The best available iron abundance measurements for early type galaxies are predominantly from the ROSAT PSPC, although a few have been measured by BBXRT and Ginga. In determining the abundances using the ROSAT PSPC observations, we extracted the x-ray spectrum both over an extended region containing most of the x-ray flux and in annular regions. Although the emission is sharply peaked toward the galaxy center, due to the possibility of strong radiative cooling in this region as well as the possible presence of a nuclear source, for each galaxy, we attempted to measure the abundance after omitting the core of the galaxy. In modelling the x-ray spectra, we used an optically thin thermal spectrum and allowed the gas temperature and the hydrogen absorption, as well as the gas temperature, to be free parameters. Figure 5 provides a summary of available iron abundance measurements.

While the most luminous and hottest galaxies have approximately solar abundances, the less luminous and cooler galaxies exhibit lower abundances, often providing only upper limits on the iron abundance. While there are several explanations, it is possible that the low iron abundances reflect low stellar iron abundances and a low SN Ia rate. Measurements of the abundances of other heavy elements in the ISM are crucial evolution. In particular, radial abundance measurements can distinguish between scenarios for the origin of Type Ia progenitor stars such as those proposed by Ciotti et al. (1991) which imply a rapid decline in the supernova rate with time compared to those used by Loewenstein and Mathews (1987) and David et al. (1990) who derived a considerably weaker dependence of the supernova rate on time.

A theoretical study of the dependence of the heavy element abundances on the supernova rate was made by Loewenstein and Mathews (1991). They characterized the time dependence of the type Ia supernova rate and its present epoch value as $R_{SN} = 2.2 f(t/t_H)^{-s}$ where R_{SN} is the number of type Ia supernovae per 100 years per $10^{11} L_\odot$, t_H is the present time, f is the ratio of the present epoch supernova rate to that given by Tammann (1974), and s parameterizes the dependence of the supernova rate with time. In Figure 4, we superpose the radial behavior of the heavy element abundance from galaxy models for a $10^{11} L_\odot$ galaxy with a heavy halo along with the abundance measurements derived from the NGC1399 and NGC4472 observations.

As is apparent from Figure 4, models with much higher supernova rates at earlier epochs conflict with the NGC4472 and NGC1399 observations. In particular, models in which the supernova rate is characterized by $s \geq 1.5$ predict heavy element abundances greater than observed. The models proposed by D'Ercole et al. (1989; see also Ciotti et al. 1991) to explain the large range in x-ray luminosity for galaxies with similar optical luminosity, predict iron abundances considerably in excess of solar and disagree with the observations. The models which adequately describe the observations have constant or slowly varying supernova rates ($|s| < 1$) and require relatively modest values of the present epoch supernova rate. In particular the model with $s = 0$ and a present epoch supernova rate equal to van den Bergh and Tammann's (1991) value ($f \sim 1$) produces iron abundances greater than observed, while the model with $f = 1/4$ gives iron abundances in good agreement with the observations. Thus for existing models of elliptical galaxies (i.e., those with heavy halos), the observations suggest a constant or slowly varying supernova rate ($|s| < 1$) and a present epoch supernova rate significantly less than $f = 1$ (consistent with the lower range of values derived by van den Bergh and Tammann 1991 and in agreement with the values derived by Cappellaro et al. 1993).

These conclusions remain tentative since the models may not be fully realistic and affects not included in the calculations could alter the derived heavy element abundance distributions. For example, the mixing of stellar and supernova ejecta into the hot coronal gas may be incomplete and thus the heavy elements produced from each process may not be injected completely into the corona. Further complications are the non-solar abundance ratios and the dispersion in these ratios (Worthey et al. 1992). With more detailed modelling, observations of additional galaxies with a range of optical luminosity (mass) and over a range of redshift (look back time), and observations having higher spectral resolution to measure abundances of individual heavy elements, we will better understand the properties of elliptical galaxies, the evolution of their hot coronae, and the evolution of their stars and supernovae which determine the mass and energy balance of the hot coronae.

5. Heavy Element Abundances in Hot X-ray Coronae

We measured the total iron abundance and, if possible, its distribution in the hot corona for sixteen early type galaxies observed with the ROSAT PSPC. While the

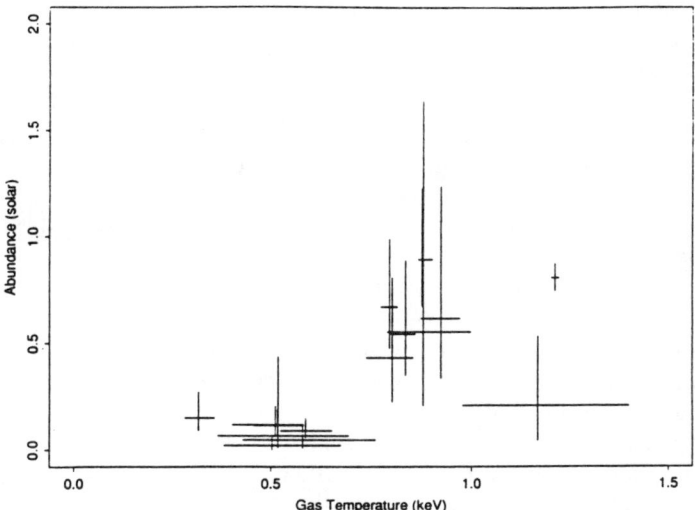

Figure 5 shows the iron abundance measured in PSPC observations of early type galaxies plotted as a function of the temperature of the corona. The hotter galaxies (1 keV) have solar abundances of iron, while the cooler galaxies have lower abundances.

supernova rate in early type galaxies remains uncertain, the stellar component should provide a lower limit to the iron abundance. Thus, our expectation, would be that the iron abundance should be roughly solar for the integrated spectrum and that there should be a decreasing gradient with increasing radius from the galaxy center (see section 4. and Forman et al. 1993). We find that while some galaxies do seem to confirm these general expectations, others do not. In this section, we summarize the available x-ray measurements of the iron abundance in the hot coronae of early type galaxies and discuss their implications.

The best available iron abundance measurements for early type galaxies are predominantly from the ROSAT PSPC, although a few have been measured by BBXRT and Ginga. In determining the abundances using the ROSAT PSPC observations, we extracted the x-ray spectrum both over an extended region containing most of the x-ray flux and in annular regions. Although the emission is sharply peaked toward the galaxy center, due to the possibility of strong radiative cooling in this region as well as the possible presence of a nuclear source, for each galaxy, we attempted to measure the abundance after omitting the core of the galaxy. In modelling the x-ray spectra, we used an optically thin thermal spectrum and allowed the gas temperature and the hydrogen absorption, as well as the gas temperature, to be free parameters. Figure 5 provides a summary of available iron abundance measurements.

While the most luminous and hottest galaxies have approximately solar abundances, the less luminous and cooler galaxies exhibit lower abundances, often providing only upper limits on the iron abundance. While there are several explanations. it is possible that the low iron abundances reflect low stellar iron abundances and a low SN Ia rate. Measurements of the abundances of other heavy elements in the ISM are crucial to understanding the hot coronae in these cooler galaxies.

This work is supported by the Smithsonian Astrophysical Observatory and NASA contracts and grants NAS8-39073 and NAG8-1536.

References

Bahcall, J., and Sarazin, J. 1977, Ap. J. (Letters), 213, L99.
Binggeli, B., Tammann, G., and Sandage, A. 1987, AJ, 94,251.
Cappellaro, E., Tunatto, M., Benetti, S., Tsvetkov, D., Bartunov, O., and Makarova, I. 1993, A&A, 273, 383.
Ciardullo, R., Rubin, V., Jacoby, G., Ford, H., and Ford, W., 1988, Ap. J. 95, 483.
Ciotti, L., D'Ercole, A., Pellegrini, S., and Renzini, A. 1991, Ap. J., 376, 380.
David, L., Forman, W., and Jones, C. 1990, ApJ, 359, 29.
David, L., Forman, W., and Jones, C. 1991, ApJ, 369, 121.
David, L, Jones, C., Forman, W., and Daines, S. 1994, Ap.J., (in press).
D'Ercole, A., Renzini, A., Ciotti, L., and Pellegrini, S. 1989, Ap.J. 341, L9.
Donnelly, R. H., Faber, S. M., and O'Connell, R. W. 1990, ApJ, 354, 52.
Faber, S. and Gallagher, J. 1976, Ap. J. 204, 365.
Fabian, A. C., 1993, AARA, in press.
Fabian, A.C., Thomas, P., Fall, M., and White, R. 1986, MNRAS, 221, 1049.
Fabricant, D., Lecar, M., and Gorenstein, P. 1980, Ap. J., 241, 552.
Fabricant, D., and Gorenstein, P. 1983, Ap. J. 535, 546.
Forman, W, Jones, C, and Tucker, W. 1985, Ap. J., 293, 102.
Forman, W, Schwarz, J, Jones, C, Liller, W, and Fabian, A 1979, Ap.J.(Lett),234, L27
Huchra, J. 1990, in Clusters of Galaxies eds. Oegerle, Fitchett, and Danly, Cambridge University Press.
Lauer, T. 1985, Ap. J., 292, 104
Loewenstein, M. 1992, ApJ, 384, 474
Loewenstein, M. and Mathews, W. 1987, ApJ, 319, 614
Loewenstein, M. and Mathews, W. 1991, ApJ, 373, 445
Marston, A., Elmegreen, D., Elmergreen, B., Forman, W., Jones, C., and Flanagan, K., 1994, submitted to Ap. J.
Mathews, W. 1978, Ap. J. 219, 413
Mathews, W. and Baker, J., 1971, Ap. J., 170, 241
Mathews, W. and Loewenstein, M. 1986, ApJ (Letters), 306, L7
Mould, J., Oke, J., de Zeeuw, P., and Nemec, J. 1990, A. J., 99, 1823
Nulsen, P.E.J., Stewart, G., and Fabian, A.C. 1984, MNRAS, 208, 185
Primini, F., Forman, W., and Jones, C., 1993, Ap. J. 410, 615
Sarazin, C., and White, R. E. 1988, Ap. J., 331, 102
Soifer, B. T., Rice, W. L., Mould, J. R., Gillett, F. C., Rowan-Robinson, M., and Habing, H. J., 1986, Ap. J. 304, 65
Tammann, G 1974, Supernovae and their Remnants ed Cosmovici, Dordrecht: Reidel
Trumper, J., et al. 1991, Nature, 349, 579.
Thomas, P, Fabian, A, Arnaud, K, Forman, W, and Jones, C, 1986, MNRAS, 222, 655.
van den Bergh, S. and Tammann, G. 1991, Ann. Rev. Astron. and Astrophys., 29, 363.
Wang, Q., Hamilton, T., Helfand, D. J., and Wu, X., 1990, Ap. J. 374, 475.
Worthey, G., Faber, S., & Jesus Gonzalez, J. 1992, Ap. J., 398, 69.

ROSAT OBSERVATIONS OF STARBURST GALAXIES

Timothy M. Heckman

Department of Physics & Astronomy, Johns Hopkins University, Baltimore, MD 21218

heckman@pha.jhu.edu

ABSTRACT

I report preliminary results of the analysis of Rosat data for a sample of about two dozen starburst galaxies spanning a range of about four orders-of-magnitude in starburst luminosity. The principal conclusions are: 1) The PSPC and HRI images show that - while the X-ray emission has its highest surface-brightness in the central starburst - most of the X-ray emission arises from far outside the starburst itself. In cases in which the starburst galaxy is viewed nearly edge-on, the X-ray emission extends out many kpc along the galaxy's minor axis. 2) Acceptable fits to the PSPC spectra require two spectral components. One is a 'hard' component that can be represented as either a nonthermal powerlaw or hot ($\gg 1$ keV) thermal emission. The second is a 'soft' thermal component (kT $\sim$ several hundred eV). Both components make comparable contributions in the Rosat band. The soft component is probably produced by ambient gas in the disk and halo of the starburst galaxy, which has been shock-heated by a starburst-driven galactic 'superwind'. The physical origin of the hard component is unclear.

1. INTRODUCTION

Starburst galaxies are one of the most fascinating phenomena in the universe. They are major constituents of the local universe (cf. Soifer et al 1987; Weedman 1991), and as such, deserve to be understood in their own right. Starbursts are even more important when placed in the broader context of contemporary stellar and extragalactic astrophysics. They are ideal laboratories in which to study massive stars, the processes involved in the formation and evolution of galaxies, and (through the galactic winds they blow) an important mechanism by which the intergalactic medium was probably heated and enriched in metals.

X-ray observations of starburst galaxies are particularly important (see Petre 1994 for a recent review). First, they provide unique insight into the nature of the starburst phenomenon itself (both its stellar and interstellar components). Massive X-ray binaries are expected to make a significant contribution to the X-ray emission of starbursts, particularly above a few keV (cf. Griffiths & Padovani 1990). Nonthermal X-ray emission in starbursts might also arise via Inverse Compton scattering of soft starburst photons off the radio-synchrotron-emitting relativistic electrons that have been created by supernova-driven shocks (cf. Schaaf et al 1989). At the softer energies probed by Rosat, emission from hot gas should be important. This thermal emission may be associated with individual young supernova remnants (cf. Bregman & Pildis 1992) or with a pervasive hot interstellar medium heated by the collective effect of stellar winds and supernovae (e.g. McKee & Ostriker 1977), which can escape the starburst in the form of a galactic 'superwind' (cf. Chevalier & Clegg 1985; Heckman, Lehnert, & Armus 1993).

The study of X-ray emission from starbursts also allows us to better assess the relevance of starbursts to a variety of cosmogonic and cosmological issues. They are likely to be significant contributors to the cosmic X-ray background (cf. Griffiths & Padovani 1990), so it is important to compare the X-ray spectra of starbursts to that of the X-ray background itself. Moreover, galactic 'superwinds' driven from starbursts are potentially important sources for the heating and chemical enrichment of the intergalactic medium (e.g. Larson & Dinerstein 1975; DeYoung 1978), and may also significantly affect the chemical evolution and even formation of galaxies (cf. Ostriker & Cowie 1981; White & Frenk 1991; Wyse & Silk 1985).

These 'superwinds' have been studied at radio and optical wavelengths - see Heckman, Lehnert, & Armus (1993) for a recent review. However, X-ray data are vital to understanding the physics of the outflow. This is because the characteristic temperature of the hot gas that drives the wind (and which contains most of the thermal-plus-kinetic energy injected by the starburst) should be of-order 10^8 K (cf. Chevalier & Clegg 1985), while the characteristic temperature of the halo gas that is shock-heated by the wind should lie in the range 10^6 to 10^7 K (cf. Suchkov et al 1994). Note that this latter temperature range is an ideal match to the Rosat bandpass.

2. THE SAMPLE

My colleagues Michael Dahlem (JHU), Claus Leitherer (STScI), and I have defined a far-infrared-selected sample of starburst galaxies which we are investigating with Rosat. These galaxies are all brighter than 30 Jy at 60 microns, and are 'warm' in the far- IR (ratio of 60 micron and 100 micron flux densities > 0.4). My collaborators (L. Armus, S. Eales, G. Fabbiano, D. Gilmore, M. Lehnert, C.M. Urry, W. Waller, and K. Weaver) and I have Rosat PSPC and/or HRI data for about a dozen of these galaxies. Dahlem & I have also retrieved Rosat archival data for an additional dozen members of the sample. This overall sample with Rosat data spans a range of about 10^4 in starburst luminosity ($\sim 10^8$ to $\sim 10^{12}$ $L_\odot$).

We are presently undertaking a two-pronged research program, involving a detailed multi-wavelength investigation of the individual galaxies with the best Rosat data and a statistical investigation of the entire sample. In this contribution I will report some of our preliminary results.

3. RESULTS

3.1 Rosat Images

Both the Rosat PSPC and HRI data provide interesting and important morphological information about the X-ray emission from starburst galaxies.

Our images show that the X-ray emission typically extends over scales ranging from several kpc in the low-luminosity starbursts to nearly 100 kpc in the highest-luminosity ones (Armus, Heckman, & Weaver 1994; Dahlem, Heckman, & Fabbiano 1994; Heckman et al 1994). The crucial point is that the region producing the X-ray emission is far larger than the starburst region itself (whose typical size is several hundred pc to a kpc). Thus (at least in the Rosat band) the dominant contribution to the X-ray emission can not come from sources that are confined to the starburst (e.g.

high mass X-ray binaries or young 'ultraluminous' supernova remnants). That is not to say that such discrete sources make a negligible contribution (they can be clearly discerned in the nearest starbursts like M82 and NGC253 - cf. Watson, Stanger, & Griffiths 1984; Fabbiano & Trinchieri 1984), only that some other much more dispersed source must also be important.

Rosat images of starburst galaxies whose stellar disks are oriented nearly edge-on are particularly revealing (Armus, Heckman, & Weaver 1994; Dahlem, Heckman, & Fabbiano 1994). Here it is clear that the X-ray emission extends far out of the disk of the galaxy (many kpc). This seems to be a common, if not ubiquitous phenomenon in starburst galaxies viewed at suitably high inclination (see Fabbiano, Heckman, & Keel 1990; Fabbiano & Trinchieri 1984; Watson, Stanger, & Griffiths 1984 and the contributions by Bregman, by Dahlem et al, and by Walterbos et al elsewhere in this volume).

We have also compared the morphologies of the X-ray and Hα nebulae. There is often a striking correspondence between the two phenomena (Armus, Heckman, & Weaver 1994; Dahlem, Heckman, & Fabbiano 1994). In a number of cases it appears that the Hα emission lies along the 'walls' of the X-ray nebula (cf. the case of NGC253 in McCarthy, Heckman, & van Breugel 1987 and NGC1569 in Heckman et al 1994).

3.2 Rosat Spectra

While the Rosat PSPC offers only limited spectral and spatial resolution by the standards of optical or radio spectroscopy, the PSPC data we have analysed have provided vital information concerning the physical state of the X-ray emitting material in starburst galaxies.

In most cases studied to date in which more than about 500 photons have been detected, the PSPC spectra can not be adequately fit by a single-component thermal or nonthermal model. Rather, at least two components are clearly present. These two components make roughly comparable contributions in the Rosat band.

The first is a 'hard' component that dominates above about a keV, and can be adequately fit by either a nonthermal powerlaw with a photon index in the range 1.5 to 2.0 or by a very hot (> several keV) thermal model. The nature of this component is not well- constrained by the Rosat data. The second 'soft' spectral component can be fit as a warm 'Raymond-Smith' plasma with kT ranging from as low as 100 eV in some galaxies, up to nearly 1 keV in others.

Comparisons of the PSPC images of the X-ray emission in the energy ranges dominated by the 'soft' and 'hard' energy bands show that their morphologies can be significantly different (cf. Pietsch 1993; Armus, Heckman, & Weaver 1994; Dahlem, Heckman, & Fabbiano 1994; and the contributions by Bregman and by Walterbos et al in this volume). This strengthens the identification of the two spectral components as physically distinct phenomena (rather than as a single complex one). The soft component also tends to be the more spatially-extended one. I will briefly discuss the implications of this below.

4. INTERPRETATION

We believe that the warm thermal X-ray emission (the 'soft' component) is almost certainly ambient gas in the disk and halo of the starburst that has been shock-heated up to temperatures of 10^6 to 10^7 K by an outflowing superwind driven by the starburst. Such temperatures imply shock speeds of roughly 300 to 900 km s^{-1}, similar to the range in Hα velocities seen in several well-studied starburst outflows (cf. Bland & Tully 1988; Heckman, Armus, & Miley 1990).

The prodigious supply of thermal and kinetic energy injected by supernovae and stellar winds in the starburst will drive a wind which will in turn blow a rapidly expanding bubble into the halo of the starburst galaxy. The physics of the X-ray emission from this 'superbubble' has recently been discussed in some detail by Suchkov et al (1994) based on two-dimensional numerical hydrodynamical simulations incorporating both a disk and halo component of interstellar gas.

Assuming here (for simplicity) a spherically-symmetric abiabatic bubble driven into a halo of constant gas density by kinetic energy that is injected at a uniform rate leads to the following simple relations for the predicted temperature, size, and luminosity (in the Rosat band) of the shocked bubble's outer wall:

$$kT \sim 600\ \dot{E}_{42}^{2/3}\ n_{-3}^{-2/3}\ r_{22}^{-4/3}\ \text{eV}$$
$$L_X \sim 5\times 10^{39}\ \dot{E}_{42}^{3/5}\ n_{-3}^{7/5}\ t_7^{9/5}\ \text{erg s}^{-1}$$

where the energy injection rate is in units of 10^{42} erg s^{-1}, the halo density is in units of 10^{-3} cm^{-3}, the radius of the bubble is in units of 10^{22} cm ($\sim$3.2 kpc), and the age of the bubble is in units of 10^7 years. Note that $\dot{E}$ is expected to be about 2% of the starburst bolometric luminosity (Leitherer & Heckman 1994). Comparison of these relations to the Rosat data on our sample of starbursts shows a satisfactory agreement (given the over-simplicity of the model).

The nature of the 'hard' component is less clear. X-ray spectra at higher energies ($\gg$keV) of a few bright starbursts obtained with BBXRT (Petre 1994) and ASCA (R. Petre, private communication) show that this hard source is best fit by thermal emission with kT $\sim$ 5 to 10 keV and an unusually weak Fe K-α line compared to hot gas with solar abundances in ionization equilibrium. The physical origin of this emission is not understood (see Petre 1994).

Thus, while many issues remain to be settled, it seems clear that the full analysis of the wonderful set of data provided by Rosat (together with the analysis of high-spectral resolution data on selected starbursts provided by ASCA) will allow considerable progress to be made in the near future towards understanding the origin of X-ray emission in starbursts, the nature of the starburst phenomenon, and the role of starbursts in the universe.

REFERENCES

Armus, L., Heckman, T., and Miley, G. 1990, ApJ, 364, 471
Armus, L., Heckman, T., and Weaver, K. 1994, in preparation
Bland, J. and Tully, R.B. 1988, Nature, 334, 43
Bregman, J. & Pildis R. 1992, ApJL, 398, L107
Chevalier, R.A., and Clegg, A.W. 1985, Nature, 317, 44
Dahlem, M., Heckman, T., and Fabbiano, G. 1994, in preparation

DeYoung, D.S. 1978, ApJ, 223, 47
Fabbiano, G., Heckman, T.M., and Keel, W.C. 1990, Ap.J., 355, 442
Fabbiano, G., and Trinchieri, G. 1984, ApJ, 286, 491
Griffiths, R., and Padovani, P. 1990, ApJ, 360, 483
Heckman, T., Armus, L., and Miley, G. 1990, ApJS, 74, 833
Heckman, T., Lehnert, M., and Armus, L. 1993, in "The Environment & Evolution of Galaxies", ed. J.M. Shull and H. Thronson, Jr., (Kluwer: Dordrecht), p.455
Heckman, T., Dahlem, M., Fabbiano, G., Gilmore, D., Lehnert, M., and Waller, W. 1994, in preparation
Larson, R.B., and Dinerstein, H.L. 1975, PASP, 87, 911
Leitherer, C., and Heckman, T. 1994, in preparation
McCarthy, P., Heckman, T.M., and van Breugel, W. 1987, AJ, 93, 264
McKee, C., and Ostriker, J. 1977, ApJ, 218, 148
Ostriker, J., and Cowie, L. 1981, ApJL, 243, L127
Petre, R. 1994, in "The Nearest Active Galaxies", ed. J. Beckman, in press
Pietsch, W. 1993, in "Physics of Nearby Galaxies: Nurture or Nature?", ed. T. Thuan, C. Balkowski, and J. Van (Paris: Editions Frontieres), p. 67
Schaaf, R., Pietsch, P., Biermann, P., Kronberg, P., and Schmutzler, T. 1989, ApJ, 336, 732
Soifer, B.T., Sanders, D., Madore, B., Neugebauer, G., Lonsdale, C., Persson, S.E., and Rice, W. 1987, Ap.J., 320, 238
Suchkov, A., Balsara, D., Heckman, T., and Leitherer, C. 1994, ApJ, in press
Watson, M., Stanger, V., and Griffiths, R. 1984, ApJ, 286, 144
Weedman, D. 1991, in Massive Stars in Starburst Galaxies, ed. C. Leitherer, N. Walborn, T. Heckman, and C. Norman (Cambridge: Cambridge University Press), p. 317
White, S., and Frenk, C. 1991, ApJ, 379, 52
Wyse, R., and Silk, J. 1985, ApJL, 296, L1

ROSAT HRI OBSERVATIONS OF MAGELLANIC CLOUD SUPERNOVA REMNANTS

John P. Hughes
Harvard-Smithsonian Center for Astrophysics

Email ID
jph@cfa.harvard.edu

ABSTRACT

Analysis of deep *ROSAT* high resolution imager (HRI) observations of two oxygen-rich supernova remnants (SNRs) in the Magellanic Clouds is described. For N132D, I exploit the limited spectral information provided by the HRI to investigate arcsecond scale spectral variations. I find that there is a region of harder X-ray emission near the southern limb and regions of softer emission near the center and northwestern limb. The remnant is believed to be interacting with a molecular cloud and the harder emission to the south is explained as a result of increased absorption along the line-of-sight there. I argue that the softer emission comes from X-ray emitting material with an enhanced abundance of oxygen. For the second SNR, E0102.2−72.2, the spatial structure is investigated in detail using two-dimensional image fitting techniques. Evidence is found for a ring-like and a spherically symmetric shell-like component both of which were modeled as homogeneous regions. In addition, a significant fraction of the observed flux (∼11%) must come from a resolved clumped component. A comparison with optical and radio imagery is made to provide a physical basis for the components identified in the X-ray analysis. The mass of X-ray emitting gas in the remnant is estimated and a value of ∼75 $M_\odot$ was determined. The dominant uncertainty on this quantity is the extent of unresolved clumping in the X-ray gas. Such clumping would tend to reduce the mass estimate by $f^{1/2}$, where f is the mean volume filling factor of the gas.

1. Introduction

N132D and E0102.2−72.2 are the brightest supernova remnants (SNRs) in soft X-rays in each of the Large and Small Magellanic Clouds (LMC and SMC), respectively. They belong to the class of oxygen-rich remnants, which include such Galactic examples as Cassiopeia A and G292.0+1.8. SNRs in this class show optical spectra with strong red- and blue-shifted oxygen emission lines indicating matter velocities of several 1000 km s^{-1}. Since Hα is usually not detected from the high velocity filaments, it is likely that they are metal-rich fragments of the stellar ejecta from the core collapse of a massive star. Studies of the stellar ejecta in these remnants are important to our understanding of the evolution of massive stars and the associated nucleosynthesis which occurs in these systems.

2. N132D

The high velocity oxygen-rich filaments in N132D appear to be distributed in a ring-like morphology with a diameter of ∼6 pc (Lasker 1980) and expansion

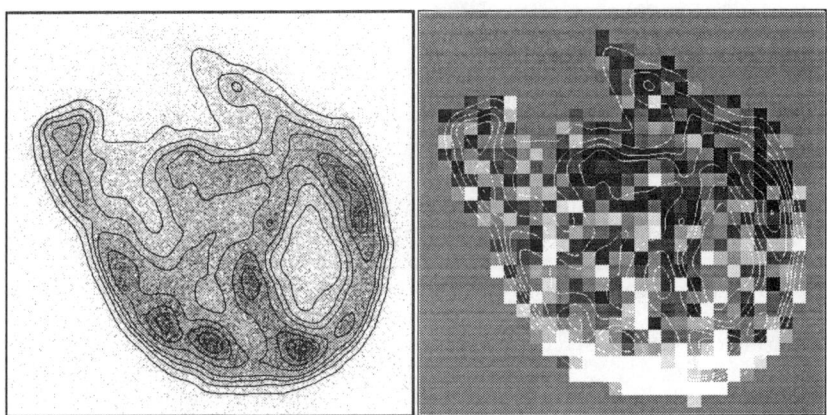

FIG. 1.–(Left) *ROSAT* HRI image of N132D where the grayscale presents the detected events, and the contours show the data after two iterations of the Lucy-Richardson deconvolution algorithm. Contours are shown at values of 0.5, 1, 1.5, 2, 3, 4, 5, and 6 counts s^{-1} arcmin^{-2}. (Right) *ROSAT* HRI pulse height centroid significance map displayed from values of -3 to 3 in units of σ. Each panel is 128″ square, which corresponds to a linear size of 31 pc assuming a distance of 50 kpc to the LMC.

velocities ($\sim$2250 km s^{-1}) implying an age of 1300 yr assuming undecelerated motion. Surrounding these filaments is a relatively quiescent limb-brightened shell of emission with a diameter of about 21 pc (for a distance to the LMC of 50 kpc). The inferred oxygen abundance is high in the central filaments, 10-100 times the abundance in the outer shell. A dual morphology is also seen in the X-rays. Analysis of the X-ray surface brightness variations in the outer shell suggested that the supernova explosion occurred in a low density cavity in the interstellar medium formed by the H II region of the precursor star (Hughes 1987). The most recent X-ray analysis of N132D was the nonequilibrium ionization (NEI) modeling of spectral data from the *Einstein Observatory* reported by Hwang *et al.* (1993) (hereafter referred to as HHCM), in which it was concluded that, for the entire remnant, oxygen was overabundant relative to iron by a factor of at least 1.9 compared to the solar ratio. I utilize the results from this work in my interpretation of the spectral information from the *ROSAT* HRI.

N132D was observed by the *ROSAT* HRI for some 2.7×10^4 s in 1991 February. Within a 100″ radius of its center there were 123911 total counts of which about 1450 were background events for a source counting rate of 4.564 ± 0.013 s^{-1}. Fig. 1 (left-hand panel) shows the *ROSAT* HRI image of N132D. The two spatial components, a limb-brightened clumpy shell, as well as excess emission near the center, are evident.

2.1. ROSAT Pulse Height Analysis

The *ROSAT* HRI has 15 pulse height (PH) channels with the potential to provide some limited energy resolution. Calibration data taken at the MPE Panter Calibration Test Facility before launch show that the centroid of the HRI PH distribution increases from 3.3 to about 5.4 as the energy of the incident X-rays varies from 0.183 to 1.49 keV (see Table 6 in David *et al.* 1993). For this observation of N132D, the PH centroid, averaged over the whole SNR, was 4.5578 ± 0.0054.

Maps of the PH centroid and its statistical error were made for N132D using 4″ square pixels. Next a significance map was determined by subtracting the mean centroid value for the entire remnant from the PH centroid in each pixel and then dividing by the statistical error on the centroid in each pixel. The significance values ranged between plus and minus 4 σ (for the 499 image pixels with counts between 68 and 676). If there were no intrinsic spatial variations in the PH centroid across the remnant, then, by construction, the distribution of significance values should be gaussian with mean $\mu = 0$ and width $\sigma = 1$. In fact a χ^2 test showed that this distribution could be rejected at a confidence level $\gg$99.95%. The observed distribution was considerably broader, $\sigma = 1.30$, strongly suggesting that the distribution of PH centroid values shows some intrinsic variation with position. Since the data were taken in "wobble" mode, spatial variations of the HRI gain should be averaged over, although this point needs additional study. I note for completeness that there was no correlation between the counts in a given pixel and the significance value there.

The PH centroid significance map is shown in Fig. 1 (right-hand panel). Only pixels containing 50 counts or more are plotted. The most dramatic feature of this image is the obvious structure which appears in it. Most of the high significance values (implying higher mean photon energies) lie together in the southern region of the remnant, while most of the low significance values (lower energies) lie in two regions: one near the center and another near the northwestern limb of the remnant. I determined the average PH centroid by summing the data from these various regions and found the following values: northwestern 4.404 ± 0.015, central 4.363 ± 0.022, southern 4.708 ± 0.015. These values differ from the mean for the entire remnant by about 10 times the statistical error.

2.2. Discussion

By interpolating the calibration data, I can relate an observed PH centroid value to a mean detected photon energy. For the softer regions (northwestern and central) the HRI PH centroid implies a mean photon energy of 0.83 keV, for the harder region (southern) the implied mean photon energy is 1.00 keV, and for the remnant as a whole, 0.92 keV. The latter value can be compared to the mean photon energy predicted from the best-fit NEI model of HHCM. Including the efficiency of the HRI and the effective area of the X-ray telescope, I obtain a predicted value for the mean photon energy of 0.92 keV. Although this excellent agreement in predicted versus observed mean photon energy for the entire remnant is very encouraging, until considerably more attention is paid to the known temporal variations of the HRI gain, the agreement should be considered largely fortuitous. On the other hand, the observed spatial variations in PH centroid values across N132D will be considerably less sensitive to temporal gain variations and thus the conclusions drawn below should be reliable.

A few years ago, strong CO emission from a molecular cloud was discovered just south of N132D and it was suggested that there was an interaction between the SNR and the cloud (Hughes, Bronfman, & Nyman 1991). Fig. 2 shows the map of molecular emission superposed on the *Einstein* HRI image. These data provide a natural explanation for the observed hardening of the X-ray spectrum toward the south: I propose that there is increased X-ray absorption along the line-of-sight there. I can reproduce the observed mean photon energy of 1.00 keV in this region if I increase the column density in the HHCM NEI model (leaving all other parameters fixed) to a value of $\sim 2.5 \times 10^{21}$ cm^{-2}. This is a factor of 4 more than the best-fit value of column derived for the entire remnant.

As the explanation for the central region of softer X-ray emission, I propose

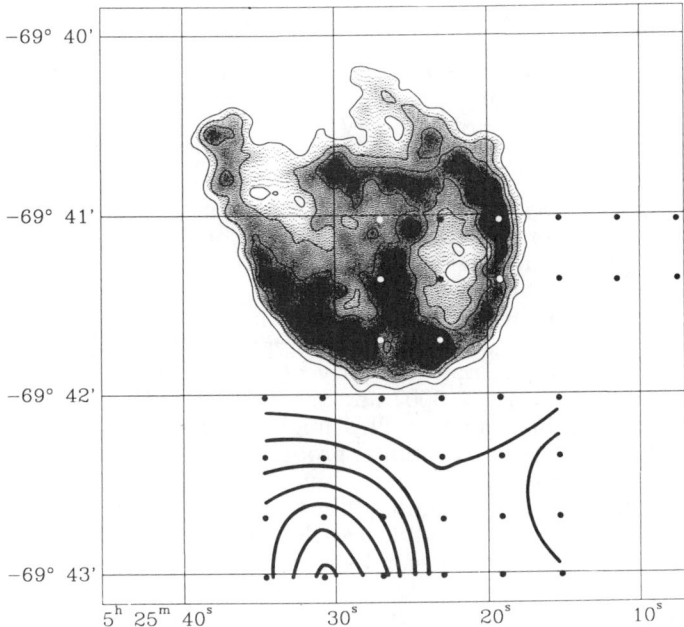

FIG. 2.– *Einstein* HRI image of N132D overlaid with contours of velocity-integrated CO emission. The grid pattern for the CO observations is shown. The contour levels are 3, 5, 7, 9, 11, 13, and 15 K km s^{-1} for a velocity integration range from 260 km s^{-1} to 270 km s^{-1}.

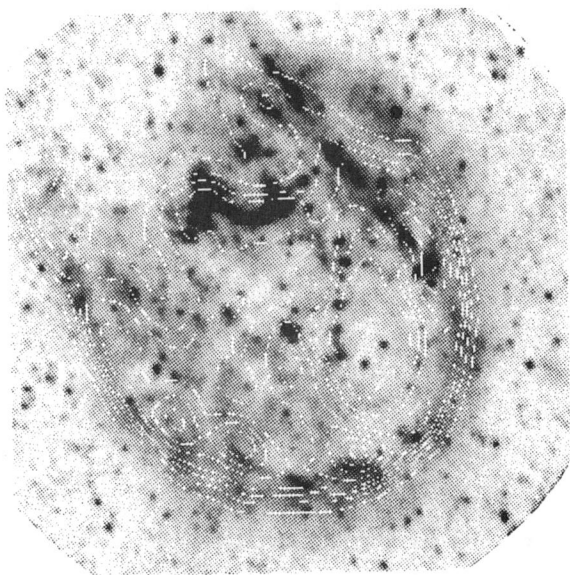

FIG. 3.– Optical image of N132D in the light of [O III] λ 5007. This map shows only the slow-moving ($v \lesssim 150$ km s^{-1}) material.

enhanced oxygen emission. First, I point out that the mean photon energy for an X-ray emitting plasma composed of pure oxygen would be 0.76 keV, assuming the same spectral parameters (*i.e.*, column density, temperature, ionization state) as the HHCM best-fit NEI model. Thus it is possible to explain the observed mean photon energy of 0.83 keV from the central regions of N132D by enhancing the oxygen abundance there by a factor of 10 or more (once again keeping the other model parameters fixed). Given the good positional agreement between the softer X-ray emission and the known oxygen-rich optical filaments, I believe that it is reasonable to draw the conclusion that oxygen is enhanced in the central X-ray emitting regions of N132D.

An optical image of N132D showing the slow-moving material is given in Fig. 3 (J. Morse 1993, private communication). The bright linear feature in the northwest apparently has a soft counterpart in the PH significance map (Fig. 1). Since this feature appears fairly bright in Hα emission, it is unlikely that its brightness in [O III] λ 5007 and the softness of its X-ray spectrum are a result of oxygen enhancement. This might suggest that excitation or NEI effects are influencing its observational properties. Additional analysis may aid in understanding this curious feature, but it is likely that the poor spectral capabilities of the HRI will ultimately limit what we can learn.

This work has demonstrated the capabilities of the *ROSAT* HRI for providing spectral information on arcsecond spatial scales. With high enough signal-to-noise it is possible to observe statistically significant spectral variations with position across extended sources. Interpretation of these variations in terms of interesting physical source quantities requires detailed knowledge of the average X-ray spectrum of the object and is (probably) restricted to constraining a single quantity at a time. Nevertheless, this capability, limited though it is, is currently unique and further effort should be devoted to its calibration.

3. E0102.2−72.2

E0102.2−72.2 was discovered as a strong soft X-ray source during a survey of the SMC by the *Einstein Observatory* (Seward & Mitchell 1981). Its identification as an oxygen-rich SNR was made by Dopita, Tuohy, & Mathewson (1981) and a velocity map of the remnant was presented by Tuohy & Dopita (1983). These data showed that the oxygen-rich material covers a velocity range of -2500 km s^{-1} to $+4000$ km s^{-1}. An *Einstein* HRI observation of the remnant (Inoue, Koyama, & Tanaka 1983) showed a strongly limb-brightened shell of emission with a radius of about $20''$ (~ 6 pc assuming a distance to the SMC of 57.5 kpc). Analysis of these data indicated that the geometry of the X-ray emitting region of the remnant was more consistent with being ring-like than shell-like (Hughes 1988). Recently the US/Japanese satellite *ASCA* obtained an X-ray spectrum of the remnant which showed strong Kα line emission from O, Ne, and Mg with evidence for weaker emission from Si, S, and Fe (Hayashi *et al.* 1994). A high resolution radio map (Amy & Ball 1993) shows a strongly limb-brightened shell with an angular diameter of $\sim 40''$. There is evidence from this image for a relatively strong compact radio component near the center of E0102.2−72.2.

The *ROSAT* HRI observed E0102.2−72.2 for 2.0×10^4 s in 1991 November. Within a radius of $40''$ there were 19500 total counts of which about 460 were background events for a source counting rate of 0.958 ± 0.007 s^{-1}. An image of the remnant, after two iterations of the Lucy-Richardson deconvolu-

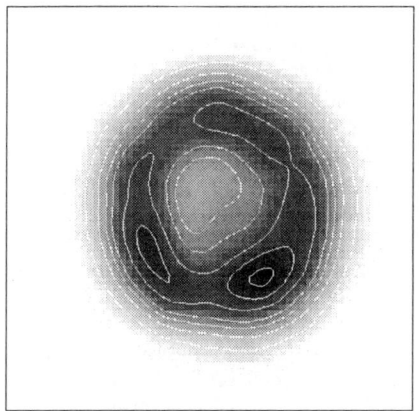 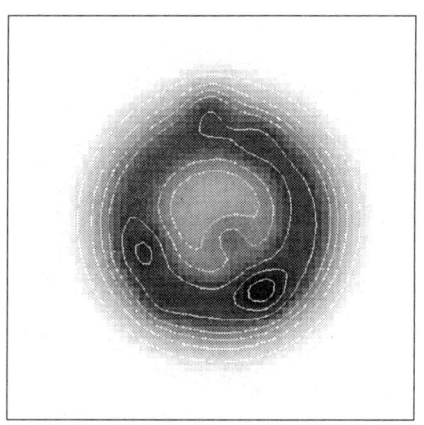

FIG. 4.–(Left) Deconvolved *ROSAT* HRI image of E0102.2−72.2. Contours are shown at values of 0.5, 1, 1.5, 2, 3, 4, and 5 counts s^{-1} arcmin^{-2}. (Right) The best-fit model for E0102.2−72.2 including both ring-like and shell-like components, plus six individual spherical clumps. The contour levels correspond to the same values as in the deconvolved data. Each panel is 64″ square, which corresponds to a linear size of 18 pc assuming a distance of 57.5 kpc to the SMC.

tion algorithm, is shown in Fig. 4 (left-hand panel). A strongly limb-brightened, apparently clumpy, shell is evident.

3.1. Spatial Model Fits

I performed spatial fits to the two-dimensional X-ray image in order to investigate the intrinsic geometric structure of the remnant. The fits were done using software specifically developed for the analysis of sparsely filled images. It employs a maximum likelihood statistical estimator derived for Poisson distributed data and parameter estimation is done using the downhill simplex method. Numerous possible geometric models, such as spherical shells, rings, spherical clumps, and so on, can be projected onto the sky and convolved with the *ROSAT* HRI point spread function for comparison to the data.

For ring models, I assumed that the plane of the ring was the same as the plane of the sky. The geometric parameters included position on the sky, the inner and outer radii of the ring, the opening angle of the ring, and a normalization. As can be seen in Fig. 4, the surface brightness of the ring appears to vary azimuthally from an average of 5.25 counts s^{-1} arcmin^{-2} in the southwest to 2.88 counts s^{-1} arcmin^{-2} in the northeast. In order to model this brightness variation I allowed for a relative azimuthal variation in the mean plasma density in the ring using $n(\theta) = [1 + \cos(\theta - \theta_0)]/2 + n_0[1 - \cos(\theta - \theta_0)]/2$. The relative density would vary smoothly, according to this function, from a maximum of 1 at the azimuthal reference angle, θ_0, to a minimum of n_0 at angle, $\theta_0 + 180°$. The surface brightness then scales as the square of the relative density times the geometric line-of-sight through the ring. (In the final model results presented below, the best-fit minimum density was ∼0.8.) In total, eight parameters were needed to fully specify this type of component. Spherical shells were parameterized in an identical manner, with the ring opening angle fixed at the value needed to produce a complete shell (90°). Finally, clumps were modeled as constant density spherical regions; each required four parameters: position on the sky, clump radius, and a normalization.

Initially, single component models were tried. Neither pure shell nor ring models provided decent fits. In the former case, the surface brightness in parts of the central ~6″–8″ radial region was overpredicted (by factors of 1.6) while in the latter it was significantly underpredicted (by up to factors of 3). For any single model fit our maximum-likelihood estimator did not yield an absolute goodness-of-fit criterion (although relative differences in the likelihood function between model fits can be accurately assigned confidence levels of relative probability, a feature I utilize below). To estimate the absolute goodness of any single best fit I compared radial averages of the model and data in quadrants and calculated χ^2, taking care only to include radial bins with 25 or more counts in the data. For the single shell fits I obtained $\chi^2 = 718.5$ for $\nu = 151$ degrees of freedom. For the pure ring models the best fit yielded $\chi^2 = 796.4$ for $\nu = 150$.

It is not surprising that these simple models were such poor descriptions of the data. However, what was unexpected was the significant improvement in the quality of the fit obtained when a model with both a ring-like and a shell-like component was used. For simplicity I assumed the two components were concentric and showed the same azimuthal density variation. In this case, then, relative to the pure ring model, only three additional parameters: the radius, thickness, and normalization of the shell-like component, were required for a complete specification of the model. When the radially-averaged data and model were compared, the best-fit two-component model yielded $\chi^2 = 287.2$ for $\nu = 147$, a tremendous improvement over the single component models. Nevertheless, for both observational and theoretical reasons, it is not expected that such a smooth, homogeneous, and unclumped model should accurately describe the structure of SNRs. Indeed, an examination of the difference image made by subtracting the data and this model revealed features on spatial scales of $\lesssim 5''$ with peak surface brightnesses approaching ~1 counts s^{-1} arcmin^{-2}.

These features were modeled as discrete clumps as described above. Clumps were introduced into the model in an iterative fashion, by examining the difference image after each successive best-fit minimization and picking out new features by eye. The process was terminated when no obvious features could be discerned in the difference image. In total, six individual clumps were fit. The maximum-likelihood estimator decreased by a value of 265.5 when the 24 parameters of the clumped component were included. Since the difference of maximum-likelihood estimators is distributed like χ^2 (for degrees of freedom equal to the number of new parameters), this result is highly significant. The fitted clumps range in diametric sizes from 2.9″ up to 4.5″. These are at the resolution limit of the HRI and therefore should probably be considered only upper limits to the true clump sizes. An image of the final two-component model including the clump component is shown in Fig. 4 (right-hand panel). Comparison with the data image shows peak surface brightness residuals at the level of ~0.6 counts s^{-1} arcmin^{-2}. Fig. 5 (left-hand side) shows the radially averaged data and model including a breakdown of the model into its various components. On the right-hand side of Fig. 5, residuals are shown for the four quadrants of the image. In general, as these various representations show, the agreement between the data and model is quite good.

3.2. Discussion

Detailed analysis of the *ROSAT* HRI X-ray image of E0102.2−72.2 shows evidence for both a ring-like component and a spherically-symmetric shell component. Other young oxygen-rich SNRs in the Galaxy, such as Cassiopeia A

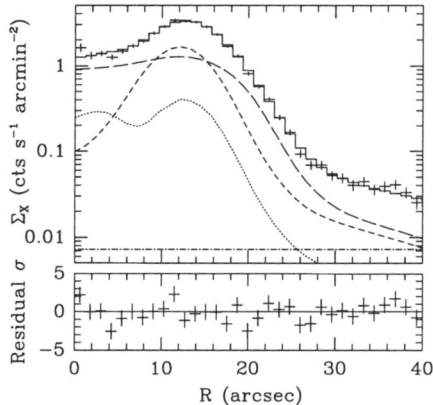 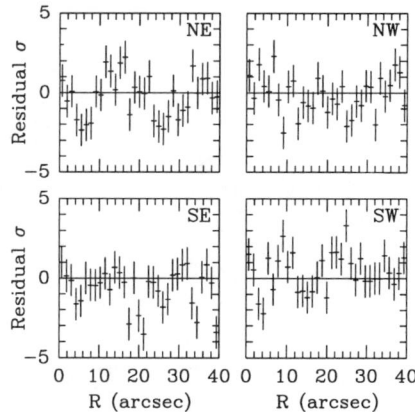

FIG. 5.–(Left) (Top) Radially-averaged X-ray surface brightness of E0102.2−72.2 compared to the best-fit spatial model. The several components to the model are shown separately. The short-dash curve is the ring-like component, the long-dash curve is the spherical shell component, the dotted curve is the clumped component, and the background is the straight line near the bottom of the panel. (Bottom) The difference between the observed data and the best-fit model in units of the statistical error in each radial bin. (Right) The difference between the observed data and the best-fit model for the four quadrants of the image as indicated.

and G292.0+1.8, have also been suggested to show such a dual morphological structure. For E0102.2−72.2 roughly 39% of the HRI emissivity comes from the ring, 50% from the spherical shell, and ∼11% from a clumped component. Given the limited statistics and the instrumental spatial resolution, the latter is merely a lower limit to the clumped component.

Fig. 6 (left-hand panel) shows a schematic cross-sectional view of the best-fit model for the X-ray emission. It is interesting to note that the inner edges of both the ring and shell components were determined (by the fitting procedure itself) to be nearly equal. Most of the clumps appear to be associated with the ring-like component. It is plausible to identify the ring component with the ejecta and its evident clumpiness is consistent with the onset of the well-known instabilities (Rayleigh-Taylor, Kelvin-Helmholtz) which tend to disrupt and fragment SN ejecta. The spherical shell (or some part of it) would arise from the swept-up circumstellar medium. This hypothesis would suggest that the integrated X-ray spectrum of the remnant should show evidence for two spectral components: one with low abundances as appropriate to gas in the SMC (∼20% cosmic) and another with enhanced abundances as from nucleosynthesis in a massive star (see mass estimate below). The *ASCA* spectrum of E0102.2−72.2 provides some qualitative support for this picture (Hayashi *et al.* 1994).

Comparison with optical imagery provides further insight into the physical basis for the components identified in the X-ray analysis. In Fig. 6 (right-hand panel), I show the radially averaged surface brightness profile of E0102.2−72.2 in the light of [O III] λ 5007 from data taken at the CTIO 4-m in 1986 November. The stellar component was subtracted from this profile using a continuum band image, centered at 6100 Å, of the same region of the sky. The optical emission shows a local peak at the inner edge of the X-ray ring component (12.3″) and then a gradual decrease as it is traced further out. By the time the outer edge of the ring is reached (17.2″), the optical emission has dropped to a minimum value. I believe that we are seeing cool, dense clumps of oxygen-rich ejecta en-

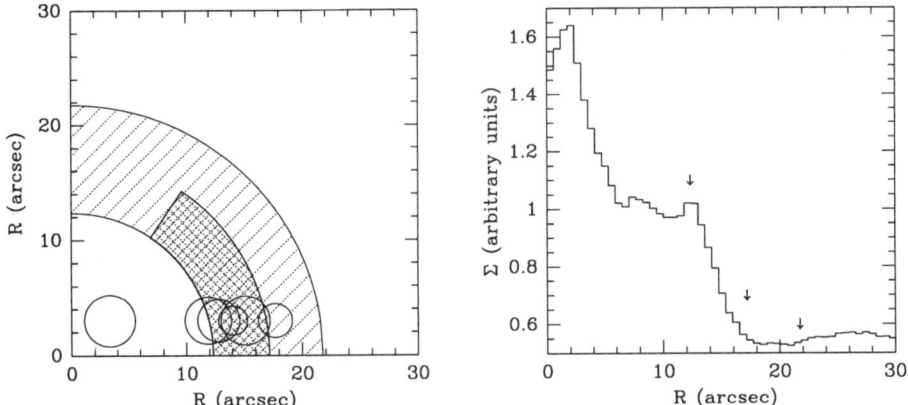

FIG. 6.—(Left) Cross-sectional view of the geometric model which best fits the ROSAT HRI image. The plane of the sky lies along the horizontal axis and runs in and out of the page. The viewing direction lies along the vertical axis. The ring component is shown as the doubly hatched region, while the spherical shell component is shown singly hatched. The radial positions of the six fitted clumps are drawn to scale as circles near the bottom of the figure. (Right) Radially-averaged surface brightness of E0102.2−72.2 in the light of [O III] λ 5007 after subtraction of the stellar component. The arrows denote, in increasing order from the center, the location of the inner edge of the ring component, the outer edge of the ring, and the outer edge of the spherical component.

tering the reverse shock and then gradually being destroyed in the high pressure environment of the hot X-ray emitting ejecta.

Beyond the blast wave (as indicated by the outermost extent of the X-ray emission $\gtrsim 22''$), there is an optical halo, presumably a fossil H II region photoionized by the precursor star, the UV flash of the SN, or the UV radiation from the SNR itself (Dopita et al. 1981). I confirm that this halo is indeed suppressed in the X-ray emitting region extending from the outer edge of the ring component to the location of the blast wave itself, as first suggested by Dopita et al. (1981).

The correlation between the X-ray and radio data is not particularly good in detail. The radio shell appears to be somewhat asymmetric, with its outer diameter varying around the remnant azimuthally by some $\pm 5''$. This level of asymmetry is just not present in the X-ray data. Furthermore the brightest part of the radio shell is toward the northeast, where the X-ray shell shows its minimum brightness. There is, however, a weak X-ray clump near the position of the relatively strong, compact central component found in the radio map which Amy & Ball (1993) speculate might be a plerionic component. This would be an exciting discovery, since, if true, it would signal the presence of a pulsar in the remnant. The X-ray clump contains only about 1.7% of the total SNR flux for an HRI counting rate of ~ 0.016 s^{-1}. The only known oxygen-rich shell remnant with a plerionic core component is the LMC SNR E0540−69.3, which has a ROSAT HRI rate of 0.187^{-1} (Seward & Harnden 1994). After correcting for the difference in distances to the two remnants, the inferred X-ray intensity for the putative plerionic clump in E0102.2−72.2 is about a factor of 10 less than the intensity of the known plerion in E0540−69.3. Using the empirical relationship between the spin-down energy loss rate for a pulsar, $\dot{E}$, and the X-ray luminosity of its associated plerion (Seward & Wang 1988) I estimate $\dot{E}$ to be $\sim 10^{37.5}$ erg s^{-1} for the putative pulsar in E0102.2−72.2. The period can also be estimated; a value of ~ 0.14 s is indicated, assuming an age of 1000 yr and the $\dot{E}$ given above. Although these values are fully consistent with those observed for other known young pulsars, it is still quite possible that the compact central component is merely a bright spot in the SNR shell seen projected onto the center of the remnant. A definitive result requires the detection of pulsations (this is difficult in the X-ray band because of the limited number of detected events) or the measurement of a flat radio spectrum from the core component.

Finally I turn to an X-ray–derived estimate for the mass of the SNR. In addition to the geometric model for the structure of the emission region, this estimate requires a spectral model to relate the observed count rates to emission measure. For this I use results from fits of a NEI model (see Hughes & Singh 1994) to the *ASCA* data for E0102.2−72.2 (Hayashi *et al.* 1994). I estimate the densities in the remnant to be $n_S \sim 3.7$ cm^{-3} for the shell component and $n_R \sim 6.0$ cm^{-3} for the ring component (after accounting for the physical overlap of the geometric models). For simplicity I ignore the contribution of the fitted clumps to the estimate. This leads to a total mass of

$$M \sim (15 f_R^{1/2} + 60 f_S^{1/2}) D_{57.5 \,\rm kpc}^{5/2} \; M_\odot.$$

Note the strong dependence of the derived mass on distance. This demonstrates the power of studying SNRs in the Magellanic Clouds where the uncertainty in distance introduces less than a 25% error in the derived mass. The result also explicitly includes the dependence on small scale clumping of the gas (*i.e.*, at spatial scales below the instrumental resolution), parameterized by the volume filling factors, f_R and f_S. Even for the rather extreme value of 0.1, the progenitor of E0102.2−72.2 clearly was a massive star.

4. References

Amy, S. W., & Ball, L. 1993, ApJ, 411, 761
David, L. P., Harnden, F. R., Jr., Kearns, K. E., & Zombeck, M, V, 1993, The ROSAT High Resolution Imager, U.S. ROSAT Science Data Center report
Dopita, M. A., Tuohy, I. R., & Mathewson, D. S. 1981, ApJ, 248, L105
Hayashi, I., Koyama, K., Ozaki, M., Miyata, E., Tsunemi, H., Hughes, J. P., & Petre, R. 1994, PASJ, to appear June 1994
Hughes, J. P. 1987, ApJ, 314, 103
Hughes, J. P. 1988, in IAU Colloquium 101, The Interaction of Supernova Remnants With the Interstellar Medium, 125
Hughes, J. P., Bronfman, L., & Nyman, L. 1991, in Supernovae, The Tenth Santa Cruz Summer Workshop in Astronomy and Astrophysics, 679
Hughes, J. P., & Singh, K. P. 1994, ApJ, 422, 126
Hwang, U., Hughes, J. P., Canizares, C. R., & Markert, T. H. 1993, ApJ, 414, 219 (HHCM)
Inoue, H., Koyama, K., & Tanaka, Y. 1983, in IAU Symposium 101, Supernova Remnants and Their X-Ray Emission, 535
Seward, F. D., & Harnden, F. R. 1994, ApJ, in press
Seward, F. D., & Mitchell, M. 1981, ApJ, 243, 736
Seward, F. D., & Wang, Z. R.. 1988, ApJ, 332, 199
Tuohy, I. R., & Dopita, M. A. 1983, ApJ, 268, L11

DIFFUSE X-RAY EMISSION FROM H II COMPLEXES: STELLAR WINDS AND SUPERNOVA REMNANTS

You-Hua Chu
Astronomy Dept., Univ. of Illinois, 1002 W. Green St., Urbana, IL 61801

Email ID
chu@dorado.astro.uiuc.edu

ABSTRACT

H II complexes containing large numbers of OB stars are often bright, diffuse, soft X-ray sources. To investigate the roles played by stellar winds and supernovae in heating the interstellar gas in complex regions, it is necessary to start with simpler objects, such as wind-blown bubbles and superbubbles. This paper reports preliminary results of *ROSAT* PSPC observations of these objects.

1. Introduction

Any massive star that emits H-ionizing photons will also blow fast stellar wind during most of its lifetime and explode as a supernova at the end. Both fast stellar winds and supernovae can heat the ambient interstellar matter to X-ray emitting temperatures. Therefore, H II complexes containing large numbers of massive stars are expected to be diffuse, soft X-ray sources. Indeed, *Einstein* IPC observations have shown diffuse X-ray emission in a wide range of interstellar objects: from simple wind-blown bubbles (Fabian & Stewart 1983; Leahy 1985; Bochkarev 1988) to superbubbles (Chu & Mac Low 1990; Wang & Helfand 1991a), and from small, Orion-type H II regions (Ku & Chanan 1979) to giant H II regions, such as the Carina Nebula (Seward & Chlebowski 1982) and 30 Doradus (Chu & Mac Low 1990; Wang & Helfand 1991b).

What astrophysics can we learn from the diffuse X-ray emission in H II complexes? Do stellar winds and supernova remnants (SNRs) play equal roles in producing hot plasma? What contribution has *ROSAT* made in this exciting new field? I will start by describing simple wind-blown bubbles, move on to more complex objects, and finish with giant H II regions.

2. Stellar Wind-Blown Bubbles

Stellar wind-blown bubbles represent the simplest examples of interaction between fast winds and the ambient media. There are two popular bubble models, assuming either energy conservation, in which case the shocked stellar wind forms a pressurized layer to push the ambient medium outwards (*cf.* Weaver *et al.* 1977), or momentum conservation, in which case the shocked stellar wind cools rapidly and mixes with the ambient medium (*cf.* Steigman *et al.* 1975). The dynamical parameters derived from optical observations, such as energy and momentum conversion factors (Treffers & Chu 1982; Chu 1982), seem to be more in accord with the momentum-conserving model. However, it has also been argued that energy conservation should not be ruled out because bubbles may be enveloped by neutral, expanding shells the kinetic energy of which may account for the missing energy (Van Buren 1986).

X-ray emission from a bubble can be used to derive the amount of hot gas and thermal energy in its interior. Therefore, X-ray observations could critically test the bubble models, and decisively tell us how stellar winds deposit energy into their ambient media. *Einstein* IPC observations detected several "bubbles," although some of them did not have optically identifiable ring morphologies. *ROSAT* PSPC observed these and additional bubbles. The *Einstein* and *ROSAT* results are summarized in Table 1.

Table 1. *Einstein* and *ROSAT* observations of Bubbles

Objects	*Einstein*	*ROSAT*
Orion Nebula	diffuse emission[a]	resolved into stars[f,i]
Rosette Nebula	diffuse emission[b]	resolved into stars[d,g]
S155/CepOB3	diffuse emission[c]	resolved into stars[h]
NGC 2359	—	not detected[d,i]
NGC 3199	not detected[d]	not detected[d]
NGC 6164-5	—	central star detected[i]
NGC 7635	—	central star detected[i]
S 308	not detected[d]	near circular rib[d,i]
NGC 6888	diffuse emission[e]	diffuse emission[j]

Notes: a) Ku & Chanan 1979; b) Leahy 1985; c) Fabian & Stewart 1983; d) archival data; e) Bochkarev 1988; f) Caillault 1993; g) our data; h) Naylor 1993; i) Wendker 1993; j) Wrigge et al. 1994.

The two most significant results from the *ROSAT* PSPC observations of bubbles are: 1) resolving three of *Einstein*'s diffuse sources into point sources associated with early-type stars and/or pre-main sequence stars; 2) providing the first useful X-ray spectrum of shocked stellar wind in a bubble, *i.e.*, NGC 6888. The X-ray emission from NGC 6888 is characterized by a low plasma temperature of 0.1–0.2 keV (Wrigge et al. 1994). The X-ray luminosity of NGC 6888, while being an order of magnitude lower than that expected in Weaver et al.'s (1977) model, can be explained if the bubble is formed by the current Wolf-Rayet wind sweeping up the circumstellar material shed by its progenitor in a red supergiant wind (Garcia-Segura & Mac Low 1993).

The *ROSAT* PSPC is the most sensitive soft X-ray imaging instrument available in the 20th century for observations of bubbles. One cannot help being disappointed by the fact that only NGC 6888 and possibly S308 are detected. More *ROSAT* PSPC observing time should have been dedicated to bubbles. RCW 58, having a similar formation mechanism and interstellar extinction as NGC 6888, is the most promising bubble for future soft X-ray imaging observations.

3. H II Complexes and Superbubbles in the Large Magellanic Cloud

For X-ray studies of H II complexes and superbubbles, the Large Magellanic Cloud (LMC) has distinct advantages over our Galaxy. The almost face-on inclination of the LMC minimizes both absorption and confusion along the line of sight. The LMC is also at a nice distance, 50 kpc, so that the angular sizes of its H II complexes and superbubbles match perfectly the PSPC's field of view.

Superbubbles are simply large bubbles blown by multiple winds and supernovae from OB associations or clusters. The X-ray emission from a superbubble can be used to derive the hot gas content and to constrain bubble models. However, unlike smaller bubbles, superbubbles are prone to additional excitation by SNRs near the shell walls to produce excess X-ray emission (Chu & Mac Low 1990; Wang & Helfand 1991a). Consequently, the astrophysical significance of X-ray studies of superbubbles depends on the observed X-ray luminosities.

For the faintest superbubbles, the observed X-ray luminosities may be used to rule out certain bubble models. For example, several LMC superbubbles are not detected by the *ROSAT* PSPC with 10–15 ksec exposures; the upper limits on their luminosities are much lower than those predicted by Weaver *et al.*'s (1977) energy-conserving models (Chang *et al.* 1994). These are probably the first confirmed X-ray dim superbubbles.

X-ray bright superbubbles are most likely excited by SNRs interacting with their shell walls. For these objects, the most significant results from the *ROSAT* observations include useful spectra and improved statistics. Most superbubbles show very soft X-ray spectra that can be characterized by Raymond-Smith plasma temperatures of 0.1–0.4 keV (Chu *et al.* 1993), although some may be as hot as 0.8 keV, *e.g.*, 30 Dor C. The archival *ROSAT* PSPC data of the LMC show many X-ray bright superbubbles. The census of these superbubbles will greatly complement the SNR statistics in the LMC.

Several LMC H II complexes have been observed with the *ROSAT* PSPC with ≥15 ksec exposure. Diffuse emission is often detected and correlates well with the interiors of shell structures. The data are being analyzed, but it is premature to report quantitative results at present.

4. Giant H II Regions

30 Doradus in the LMC is the nearest giant H II region. Its bright, diffuse X-ray emission was detected by the *Einstein* IPC. The *ROSAT* PSPC observations yielded useful spectra with a much improved spatial resolution. Figure 1 shows a pair of Hα and PSPC images of the 30 Dor field. Four outlying regions of SNR activities are marked: 30 Dor B and 0540-69.3 are two known SNRs; 30 Dor C is an X-ray bright superbubble; the Honeycomb Nebula is a sheet of neutral cloud shocked by a SNR in a low density cavity (Chu, Dickel, & Staveley-Smith 1994).

Both point sources and diffuse sources exist in 30 Dor. The two point sources near the core are coincident with two tight groups of stars – R136 and R140. The brighter source, at R136, shows relatively hard X-ray spectrum; the spectral fits of Raymond-Smith models give plasma temperatures of 1–2 keV. It is likely that one of the member stars in R136 is an X-ray binary.

The diffuse X-ray emission correlates very well with both the shell structures shown in Fig. 1 and the kinematic features shown in high-dispersion echelle Hα spectra (Chu & Kennicutt 1994). Some X-ray sources can be identified with fast expanding ($V_{exp} > 100$ km s^{-1}) shells; for example, the X-ray ring to the east of R136 is coincident with a shell expanding at 150–180 km s^{-1}. The diffuse X-ray emission from large shells can be explained in the same way as for X-ray bright superbubbles, *i.e.*, SNRs shocking the shell walls. This explanation is further supported by the nebular kinematics; for example, the large elliptical shell at position angle of ∼20° and the bright shell to the south of R136 both show high-velocity, $\Delta V > 100$ km s^{-1}, shocked material superposed on a shell expanding at 30–35 km s^{-1} (Chu & Kennicutt 1994). SNRs clearly play a more important role than stellar winds for the diffuse X-ray emission from 30 Dor.

This conclusion is also supported by the absence of correlation between diffuse X-rays and the distribution of massive stars in 30 Dor (Parker & Garmany 1993).

30 Dor is by no means unique in its X-ray emission, as NGC 604 in M33 (Long et al. 1993) and 5 giant H II regions in M101 (Murphy & Chu 1993) are all similarly bright in X-rays. However, it is much harder to analyze the detailed nature of X-ray emission from distant regions that are not spatially resolved by the *ROSAT* PSPC, although the sources in M101's giant H II regions indeed appear to be combinations of X-ray binaries and SNRs.

5. Conclusion

H II complexes are often bright, diffuse X-ray sources, and the X-ray surface brightness correlates well with shell structures. SNRs hitting the inner shell walls of superbubbles can produce excess X-ray emission, and this mechanism can explain most of the diffuse X-ray emission in H II complexes. X-ray observations of H II complexes can effectively uncover hidden SNRs.

YHC acknowledges the support of NASA grant NAG 5-1900. She also thanks her collaborators for moral support and useful discussions.

References

Bochkarev, N. G. 1988, Nature, 332, 518
Caillault, J. P. 1993, in this volume
Chang, H.-W., Su, Y.-L., Chu, Y.-H., & Mac Low, M.-M. 1994, in preparartion
Chu, Y.-H. 1982, ApJ, 254, 578
Chu, Y.-H., Dickel, J. R., & Staveley-Smith, L. 1994, in preparation
Chu, Y.-H., & Kennicutt, R. C. 1994, ApJ, in press
Chu, Y.-H., & Mac Low, M.-M. 1990, ApJ, 365, 510
Chu, Y.-H., Mac Low, M.-M., Garcia-Segura, G., Wakker, B., & Kennicutt, R. C. 1993, ApJ, 414, 213
Fabian, A. C., & Stewart, G. C. 1983, MNRAS, 202, 697
Garcia-Segura, G., & Mac Low, M.-M. 1993, in ASP Conf. Series 35, Massive Stars: Their Lives in the Interstellar Medium, p. 354
Kennicutt, R. C., & Hodge, P. W. 1986, ApJ, 306, 130
Ku, W. H.-M., & Chanan, G. A. 1979, ApJ, 234, L59
Leahy, D. A. 1985, MNRAS, 217, 69
Long, K. S., Gordon, S. M., Blair, W. P., & Charles, P. 1993, in this volume
Murphy, R., & Chu, Y.-H. 1993, in this volume
Naylor, T. 1993, abstract of an AO4 ROSAT HRI proposal, in MIPS
Parker, J. W., & Garmany, C. D. 1993, AJ, 106, 1471
Seward, F. D., & Chlebowski, T. 1982, ApJ, 256, 530
Steigman, G., Strittmatter, P. A., & Williams, R. E. 1975, ApJ, 198, 575
Trefferes, R. R., & Chu, Y.-H. 1982, ApJ, 254, 569
Van Buren, D. 1986, ApJ, 306, 538
Wang, Q., & Helfand, D. J. 1991a, ApJ, 373, 497
Wang, Q., & Helfand, D. J. 1991b, ApJ, 370, 541
Weaver, R., et al. 1977, ApJ, 218, 377
Wendker, H. J. 1993, personal communication
Wrigge, M., Wendker, H. J., Wisotzki, L. 1994, A&A, in press

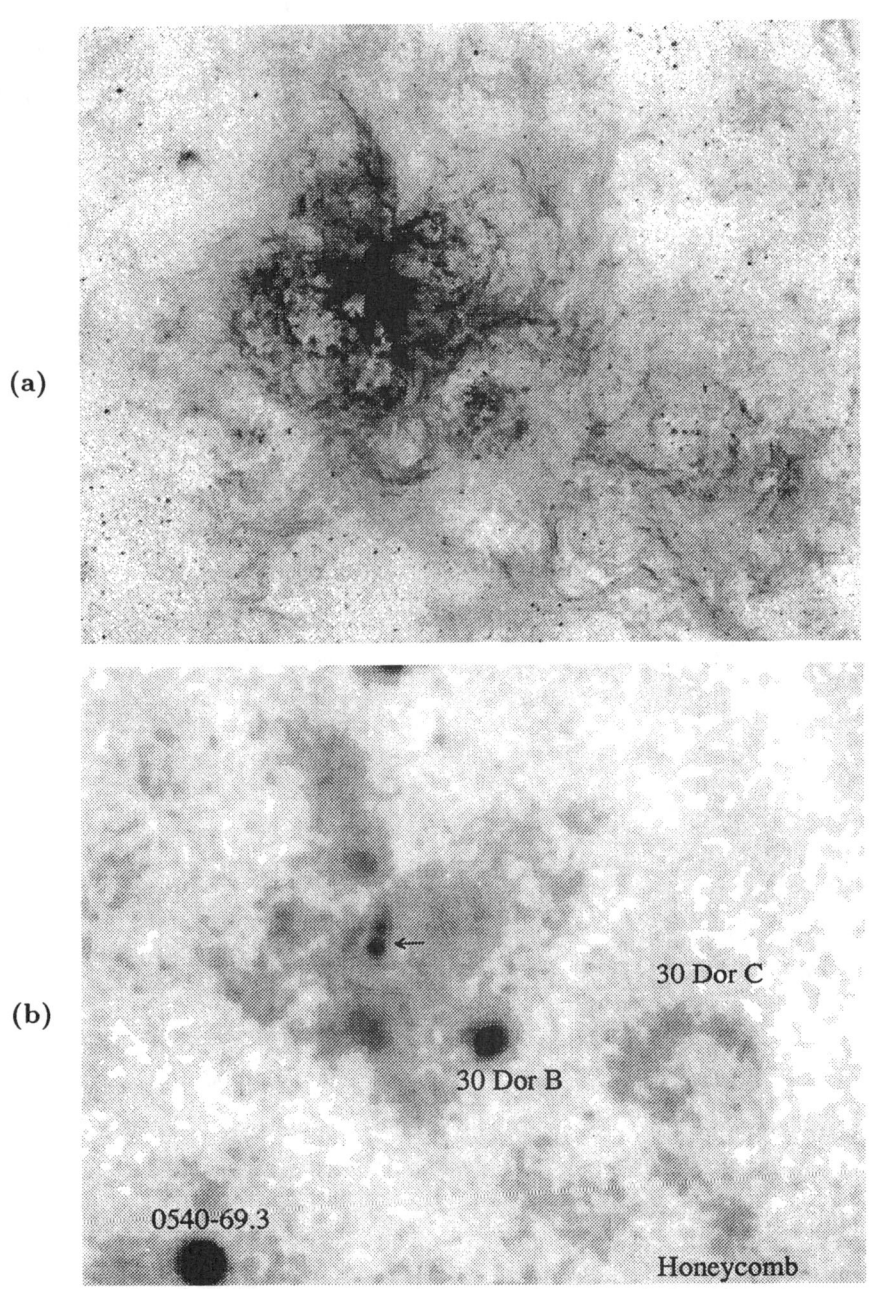

Figure 1. a) Hα image of 30 Dor, reproduced from a Curtis Schmidt plate taken by Kennicutt & Hodge (1986). b) *ROSAT* PSPC image of 30 Dor, rp500131. The picture size is $34' \times 27'$. The arrow points to the source at R136.

THE CARINA NEBULA IN X-RAYS

M. F. Corcoran[1,2], J. Swank[2], G. Rawley[3], R. Petre[2], J. Schmitt[4], C. Day[1,2]

ABSTRACT

New *ROSAT* PSPC and HRI observations of the Carina Nebula region are used to examine the X-ray emission from the discrete sources as well as the diffuse hot gas in the Carina Nebula near Eta Carina. The spectral and spatial response of the PSPC allow analysis of the 0.1 - 2.4 keV spectra of all the bright point sources in the region and an examination of the spectral variation of the diffuse X-rays. Analysis of the diffuse emission shows that most of the emission comes from gas at a temperature of a few million degrees, but also indicates the presence of hot (4E7 K) diffuse gas in an 11 pc region around Eta Car. Eta Car shows a 3 component spectrum with temperatures of 1E6, 6E6 and 4E7 K. A 0.1 × 0.2 pc shell of X-ray emitting gas around Eta Car has been resolved by the HRI, and comparison of the *Einstein* HRI and *ROSAT* HRI images supports the suggestion that the soft components originate in the shell (which is unresolved by the PSPC) while the hottest gas is produced less than an arcsec from the optical star. Spectra of the bright O stars are characterized by 2 temperature components (1E7 K and 2E6 K) and generally show absorption columns larger than the interstellar column. PSPC and *ROSAT* HRI observations are used to examine the L_x/L_{bol} relation for hot stars in the field through the early-B spectral class.

1. INTRODUCTION

The Carina Nebula region is one of the most interesting massive star forming regions of the Galaxy, containing bright nebulosities, faint dust lanes, open clusters with dozens of massive hot stars in various evolutionary stages (including 6 O3 stars and 3 Wolf-Rayet stars), and the peculiar object Eta Car. Eta Car is surrounded by a nebulosity called the homunculus which apparently was ejected from the star during an eruptive event in the 1840s. The nature of Eta Car is unknown but the star is generally thought to be a single very massive star close to the Humphreys-Davidson instability limit (Davidson 1971) although binary and multiple-star scenarios have also been proposed (Warren-Smith et al. 1979).

Observation of the Carina Nebula region with early non-imaging X-ray detectors (*OSO*-8, Becker et al. 1976, *Uhuru*, Forman et al. 1978) revealed at least one source of X-rays, and indicated the presence of a hard component at a temperature of a few keV. Observation with the imaging instruments on the *Einstein* Observatory (Seward et al. 1979, Seward and Chlebowski 1982, Chlebowski et al. 1984) showed various contributors to the X-ray flux: point source emission from the hot stars, point source and extended emission from Eta Car, and soft emission from a diffuse component generally associated with the optical nebulosities. The diffuse emission and the

---

[1] USRA, Code 668, Goddard Space Flight Center, Greenbelt, MD 20771
[2] LHEA, Code 666, Goddard Space Flight Center, Greenbelt, MD 20771
[3] Applied Research Corporation, 8201 Corporate Dr., Landover, MD 20785
[4] MPE, 8046 Garching bei Munchen, Germany

emission from most of the hot stars is produced by a thermal plasma of temperature near 1 keV. Eta Car however showed a complex, multi-temperature spectrum. Spectra of Eta Car obtained with the IPC and the SSS (Chlebowski et al. 1984) detected thermal emission at temperatures of a few hundred eV and 4 keV, which indicated that at least some of the high temperature emission observed by *OSO*-8 and *Uhuru* originated very near Eta Car. A more recent *Ginga* observation (Koyama et al.1990) of the Carina Nebula region determined a temperature for the hot gas of 4.1 keV, while a comparison of the *Ginga* and SSS fluxes suggested that some of the hot emission originated over an extended region of a few tenths of a degree in size. The Carina Nebula was also the target of a BBXRT pointing, which provided the first moderate resolution measure of the iron line region.

2. THE *ROSAT* OBSERVATIONS

The Carina Nebula was observed by both the PSPC and the HRI on *ROSAT*. The PSPC observations took place in June and December 1992 for a total of 37 ksec. The HRI observation in hand consists of a 12 ksec observation obtained in July 92. Another 38 ksec of HRI time awaits scheduling and processing. The purpose of these observations was to determine temperature variations in the diffuse emission, to determine the extent of the 4 keV component, to examine the X-ray luminosity function of the hot stars and look for variability in the point sources, and to resolve the extent and spectral distribution of the hot gas in and around Eta Car. The PSPC greyscale image is shown in Fig. 1.

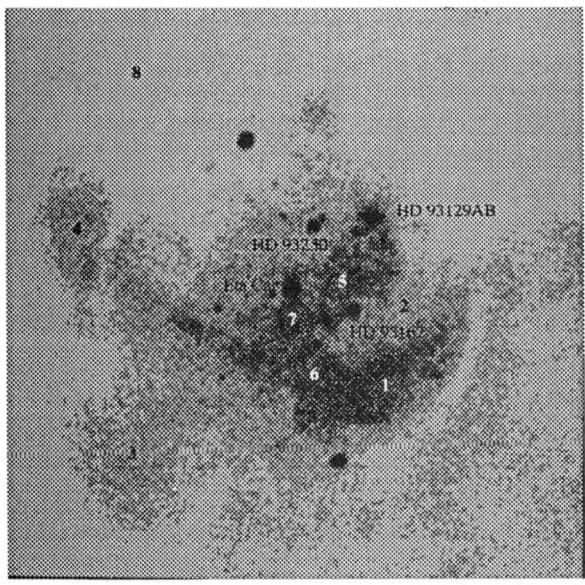

Figure 1. PSPC image of Eta Car. North is to the top, east to the left. Several bright stars and diffuse emission are apparent. Bright stars and regions of diffuse emission used in the spectral analysis are marked.

3. ANALYSIS OF THE DIFFUSE EMISSION

We extracted spectra from 7 regions within the diffuse emission indicated in Fig. 1. We used the spectrum from region 8 (which is outside most of the diffuse emission) as a measure of the background (a combination of scattered solar X-rays, charged particles and cosmic background not associated with the Carina nebula). We modeled the spectrum from region 8 and included this model as a fixed background component in our spectral modeling of the other 7 regions. Derived temperatures, columns and emission measures are given in Table 1. The average number density n is $n = \sqrt{EM/V}$, where EM is the emission measure and V is the volume of the hot gas. Because we do not know the depth of the hot gas along the line of sight, the volume of the hot gas is not known and thus we cannot determine the density of the hot gas accurately. In order to crudely estimate densities, we assumed that the hot gas uniformly filled the volume obtained by revolving the spectral extraction region (which was either a circular or elliptical region) around one axis. The derived densities are given in table 1.

Table 1. Analysis of Diffuse Emission Spectra.

#	N_H	Log(T)	Log(EM)	Density	Surface Brightness	χ^2/Nbins
	(10^{22} cm^{-2})	(K)	(cm^{-3})	(cm^{-3})	cgs	
1	0.26	6.62	56.82	0.66	1.38E-13	57/33
2	0.45	6.37	56.85	0.90	3.11E-14	28/33
3	0.64	6.21	58.11	2.67	4.61E-14	31/33
4	0.63	6.43	57.32	1.08	4.94E-14	43/33
5	0.76	6.21	57.94	7.17	8.02E-14	46/34
6	0.20	6.62	56.18	0.72	9.97E-14	35/33
7	0.56	6.43	58.08	0.62	9.59E-14	219/33
2 Temperature Fits						
7	0.20	7.65	56.66	0.12	1.05E-13	37/33
	0.21	6.58	57.19	0.14		
1	0.24	6.91	55.63	0.17	1.39E-13	55/33
	0.20	6.63	56.67	0.16		
Total	0.54	6.93	57.27	0.01	7.20E-14	69/31
	0.20	6.50	57.73	0.04		

Most regions were adequately fit by 1 temperature models with a temperature of a few million degrees. An exception to this is region 7, a 7 arcmin radius circle centered near Eta Car (which excluded Eta Car, HD 93162, HD 93204 and HD93205) which required an additional hot component with T ≈ 4.1E7 K. Thus the PSPC confirms the presence of very hot gas in an extended region of radius at least 7 arc minutes (11 pc at the distance of the Carina Nebula, taken as 2600 pc). However, the surface brightness we derive by extrapolating the ROSAT spectrum into the 2-10 keV Ginga band is 9.3 × 10^{-14} ergs s^{-1} cm^{-2} arcmin^{-2} which is almost a factor of 2 smaller than the Ginga value (1.6 × 10^{-13} ergs s^{-1} cm^{-2} arcmin^{-2}). However, this discrepancy may be due to uncertainties in the extrapolation of the ROSAT spectrum to the Ginga bandpass. Alternatively it may indicate that other regions of high temperature gas exist beyond the vicinity of Eta Car.

4. EXTENDED EMISSION AROUND ETA CAR

A 15 × 20 arcsec (0.2 × 0.3 pc) shell of X-ray emitting gas surrounding the homunculus was imaged by both the *Einstein* and *ROSAT* HRI. Fig. 2 shows the *ROSAT* HRI image including optical contours of the homunculus. The X-ray shell is aligned with the axes of symmetry of the homunculus. Analysis of the Eta Car PSPC spectrum (which includes the emission from the extended shell) indicates the presence of soft components with a temperatures of 1 and 6 million degrees as well as a much hotter, highly absorbed component with a temperature near 45 million degrees. If the soft emission originates in the extended shell, then the velocities implied by the X-ray temperatures are about 200 km/s. If the gas was ejected from Eta Car, this velocity and distance imply that the gas was ejected more than 100 years ago. Assuming that the hot gas forms an elliptical shell with semi-major axes of 0.10, 0.10, 0.12 pc with a thickness of 0.02 pc, the density of the shell is 157 cm^{-3} and the total mass in the shell is 0.01 solar mass. Fig. 3 shows a comparison of the *Einstein* and *ROSAT* HRI images. A point source of emission at the position of the optical center of light of Eta Car is apparent in the *Einstein* image but not in the *ROSAT* HRI image. This is consistent with the suggestion made by Chlebowski et al. of the existence of a hard absorbed source less than an arcsec from the location of Eta Car. Because the source is hard (T $\approx$ 4 keV) and absorbed ($N_H \approx 10^{22}$ cm^{-2}) it is not visible in the *ROSAT* band, which cuts off at about 2.4 keV. However, this hard source would be detectable by *Einstein* since the *Einstein* bandpass extends to 4.5 keV.

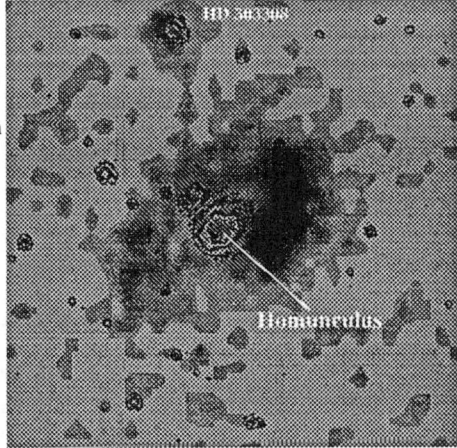

Figure 2. *ROSAT* HRI map of Eta Car and optical image of homunculus. The bright source to the northeast of the homunculus is HDE 303308 which is about 1 arcminute from Eta Car.

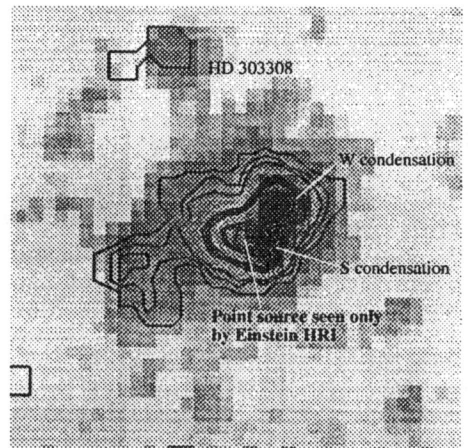

Fig. 3. Comparison of Einstein and ROSAT HRI maps of Eta Car. The S and W condensations are sources of soft X-rays visible in both the Einstein and ROSAT maps. The point source at the optical center of the homunuculus is hard and only visible in the Einstein map.

5. OTHER POINT SOURCES

Results of modeling the X-ray spectra of the 5 brightest O stars are given in Table 2. The PSPC detected close to 20 stars as X-ray sources. We find that X-ray luminosity is correlated with bolometric luminosity with $2 \times 10^{-7} < L_X/L_{bol} < 2 \times 10^{-6}$, which is somewhat larger than the L_X/L_{bol} ratio derived from analysis of the IPC detections.

Table 2. Spectral Analysis of the Bright Stars

Star	PSPC rate cts s^{-1}	NH 10^{22} cm^{-2}	Log(T) K	Log(EM) cm^{-3}	χ^2_ν	Log(L_X)[1] (cgs)
HD93250		0.21	7.20	56.14		
		0.32	6.57	56.09	0.93	33.64
HD93129	0.05	0.30	7.67	56.58		
		0.20	6.89	55.84	1.04	33.73
HD93162	0.15	0.41	7.30	56.73		
		0.42	6.91	56.00	1.48	33.48
HD93403	0.03	0.36	7.64	56.19		
		0.22	6.96	55.85	0.96	33.07
HD93205	0.03	1.47	7.29	56.30		
		0.20	6.39	55.89	1.44	33.59

[1]L_X has been corrected for ISM + circumstellar absorption

References
Becker, R. H., et al., 1976, ApJ, 209, L68.
Chlebowski, T., et al., 1984, ApJ, 281, 665.
Davidson, K., 1971, MNRAS, 154, 415.
Forman, W., et al., 1978, ApJS, 38, 357.
Koyama, K., et al., 1990, ApJ, 362, 215.
Seward, F., and Chlebowski, T., 1982, ApJ, 256, 530.
Seward, F., et al., 199, ApJ, 234, L55.
Warren-Smith, R. F., et al., 1970, MNRAS, 187, 761.

HALOS, STARBURSTS, AND SUPERBUBBLES IN SPIRALS

Joel N. Bregman
Department of Astronomy, University of Michigan, Ann Arbor, MI 48109-1090
jbregman@astro.lsa.umich.edu

ABSTRACT

Detectable quantities of interstellar material are present in the halo of the Milky Way galaxy and in a few edge-on spiral galaxies, largely in the form of neutral atomic gas, warm ionized material, and cosmic rays. Theoretical and observational arguments suggest that million degree gas should be present also, so sensitive ROSAT observations have been made of the large nearby edge-on spiral galaxies for the purpose of detecting hot extraplanar gas. Of the six brightest non-starburst edge-on galaxies, three exhibit extraplanar X-ray emission: NGC 891, NGC 4631, and NGC 4565. In NGC 891, the extended emission has a density scale height of 7 kpc and an extent along the disk of 13 kpc in diameter. This component is close to hydrostatic equilibrium, has a luminosity of 4.4×10^{39} erg s^{-1}, and a mass of 10^8 M$_\odot$. Extended and structured extraplanar hot gas is seen around the interacting edge-on spiral NGC 4631, with X-ray emission associated with a giant loop of Hα and HI emission; spurs of X-ray emission extending from the disk are seen also. Hot gas is expected to enter the halo through superbubble breakout, and a search for superbubbles in normal spiral galaxies have shown that these phenomena are present, but of low surface brightness and are detected in only a few instances.

Unlike the normal spiral galaxies where the gas is bound to the systems, the hot gas in starburst galaxies is being expelled. In M82, the X-ray emission lies in loosely-defined cones that extend to a projected radius of 6.5 kpc. The decrease of the surface brightness with position from the nucleus is consistent with adiabatic expansion of gas flowing from the central region of the galaxy.

1. The Theoretical Expectations For Hot Gas In And Around Spiral Galaxies

An abundance of hot gas is the expected manifestation of an active star formation region. During their brief life, high mass stars (>5 M$_\odot$) drive strong stellar winds at speeds in the thousand km s^{-1} range, which produces a hot bubble of gas in the vicinity of the star. At the end of their lives, these high mass stars become supernovae, depositing into the interstellar medium 3-4 times more energy than during the wind phase, along with a considerable amount of processed materials. Usually, high-mass stars occur in associations where there might be 30 supernova progenitors, and the mean time between supernova explosions in the association is less than the cooling time of the hot material. This leads to the collective growth of a hot bubble (or "superbubble"), which can become large enough to break out of the disk (review of these phenomena by Tenorio-Tagle and Bodenheimer 1988).

Breakout of hot bubbles depends upon the number of supernova per unit time (the input "luminosity", L_*) and the mean density of the gas, but there are three possible scenarios. If the input energy rate L_* is too low or (and) too short in duration, the bubble radiates its thermal energy before becoming large enough to break out of the disk. In this event, one might see bubbles of X-ray emitting material that were roughly spherical and confined to the disk of a spiral galaxy. When the input energy rate L_* and the total energy released over time is sufficient for the bubble size to exceed the thickness of the cold disk of gas, the bubble grows in an irregular fashion, becoming elongated in the direction perpendicular to the disk. When the growth of the bubble in the vertical direction begins to accelerate, the cold confining shell becomes unstable to Rayleigh-Taylor instabilities, and the hot gas escapes the bubble, expanding into the halo or extraplanar region of the galaxy (Tomisaka and Ikeuchi 1986; MacLow, McCray, and Norman 1989).

The fate of hot gas that has escaped the disk depends upon its thermal energy at the time of breakout and the depth of the potential well of the galaxy (Chevalier and Oegerle 1979; Bregman 1980). If the sum of the thermal, kinetic, and gravitational energies is negative, the gas remains bound to the galaxy and will try to establish hydrostatic equilibrium with the galactic gravitational field. The timescale for this to occur is a sound-crossing time over a length scale of the pressure scale height that the gas would have in hydrostatic equilibrium. Gas that cools slowly (cooling time longer than a sound-crossing time) will have a thickness, perpendicular to the disk that is given by hydrostatic theory, for which the length scale depends only on the temperature. This is a simple prediction that can be tested by ROSAT observations, as will be discussed. However, if the cooling time is shorter than a sound-crossing time, the thickness of the X-ray emitting material will be smaller than the hydrostatic value, and the X-ray emission will probably appear to be irregular or structured in appearance (either of these two possibilities is often referred to as galactic fountains).

Gas with positive total energy will escape freely, being unbound to the system. Some thermal energy is converted to kinetic energy and the temperature of the gas drops rapidly through adiabatic expansion. Both of these mechanisms lower the amount of radiated X-ray emission, compared to gravitationally bound gas, so systems with unbound outflows (galactic winds) would be more difficult to detect unless L_* was very high. Several of these theoretical expectations are clear and easy to test, and this has been a goal of X-ray astronomy.

2. The Observational Picture Prior To ROSAT

There were abundant X-ray observations of individual supernova remnants prior to ROSAT, but the observations of individual bubbles were less plentiful. Only a few superbubbles were known, and all but one in our Galaxy. In our galaxy, the Sun seems to lie in a hot low density region of size 100 pc (Fried et al. 1980; Cox and Reynolds 1987 and references therein). It is generally easier to study superbubbles that we do not live in, such as the Cygnus superbubble (Cash et al. 1980), which has a diameter of 450 pc, a luminosity of 5×10^{36} erg s^{-1}, and a temperature of 2×10^6 K. Singh et al. (1987) found a large superbubble in the LMC (1.0×1.4 kpc) associated with a shell of optical nebulosity (Shapley III) with an X-ray luminosity of 3.4×10^{37} erg s^{-1}; this lies in an HI hole (also, see Wang and Helfand 1991a,b).

On theoretical grounds, it is expected that a neutral shell of gas surrounds a superbubble. Giant HI shells and partial shells with radii of 100–1000 pc are found with relative ease (compared with X-ray discoveries) and have been seen in our galaxy by Heiles (1979, 1984) and Colomb, Poppel, and Heiles (1980). In M31 (Brinks and Bajaja 1986) and in M33 (Deul 1988; Deul and den Hartog 1989), "holes" in the HI disk are found with sizes exceeding 100 pc. It has been suggested that these shells are the HI signatures of superbubbles (e.g., McCray and Kafatos 1987), although for a convincing case to be made, one should also find a bubble of hot X-ray emitting gas within the HI shells.

Regarding the breakout of gas around normal galaxies, previous X-ray observations have been disappointing. None of the normal edge-on galaxies observed with the Einstein Observatory displayed detectable emission (Bregman and Glassgold 1982; Fabbiano 1989). Furthermore, spiral disks, when viewed face-on, showed little evidence of diffuse emission that might be associated with an unresolved ensemble of hot gas bubbles.

The only spiral galaxies that displayed extended X-ray emission were starburst galaxies. In these systems, the energy input rate L_* is estimated to be very high (0.1 SN yr^{-1}) so that hot gas is produced in abundance and with enough energy to flow outward as a wind. The best examples of these starburst galaxies are M82 and NGC 253, and we will concentrate on the former as an example. The X-ray map made with the HRI on the Einstein Observatory (Watson, Stanger, and Griffiths 1984) shows several point sources distributed through the star-forming region (about 30″ in size) as well as a diffuse X-ray halo. This X-ray halo is extended along the minor axis (3′ toward the SE, 1.7–2.5′ toward the NE) much more so than along the major axis (half-width < 1.4′). Deeper, low resolution imaging with the Einstein IPC suggests that the X-ray halo may extend as far as 9′ from the nucleus (Fabbiano 1989).

The recent observations with the ROSAT HRI and PSPC have revised our perspective of these phenomena, especially in the areas of edge-on spiral galaxies and starburst galaxies.

3. The New Perspective Provided By ROSAT

3.1 Superbubbles

Observations of superbubbles in our Galaxy have been assisted greatly by the ROSAT All-Sky Survey, which have been discussed in detail in this meeting by Snowden (1994) and Plucinsky et al. (1994). For normal external spiral galaxies, several of the presentations at this meeting indicate that diffuse emission is present, often in spiral arm regions where star formation activity is high (M33, Long et al. 1994; NGC 6946, Schlegel 1994; M101, Murphy and Chu 1994; the LMC, Wang 1994). Individual supernovae are seen occasionally, especially in the nearby systems, and several star forming associations are detected, sometimes extended on the size scale expected from superbubbles (M33, Schulman and Bregman 1994; M101, Murphy and Chu 1994). Most of these superbubbles are weak detections, so the ROSAT observations will serve to determine the high end of their luminosity function.

3.2 Extraplanar Gas Around Spiral Galaxies With Normal Star Formation

The most direct and unambiguous approach for detecting halo or extraplanar gas is to observe edge-on galaxies without warps, or at least without warps that are projected onto the inner part of the galaxy, where extraplanar gas is expected to be most common. Also, one wishes to observe "normal" galaxies, as opposed to interacting systems, a qualification that will simplify the interpretation, when emission is detected. Of the nine edge-on galaxies that have been detected, three are starburst systems (NGC 3079, NGC 2146, and NGC

3628), another is interacting with neighboring systems (NGC 4631), and the remaining five are nearly normal galaxies (NGC 891, NGC 4244, NGC 5907, NGC 4565, and NGC 5529). These galaxies are the nearest edge-on galaxies and are typically $5'$–$15'$ in optical diameter.

Many of these edge-on galaxies were observed, beginning 15 years ago, in the radio region in an effort to search for radio halos and evidence that magnetic fields and relativistic particles could rise above the thin disk of cold interstellar gas. Some, such as NGC 891 and NGC 4631 have prominent radio halos extending a few kiloparsecs above the disk (Sancisi and Allen 1979; Rupen 1991). Many of these same galaxies were observed in the optical region in the light of Hα, and considerable diffuse emission was detected in NGC 891 and NGC 4631, but not in NGC 4244, NGC 5907, or NGC 4565 (Rand, Kulkarni, and Hester 1990).

Many, if not most galaxies fail to show extraplanar X-ray emission. PSPC observations indicate null detections of X-rays from NGC 4244 and NGC 5529, while emission from the plane is detected from NGC 5907 (Frenk, private communication; Pietsch 1993). However, three systems show evidence for extraplanar gas – NGC 891, NGC 4631, and NGC 4565. The case of NGC 4631 is discussed separately in these proceedings by Walterbos (1994), but to summarize the morphological appearance, the extraplanar gas is semi-spherical, but with localized bright regions, possibly caused by filaments. The luminosity is 6×10^{39} erg s^{-1}, and the temperature of the gas is 3.5×10^6 K, although there may be a cooler gas component, which is difficult to quantify. A second galaxy with diffuse emission is NGC 4565 (Pietsch 1993), but these data have been obtained only recently and a detailed analysis is not yet complete.

The third galaxy with detected extraplanar X-ray emission is NGC 891, a spiral galaxy that is nearly an exact twin of our own Milky Way galaxy, with a rotational velocity of 225 km s^{-1}, an inclination of 88.6–90°, and a distance of 10 Mpc. NGC 891 has a more extensive Hα halo than any of the other normal or near-normal galaxies, and there is a close coincidence between the Hα halo and the radio halo. Our 25 ksec PSPC observation of NGC 891 revealed a few point sources in the disk (the brightest being SN 1986J; Bregman and Pildis 1992) plus extended diffuse emission with a FWHM of $100''$ (4.8 kpc) and a maximum detectable halo size of twice that value (Bregman and Pildis 1994). The diffuse emission extends 130–$140''$ (6.5 kpc) along the disk relative to the center of the galaxy (Fig. 1), and there is an additional "plume" of emission extending about $130''$ (6.1 kpc) above the disk and from the nuclear region (this plume has no equivalent Hα or radio feature). This X-ray emission has a similar, if not somewhat larger extent when compared to the Hα and radio halo, suggesting that the various types of interstellar medium are cospatial in a general sense.

Spectral fits to the data require a two temperature fit, where the lower temperature is approximately 0.3 keV (3.6×10^6 K; associated luminosity is 4.4×10^{39} erg s^{-1}) while the higher temperature is > 2 keV ($>2.4 \times 10^7$ K; associated luminosity of 5.2×10^{39} erg s^{-1}). We interpret this lower temperature component as originating from the hot gas while the higher temperature component originates from an ensemble of X-ray binary stars in the disk. For a filling factor of unity, the hot gas has a density scale height of 7 kpc and a density near the disk of 2×10^{-3} cm^{-3}, which leads to a total hot gas mass of 1×10^8 M$_\odot$. This gas is most likely bound to NGC 891 in that the density scale height is the value expected for gas in hydrostatic equilibrium.

168 Halos, Starbursts, and Superbubbles in Spirals

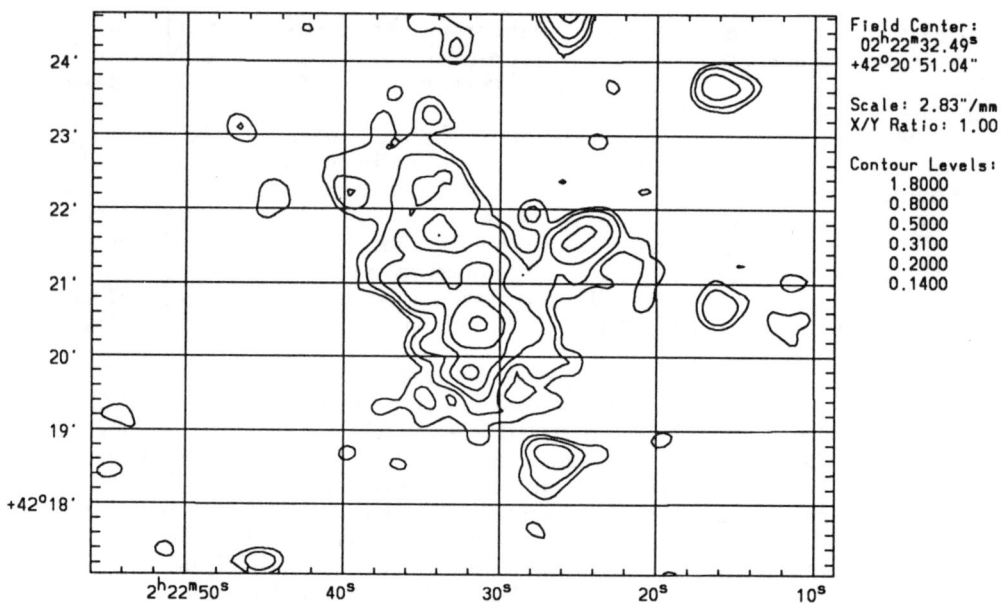

Figure 1. The PSPC contour map of NGC 891 after removal of the point source SN 1986J. The midplane of this edge-on galaxy is aligned N/NE - S/SW and the X-ray emission has a FWHM perpendicular to the disk of 100'' (5kpc).

These observations reveal a galactic fountain in the hydrostatic limit. That is, the cooling time of the gas is longer than a sound-crossing time, so it is able to approach hydrostatic equilibrium before cooling. The pressure in the hot gas is $P/k \sim 14,000$ K cm^{-3}, which is similar to the value expected in the disk, so there appears to be pressure balance between the halo and disk, which would be expected if the flow time for hot material out of the disk occurred on a time shorter than the cooling time (the porosity of the disk is high enough to permit hot gas to flow into the halo easily). The cooling rate of the hot gas is approximately 0.1 $M_\odot$ yr^{-1}, which would recycle the disk gas in about a Hubble time.

It is gratifying that ROSAT has been able to, for the first time, detect galactic fountains in disk galaxies. However, it brings to light an important problem – why are these halos so faint (low luminosity)? The energy input from the ensemble of supernovae in NGC 891 (similar arguments can be made for NGC 4631) is 1×10^{42} erg s^{-1}, and most of the energy is expected to go into thermal energy. However, the observed X-ray luminosity from halo gas is $<$ 1% of the energy released by supernovae. We can rule out the possibility that 10^6–10^7 K gas is trapped in the disk since it would produce face-on spirals with very luminous diffuse emission, but face-on disks are faint as well (Fabbiano 1989). The other possibility is that hot gas is indeed produced, but it radiates most of its thermal energy at a temperature $< 10^6$ K, where ROSAT would have difficulty detecting such emission.

3.3 Starburst Galaxies

A leading issue in the study of starburst galaxies is how the production of hot gas evolves in time, as this is expected to have profound consequences on star formation in the core and on the evolution of the superwind. Regarding star formation, the question can be raised as to whether the production of hot gas compresses the existing cold material, leading to an enormous enhancement of star formation. That is, does the hot gas initiate the starburst event, or it the hot gas merely a byproduct, having little influence on the star formation scenario? The second issue addresses the breakout of the hot material and its interaction with its environment. Some models suggest that, after breakout, the hot material flows along a column aligned with the minor axis, while other models suggest that the flow will be spherically symmetric (Chevalier and Clegg 1985; Tomisaka and Ikeuchi 1988). Since thorough discussions of a variety of starburst galaxies were presented in this conference by Heckman (1994) and Dahlem (1994), this contribution will describe results and interpretations of two other famous starburst galaxies, M82 and IC 342.

The early evolution of the hot gas in a starburst is expected to be spherical in nature (while still confined to the disk) and near pressure equilibrium with its surrounding. The thickness of the disk is typically 300 pc, so at a distance of the nearest starburst galaxies (3–5 Mpc), the size of a young hot bubble would be only 10–20″, which demands the use of the HRI if morphology is to be studied.

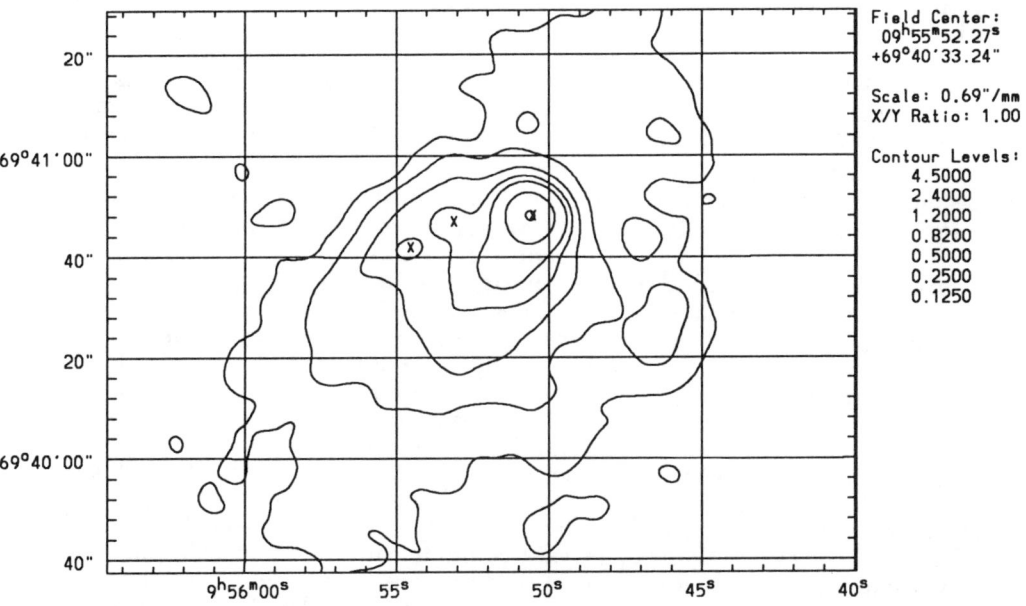

Figure 2a. The ROSAT HRI contour map of the center of M82 shows three point sources (marked by X) plus diffuse emission extending along the minor axis of the galaxy, which is in the N/NW direction.

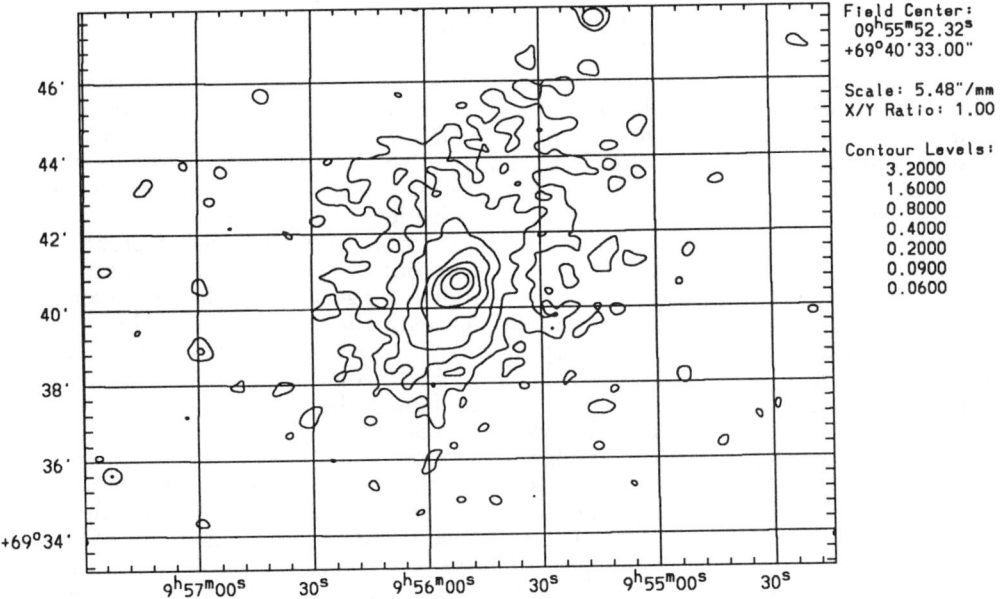

Figure 2b. The ROSAT HRI map of M82, smoothed to 20″ resolution. The emission is seen to extend to at least 6′ to the north and to nearly 5′ to the south.

Our HRI observation of the starburst galaxy IC 342 reveals X-ray emission from the nuclear region that is fit with a point source plus extended diffuse emission of radius 200 pc (Bregman, Cox, and Tomisaka 1993). The total luminosity from this central region is 1.5×10^{39} erg s^{-1}, of which half is due to the diffuse emission. For a filling factor of unity, the density in this diffuse gas is 0.2 cm^{-3}, and the pressure is $P/k \sim 10^6$ K cm^{-3}, which places this material close to the pressure of the molecular gas (Turner and Hurt 1992).

Based upon the bubble size, luminosity, and density, its age was inferred from theory and found to be quite young, approximately 2×10^5 yr. It is still confined to the disk and considerably smaller in extent than the intense molecular emission, which forms a bar nearly an arcminute in length (1.3 kpc). This may suggest that star formation occurs over a region large compared to the X-ray emission, in which case the X-ray emission does not drive the starburst event, although it may enhance the star formation rate locally.

In M82, we find a more evolved starburst galaxy in which the hot gas bubble has broken out of the disk and is flowing away from the galaxy. Our HRI observation (Fig. 2a,b; Bregman, Schulman, and Tomisaka 1994), and the PSPC observation (Watson, private communication) reveal a bright core of X-ray emission, which extends in both directions along the minor axis to a maximum distance of about 7′ (7 kpc) from the nucleus. The HRI observation of the central region (Fig. 2a) shows three point sources present plus intense diffuse emission (point sources in M82 and their variability are discussed in

Collura et al. 1994). For the inner 1′ (1 kpc), the X-ray and Hα emission are closely correlated, and for a temperature near 1 keV, the hot gas pressure is $10^7.5$ K cm^{-3}, which is comparable to that deduced from the optical emission lines (Heckman, Armus, and Miley 1990), suggesting near pressure equilibrium between the two phases.

The extended emission is largely confined to cones that are nearly along the minor axis, although these are not cones of constant opening angle (Fig. 2b). The half-width at half-maximum of the cone, measured in degrees, decreases from 55–60° in the inner 0.5′ to 20–30° at a distance of about 1.5′ from the center. At distances beyond about 2′, the emission is consistent with having a constant opening angle. This change in the shape of the jet is consistent with it being confined by some surrounding medium close to the galaxy, but freely expanding further out.

Jet confinement and free expansion make different predictions for the change of X-ray surface brightness along the jet. Neglecting bandpass effects, the projected X-ray surface brightness, $S_x(r)$, in a freely expanding jet will depend upon the distance along the jet as $S_x(r) \propto r^{-3}$. In contrast, for a confined jet, the surface brightness is related to the width of the jet (d) by $S_x \propto d^{-3}$, so S_x can decrease more slowly than r^{-3}. For the case of M82, the surface brightness in the southern side decreases as expected for an adiabatically expanding flow in a partially confined jet. This may be true in the northern jet as well, although the analysis is complicated from absorption effects by the M82 disk and the morphology of the X-ray jet. Finally, adiabatically expanding gas should lead to a temperature drop, and increasingly softer emission is found outward along the jet (Watson, private communication; also seen in other starburst galaxies; see Heckman 1994). These observations support the outflow picture predicted from hydrodynamical considerations and indicate that a combination of confinement and free expansion offers the best explanation for the phenomenon.

JNB would like to acknowledge support for these research activities by NASA grants NAGW-2135 and NAG5-1955.

References

Bregman, J.N. 1980, ApJ, 236, 577.
Bregman, J.N., Cox, C.V., and Tomisaka, K. 1993, ApJ, 415, L79.
Bregman, J.N., and Glassgold, A.E. 1982, ApJ, 263, 564.
Bregman, J.N., and Pildis, R.A. 1992, ApJ, 398, L107.
Bregman, J.N., and Pildis, R.A. 1994, ApJ, in press.
Bregman, J.N., Schulman, E., and Tomisaka, K. 1994, ApJ, submitted.
Brinks, E., and Bajaja, E. 1986, A&A, 169, 14
Brinks, E., and Shane, W.W. 1984, A&A Sup, 55, 179.
Cash, W., Charles, P., Bowyer, S., Walter, F., Garmire, G., and
Riegler, G. 1980, ApJ, 238, L71.
Chevalier, R.A., and Clegg, A.W. 1985, Nature, 317, 44.
Chevalier, R.A., and Oegerle, W.R. 1979, ApJ, 227, 398.
Collura, A., Reale, F., Schulman, E., and Bregman, J. 1994, ApJ, in press.
Cox, D.P., and Reynolds, R.J. 1987, ARAA, 25, 303.
Dahlem, M. 1994, in "The 1st Annual ROSAT Science Symposium and Data Analysis Workshop", held Nov. 8-10, 1993 at the Univ. of Maryland, in press.
Deul, E.R., and van der Hulst, J.M. 1986, A&A Sup, 67, 509.

Fabbiano, G. 1989, ARAA, 27, 87.
Fried, P.M., Nousek, J.A., Sanders, W.T., and Kraushaar, W.L. 1980, ApJ, 242, 987.
Heckman. T.M 1994, in "The 1st Annual ROSAT Science Symposium and Data Analysis Workshop", held Nov. 8-10, 1993 at the Univ. of Maryland, in press.
Heckman, T., Armus, L., and Miley, G.K. 1990, ApJSS, 74, 833.
Heiles, C. 1979, ApJ, 229, 533.
Heiles, C. 1984, ApJ Suppl., 55, 585.
Long, K.S., Gordon, S.M., Blair, W.P., and Charles, P.A. 1994, in "The 1st Annual ROSAT Science Symposium and Data Analysis Workshop", held Nov. 8-10, 1993 at the Univ. of Maryland, in press.
MacLow, M.-M., McCray, R., and Norman, M.L. 1989, ApJ, 337, 141.
McCray, R., and Kafatos, M. 1987, ApJ, 317, 190.
Murphy, R., and Chu, Y.-H. 1994, in "The 1st Annual ROSAT Science Symposium and Data Analysis Workshop", held Nov. 8-10, 1993 at the Univ. of Maryland, in press.
Pietsch, W. 1993, in the "The ISM In Galactic Halos: Current Views", held at STScI, August 10-12, 1993.
Plucinsky, P.P., Snowden, S.L., Aschenbach, B., Egger, R., Edgar, R., and McCammon, D. 1994, in "The 1st Annual ROSAT Science Symposium and Data Analysis Workshop", held Nov. 8-10, 1993 at the Univ. of Maryland, in press.
Rand, R.J., Kulkarni, S.R., and Hester, J.J. 1990, ApJ, 352, L1.
Rupen, M.P. 1991, AJ, 102, 48.
Sancisi, R., and Allen, R.J. 1979, A&A, 74, 73.
Schlegel, E.M. 1994, in "The 1st Annual ROSAT Science Symposium and Data Analysis Workshop", held Nov. 8-10, 1993 at the Univ. of Maryland, in press.
Schulman, E., and Bregman, J.N. 1994, in "The 1st Annual ROSAT Science Symposium and Data Analysis Workshop", held Nov. 8-10, 1993 at the Univ. of Maryland, in press.
Singh, K.P., Nousek, J.A., Burrows, D.N., and Garmire, G.P. 1987, ApJ, 313, 185.
Snowden, S.L. 1994, in "The 1st Annual ROSAT Science Symposium and Data Analysis Workshop", held Nov. 8-10, 1993 at the Univ. of Maryland, in press.
Tenorio-Tagle, G., and Bodenheimer, P. 1988, ARAA, 26, 145.
Tomisaka, K., and Ikeuchi, S. 1986, PASJ, 38, 697.
Tomisaka, K. and Ikeuchi, S. 1988, ApJ, 330, 695.
Turner, J.L., and Hurt, R.L. 1992, ApJ, 384, 72.
Wang, Q.D. 1994, in "The 1st Annual ROSAT Science Symposium and Data Analysis Workshop", held Nov. 8-10, 1993 at the Univ. of Maryland, in press.
Wang, Q., and Helfand, D.J. 1991a, ApJ, 373, 497.
Wang, Q., and Helfand, D.J. 1991b, ApJ, 379, 327.
Watson, M.G., Stanger, V., and Griffiths, R.E. 1984, ApJ, 286, 144.

THE SOFT X-RAY HALO OF THE SPIRAL GALAXY NGC4631

René A.M. Walterbos & Michael F. Steakley
New Mexico State University, Astronomy Department, Las Cruces, NM

Email ID
rwalterb@nmsu.edu

Q. Daniel Wang
University of Colorado, JILA/CASA, Boulder, CO

Colin.A. Norman
Johns Hopkins University, Dept. of Physics. & Astron., Baltimore, MD

Robert Braun
Netherlands Foundation for Research in Astronomy, Dwingeloo, NL

ABSTRACT

We present ROSAT PSPC observations of the close to edge-on spiral galaxy NGC4631. This vigorously star forming galaxy shows extented X-ray emission perpendicular to the plane, out to about 6 to 8 kpc. The spatial extent is largest at soft X-ray energies. The total X-ray luminosity of hot gas can be easily supplied by star formation in the disk, and it is likely that the halo is due to outflow of hot gas from the inner disk. Spectral analysis of the X-ray data shows that part of the halo emission may be quite cool, well below 10^6 K. We briefly discuss implications of these results.

1. Introduction

Ever since Spitzer (1956) predicted that halos of spiral galaxies might contain hot gas, this component of the interstellar medium (ISM) has stimulated much interest. Models of the interaction of supernovae with the ISM (Cox & Smith 1974, McKee & Ostriker 1977), and of the disk-halo interface (*e.g.* Bregman 1980, Norman & Ikeuchi 1989) have demonstrated that it is quite likely for such a hot component to exist; the surprising result is actually that it has been quite difficult to find it! (*e.g.* Fabbiano 1989, Fabbiano & Trinchieri 1987). NGC4631, a close to edge-on Sc spiral galaxy, may be the most likely target for finding a hot halo. It has vigorous star formation, as judged from its far-infrared luminosity (Rice *et al.* 1988) and its extensive radio halo (*e.g.* Hummel & Dettmar 1990). Its central 4 kpc of the disk shows a few long Hα filaments, indicative of disk-halo interaction, and possibly a thick diffuse ionized gas layer (Rand *et al.* 1992, Walterbos *et al.* 1993, 1994). In addition, the galaxy is only 7.5 Mpc distant, and its high Galactic latitude implies a low foreground HI column. The latter is crucial for the detection of X-rays below 0.4 keV, the range where hot gas is expected to emit most of its energy. However, NGC4631 may not be a proto-typical late-type spiral because it is interacting with two neighboring galaxies, producing extensive HI tidal tails (Rand & van der Hulst 1993), a radio halo, and enhanced star formation in the disk.

© 1994 American Institute of Physics

The Soft X-ray Halo of the Spiral Galaxy NGC4631

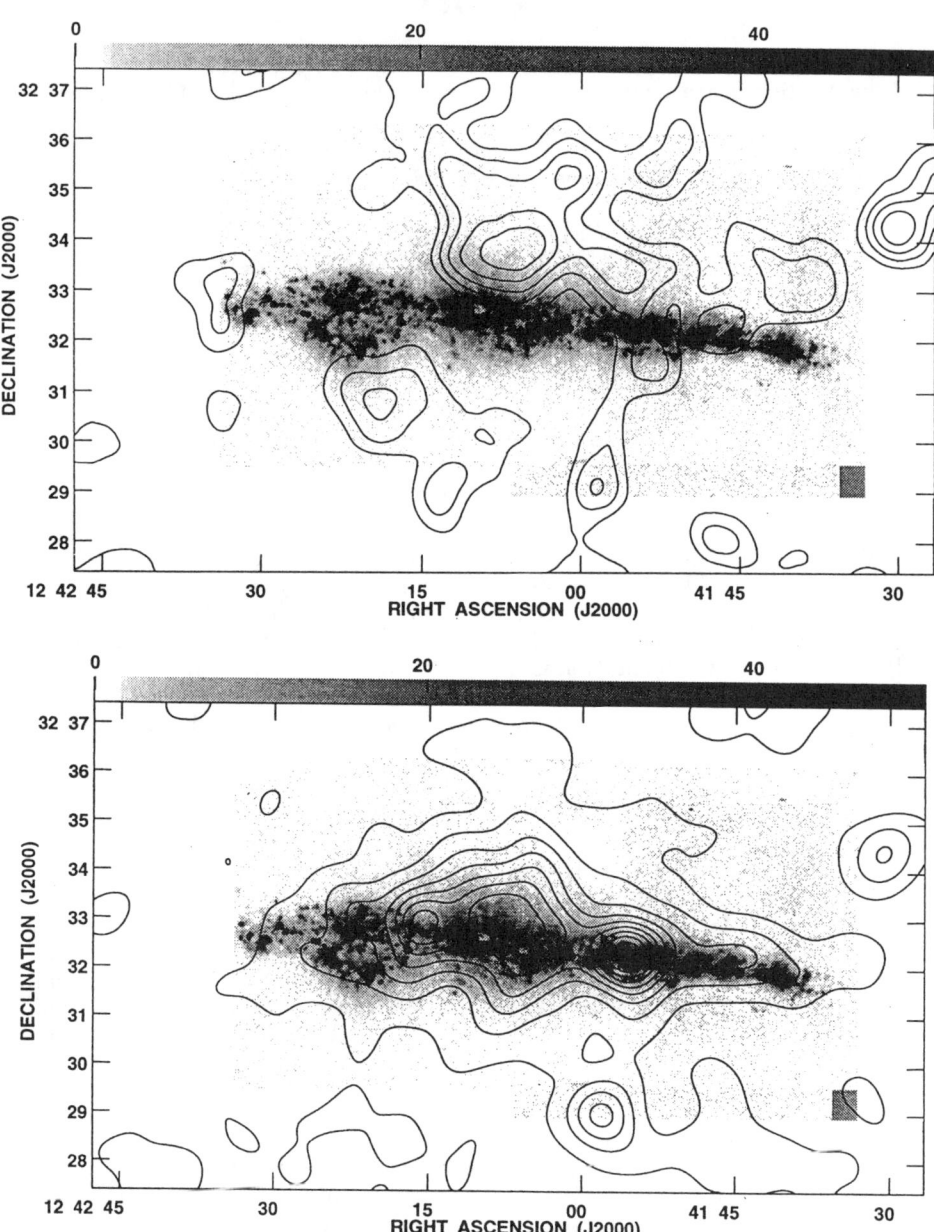

Figure 1. (Top) PSPC image of NGC4631 in the range 0.2-0.4 keV, superposed on an Hα image (grey scale; note that the spatial alignment is still somewhat uncertain). Contours at 1, 2, 3, 4, and 5 times 10^{-6} counts s^{-1} pix^{-1} above background. (Bottom) The X-ray emission in the range 0.5-0.9 keV. Contours start at 2 and increase in steps of 2 (same units). One arcmin corresponds to 2.2 kpc. Data smoothed to 45 arcsec.

2. Observations and results

We obtained a 23 ksec ROSAT PSPC observation of NGC4631, of which about 16 ksec was left after editing the data to delete periods with enhanced non-cosmic background count rates (Snowden et al. 1994). Complete results of this observation are presented by Wang et al. (1994) and by Walterbos et al. (1994). Data below 0.2 keV suffer badly from electronic ghosts, so these were not used in creating images, but only in the spectral analysis. About 1200 counts were detected, 40% each from disk and halo, and 20% from a bright source west of the nucleus (Fig. 1) also seen by Einstein (Fabbiano & Trinchieri 1987). This source appears to coincide with a giant HI supershell (Wang et al. 1994). We plot X-ray contours over an Hα image in Fig. 1 for the soft and medium energy ranges. The X-ray emission, although weak (the highest contour in Fig. 1 (top) is about 5 sigma), is clearly extended above the disk, out to 6 or 8 kpc. The extra-planar X-ray emission is prominent above the region of most intense Hα emission, near the "double worm" in Hα (Rand et al. 1992). The halo emission appears asymmetric, which may be intrinsic or due to the orientation at which we see NGC4631 (Walterbos et al. 1994). The medium energy image is also significantly extended in z-direction. The hard X-ray image (not shown) is more confined to the disk, which implies that the extra-planar *medium* X-rays are also likely due to hot gas, not to a population of hard X-ray binaries.

3. Spectral analysis and discussion

The spectra for the halo and disk regions are dramatically different (Fig. 2). A two-component fit to the disk spectrum is required, with one component hotter than 2.4×10^7 K, probably due to hard X-ray binaries, and a cooler component of 3.5×10^6 K, which we tentatively identify with hot gas in the disk. A foreground column of 10^{21} atoms cm^{-2} was assumed in the fit; the derived temperatures are not extremely sensitive to this assumption. The halo spectrum appears more complicated. A single temperature fit (middle panel) works well, and indicates a temperature of 2.8×10^6 K, similar to the temperature of hot gas in the disk. The total luminosity for the hot gas (disk and halo) is about $5 - 7 \times 10^{39}$ erg s^{-1}. This luminosity is easily supplied by the energy input from supernovae, which amounts to about 10^{42} erg s^{-1}. The hard disk component contributes about 4×10^{39} erg s^{-1}.

The single temperature fit to the halo gas is appealing but there may be a problem: the fit requires a foreground column of less than 3.3×10^{19} atoms cm^{-2}, which does not agree with the observed foreground HI column of 1.2×10^{20} atoms cm^{-2}. There is likely also a substantial H$^+$ foreground column (*e.g.* Reynolds 1991). If we assume a foreground column of 1.8×10^{20} atoms cm^{-2}, a single temperature fit to the halo spectrum is rejected at the 95% confidence level (top panel). In that case, a significant fraction of the halo gas must be cool, about 3×10^5 K. Emission from this gas would be heavily absorbed, so the excess in Fig. 2 would imply an order of magnitude larger (compared to the several million K gas) X-ray luminosity. If correct, this might be where a large fraction of the supernovae energy goes. It would also explain the previous lack of success in detecting hot gas in spirals. Only the favorable conditions for NGC4631 give us a hint of this possibility; it may be hard to find other galaxies where ROSAT data could confirm the presence of large amounts of "cool" hot gas.

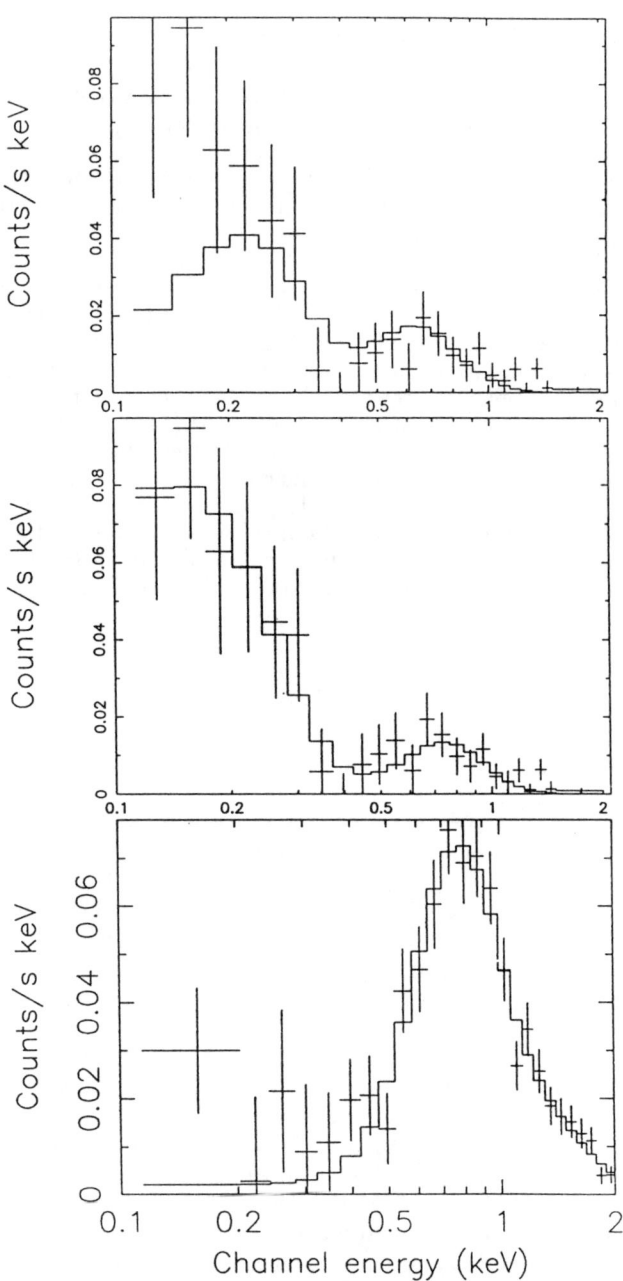

Figure. 2. Spectral fits to the halo (top and middle) and disk X-ray data (bottom). The two halo fits correspond to the two options discussed in the text; the halo is clearly quite soft. The disk spectrum is harder, partially due to absorption but also to a contribution of hard X-ray sources.

A gratifying result is that we have detected X-ray emission from hot gas in both the disk and halo of NGC4631, confirming a long-suspected ISM component, and models of disk-halo interaction (Bregman 1980, Norman & Ikeuchi 1989). On the other hand, the strong tidal interaction does not make NGC4631 a typical spiral, so extrapolation of our results to other galaxies may not be entirely appropriate. Outflows of hot gas have been seen in starburst galaxies (see Bregman and Heckman, these proceedings) and although NGC4631 has not typically been placed in this class, the concentration of halo gas near the central disk might indicate that also here it is especially the central (but not necessarily nuclear) region that is responsible for powering the halo. Thus it remains uncertain how many "normal" spirals will have hot halos. The other good candidate, NGC891 (Bregman, these proceedings, and Bregman & Pildis 1993) shows a thick disk of hot gas, but not as extensive as NGC4631, possibly because of the large foreground column which makes it impossible to see a soft X-ray halo. Apart from these two systems, which both show high star formation rates, even ROSAT may have a hard time detecting extended soft X-ray *halos* in other "normal" (*i.e.* non-starburst) edge-on spirals.

Acknowledgements

We are grateful to Richard Rand for making his integrated HI image of NGC4631 available to us. This work has been supported by grants from NASA (NAG 5-1924) and NSF (AST9123777) to RAMW.

References

Bregman, J.N., 1980, ApJ 236, 577
Bregman, J.N., Pildis, R.A., 1993, ApJ , in press
Cox, D.N., Smith, B.W., 1974, ApJL 189, L105
Fabbiano, G., 1989, ARA&A 27, 87
Fabbiano, G., Trinchieri, G., 1987, ApJ 315, 46
Hummel, E., Dettmar, R.-J., 1990, A&A 236, 33
McKee, C.F., Ostriker, J.P., 1977, ApJ 218, 148
Norman, C.A., Ikeuchi, S., 1989, ApJ 345, 372
Rand, R.J., Kulkarni, S.R., Hester, J.J.: 1992, ApJ 396, 97
Rand, R.J., van der Hulst, J.M., 1993, AJ 105, 2098
Reynolds, R.J., 1991, in The Interstellar Disk-Halo Connection, IAU Symp. 144, ed. J.B.G.M. Bloemen, (Dordrecht: Kluwer), p 67
Rice, W.A., Lonsdale, C.J., Soifer, B.T., Neugebauer, G., Koplan, E.L., Lloyd, L.A., de Jong, T., Habing, H.J., 1988, ApJS 68, 91
Snowden, S.L., McCammon, D., Burrows, D.N., Mendenhall, J.A., 1994, ApJ , in press
Spitzer, L., 1956, ApJ 124, 20
Walterbos, R.A.M., Braun, R., Norman, C.: 1993, in Evolution of Galaxies and Their Environment, eds. Hollenbach, Thronson, & Shull, NASA Conf. Publ. 3190, p326
Walterbos, R.A.M., Steakley, M.F., Wang, Q.D., Norman, C.A., Braun, R., 1994, to be submitted to ApJ
Wang, Q.D., Walterbos, R.A.M., Steakley, M.F., Norman, C.A., Braun, R., 1994, submitted to ApJ

COSMOLOGICAL IMPLICATIONS OF ROSAT OBSERVATIONS OF GROUPS AND CLUSTERS OF GALAXIES

L. P. David, C. Jones, and W. Forman
Smithsonian Astrophysical Observatory

Email ID
david@cfa.harvard.edu

ABSTRACT

We have used ROSAT PSPC observations to determine the total gravitating mass in a sample of 9 groups and clusters of galaxies. All of these clusters (spanning a range in gas temperatures of 1.0 to 8 keV) have mass-to-light ratios of $100-150 M_\odot/L_\odot$. This indicates that rich clusters do not contain any more dark matter than individual galaxies. We also find that the ratio of total mass-to-luminous mass (hot gas and stars) uniformly decreases from galaxies, to groups, to clusters. Only 10% of the gravitating mass in galaxies and groups can be accounted for while 30% is observable in rich clusters. This indicates that the universe becomes "brighter" on larger scales not "darker".

1. Introduction

The distribution of matter in the universe and its composition is one of the most fundamental problems in astrophysics. Such information places strong constraints on theories concerning the origin of structure. X-ray astronomy is an excellent tool for probing the universe on scales from galaxies to rich clusters ($10^{12} - 10^{15} M_\odot$). A convenient way of parameterizing the amount of dark matter on different scales is the mass-to-light ratio (see Tremaine 1992 for a recent review). Typical mass-to-light ratios are $3 M_\odot/L_\odot$ in the solar neighborhood, $10 M_\odot/L_\odot$ in the luminous portions of galaxies, and $100 M_\odot/L_\odot$ for the total mass in a galaxy. Estimates of the gravitating mass in clusters of galaxies using galaxy redshifts give $M/L \sim 300 M_\odot/L_\odot$ (assuming an isotropic velocity dispersion). All of these values fall far short of that required if the mass density of the universe is to equal to the critical density ($M/L = 1200 h_{50} M_\odot/L_\odot$; Davis and Huchra 1982). The common assumption to reconcile this discrepancy is that galaxies preferentially form in regions of high density. Such biased galaxy formation scenarios predict that M/L should increase with scale (Blumenthal et al. 1984).

Analysis of x-ray observations of groups and clusters has substantially changed this picture. We will show that galaxies, groups, and clusters all have mass-to-light ratios between $100 - 150 M_\odot/L_\odot$ (we use $H_0 = 50$ km s^{-1} Mpc^{-1} throughout). This indicates that rich clusters do not contain any more dark matter than galaxies, and that the distribution of mass is not biased with respect to the light. We also will show that the fraction of the gravitating mass that is observable (hot gas and galaxies) uniformly increases from galaxies to rich clusters.

2. Mass-to-Light Ratios

The gravitating mass of a cluster can be determined most reliable from x-ray observations. Optical spectroscopic observations provide information only about the dispersion in the component of a galaxy's velocity that is projected along the line-of-sight. In order to estimate the mass of a cluster, assumptions must be made about the distribution of galaxy orbits. Merritt (1987) has shown that even in the well studied Coma cluster, mass estimates can vary by a factor of 7 depending on whether the galaxies' orbits are mostly circular or mostly radial. However, since the hot gas in clusters is collisional, the mass of a cluster can be derived from the equation of hydrostatic equilibrium

$$M_{tot}(<r) = -\frac{kTr}{\mu m_p G}\left(\frac{d\ ln\rho_{gas}}{d\ lnr} + \frac{d\ lnT}{d\ lnr}\right) \quad (1)$$

where the gas temperature and density distributions can be deduced from x-ray observations. The primary difficulty in determining the binding mass of clusters from x-ray missions prior to ROSAT had been the lack of spatially resolved X-ray spectra, required to determine the gas temperature profile.

We have compiled a sample of 9 groups and clusters that were observed by the PSPC. Seven of these have information on the total visual light in the cluster (Arnaud et al. 1992). In the groups and cool clusters ($kT \lesssim 3$ keV), we unambiguously determine the temperature profile of the gas. In general, the gas is essentially isothermal beyond the central cooling flow regions in groups and cool clusters (Ponman and Bertram 1993, David et al. 1994). For the cool clusters in our sample, the gravitating masses are determined from the best fit β model (which gives the density distribution of the gas), and the best fit power-law temperature profile. The gravitating masses of the hotter clusters are determined by assuming an isothermal profile. Henry et al. (1993) have shown that the PSPC observations of A2256 ($kT \approx 7$ keV) are consistent with an isothermal distribution. Hughes (1992) also has shown that the temperature distribution in the Coma cluster is best fit by a isothermal distribution within several core radii, with an adiabatic profile at larger radii. Since we calculate M/L at 1 Mpc in each cluster, we should be well within the isothermal region.

Our analysis shows that M/L is essentially a constant in galaxies, groups, and rich clusters (see Figure 1). This implies that rich clusters do not contain any more dark matter than galaxies. The dark matter in clusters can result simply from stripping dark galactic halos as clusters form through hierarchical merging. This result also indicates that mass follows light over scales from $10^{12} M_\odot$ to $10^{15} M_\odot$. The dashed lines in Fig. 1 are the mass-to-light ratios derived using the velocity dispersions in Zabludoff, Huchra, and Geller (1991), and assuming an isotropic distribution of orbits. In cases where there is good agreement, only galaxies within a region comparable to the extent of the x-ray emission are used to determine the velocity dispersion (typically 30'). In cases where there is a large discrepancy, the galaxy redshifts are accumulated over many degrees. This comparison shows that, in general, the central galaxies in a cluster are virialized with an isotropic velocity dispersion, while the outer galaxies are either on predominantly radial orbits, or are just now falling into the cluster for the first time. Such a situation is evident in the numerical simulations of Evrard (1990) in which the anisotropy parameter increases significantly with radius in clusters at the present time. Previous reports of large mass-to-light ratios in clusters based on velocity dispersions, are primarily due to the inclusion of galaxies which are not yet virialized or are on predominantly radial orbits.

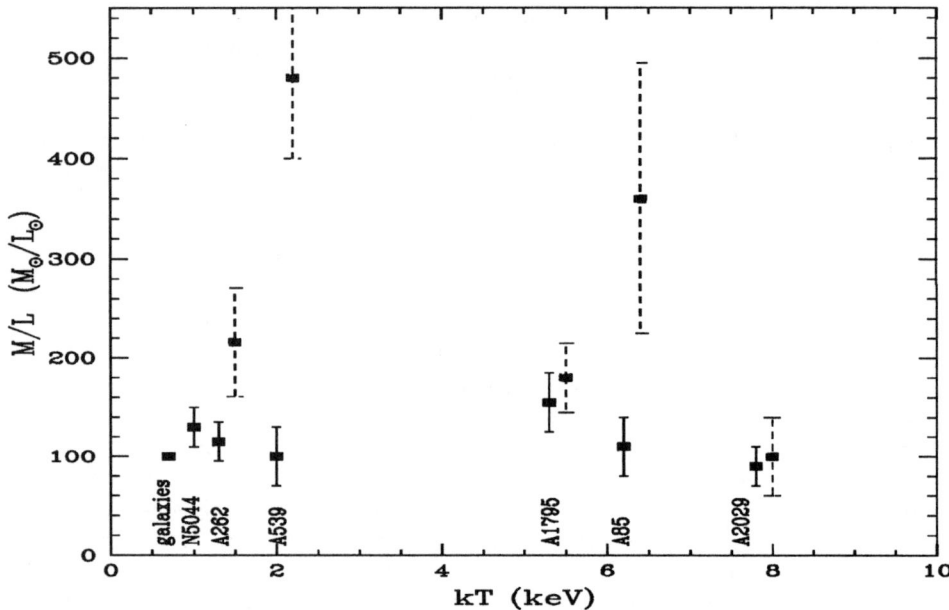

Figure 1: Mass-to-light ratios for our sample of groups and clusters. Solid lines give M/L based on x-ray observations. Dashed lines give M/L based on galaxy redshifts with the assumption of an isotropic velocity dispersion.

3. Ratio of Mass-to-Luminous Mass

We define the luminous mass in clusters as both the hot x-ray emitting gas and the optically luminous portions of galaxies (determined using $M/L = 8 M_\odot/L_\odot$). The ratio of the total mass-to-luminous mass M/M_{lum} is shown in Fig. 2 for our entire sample. Instead of plotting each object as a single point we plot M/M_{lum} as a function of encircled mass (or equivalently radius) in each object. The varying lengths of the curves in Fig. 2 reflect the fact that the x-ray emission is observable to larger radii in clusters compared to groups. It is obvious from Fig. 2 that there is a strong continuity in M/M_{lum} over scales of $10^{12} M_\odot$ to $10^{15} M_\odot$. This indicates that the hierarchical merging process does not completely erase all prior segregation of matter. The decrease with radius of M/M_{lum} for a given object occurs because the hot gas is the most extended mass component. The decrease of M/M_{lum} from groups to rich cluster arises primarily because the x-ray emission is observable to greater radii in rich clusters. On average, the gas fraction increases from 2% in galaxies, to 10% in groups, to 20-30% in rich clusters.

4. Discussion

The above analysis shows that mass-to-light ratios of galaxies, groups, and

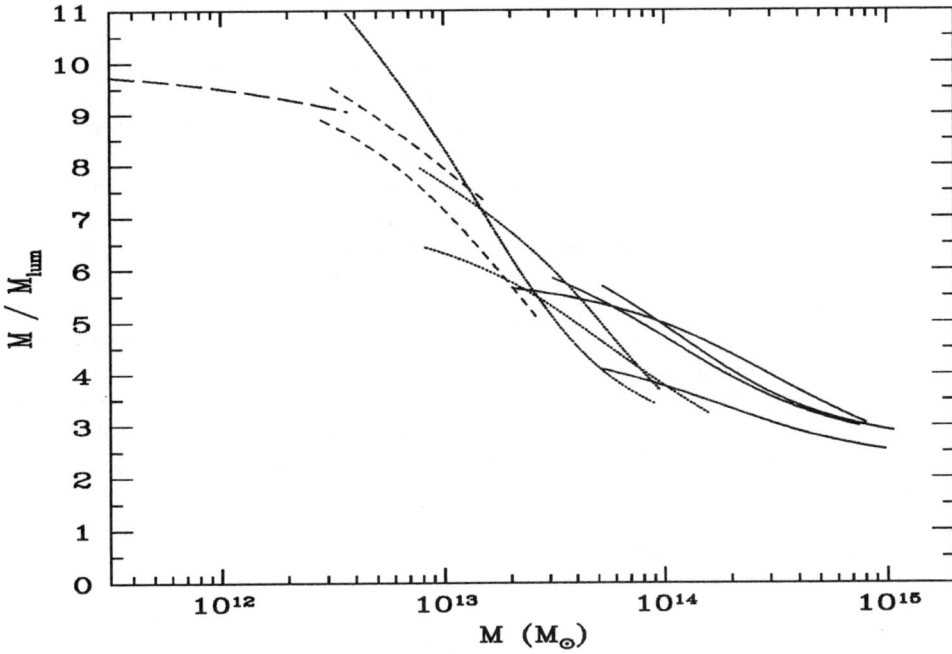

Figure 2: Ratio of total mass-to-luminous mass in the elliptical galaxy NGC 4636 (long dashed line), the groups NGC 5044 and HC62 (short dashed lines), the cool clusters A262, A539, A2589 (dotted lines), and the hot clusters A1795, A85, A2029, A478 (solid lines). For the clusters which do not have photometric results (HC62, A2589, and A478) we assume that 10% of the total mass is contained in the luminous portion of galaxies.

clusters are constant at $100-150 M_\odot/L_\odot$, while M/M_{lum} decreases from 10 to 3 in these objects. Combining these results indicates that the mass follows the light on scales of 10^{12} to $10^{15} M_\odot$, but that the hot gas is more uniformly distributed than the mass and is anti-biased with respect to the light. This could result by heating the gas after it is accreted into clusters (David, Forman, Jones 1991; Metzler and Evrard 1993). Recent numerical simulations by Cen and Ostriker (1993) have shown that heating by galactic winds and photoionization after galaxy formation produces a hot gas component that is anti-biased with respect to the mass.

If the mass continues to follow the light on scales larger than $10^{15} M_\odot$ (i.e., no hot dark matter) than the mass-to-light ratios of our sample of groups and clusters imply that $\Omega = 0.1$. Standard big band nucleosynthesis calculations limit the baryonic component of the universe to (Walker et al. 1991)

$$0.04 < \Omega_{baryons} h_{50}^2 < 0.06 \qquad (2)$$

Our results show that at least 40% of the mass in clusters must be baryonic (this is consistent with the lower limit of 30% obtained from the x-ray observations), and that less than 60% is nonbaryonic.

Prior to the era of x-ray astronomy, the common lore was that the universe becomes "darker" on larger scales. The discovery of large amounts of hot gas in rich clusters (up to 6 times the mass in galaxies; David et al. 1990) and more reliable mass estimates using x-ray observations, has shown that the universe actually becomes "brighter" on larger scales. Only 10% of the gravitating mass in galaxies can be accounted for, while 30% of the mass in clusters is observable. This fraction can only increase with observations by more sensitive x-ray telescopes (i.e., ASCA and AXAF).

The above estimate on Ω assumes that only baryons and cold dark matter exist. The presence of hot dark matter (which is only gravitationally bound on scales greater than rich clusters) could yield mass-to-light ratios consistent with $\Omega = 1$. The fluctuations in the cosmic microwave background detected by COBE show that the amplitude of density fluctuations on a scale of 10° corresponds to a biasing parameter of unity (Wright et al. 1992). This result is in conflict with the standard biased cold dark matter (CDM) scenario. Alternative models include: 1) a mixture of hot and cold dark matter, 2) a low density CDM model with a non-zero cosmological constant, or 3) a tilted CDM model. There are several observations requiring values of Ω near unity on larger scales. A comparison of the QDOT IRAS galaxy redshift survey with measured peculiar velocities indicates that $0.6 < \Omega < 1.1$ on scales of 100 Mpc (Kaiser et al. 1991). Values of Ω near unity are also required to account for the large fraction of clusters with substructure (Richstone et al. 1992).

In conclusion, comparing our results on groups and clusters with results favoring a high Ω on large scales, we arrive at a scenario in which approximately 90% of the mass in the universe is hot, 5% is baryonic, and 5% is cold nonbaryonic dark matter.

References

Arnaud, M., Rothenflug, R., Boulade, O., Vigroux, L., and Vangioni-Flam, E. 1992, A&A, 254, 49.
Blumenthal, G.R., Faber, S.M., Primack, J.R., Rees, M.J. 1984, Nature, 311, 517.
Cen, R., and Ostriker, J. 1993, Ap.J., 417, 404.
David, L.P., Arnaud, K.A., Forman, W., and Jones, C. 1990, Ap.J., 356, 32.
David, L, Jones, C., Forman, W., and Daines, S. 1994, Ap.J., (in press).
Davis, M., and Huchra, J.P. 1982, Ap.J., 254, 437.
Evrard, A. 1990, Ap.J., 349.
Henry, J., Briel, U., and Nulsen, P. 1993, (preprint).
Hughes, J. 1989, Ap.J., 337, 21.
Kaiser, N., Efstathiou, G., Ellis, R., Frenk, C., Lawrence, A., Rowan-Robinson, M., and Saunders, W. 1991, MNRAS, 252, 1.
Merritt, D. 1987, Ap.J., 313, 121.
Metzler, C., and Evrard, A. 1993 (preprint).
Ponman, T.J., and Bertram, D. 1993, Nature, 363, 51.
Richstone, D., Loeb, A., and Turner, E. 1992, Ap.J., 393, 477.
Tremaine, S. 1992, Physics Today, 28.
Walker, T.P., Steigman, G., Schramm, D.N., Olive, K.A., and Kang, H. 1991, Ap.J., 378, 186.
Wright et al. 1992, Ap.J. Lett., 396, L13.
Zabludoff, A., Huchra, J., and Geller, M. 1991, Ap.J. Supp., 74, 1.

CLUMPED X-RAY EMISSION AROUND RADIO GALAXIES IN CLUSTERS: NEW TOOLS FOR INVESTIGATING CLUSTER EVOLUTION

J. Burns, K. Roettiger, J. Pinkney, C. Loken, & S. Doe
Department of Astronomy, New Mexico State Univ., Las Cruces, NM 88003

Email ID
jburns@nmsu.edu

F. Owen
National Radio Astronomy Observatory, P.O. Box O, Socorro, NM 87801

W. Voges
Max-Planck-Institut für Extraterrestrische Physik,
D-85748, Garching, Germany

R. White
NASA Goddard Space Flight Center, Code 932, Greenbelt, MD 20771

ABSTRACT

New ROSAT X-ray imaging of clusters with extended radio galaxies is described. We present evidence for a strong correlation between the positions of radio galaxies and clumped X-ray emission within galaxy clusters. This correlation is interpreted as the result of mergers between clusters. We argue that radio galaxies are good probes of the dynamical state of clusters. In particular, wide-angle tailed radio sources may be the best tool that presently exists for sampling large-scale gas motions in clusters which have undergone mergers. We also present hybrid Hydro/N-body simulations of cluster mergers which show that (1) cluster gas cores are heated via shocks, (2) cooling cores are likely destroyed during mergers, (3) bulk gas motions of ≈ 1000 km/s persist for about 5 Gyr following the mergers, and (4) the intracluster gas may not be in hydrostatic equilibrium for clusters with significant X-ray substructure.

1. Introduction

Recent observational evidence suggests that galaxy clusters have undergone significant evolutionary changes during a relatively recent epoch. Both the cluster X-ray luminosity function and the numbers of cluster cooling flows have changed markedly from $z \approx 0.3$ to the present. At about the 3σ significance level, there appear to be fewer X-ray luminous clusters at $z > 0.3$ based upon samples gleaned from the *Einstein* and EXOSAT cluster archives (Henry et al. 1992; Edge et al. 1990); the absence of many X-ray detections for $z > 0.3$ clusters in the ROSAT North Ecliptic Pole survey is qualitatively consistent with the claimed evolution (Henry 1994). Similarly, it appears that the numbers of cooling flows have declined over this same time period (based upon $H\alpha$ surveys of moderate distance clusters (Donahue et al. 1992), where central $H\alpha$ emission is interpreted as thermal instabilities in cooling inflows). Although evolution in clusters is not unexpected, the very recent epoch of such evolution presents a challenge to theories of cluster and large-scale structure formation.

The abundance of observed substructure in galaxy clusters (*e.g.*, Beers et al. 1992) is consistent with late, possibly on-going evolution in the cluster potential wells. Here, ROSAT has contributed substantially to the recognition of X-ray substructure in clusters such as Coma (White et al. 1993) and A2256 (Briel et al. 1991). But, even the *Einstein* archives have revealed that 40%-70% of rich clusters have statistically significant evidence for substructure (Jones & Forman 1992; Mohr et al. 1993; Burns et al. 1994). In this paper, we describe a newly-discovered strong correlation between the positions of radio galaxies and X-ray substructure peaks within the intracluster medium (ICM). We make the case that radio-loud clusters are good hunting-grounds for clusters which are in a still-evolving dynamical state. Hybrid hydro/N-body simulations are used to model both the gas and the dark matter during the merger between clusters; results from these simulations are used to explain some of the newly discovered characteristics of clusters in the radio and the X-ray.

2. Radio Galaxies & X-ray Substructure in Rich & Poor Clusters

Over the past dozen years, we have been investigating the radio properties of rich and poor clusters using the VLA at 20-cm with $\approx 15''$ resolution (*e.g.*, Zhao et al. 1989; Owen et al. 1992, 1993; Burns et al. 1987). Extended synchrotron plasmas from radio galaxies are strongly influenced (via thermal confinement and ram pressure bending) by the surrounding medium. In order to understand the origin and evolution of the radio emission, one must also examine the hot ICM that surrounds the galaxies. So, we have undertaken two correlative studies of the X-ray and radio properties of northern Abell clusters to investigate the relationship between the cluster gas and the radio plasma.

First, we correlated our statistically complete sample of VLA-observed Abell clusters that have radio sources with diameters $> 2'$ and the *Einstein* archives (Burns et al. 1994). We found 41 radio-loud clusters with IPC images. Over 70% of these clusters appear to have statistically significant X-ray substructure. But, even more interesting was the discovery of a surprisingly strong correlation between the positions of the radio galaxies and X-ray peaks within the clusters. About 75% of the radio galaxies have X-ray peaks within $5'$ and 46% have X-ray peaks within $1'$ of the radio galaxy positions. Most of the X-ray features are resolved and are likely caused by overdense hot gas clumps in the ICM. A few representative examples of the X-ray/radio coincidences are shown in Fig. 1. A histogram of the X-ray/radio separations along with the expected distribution for random chance coincidences is shown in Fig. 2. These figures indicate a strong and highly significant relationship between radio galaxies and X-ray subclumps in the ICM.

Second, since the *Einstein* archives do not contain a statistical sample of rich clusters, we have begun a correlative study of radio galaxies from our VLA survey with X-ray emission from the ROSAT all-sky survey (RASS; Voges 1992). For a subsample of Abell clusters with $0.06 < z < 0.09$, we find that 76% of the 90 radio sources have RASS X-ray peaks within $5'$ of the radio galaxy positions. This result is in excellent agreement with that found from the *Einstein*/VLA sample.

The above correlation between radio galaxies and X-ray subclumps for rich clusters appears to extend to poor clusters as well. Using a percolation analysis of Zwicky galaxies, Burns et al. (1987) defined a statistical sample of 130 nearby ($z < 0.05$), relatively compact poor clusters that were imaged with the VLA. In

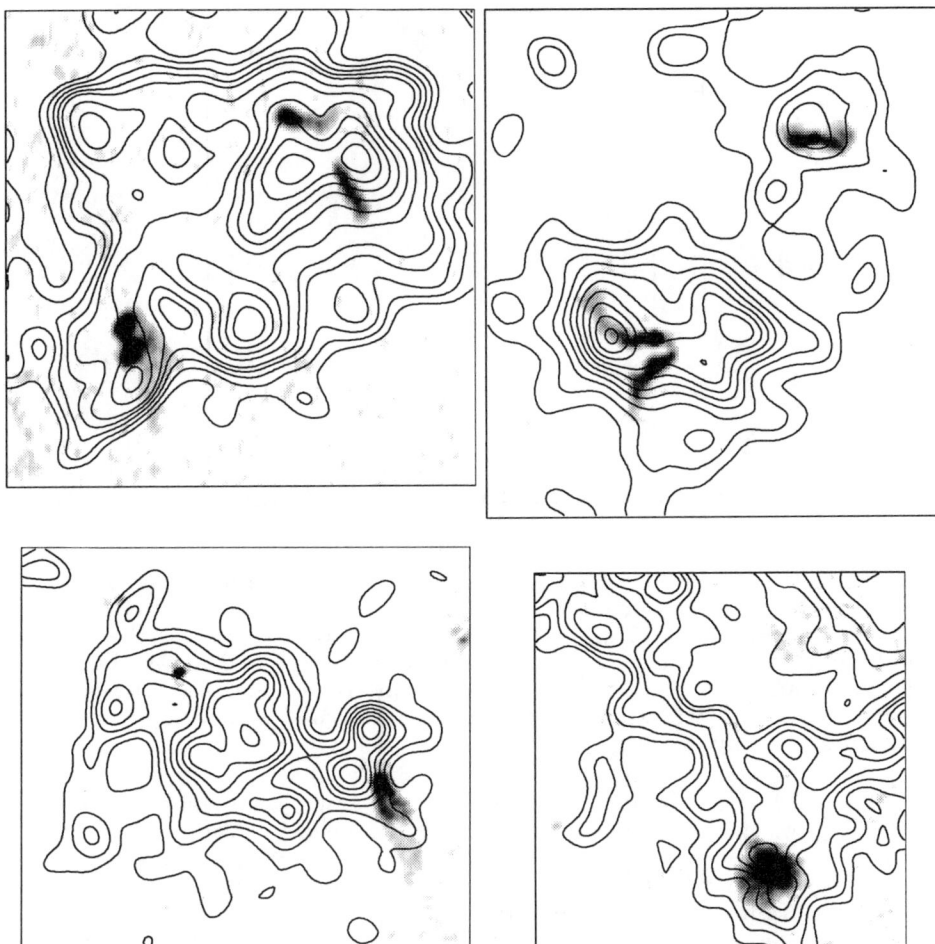

Figure 1. Overlays of *Einstein* X-ray images (contours) onto VLA radio maps (grey-scales) of Abell clusters. The clusters (& the angular sizes of the images) are: A514 (upper left; 15' × 15'), A1569 (upper right; 14' × 15'), A1589 (lower left; 14' × 14'), and A2147 (lower right; 14' × 14'). Note the X-ray clumps or X-ray extensions near the radio galaxies. In the case of A2147, the cluster center is just off the image to the NW and there is an interesting X-ray extension toward the amorphous radio source to the south.

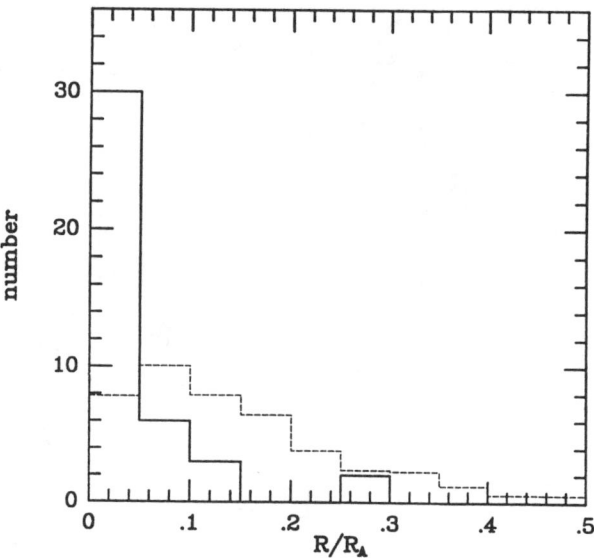

Figure 2. Distribution of separation distances between radio galaxies and the nearest significant X-ray peak. The distances are in units of Abell cluster radii ($R_A = 2$ Mpc for $H_0 = 75$ km/s/Mpc). The dashed histogram is the result of 1000 Monte Carlo simulations assuming that an average of 2 X-ray clumps per cluster are randomly placed with respect to the radio galaxies. The observed and Monte Carlo distributions are drawn from the same parent population with a probability of only 0.4%.

Fig. 3, ROSAT PSPC images are superposed on new VLA maps (from Batuski et al. 1994) of two poor clusters which contain head-tail radio sources. The X-ray, radio, and optical properties of these clusters are all interesting. The presence of U-shaped head-tail radio sources suggests that the radio galaxies are experiencing ram pressures comparable to that in rich clusters – a surprising result since the lower average gas densities and velocity dispersions in poor groups should give rise to average ram pressures that are ≈100 times less than that in rich clusters. In addition to the head-tail sources, nearly all the other Zwicky galaxies in these clusters are radio-loud as well (albeit weak and generally compact). The X-ray emission is clumped around the radio galaxies and is generally asymmetric. Finally, the velocity differences between some of the galaxies is greater than what one expects for a bound group (e.g., ΔV_r between NGC 4061 and 4065 in N79-299A is 850 km/s). These characteristics suggest that we are dealing with very young systems that are still in an early stage of collapse and formation. We are in the process of examining the X-ray properties of all 130 clusters in the sample from the RASS; to date, we detect X-ray emission in 45% of the poor clusters.

The above new results coupled with observations discussed in §1 are beginning to paint a somewhat revised picture of the origin and evolution of galaxy clusters. It seems that rich clusters grew hierarchically via the mergers of poor groups. Given the abundance of observed X-ray and optical substructure in clusters, it appears that this merging activity continues even today. The rate

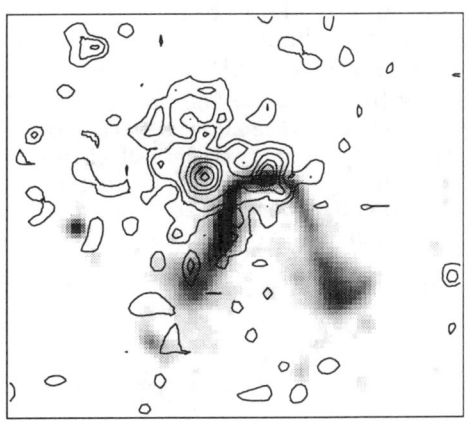 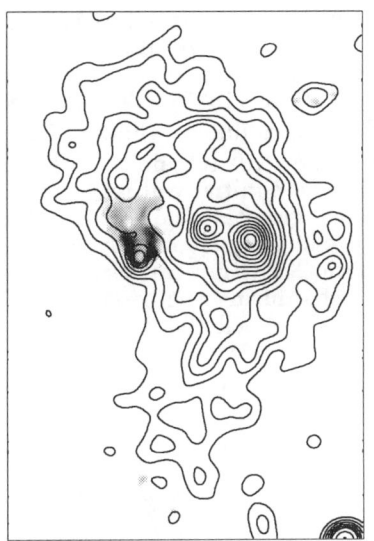

Figure 3. Overlays of ROSAT X-ray images (contours) onto VLA radio maps (grey-scales) of poor clusters. Left-most image is of N79-299A (image size is $8' \times 7'$); the X-ray peaks coincide with NGC 4065 (eastern peak) and NGC 4061 (western peak & head-tail radio galaxy). Right-most image is S49-132 (image size is $15' \times 22'$); the eastern X-ray peak coincides with the tailed radio galaxy NGC 7503.

of merging in clusters is a strong constraint on initial conditions for large-scale structure formation. As a result of mergers, the potential wells of clusters deepen with time and the X-ray luminosities increase, as is observed. Also, as discussed in the next section, merging probably destroys cooling flows which would explain the relative dearth of X-ray cooling inflows at recent times. Finally, radio galaxies seem to live within X-ray clumps inside of galaxy clusters. We suggest that these clumps are the gaseous residue from recent mergers between clusters. The radio emission is shaped by and may even be triggered by this local gaseous environment. So, radio galaxies may serve as important beacons of cluster merger activity. The new X-ray, radio, and optical observations are all consistent with a picture of clusters as relatively young, dynamically-evolving systems that have experienced mergers.

3. Hydro/N-Body Simulations of Cluster Mergers

If merging between clusters is as important as suggested by the new X-ray and radio data, then we must develop better physical insight into the physics of the merger process. To do so, we have constructed a new hybrid Eulerian hydrodynamics + N-body code that allows us to simulate the evolution of both the gas as well as dark matter during cluster mergers. Our computer program combines the Hernquist N-body Treecode with a robust, well-tested finite difference hydrodynamics code called ZEUS-3D, which has been used extensively to investigate astrophysical jets (*e.g.*, Burns *et al.* 1991).

The first results for the merger of two idealized spherical clusters, initially in hydrostatic equilibrium with adiabatic ICMs, were reported in Roettiger *et al.* (1993). We have recently re-run this simulation to include cooling (bremsstrahlung + lines). One cluster was poor ($6 \times 10^{13} M_\odot$, $r_c = 125$ kpc, $T = 2.5 \times 10^7$ K) and the other was rich ($5 \times 10^{14} M_\odot$, $r_c = 250$ kpc, $T = 10^8$ K); they were initially at rest, separated by 6 Mpc, and were allowed to fall together under gravitational forces. Upon collision of the cores, the small cluster was travelling supersonically relative to the sound speed of the rich cluster. Synthetic X-ray images, using the ROSAT bandwidth, were constructed at several epochs during the evolution. One of these images is compared with *Einstein* data for Abell 168 in Fig. 4.

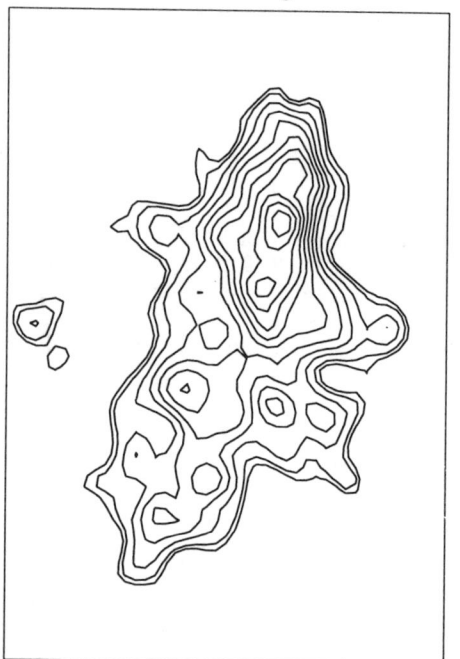 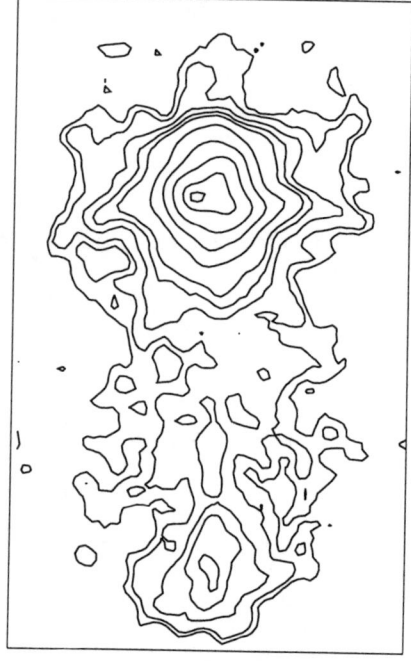

Figure 4. Comparison of *Einstein* X-ray image of Abell 168 (left) with a synthetic X-ray image generated from our Hydro/N-body simulation (right). Ulmer *et al.* (1992) have suggested that the unusual X-ray morphology of A168 is the result of a merger. The merger simulation shows a dispersed subcluster which has exited near the bottom nearly 3.5 Gyrs after the merger event (coincident cores). The synthetic image has been scaled to a comparable dynamic range as the A168 image and Gaussian noise has been added.

Several important conclusions can be drawn from these simulations. First, the gaseous core of the rich cluster is heated and expands during the merger, largely caused by the development of strong shocks. These shocks continue to "stir" the cluster gas, causing large scale turbulence in the ICM to persist for much longer than the sound crossing time. Thus, gaseous substructure exists for long periods of time (> 5 Gyr) following the merger event. Second, relatively high bulk velocities within the ICM (nearly 1000 km/s over scales of

> 400 kpc) are sustained for up to 5 Gyr following the merger. Third, the gas and the dark matter (and presumably galaxies) are only loosely spatially correlated for the disrupted poor group following the initial head-on collision. A "spray" of generally unbound dark matter particles emerges on the opposite side of the cluster with a relatively high velocity dispersion. These properties resemble those found in several clusters, including A168 in Fig. 4 and the Coma cluster. We conclude that these are *post-merger* rather than pre-merger clusters. Fourth, applying hydrostatic equilibrium to determine masses of clusters with substructure is dangerous. We used the synthetic X-ray images to generate surface brightness profiles and to deproject the distributions of ICM density and temperature, as is commonly done with real X-ray images (*e.g.*, Arnaud 1988). If we assume that all the gas is bound to the merged cluster and in hydrostatic equilibrium, our calculated mass underestimates the real total mass by factors of 2-5. Clearly, one must be careful in applying equilibrium assumptions to clusters with substructure since our models demonstrate that the ICM in merged clusters can exist in a dynamic state for 5 Gyr following the merger.

Several new sets of simulations are underway which should further illuminate the physics of cluster merging. In the first set, we are using the above hybrid code to explore the effects of different cluster masses and infall velocities on the resulting X-ray gas properties. In the second set, we are utilizing a new two-level grid code (Anninos *et al.* 1994) to produce more realistic initial conditions from a large-scale structure simulation (on the first grid) which, in turn, will be used to study the merger of two proto-clusters at higher resolution (on the second grid).

4. Probing Gas Dynamics in Merged Clusters with Wide-Angle Tailed Radio Sources

How can we test our simulation results that predict relatively high bulk velocities in the ICM in merged clusters? Current X-ray spectroscopic resolutions are generally insufficient to detect such motions using spectral lines. However, indirect probes of the ICM may exist via tailed radio sources which extend over large regions within clusters. In particular, wide-angle tailed (WAT) radio sources, associated with giant centrally-located D or cD galaxies, have typical radii of 100-500 kpc (see Fig. 5). The bulk speed of the radio plasma in the diffuse tails is thought to be transonic, and thus very susceptible to pressure gradients in the ICM – the bending of the tails reflect motions in the ICM much like smoke buffeted by the local wind as it rises from a fireplace chimney.

Pinkney *et al.* (1993) have argued that WAT clusters are the products of relatively recent mergers. Spectroscopic measurements of galaxies in clusters with cDs reveal that about a third of these supergiant galaxies have apparent peculiar velocities of 200-400 km/s relative to the cluster means (*e.g.*, Hill & Oegerle 1993; Bird 1994). It is unlikely that the cDs are actually moving through the cluster cores at these speeds since the resulting displacements from the cluster centroids would tidally strip the cD halos (*e.g.*, Merritt 1984). Instead, we believe that this apparent peculiar motion reflects velocity substructure resulting from a cluster/subcluster merger (Pinkney *et al.* 1993). Even if this velocity represents real motion of the WAT galaxy through the ICM, the resulting ram pressure is still insufficient to bend the radio jets and tails (Eilek *et al.* 1984) to their observed curvature (see Fig. 5). Another reason to believe that WAT clusters have undergone mergers is the lack of strong (*i.e.*, $\approx$100 $M_\odot$/yr) cooling inflows that are observed in 75%-90% of morphologically similar clusters (Edge

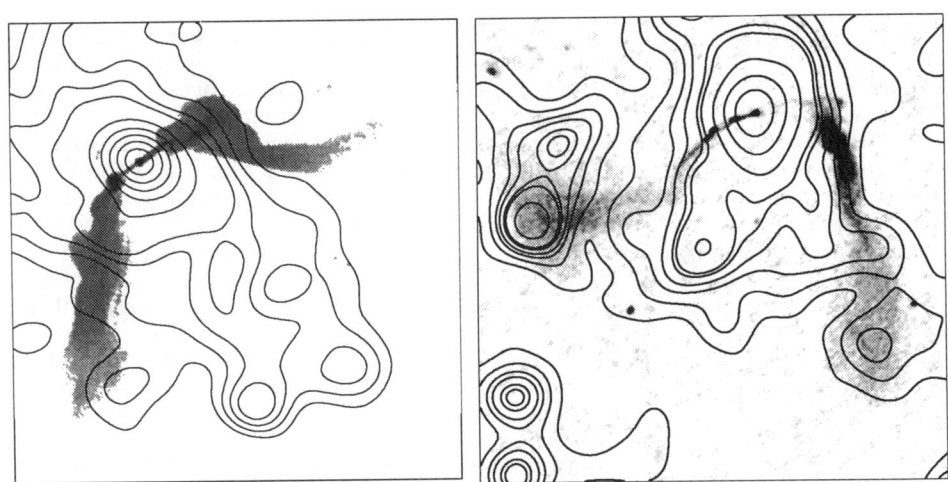

Figure 5. Overlays of ROSAT X-ray images (contours) onto VLA radio maps (grey-scales) of two clusters with Wide-Angle Tailed (WAT) radio galaxies. Left image is A2634/3C 465 (image size is $9' \times 9'$). Right image is 1919+479 (image size is $10' \times 10'$); 1919+479 is one of the largest known WATs with a linear diameter of 1.2 h_{75}^{-1} Mpc. Note the extension of X-ray emission between the radio tails.

et al. 1992). As described in §3, mergers heat the gas in cluster cores and would likely destroy a pre-existing cooling flow. Finally, we have found asymmetries and elongations in the central X-ray surface brightness in most of a dozen WAT clusters observed to date with the ROSAT PSPC as shown in Fig. 5. The X-ray extensions usually occur along a line stretching between the radio tails. Although such asymmetries appear to be uncommon for most non-WAT cD clusters, our numerical simulations indicate that such X-ray extensions are expected for clusters which have undergone a merger.

A cluster/subcluster merger would appear to solve the long-standing problem of bending the jets/tails in WATs. We believe that the jet curvature is not produced by motion of the radio galaxy through the ICM, but rather by bulk motion of the ICM past the central WAT galaxy. Numerical simulations indicate that 1000 km/s bulk motions within the ICM occur in post-merger clusters for timescales well in excess of the typical synchrotron lifetimes of the radio sources ($\approx 10^8$ yr). The resulting ram pressure is sufficient to bend the radio jets/tails (*e.g.*, Burns 1986; Loken *et al.* 1994). If this idea is correct, then WATs can serve as important tracers of the ICM gas pressure and dynamics in merged clusters.

5. Conclusions

We present the following conclusions from this work:
(1) New X-ray, radio, and optical data suggest that clusters of galaxies are more complex, dynamically-evolving structures than previously believed. The galaxies and gas in post-merger clusters remain in a nonequilibrium state

for relatively long time periods (≈5 Gyr) following the merger. Thus, clusters with evidence for X-ray or optical substructure may not be in either virial or hydrostatic equilibrium. Applying such assumptions to these clusters may result in seriously incorrect total mass estimates.

(2) It appears that radio galaxies live within X-ray subclumps which may be the by-product of cluster/subcluster mergers. The local gaseous environment strongly influences the radio properties of cluster galaxies. Such clumped X-ray emission is seen around radio galaxies in both rich and poor clusters. This might explain why the radio luminosities and radio morphologies of extended radio sources are so similar for radio galaxies inside and outside of rich clusters.

(3) Mergers may be responsible for producing the unusual properties of WAT radio sources and the WAT clusters. The bulk motion of the ICM gas around the radio galaxy, resulting from the merger, appears sufficient to explain the bending of the WAT jets/tails. In turn, WAT radio sources provide the best available probe of gas dynamics in merged clusters.

(4) Extended radio tails in cluster radio sources also appear to be in thermal pressure equilibrium with the ICM (*e.g.*, Morganti *et al.* 1988). As a result, these radio galaxies are good predictors of X-ray luminosities for the ICMs. Potentially, these radio galaxies can serve as an important check on the evolution of the cluster X-ray luminosity function. We are in the process of performing new VLA observations, optical imaging and spectroscopy, and analyzing RASS data for a sample of rich Abell clusters with $z > 0.2$; we will use this sample to study the evolution of cluster gas and radio sources at these intermediate redshifts.

(5) Finally, radio galaxies should be viewed with renewed interest as potentially important tools for studying cluster evolution and cluster gas dynamics. They are an important complement to X-ray observations. These radio galaxies appear to be intimately connected to the local gaseous environment and are sensitive to recent cluster/subcluster mergers.

Acknowledgements

This work was supported by grants from NASA and the National Science Foundation. We thank our collaborators including Hans Böhringer, George Rhee, Jean Eilek, Mike Norman, David Batuski, and Anatoly Klypin for their input and advice.

References

Anninos, P., Norman, M., & Clarke, D. 1994, preprint
Arnaud, K. 1988, in Cooling Flows in Galaxies and Clusters, ed. A. Fabian (Cambridge: Cambridge Univ. Press), p. 31
Batuski, D., Venkatesan, V., Hanisch, R., & Burns, J. 1994, ApJ, submitted
Bird, C.M. 1994, ApJ, in press
Briel, U., Henry, J., Schwarz, R., Böhringer, H., Ebeling, H., Edge, A., Hartner, G., Schindler, S., Trümper, J., & Voges, W. 1991, A&A, 246, L10
Burns, J. 1986, Can. J. Phys., 64, 373
Burns, J., Hanisch, R., White, R., Nelson, E., Morrisette, K., & Moody, J. 1987, AJ, 94, 587
Burns, J., Norman, M., & Clarke, D. 1991, Science, 253, 522
Burns, J., Rhee, G., Owen, F., & Pinkney, J. 1994, ApJ, in press

Donahue, M., Stocke, J., & Gioia, I. 1992, ApJ, 385, 49
Edge, A., Stewart, G., Fabian, A., & Arnaud, K. 1990, MNRAS, 245, 559
Edge, A. Stewart, G., & Fabian, A. 1992, MNRAS, 258, 177
Eilek, J., Burns, J., O'Dea, C., & Owen, F. 1984, ApJ, 278, 37
Henry, J.P., Gioia, I., Maccacaro, T., Morris, S., Stocke, J., & Wolter, A. 1992, MNRAS, 386, 408
Henry, J.P., 1994, these proceedings
Hill, J. & Oegerle, W. 1993, AJ, 106, 831
Jones, C. & Forman, W. 1992 in Clusters and Superclusters of Galaxies, ed. A. Fabian (Dordrecht: Kluwer), p. 49
Loken, C., Burns, J., & Clarke, D. in The Physics of Active Galactic Nuclei, ed. G. Bicknell & P. Quinn (San Francisco: PASP), in press
Merritt, D. 1984, ApJ, 276, 26
Mohr, J., Fabricant, D. & Geller, M. 1993, preprint
Morganti, R., Fanti, R., Gioia, I., Harris, D., Parma, P., & de Ruiter, H. 1988, A&A, 189, 11
Owen, F., White, R., & Burns, J. 1992, ApJS, 80, 501
Owen, F., White, R., & Ge, J.-P. 1993, ApJS, 87, 135
Pinkney, J., Rhee, G., Burns, J., Hill, J., Oegerle, W., Batuski, D., & Hintzen, P. 1993, ApJ, 416, 36
Roettiger, K., Burns, J., & Loken, C. 1993, ApJ, 407, L53
Ulmer, M., Wirth, G., & Kowalski, M. 1992, ApJ, 397, 430
Voges, W. 1992 in Proc. of the European "International Space Year" Conf., ESA ISY 3, ed. T.D. Guyenne & J.J. Hunt (Noordwijk: ESA Publ. Div.), p. 9
White, S., Briel, U., & Henry, J. 1993, MNRAS, 261, L8
Zhao, J.-H., Burns, J., & Owen, F. 1989, AJ, 98, 64

**ORAL PRESENTATIONS
SERENDIPITY**

ROSAT Observations of Supernovae

Eric M. Schlegel
MC 668, NASA Goddard Space Flight Center, Greenbelt, MD 20771

Email ID
eric@heasfs.gsfc.nasa.gov

ABSTRACT

Supernovae have been expected to be X-ray sources for many years. This paper reviews the upper limits placed on or the observations of supernovae using $ROSAT$. Five Type II supernovae have been detected in the X-ray band, of which $ROSAT$ has detected four. In addition, an improved upper limit exists for Type Ia supernovae.

1. Introduction

Supernovae may be X-ray sources, given the quantity of thermal energy released in a supernova outburst ($\sim 10^{49}$ ergs s^{-1} of electromagnetic radiation (Kirshner 1990)). Noting the rapid decay times typical of the highest energy bands of supernovae, the observer might expect supernovae to be detectable X-ray sources only at the very earliest times in the evolution of the outburst. For example, the ultraviolet flux of SN 1987A declined by $\sim 10^3$ in three days (Kirshner et al. 1987). A decline of this magnitude occurs more quickly than satellites can typically be re-oriented for a pointed observation. However, X-ray satellites have been pointed at supernovae, resulting in the detection of several supernovae. The quality of the observations appears very likely to take a quantum leap upward in the next few years, so a review will focus attention on the questions which better signal-to-noise observations can answer. This paper is a subset of a larger review of X-ray observations of supernovae (Schlegel 1993).

Supernovae are classified on the basis of their optical spectra (Zwicky 1965) and fall into two broad categories. Type II supernovae show emission lines of hydrogen in their spectra. These supernovae are produced by the collapse of the progenitor's core. Type I supernovae show no hydrogen in their spectra. This class has been subdivided recently into Type Ia and Type Ib, with the Type Ib subclass showing strong emission lines of oxygen after about 100 days of evolution (see, for example, Porter and Filippenko 1987). The Type Ib objects are also thought to start from massive stars, so the evolution also commences with a core collapse (see, for example, Schlegel and Kirshner 1989). Type Ia supernovae, however, are believed to be detonating or deflagrating white dwarfs (see, for example, Woosley 1990)

Type II supernovae can be further subdivided based upon the optical light curve (see, for example, Kirshner 1990). SN IIL show a linear decline from outburst maximum to ~ 4.5 magnitudes below maximum at which time a new, linear, decay rate dominates. SN IIP show an approximately constant plateau extending out to about 50-100 days, followed by a rapid, ~ 1 magnitude decline, followed by the decay appropriate to radioactive ^{56}Co.

The expected X-ray emission of supernovae can be roughly divided into two temporal boxes. First, we might expect X-rays to appear promptly, accompanying the optical maximum, either due to the high-temperature flash associated with the breakout of the shock wave through the stellar surface, or to the prompt thermal detonation/deflagration of a white dwarf, or to the interaction of the initial photoionization and shock with circumstellar material. Second, we might expect X-rays to appear at late times (≥ 100 days) whenever the expanding debris field becomes optically thin to X-rays, or when the outgoing shock wave runs into the circumstellar shell from earlier phases of mass loss or when the material has thinned sufficiently that any radiation from a collapsed remnant remaining after the explosion can escape the nebula.

2. Mechanisms of X-ray Production

There is insufficient space to review the mechanisms for the production of X-rays from supernovae. The reader is referred to the larger review (Schlegel 1993).

The proposed production mechanisms for X-rays are: i) inverse Compton scattering; ii) prompt thermal burst accompanying shock breakout; iii) pulsar-driven input; iv) radioactive decay input; v) circumstellar interaction. The first mechanism has been eliminated by Chevalier (1982), as he showed that the expected flux from inverse Compton scattering is ten times lower than the detected flux from SN 1980K. A match with the observations can be made by changing the parameters (such as the effective temperature), but inverse Compton scattering then decreases the density of synchrotron electrons responsible for the radio emission.

Bandiera et al. (1984) advance the idea that a pulsar could inject magnetic energy and relativistic particles into the expanding debris field from the supernova explosion, leading to a large flux of non-thermal radiation. Since their model requires a pulsar, it only applies to core collapse supernovae as detonating white dwarfs are not expected to leave a collapsed remnant (e.g., Woosley 1990). To my knowledge, no attempt has been made to refute this hypothesis nor test it against data.

A prompt thermal burst accompanies the breakout of the shock The effective temperature at breakout is $\sim 2 \times 10^5$ K with a luminosity of $\sim 10^{45}$ ergs s^{-1} in a burst lasting $\sim 10^3$ s. The burst lasts for a time too short to re-orient a pointed-mode X-ray satellite. Only a sensitive, all-sky X-ray monitor would catch the X-ray burst accompanying shock breakout.

Circumstellar interaction is possible. When the outgoing shock encounters the wind from the progenitor, a shock is established in the material, and a reverse shock propagates back into the shocked supernova gas. The development of the circumstellar interaction has largely been the work of Chevalier (1982, 1984, 1990) and Fransson (1984). The presence of a circumstellar medium is stongly dependent upon the immediate prior evolution of the progenitor (Chevalier 1990).

The presence of radioactively decaying elements in the debris of a supernovae immediately produces a source of γ-rays, and by Compton scattering, a source of X-rays. The X-rays emerging are expected to be hard, as the Compton degradation of the γ-rays will produce a pseudo-continuum down to ~ 20keV, at which point photoelectric absorption will dominate. No flux is expected below 10keV because of the presence of absorption edges in that region (for example, edges due to Fe, Si, S, O). The time at which X-rays emerge is set by the time at which the effective optical depth (scattering plus absorption opacity) becomes approximately unity.

3. Observations of SN Ia

There have been quite a few X-ray observations of Type Ia supernovae, but to date, there have been no detections. No obvious mechanism exists for producing X-ray emission in Type Ia supernovae after the outburst but during the first tens of days when the supernova is optically bright. No circumstellar medium is expected to be present, given that the typical white dwarf lifetime prior to the detonation/deflagration of the hydrogen layer should be of sufficient length to dissipate any circumstellar medium, so late-time emission due to circumstellar interaction will not occur.

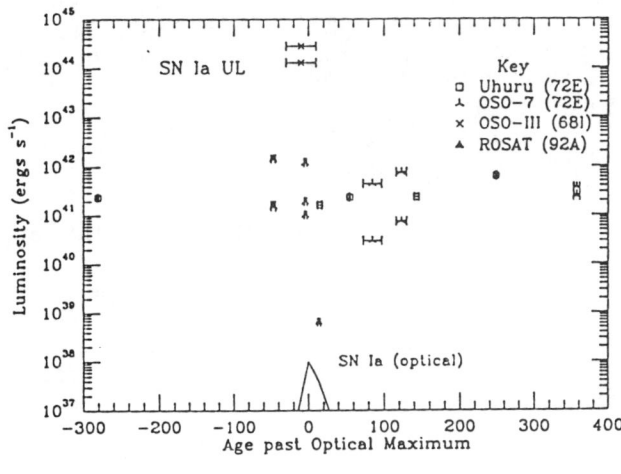

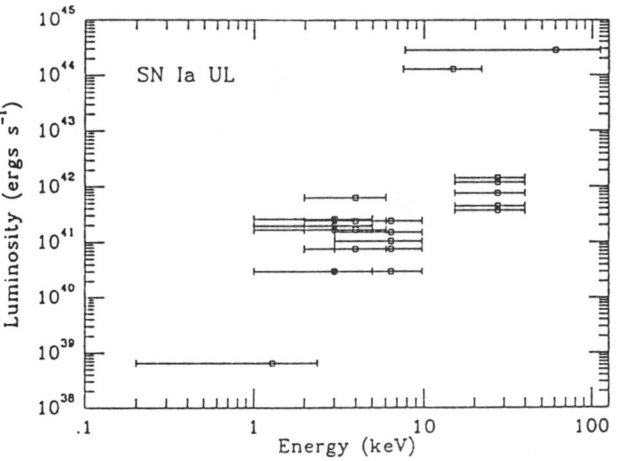

Figure 1 - Upper limits on X-ray emission from SN Ia on a) the light curve; b) the spectrum.

Previous upper limits on the X-ray emission from Type Ia supernovae have not strongly constrained possible models due to the relatively low sensitivity and high internal backgrounds of the instruments used. Figure 1 shows the temporal and spectral upper limits. The data points have been collected from the literature (Schlegel 1993). The temporal upper limit is compared to an *optical* light curve (labeled 'optical' in the figure) for reference. Note that the optical light curve has been normalized to a luminosity of 10^{38} ergs s^{-1}.

The launch of *ROSAT* created an opportunity to obtain a much lower upper limit on the X-ray emission from a Type Ia supernova. *ROSAT* has excellent X-ray optics which place over 80% of the photons into the central 10" of the point spread function (Aschenbach 1988). In addition, the *ROSAT* Position Sensitive Proportional Counter (PSPC) is characterized by a very low internal background ($\leq 2.5 \times 10^{-5}$ counts s^{-1} arcmin^2 for $\sim$95% of the observing time (e.g., Snowden and Freyberg 1993)). The overall background of the *ROSAT* PSPC then becomes a strong function of the local particle and scattered solar X-ray environment. Such a low internal background can lead to a sensitive upper limit on X-ray emission from any source. *ROSAT* was used to place an upper limit on SN 1992A in NGC 1380. Eleven counts were detected at the position of the supernova, consistent with the background rate. For an assumed distance of 16.9Mpc (Tully 1988), the 99% upper limit obtained on the 0.5-2.0keV luminosity is 2.9×10^{38} ergs s^{-1} (the 5σ upper limit is 4.7×10^{38} ergs s^{-1}) (Schlegel and Petre 1993). This upper limit is at least a factor of 100 lower than all previous upper limits (Figure 1).

4. Observations of Massive Supernovae

X-ray emission is more generally expected from massive stars due to the breakout of the shock through the stellar surface, or to the higher circumstellar density, or to Compton scattered γ-rays. Five supernovae resulting from massive stars have now been observed with four different X-ray satellites (Table 1). I concentrate here on those supernovae observed with *ROSAT*.

Table1: Supernovae Detected in X-rays

SN Name	Galaxy	Date of Optical Max	B mag at max	SN Type	Galaxy Distance	X-rays first observed
SN1978K	NGC 1313	$\sim$1978 June 10 $\sim$1978 May 25	$\sim$13 $\sim$14.5	IIL IIP	4.5Mpc	$\sim$12.1yr
SN1980K	NGC 6946	1980 Nov 5	11.5	IIL	5.1Mpc	+35d
SN1986J	NGC 891	1983 Jan?	?	IIpec	9.6Mpc	?
SN1987A	LMC	1987 May 9	3.5	IIP	50kpc	+154d
SN1993J	NGC 3031	1993 Apr 18	11.4	IIpec	3.63Mpc	+5d

a. SN 1986J in NGC 891

SN 1986J was first discovered in the radio (Rupen et al. 1987) and rapidly became the brightest radio source in NGC 891. Optically, the supernova was clearly well-evolved, showing only narrow emission lines characteristic of a supernova at least several hundred days beyond the outburst. The strong radio emission suggested an interaction with a dense circumstellar environment.

SN 1986J was observed with the *ROSAT* PSPC on 1991 August 18-20 for 24.9ksec (Bregman & Pildis 1992). The extracted spectrum showed essentially no emission below ~0.65keV. A thermal bremsstrahlung spectrum was fitted to the extracted counts, yielding essentially two equally good fits: a low kT-high N_H solution (kT ~0.5keV, log N_H ~22.3) and a high kT-low N_H solution (kT ~1.8keV, log N_H ~21.8) (Figure 2). The dual solutions exist because of the low spectral resolution of the PSPC, and because, at low temperatures, L-shell lines of Fe are very strong and, when heavily absorbed, can not be distinguished from a higher-temperature plasma.

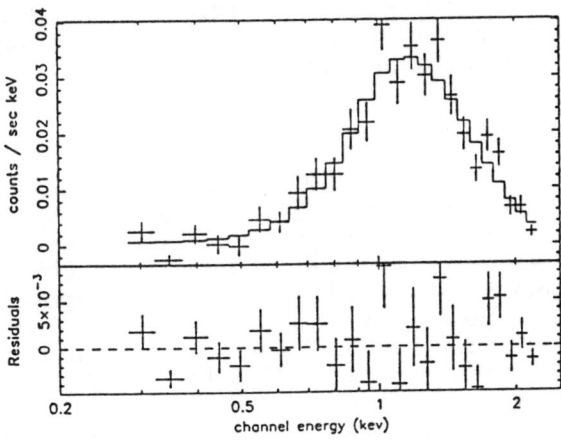

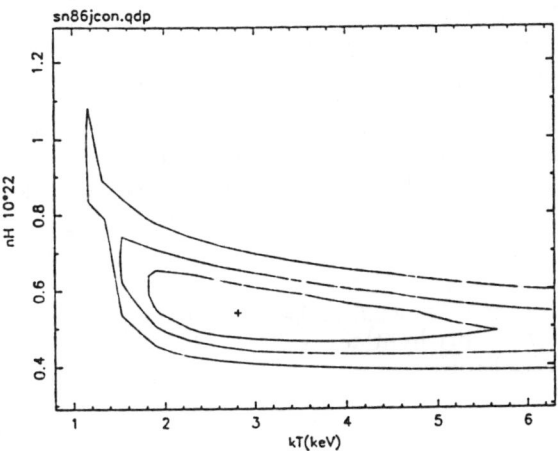

Figure 2 - a) The *ROSAT* PSPC spectrum and b) the χ^2 contours of the fit parameters for SN 1986J. Essentially no flux is detected below ~0.65keV.

The lower temperature solution is consistent with the blast wave interaction model of Chevalier & Fransson (1994). On that basis, Bregman & Pildis (1992) suggested that the X-rays from SN 1986J were produced by the inward-moving shock from the interaction of the blast wave with the circumstellar medium. The inward-moving shock produces a dense, cool (~10^7 K) shell. The model then accounts for the observed high column, N_H ~2×10^{22} cm^{-2}, about 2.6 times larger than the observed H I column used to correct the optical spectroscopy, as being partly due to the galactic column in the direction of NGC 891 and largely due to the cool, dense shell behind the reverse shock.

The detection of the radiatively cooled material behind the shock, if verified, will be the first such observation in the X-ray band. X-ray data could then be used to infer the density distribution of the circumstellar medium prior to the supernova outburst, an important quantity used in modeling the evolution of the blast wave (e.g., Chevalier & Fransson 1994; Suzuki et al. 1993). Verification of the detection of the radiatively cooled material can be achieved by observing the temporal development of the spectrum, which should show a decline in the absorption column as the shell moves through the cool material, or by detecting absorption edges in a high resolution spectral observation, such as should be possible with *Asuka*. Either way, the low energy spectrum will gradually be unveiled.

b. SN 1978K in NGC 1313

A *ROSAT* PSPC observation was made of NGC 1313 between 1991 April 24 and May 11 for 11.2ksec. Three sources were present: the nucleus (or, at least, a source consistent with the nucleus), a source about 7' south of the nucleus, and a source 11' southwest of the nucleus. The *Einstein* IPC data on NGC 1313, obtained on 1980 January 2 with an exposure of 8.3ksec showed two sources corresponding to the nucleus of NGC 1313 and the southern source. The third source in the *ROSAT* frame is now known as SN 1978K (Ryder et al. 1993a).

Serendipitously, SN 1978K lies in three other *ROSAT* pointings, so a light curve can be constructed (Schlegel et al. 1993). In two of these, SN 1978K lies about 20' off-axis. Once the counts are corrected for wobble-induced occulation by the window supports, a light curve can be produced. The 0.5-2.0keV luminosities, assuming a thermal bremsstrahlung spectrum, lie near 1×10^{39} ergs s^{-1}. The light curve is shown in Figure 3. There is a suggestion of a decline in the light curve indicating the end of this X-ray bright phase. Two additional pointed observations have been scheduled with *ROSAT* (one PSPC, one HRI) for late 1993, and a verification phase observation of NGC 1313 with *Asuka* was made in September 1993, so additional points will be added to the light curve. The full light curve will be presented in Schlegel et al. (1993).

Figure 3 - The 0.5-2.0keV luminosity light curve of SN 1978K.

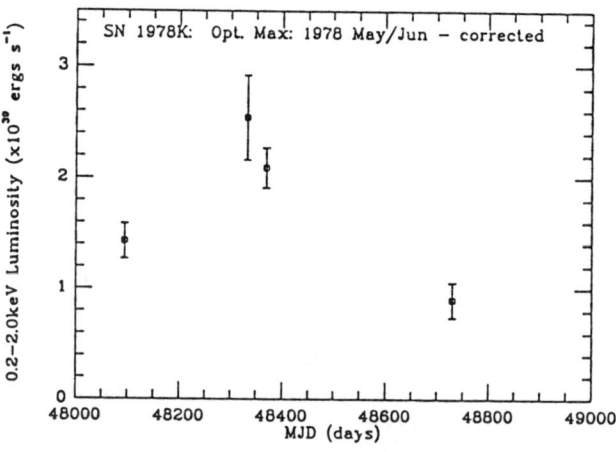

The fitted spectrum is shown in Figure 4, with no one model preferred. The thermal bremsstrahlung model gave a fitted temperature of ~0.5keV, similar to that of SN 1986J. The column to the X-ray source is quite high, lying about a factor of two above the 21-cm H I column density in the surrounding region (Ryder et al. 1993b). The difference may be the absorption of X-rays through the shell of cool material behind the reverse shock. We may observe the evolution of the shell by observing the changing column in the fitted spectrum with additional observations. The unabsorbed luminosity in this spectrum is ~9.5x10^{39} ergs s^{-1} (Ryder et al. 1993a). As SN1978K is evolving while X-ray satellites are in orbit, the long-term light curve is of considerable interest. An upper limit can be assigned to the *Einstein* IPC data as was done for SN 1992A. The resulting value is 2x10^{39} ergs s^{-1}, a factor of ~5 less than the *ROSAT* luminosity.

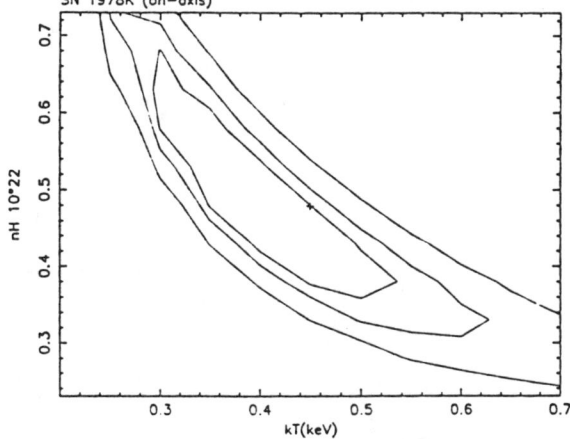

Figure 4 - a) The *ROSAT* PSPC spectrum and b) the χ^2 contours of the fit parameters for SN 1978K for the exposure of 1991 April = MJD 48370.0.

e. SN 1993J in NGC 3031 (M81)

The newest member of the X-ray detected supernova happens to be the second closest and the brightest northern hemisphere supernova since SN 1937C. The supernova was discovered on 1993 March 28 by Ripero (1993). An optical maximum occurred on 1993 March 28. The supernova declined rather quickly and then recovered to a second maximum. The supernova was classified as a Type II once the hydrogen lines became readily apparent. The progenitor appears to be a slightly reddened (E_{B-V} <0.1) K supergiant. The apparent fading of the Hα line (Filippenko and Matheson 1993) suggests the supernova is a Type IIb (Wheeler & Harkness 1990). Nomoto et al. (1993) show that the likely progenitor is a hydrogen stripped evolved star in a binary system.

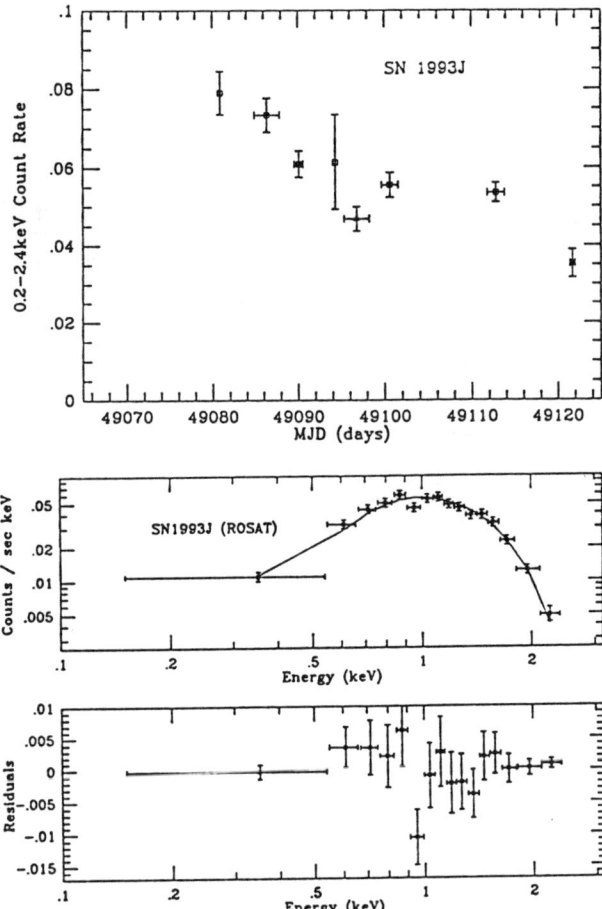

Figure 5 - a) For SN 1993J in NGC 3031, the *ROSAT* PSPC a) light curve; b) the spectrum.

SN 1993J must be accorded another accolade: it is the first supernova to be observed and detected with two X-ray satellites at the same time. The first observation was obtained with *ROSAT* (Zimmermann et al. 1993) in a 2.8ksec exposure on 1993 April 3.4 (UT). An X-ray source was visible within 10" of the optical position and ~1' north of a known X-ray binary (X6 of Fabbiano 1988). The extracted spectrum, shown in Figure 5b, is rather flat, consistent with either a power law having a photon index of -0.7 or a thermal spectrum with kT >7keV (Zimmermann et al. 1993). Assuming a distance of 3.63±0.34Mpc (Freedman et al. 1993), the corresponding luminosity in the 0.1-2.4keV band is 3×10^{38} ergs s^{-1}. The fitted column is consistent with the galactic column, implying either that no additional absorption near the source is necessary, or that the absorbing matter near the source has been completely ionized. The second case predicts the emergence of emission lines once the matter recombines. A lower limit to the bremsstrahlung kT of ~7 keV was found. Additional *ROSAT* observations were made throughout April 1993. The count rates showed a decrease of about 30% in the days between the first and last observations. implying a decay time of a few weeks to a few months. The X-ray light curve is shown in Figure 5a.

Models of the X-ray emission have been computed by Suzuki et al. (1993). The models which explain the optical light curve are those which have a thin hydrogen envelope ejected at relatively high velocities (~few x 10^4 km s^{-1}) (Nomoto et al. (1993)). Rayleigh-Taylor instabilities occuring during the propagation of the shock mix some ^{56}Ni and ^{56}Fe into the envelope. The X-rays are then produced by the outgoing shock running into the dense circumstellar medium, which for SN 1993J, is apparently extreme. The X-rays arise from thermal bremsstrahlung from the shocked ejecta. Suzuki et al. predict the softening of the X-ray spectrum as the reverse shock decays as it runs into the steep density gradient of the expanding ejecta. Zimmermann et al. (1993) show that the Suzuki et al. model predicts a faster decline than the observed one. Additional parameter space must be explored to see if the predicted decline is truly discrepant with the observations.

5. Summary Discussion

The observations detailed above point out the variety of behavior seen to date. Three supernovae have been detected in X-rays. Of the emission mechanisms considered likely as explanations for the X-ray emission of supernovae, only one is firmly established, namely, the Compton scattering of the γ-rays from radioactive decay in SN 1987A.

Circumstellar interaction almost certainly accounts for the early X-ray emission observed from SN 1993J. The most likely explanation is that the immediate surroundings of the progenitor of SN 1993J were filled with matter, perhaps as a result of the detonation occuring during a mass loss phase.

The late-time behavior of SN 1986J and SN 1978K likely requires the interaction of the blast wave with the circumstellar medium. The issue to be addressed by future observations is the exact nature of the interaction.

6. The Future

What is the future of X-ray emission from supernovae? In two words, quite good. SN 1993J has been observed with *Asuka* and the results will be available by mid-1994. SN 1987A has been detected with *ROSAT* (Gorenstein et al. 1994); it will also be studied with *Asuka*. SN 1978K has also been detected by *Asuka*. SN 1986J is a first-round target for *Asuka*.

The X-ray study of supernovae is a newly established twig of the X-ray astrophysics tree. The observations obtained will likely help most handily in increasing our understanding of the evolution of shocks in a dense circumstellar medium, thereby growing the twig into a branch.

Acknowledgements

I thank U. Zimmermann for sending the SN1993J data files in electronic form, F. Seward for a SN Ia reference, and the scientific organizing committee for inviting me to speak at this conference. Not all those researchers contributing to this area could be cited here; I hope the full paper includes everyone.

References

Aschenbach, B. 1988, Appl. Opt., 27, 1404
Bandiera, R., Pacini, F., & Salvati, M. 1984, ApJ, 285, 134
Bregman, J. N. & Pildis, R. A. 1992, ApJ, 398, L107
Chevalier, R. 1982, ApJ, 259, 302
Chevalier, R. 1984, ApJ, 285, L63
Chevalier, R. 1990, *Supernovae*, A. Petschek, (Berlin: Springer-Verlag Inc.), 91
Chevalier, R. & Fransson, C. 1994, ApJ, 000, 000
Fabbiano, G. 1988, ApJ, 325, 544
Filippenko, A. & Matheson, T. 1993, IAU Circular 5787
Fransson, C. 1984, AA, 133, 264
Freedman, W., et al. 1993, ApJ, 000, 000
Gorenstein, P., Hughes, J., & Tucker, W. 1994, ApJ, 000, L000
Itoh, H. & Masai, K. 1989, MNRAS, 236, 889
Itoh, H., Hayakawa, S., Masai, K., & Nomoto, K. 1987, PASJ, 39, 529
Kirshner, R. 1990 in *Supernovae*, ed. A. Petschek, (Berlin: Springer-Verlag), p59
Kirshner, R., Sonneborn, G., Crenshaw, D., & Nassiopoulos, G. 1987, ApJ, 320, 602
Nomoto, K., Suzuki, T., Shigeyama, T., Kumagai, S., Yamaoka, H., & Saio, H. 1993, Nature, 000, 000
Porter, A. & Filippenko, A. 1987, AJ, 93, 1372
Ripero, J. & Garcia, F. 1993, IAU Circular 5731
Rupen, M. P., van Gorkom, J. H., Knapp, G. R., Gunn, J. E., & Schneider, D. P. 1987, AJ, 94, 61
Ryder, S. D., Staveley-Smith, L., Dopita, M. A., Petre, R., Colbert, E., Malin, D. F., & Schlegel, Eric M. 1993a, ApJ, 416, 167
Schlegel, Eric M. 1993, In preparation
Schlegel, Eric M. et al. 1993, ApJ, 000, 000
Schlegel, Eric M. & Petre, R. 1993, ApJ, 412, L29
Schlegel, Eric M. & Kirshner, R. 1989, AJ, 98, 577
Snowden, S. & Freyberg, M. J. 1993, ApJ, 404, 403
Suzuki, T., Kumagai, S., Shigeyama, T., Nomoto, K., Yamaoka, H., & Saio, H. 1993, Nature, 000, 000
Wheeler, C. & Harkness, R. 1990, *Supernovae*, A. Petschek, (Berlin: Springer-Verlag Inc.), 1
Woosley, S. 1990, *Supernovae*, A. Petschek, (Berlin: Springer-Verlag), 182
Zimmermann, H. et al. 1993, AA, 000, 000
Zwicky, F. 1965, *Stellar Structure*, L. Aller & D. McLaughlin, (Chicago: University of Chicago Press), 421

DETECTION OF X-RAYS FROM SN 1987A WITH ROSAT

Paul Gorenstein, John P. Hughes, and Wallace H. Tucker
Harvard-Smithsonian Center for Astrophysics

ABSTRACT

Soft X-rays (0.5 - 2 keV) were detected with the ROSAT PSPC from the direction of SN 1987A which falls within an association in the LMC that is rich in B stars. The emission is consistent with two point sources in the LMC with total luminosity $\sim 10^{34}$ erg/s. The brighter source is identified with SN 1987A on the basis of its positional agreement. We interpret the emission as arising from the interaction of supernova ejecta with the pre-existing blue giant wind. The second source may be an unidentified Be/X-ray binary. Another possibility is that the second source is an X-ray echo from the supernova outburst. The X-ray echo interpretation requires that the supernova emitted about 10^{47} erg in a burst of soft X-rays.

1 Introduction

On February 23, 1987 SN 1987A, the nearest visible supernova in 400 years, was detected in the Large Magellanic Cloud. 131 days later the first positive detection of X-rays was made by Ginga (Dotani et al. 1987) and by detectors on the MIR Space Station (Sunyaev et al. 1987) which were sensitive in the hard X-ray band. Ginga detected X-ray emission for about 1000 days after the explosion (c.f. Inoue et al. 1992).

The ROSAT PSPC observed the SN 1987A region for 6400 seconds during 17-28 June 1990, following the period of "first light". It obtained an upper limit of 2.5×10^{34} erg/cm^2-s in the 0.3-2.4 keV band (Trümper et al. 1991). We carried out a longer, more sensitive PSPC measurement of the region in 1991 and 1992 and detected an X-ray source at the expected position.

2 Measurements

The dates and measurement times are given in Table 1. A contour diagram of the central region is shown in Figure 1. There is a very significant excess of about 60 counts above an expected background of 54 in a one square arcmin region. The emission is more extended than what is expected from a simple point source. A model consisting of two point sources was fit to the data. Their positions and intensities plus a background level were treated as seven free parameters. A single source model was also fit, but the two-point source model provides a significantly better fit. However, a single extended source cannot be excluded. The positions of SN 1987A, and the two sources, Source 1 and Source 2 are given in Table 2.

Table 1
ROSAT Observation Times of SN 1987A

Date	Total Time
6 Oct. 1991	17320 sec
30 April 1992	1520
10 May 1992	1159
12-14 May 1992	7208
	27207

© 1994 American Institute of Physics

Table 2
Source Positions (J2000)

Source	RA*	Dec*
1	5h 35m 26.3s ± 0.8s	-69° 16' 9.4" ± 4.9"
2	5h 35m 21.9s ± 1.7s	-69° 16' 26.5" ± 6.5"
SN1987A**	5h 35m 28.0s	-69° 16' 11.7"

*Including one sigma statistical errors.
**West et al., 1987, processed to epoch 2000 coordinates.

The position of Source 1 differs from that of SN 1987A by 9.3 arcsecs. The one sigma statistical error in Source 1's position is 6.6 arcsec in two dimensions. A study of a stellar association indicates there is a random systematic error of about 3 to 4 arcsecs in PSPC positions across the central region of the detector (E. Feigelson, Private Communication). The combined statistical and systematic uncertainty in the position of Source 1 is about 8 arcsecs. This is sufficient to account for its offset from SN 1987A. The position of Source 2 is 35 arcsecs from SN 1987A which is well beyond the range of uncertainty. Based upon the positional agreement, we identify Source 1 with SN 1987A. Table 3 lists the number of counts detected in three pulse height bands of the ROSAT PSPC for Source 1 and Source 2.

The best fit value of kT is 0.76 ± 0.20 keV (one sigma error) for relative abundance of 1. The value of kT is essentially the same for abundances of 1/3, and 3. In each case, there is no upper limit to the allowed temperature at the 2-sigma confidence level. At a distance of 51 kpc (Jakobsen et al. 1991, and Panagia et al, 1991), the luminosity of Source 1 is 8×10^{33} erg/s and of Source 2 in 4×10^{33} erg/s. We did not analyze the data for periodic temporal variability. The small number of counts makes it difficult to search for a pulsar without prior knowledge of the period.

Table 3
Net Counts from Source 1 and Source 2
ROSAT PSPC, 27 ksec

Pulse Height (keV)	Source 1 (SN1987A)	Source 2
0.07−0.40	6.5 (+8.6, −6.5)	0.0 (+6.1, −0.0)
0.40−1.00	26.6 (+8.7, −7.6)	12.5 (+10.8, −9.1)
1.00−2.40	17.6 (+6.2, −5.4)	10.9 (+5.2, −4.5)

3 Discussion

The source positions coincide with a B association within the LMC. We consider the possibility that the emission is due to stars. Walker and Suntzeff (1990) have published the UBV colors for the 39 brightest stars within 30 arcsec of SN 1987A. They determine the reddening correction for these stars to be E(B-V) = 0.17, corresponding to A_V = 0.51 and E(U-B) = 0.13. Adopting a distance modulus of 18.5 for the LMC, we determine the intrinsic V magnitude, colors and absolute visual magnitudes, and used the tables given in Lang (1992) to assign the spectral types and the bolometric luminosities. We use a value of log L_X/L_{bol} = (-6,-5,-6,-7) for (B1-B4,B5-B9,FGK,A) stars to obtain upper limits to the X-ray emission from the coronae of these stars (Maggio et al. 1990, and Grillo et al. 1992). The most luminous stellar coronae in the association are early

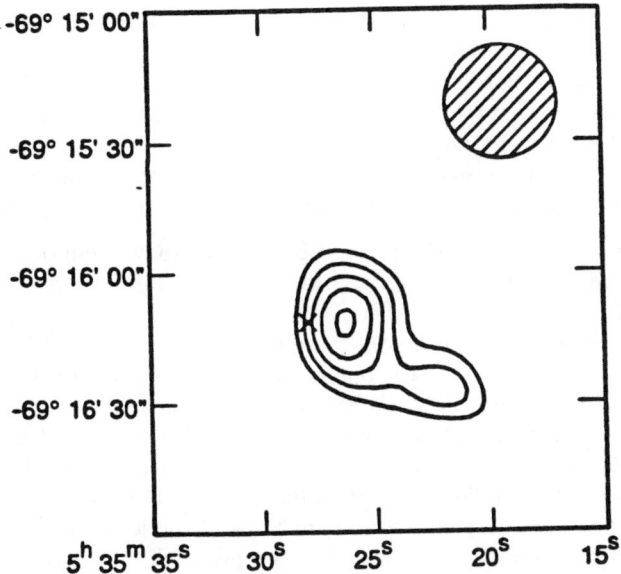

Fig. 1.— X-ray contours for the 27 ksec of observing time. The contour levels are linear and range from 1.25 e-6 to 1.88e-6 cts/s/sq-arcsec. The "X" symbol denotes the position of SN 1987A. The coordinates are epoch 2000. The round cross-hatched circle shows the angular width of the ROSAT PSPC resolution function.

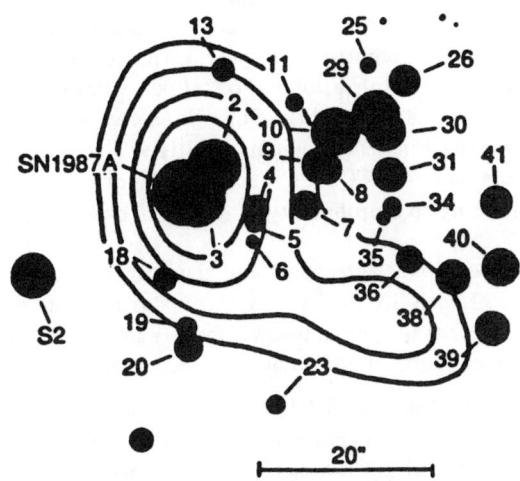

Fig. 2.— The X-ray contours superimposed upon a star field derived from a U band photograph by Walker and Suntzeff (1990) of the region near the SN 1987A. The contours were shifted to make Source 1 coincident with SN 1987A. The stars' numbers correspond to those of Fig. 3 in the Walker and Suntzeff paper.

B giants with $L_X \leq 10^{32}$ erg/s. The summed X-ray emission from all the stellar coronae should be $\leq 5 \times 10^{32}$ erg/s. The log $L_{bol}/L_\odot$ limits used are valid for main sequence or giant stars, so this conclusion is unaffected if the G and K stars are foreground objects within own galaxy. From these numbers it is apparent that none of these stars, individually, can explain the observed X-ray sources, nor can their integrated X-ray emission. An undetected cataclysmic variable cannot explain the emission since CV X-ray luminosities are in the range $10^{31\pm1}$ erg/s (Patterson and Raymond, 1985).

The only type of main sequence or giant early B type star known to produce X-rays of the intensity required is a Be/X-ray binary. Examples are X Per (Meurs et al. 1992) and Hen 715 (Grillo et al. 1992) both of which have been observed to have $L_X \sim 10^{34}$ erg/s over a period of two decades since they were first observed by the Uhuru X-ray satellite (Forman et al. 1978). The recent report by Walborn et al. (1993) that Star 3 of the stars within a few arcseconds of the supernova explosion (Sonneborn, Altner, and Kirshner, 1987) is a Be star with a suggested spectral classification of B2II raises the possibility that Source 1 is a Be/X-ray binary. However, the survey of Meurs et al. (1992) indicates that only 1 of the 114 stars OBe stars (X Per) in the Bright Star Catalog were observed to have a luminosity within a factor three of the level we measured for Source 1. If the same ratio applies to the LMC, the probability that Source 1 is a Be/X-ray binary is ~ 0.01.

It is not known whether any of the other 27 B stars of the association is a Be star, or more to the point, whether it is a Be/X-ray binary emitting at a level $\sim 10^{33.8\pm0.2}$ erg/s. Grillo et al. (1992) found one such star (Hen 715) in a sample of 1545 B stars observed by the Einstein IPC, whereas Meurs et al. (1992) found one (X Per) in a sample of 1026 OB stars observed with ROSAT. The frequency of 1 Be X-ray binary per thousand B stars is in agreement with the results of model calculations of the formation of Be stars through evolution of close binaries to a system with a neutron star and a rapidly spinning B star well inside its Roche lobe (Pols et al. 1992). Based on the above discussion, we estimate that the probability of one Be/X-ray binary in our sample of 27 B stars is ~ 0.03. From the Einstein Medium Sensitivity Survey (Gioia, et al., 1990), we can estimate that the number of QSO's sources at the observed flux is about 60 per square degree, corresponding to a probability of ~ 0.02 of observing one QSO in a 1 arc minute field of view.

On the basis of the positional agreement and the low probability of other explanations, we identify Source 1 with SN 1987A. A neutron star remnant of SN 1987A should be shrouded by the ejecta that is opaque below 5 for keVhe radius at least 10 years after the explosion (Gorenstein, Pinto, Hughes and Tucker, 1990).

We interpret the X-radiation from SN 1987A as being due to the reverse shock wave created by the interaction of the expanding ejecta with a blue giant wind, characterized by $1/r^2$ density profiles. Following Chevalier's (1982) treatment of this interaction, the luminosity of the reverse shock is $L_r = < n^2 V \Lambda(T) >_r = 4\pi R_s^3 (\Delta R_S/R_S) < n^2 \Lambda(T) >_r$ where n is the electron number density, $\Lambda(T)$ is the cooling function for a plasma of temperature T; $<>_r$ represents an average over the reverse shock wave, and R_s is the radius of the leading shock wave, given by $R_s = 1.1 v_c \tau (t/\tau)^{6/7}$ for a similarity solution for ejecta with a $1/r^9$ density profile (Arnett et al. 1989) interacting with a $1/r^2$ blue giant wind. The characteristic velocity of the supernova ejecta is v_c and τ is the characteristic time for the leading shock wave to sweep up a mass in the blue giant wind equal to the ejecta mass.

Two observational tests could determine whether the X-ray emission we detect is due to shock waves produced by explosion ejecta or to the fact that the Be system Star 3 is indeed an X-ray emitter, despite the low a priori probability. One test is detection of variability. If shock waves are responsible, the flux will decrease at a modest rate $\sim 25\%$ a per year (until the ejecta encounters the dense red giant wind). If an X-ray emitting Be star is the source, then taking X per and Hen 715 as examples, we can expect variability by factors of 3 to 7. The other test is the detection of a finite size. The size is expected to be about 1 arcsecond for the shock wave source.

As described above, there is a small probability ($\sim 3\%$) that Source 2 is a Be/X-ray binary. The models of Pols et al (1992) indicate that, in order for a neutron star companion to have been formed, the B star must be of type B2 or earlier. Stars 38, 39 or 40 would appear to be the prime suspects by virtue of their spectral type and positions.

An intriguing alternate possibility is that Source 2 represents an X-ray echo of SN 1987A. Chevalier and Emmering (1989) have presented a model in which the visible light echo observed 9" from SN 1987A (Bond et al. 1989, Couch and Mailin 1989, Crotts and Kunkel, 1991) is due to a shock wave produced by the interaction of the presupernova red giant wind with the interstellar medium. They estimate the radial distance R of the shock from the supernova to be ~ 15 light years.

The brightness of the echo requires that the supernova was a luminous source of X-radiation. Taking a column density through the red giant shell of $N_H = 3 \times 10^{18}/cm^2$ (cf. Chevalier and Emmering, 1989), the fraction f_{sc} of scattered X-ray luminosity from the shell is $\sim 3 \times 10^{-5}$ (cf. Mauche and Gorenstein, 1986, Gorenstein, 1975) and a fraction, 3×10^{-14} of the energy of the supernova X-ray burst will appear in the echo (cf. Chevalier and Emmering, 1988). An X-ray burst of the order of 10^{47} erg is required to explain our Source 2 in terms of an X-ray echo. In their study of the UV and optical emission lines of SN 1987A, Fransson and Lundqvist (1989) found that black body temperatures in the range 3×10^5 to 8×10^5 K are favored and that at least 10^{56} photons above 100 eV were required. If we assume that $\sim 3 \times 10^{56}$ photons of average energy 200 eV (8×10^5K black body) were emitted, the emitted energy in soft X-rays is 10^{47} erg, which is not inconsistent with the data, given the uncertainties in the spectrum of Source 2.

The X-ray echo hypothesis makes a number of predictions: (1) the distance of Source 2 from SN 1987A should increase with time; (2) the intensity of Source 2 should decrease by 25%/yr due to changes in the scattering angle; (3) the spectrum of Source 2 should be soft, corresponding to a black body temperature 8×10^5 K.

4 Acknowledgements

We would like to thank Robert Kirshner and Jeffrey McClintock of CfA, and Salvatore Sciortino of the University of Palermo for information on objects discussed in this paper. We also thank Mark Birkinshaw, Robert Kirshner, Frederick Seward, Harvey Tananbaum and Olaf Vancura for their comments on the manuscript. This work was supported in part by the ROSAT Guest Investigator Program, contract NAG5-2233.

5 References

Arnett, W.D., Bahcall, J.N., Kirshner, R.P., and Woosley, S.E., 1989, *ARAA* **27**, 629.
Bond, H, Gilmozzi, R., Meakes, M., and Panagia, N., 1990, *Ap.J.* **354**, L49.
Chevalier, R.A., 1982, *Ap.J.* **258**, 790.
Chevalier, R.A., 1986, *Ap.J.* **308**, 225.
Chevalier, R.A., 1992, *Nature* **355**, 617.
Chevalier, R.A., and Emmering, R.T., 1988, *Ap.J.* **342**, L105.
Chevalier, R.A., and Emmering, R.T., 1989, *Ap.J.* **342**, L75.
Couch, W., and Malin, D., 1989, *IAU Circ.* **No. 4739**.
Crotts, A., and Kunkel, W., 1991, *Ap.J.* **366**, L73.
Dotani, T., et al, 1987, *Nature* **330**, 230 (37 authors).
Forman, W., Jones, C., Cominsky, L., Julien, P., Murray, S., Peters, G., Tananbaum, H., and Giacconi, R., 1978, *Ap.J. Suppl.* **38**, 357.
Fransson, C., and Lundqvist, P., 1989, *Ap.J.* **341**, L59.
Gioia, I.M., Maccacaro, T., Schild, R.E., Wolter, A., Stocke, J.T., Morris, S.L., and Henry, J.P., 1990, *Ap.J. Suppl.* **72**, 567.
Gorenstein, P., 1975, *Ap.J.* **198**, 95.
Gorenstein, P., Pinto, P., Hughes, J., and Tucker, W., 1990, in, *SN 1987A and Other Supernovae*, ed. I.J. Danziger and K. Kjär (ESO Workshop and Conference Proceedings No. 37), P499.
Grillo, F., Sciortino, S., Micela, G., Vaiana, G. and Harnden, F.R., 1992, Ap.J. Suppl. **81**, 795.
Inoue, H., Hauashida, K., Itoh, M., Kondo, K., Takeshima, T., Yoshida, K., and Tanaka, Y., 1991, *Publ. Astron Soc., Japan* **43**, 213.
Jakobsen, P., Machetto, F., Panagia, N., 1993, *Ap.J.* **403**, 736.
Lang, K., 1992, Astrophysical Data, (Springer-Verlag, New York) p. 125ff.
Luo, D., and McCray, R., 1991, *Ap.J.* **379**, 659.
Mac Low, M., and McCray, R., 1988, *Ap.J.* **324**, 776.
Maggio, A., Vaiana, G., Haisch, B., Stern, R., Bookbinder, J., Harnden, F., Jr., and Rosner, R., 1990, *Ap.J.* **348**, 253.
Mauche, C., and Gorenstein, P., 1986, *Ap.J.* **302**, 371.
Meurs, E., et al, 1992, *Astron. Ap.* **265**, L41 (10 authors).
Panagia, N., Gilmozzi, R., Macchetto, F., Adorf, H.-M., and Kirshner, R.P., 1991, *Ap.J.*, **380**, L23.
Patterson, J., and Raymond, J., 1985, *Ap.J.* **292**, 535.
Pols, O.R., Cote, J., Waters, L. and Heise, J. 1991, *Astron. Ap.* **241**, 419
Sonneborn, G., Altner, B., and Kirshner, R.P., 1987, *Ap.J.* **323**, L35.
Stark, A.A., Gammie, C.F., Wilson, R.W., Bally, J., Linke, R.A., Heiles, C., Hurwitz, M., 1992, *Ap.J. Suppl.* **79**, 77.
Sunyaev, R., et al, 1987, *Nature* **330**, 227 (34 authors).
Trümper, J., et al, 1991, *Nature* **349**, 579 (14 authors).
Walborn, N.R., Phillips, M.M., Walker, A.R., and Elias, J.H., 1993, Pub. Astron. Society of the Pacific (In Press).
Walker, A.R., and Suntzeff, N., 1990, *Pub. Astron. Society of Pacific* **102**, 131.
Wang, L., Mazzali, P., 1992, *Nature*, 355, 58.
West R.M., Laubert, A., Jorgensen, H.E., and Shuster, H.-E., 1987, *A&A* **177**, L1.

ROSAT OBSERVATIONS OF AN OPTICAL QUASAR SURVEY FIELD

Hermann Brunner
Astronomy Institute, University of Tübingen, Germany

Email ID
brunner@ait.physik.uni-tuebingen.de

ABSTRACT

Deep (T $\sim$ 35 ksec) pointed ROSAT observations of a 2.2° × 2.2° optical quasar survey field (149 quasars; m_{lim} = 20.5; Crampton et al., 1989) have yielded a detection rate (3 σ) of $\sim$ 60 % (86 quasars; limiting sensitivity $\sim$ 5 · 10^{-15} erg cm^{-2} s^{-1} keV^{-1} at 1 keV). 46 quasars were bright enough to perform spectral power law fits. The mean energy power law index drops from $\sim$ 1.4 at z = 0 to $\sim$ 0.9 at z > 2. This is interpreted as being due to a break in the spectrum between a soft, thermal accretion disk and a hard power law component, occuring at a source frame energy around 1 keV. Mean accretion disk model parameters are derived (M = 5 · 10^8 M$_\odot$, $\dot{M}$ = 0.65 · M$_{Edd.}$, $\alpha_{visc.}$ = 0.5), using an geometrically thin α-accretion disk model (Dörrer et al., 1992, and references therein). The α_{ox} distribution and the optical number-redshift relation is modeled, using the accretion disk parameters as determined from the X-ray spectral data and assuming a constant comoving volume density (H$_0$ = 100 km/s Mpc, q_0 = 0.5) and statistical orientation of the inclination angles of the model source population.

1. X-Ray and Optical Data

We have observed a 2.2° × 2.2° field from the CFHT grating lens (grens) quasar survey (Crampton et al. 1989) in 9 overlapping ROSAT PSPC pointings of 9 ksec each. The mean exposure in the survey area is $\sim$ 35 ksec. The CFHT quasar survey has a completeness limit of $m < 20.5$. The part of the survey selected for ROSAT observations (4.85 deg^2) is centered around $13^h39^m33^s$ 27°05'. The 149 quasars in the field have redshifts ranging from 0.24 to 3.06. Galactic N_H in the field is at 1.05 × 10^{20} cm^{-2}.

2. Data Analysis

At the position of each of the 149 optical quasars, ROSAT PSPC spectra in the energy range from 0.15 to 2.4 keV were extracted from a circular area of radius 2.5'. The background was accumulated from an annulus with inner radius 2.5' and outer radius 6.25', centered on the source position. Areas in the background annulus contaminated by sources were removed, again using an extraction radius of 2.5'. In an eyeball inspection 6 of the extracted spectra were rejected in order to avoid cases where (1) an X-ray source was not centered in the detection circle, indicating that the detected X-ray source is not the optical quasar, and where (2) more than one X-ray source contributed to the flux in the extraction circle. This left 86 positions where a significant signal (> 3σ) was observed. The count rate distribution of the 86 detections is plotted in Fig. 1.

Of these detections, 46 were bright enough to permit binning the data into a minimum of 3 spectral bins with a significance of 5σ each, thus permitting to perform at least rudimentary spectral fits on these targets.

3. Detected Fraction; Distribution of the Broad-Band Spectral Index α_{ox}

The fraction of quasars detected in each redshift range is plotted in Fig. 2 a. Note the decline of the detected fraction at redshifts $z > 1.5$. The fraction of X-ray detected quasars is found to be independent of the optical magnitude of the counterparts (Fig. 2 b). This suggests that many of the as yet unidentified X-ray sources in the field are quasars with magnitudes below the optical survey cutoff.

The observed broad-band spectral index α_{ox} is found to be independent of redshift and closely clusters around 1.5. The optical quasars not detected in X-rays, on average, yield a lower limit of $\alpha_{ox} \gtrsim 1.6$ (Fig. 6; dotted line), indicating that, on average, the non detections are less X-ray bright than the quasars detected in X-rays.

α_{ox} is found to correlate closely with the optical magnitude, in the sense that optically weak quasars have a flat α_{ox}. We find $<\alpha_{ox}>\sim 1.3$ for the optically weakest quasars in the sample. Taken in conjunction with the observation that the X-ray detected fraction does not break off at magnitudes approaching the optical survey limit, this points to the existence of a class of quasars with extremely flat α_{ox} distributions ($\alpha_{ox} \lesssim 1.3$), which are too weak to be easily detected optically. Optical follow up observations to clarify this point will be performed.

With the existence of a population of quasars with $\alpha_{ox} \gtrsim 1.6$, not observed in X-rays, and a population of quasars with $\alpha_{ox} \lesssim 1.3$, not observed in the optical, the observed shape of the α_{ox} distribution may thus more reflect the different sensitivity thresholds in both spectral domains than any true property of the underlying population. Attempts to model the shape of the observed α_{ox} distribution are described below.

4. Spectral Fitting

The 46 quasar spectra were fit, both to power laws with fixed galactic N_H ($N_H = 1.05 \times 10^{20} cm^{-2}$; Stark et al., 1992), and to power laws where N_H was left free to vary. No excess N_H over the galactic value was observed in any of the targets. We find a decline of the mean X-ray power law spectral index (N_H fixed at galactic value) from $\alpha_{energy} \sim 1.4$ at redshift $z = 0$ to $\alpha_{energy} \sim 0.9$ at $z > 2$ (see Figs. 3 and 4). A physical interpretation for this decline is given below.

5. Contamination from Radio-Loud Quasars

In an optically selected sample only a small percentage of radio-loud quasars is expected. A cross-correlation of our optical/X-ray sample with the 6 cm northern sky catalogue (Becker et al. 1991; completeness limit $\sim$ 30 mJy) has yielded one radio-loud quasar in the sample. Unfortunately, the radio data available to us are not sensitive enough to decide which, if any, of the remaining quasars are radio-loud. Results derived on the Einstein EMSS (Gioia et al. 1990; Stocke et al. 1991) suggest that the fraction of radio-loud quasars is $< 10 \%$ (75 radio detections among 395 EMSS AGN, 42 of which (10.6 %) were radio-loud). As our X-ray data are based on an optically selected sample, the fraction of radio-loud quasars is expected to be lower than in the EMSS, which was X-ray selected.

Radio-loud quasars are known to be more luminous (Zamorani et al. 1981) and to have harder X-ray spectra (Wilkes and Elvis 1987) than radio-quiet quasars. We have tested whether the decline in spectral index with redshift might be due to a contamination of the sample with radio-loud quasars at high redshifts: Removing the one source known to be radio-loud (z=1.185) plus a random selection of four additional sources with $z > 1$ does not change the general trend of a declining spectral index with redshift. We conclude that the decline in spectral index is not caused by a contamination from up to $\sim 10\%$ radio-loud quasars.

6. Modeling of the X-Ray Spectra

We interpret the observed decline of the mean X-ray power law index with redshift as being due to a break in the mean X-ray spectrum, occuring close to a source frame energy of 1 keV, which is gradually shifted out of the ROSAT sensitivity window at higher redshifts. This break in the spectrum is modeled in terms of a soft accretion disk component which dominates the spectrum at source frame energies below ~ 0.6 keV, and a standard, AGN type hard power law component with spectral index $\alpha_{energy} = 0.7$ which dominates the spectrum at higher energies.

Our semi-analytical accretion disk calculations (Dörrer et al. 1992) closely follow Czerny and Elvis (1987), Wandel and Petrosian (1988) and Maraschi and Molendi (1990). The accretion disk model parameters are: the central mass, M, the mass accretion rate in units of the Eddington accretion rate, $\dot{M}_{Edd.}$, the viscosity parameter, α, and the inclination angle of the disk, Θ. Sample model spectra are shown in Fig. 5.

In order to limit the number of free parameters, the normalization of the power law component (1×10^{27} erg/sec Hz at 1 keV, source frame), the viscosity parameter ($\alpha = 0.5$) and the inclination angle of the accetion disk ($\cos \Theta = 0.9$) were fixed at typical values. For the two remaining accretion disk parameters, the central mass M and mass accretion rate $\dot{M}_{Edd.}$, model spectra for a grid of values were calculated, transformed to the observer frame for a range of redshifts ($z = 1 \dots 3$), folded through the response of the ROSAT PSPC, and re-fit to simple fixed N_H power law spectra.

The set of parameters best representing the observed decline of the power law index with redshift is:

M	$= 5 \times 10^8 M_\odot$
$\dot{M}_{Edd.}$	$= 0.65$
$\alpha_{visc.}$	$= 0.5$ (fixed)
$\cos \Theta$	$= 0.9$ (fixed)
α_{PL}	$= 0.7$ (fixed)
norm. of PL (at 1 keV, source frame)	$= 10^{27}$ erg/sec Hz (fixed)

7. Modeling of the α_{ox}-Distribution

We have also attempted to model the shape of the observed distribution of α_{ox}. For this purpose, the model parameters determined from the X-ray spectral data (see above) were used and three additional assumptions were made: (1) The comoving volume density of the model quasars was assumed to be constant and a Hubble constant of $H_0 = 100$ km/s Mpc and $q_0 = 0.5$ was used. (2) The inclination angles of the accretion disks were assumed to have random orientation. (3) 50 % of the optical emission was assumed to be due to the accretion disk. For the appropriate number of sources in each redshift shell and each inclination angle range, the X-ray and optical fluxes were calculated. An α_{ox} histogram was accumulated for the subset of sources above the detection threshold in both spectral ranges. In Fig. 6 the result of this calculation is presented for two different central masses (dashed lines). Note that the shape of the α_{ox}-distribution is mainly due to point (2), above, and the fact that the emission observed from the accretion disk is strongly dependent on the inclination angle, in the sense that disks seen face-on have the highest α_{ox} values. While the agreement is not complete, our simple model assumptions nevertheless seem to be adequate to reasonably represent the mean and width of the α_{ox} distribution.

8. Modeling of the Optical Number-Redshift Relation

The assumptions made to model the α_{ox} distribution were also used to calculate, in each redshift range, the expected number of quasars brighter than the limiting sensitivity of the optical quasar survey ($m_{vis.} < 20.5$). In Fig. 7, the resulting distribution is plotted for two different masses together with the observed quasar counts. Again, in the light of the simple model assumptions made, the agreement seems to be satisfactory.

9. Summary

We find a decline of the X-ray spectral index measured by the ROSAT PSPC detector with redshift in a sample of 46 optically selected quasars. The most probable cause for this decline is curvature in the X-ray spectrum due to emission from an accretion disk which dominates the X-ray spectrum at energies below 0.6 keV and is shifted out of the ROSAT sensitivity window at higher redshifts. We present accretion disk parameters which model the observed decline in spectral index. Model predictions for the α_{ox}-distribution and the optical number-redshift relation are given.

References

Becker, R. H., White, R. L., Edwards, A. L., 1991, ApJS, 75, 1
Crampton, D., Cowley, A. P., Hartwick, F. D. A., 1989, ApJ, 345, 59
Czerny, B., and Elvis, M., 1987, ApJ, 321, 305
Dörrer, Th., Friedrich, P., Brunner et al., 1992, 'X-ray emission from active galactic nuclei and the cosmic X-ray background', eds. Brinkmann and Trümper, Garching 1991, p. 130
Gioia, I. M., Maccacaro, T., Schild, R. E. et al., 1990, ApJS, 72, 567
Maraschi, L. and Molendi, S., 1990, ApJ, 353, 452

Stocke, J. T., Morris, S. L., Gioia, I. M. et al., 1991, ApJS 76, 813
Wandel, A. and Petrosian, V., 1988, ApJ, 329, L11
Wilkes, B. J. and Elvis, M., 1987, ApJ 323, 243
Zamorani, G., Henry, J. P., Maccacaro, T. et al., 1981, ApJ, 245, 357

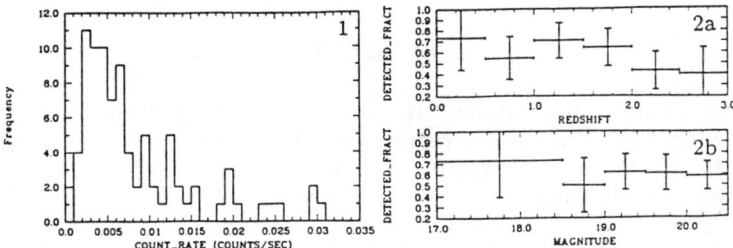

Fig. 1 — Distribution of ROSAT PSPC source count rates for 86 quasars detected in the survey area.

Fig. 2 — a: Fraction of quasars detected in X-rays as a function of redshift. b: Fraction of quasars detected in X-rays as a function of optical magnitude.

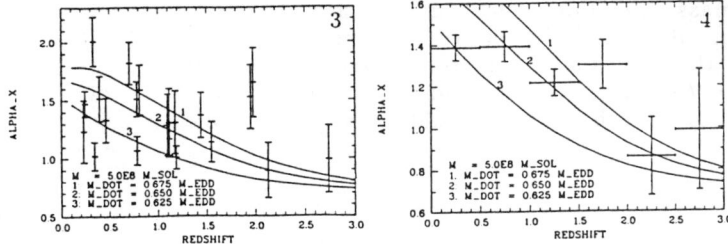

Fig. 3 — Power law spectral indices for fixed N_H fits, plotted as a function of redshift. Only sources with 1σ errors < 0.3 are plotted (21 targets). Different model predictions are shown. See text for model parameters not listed in the plot. The corresponding source frame spectra are plotted in Fig. 5.

Fig. 4 — Same as Fig. 3, but weighted means of power law spectral indices are shown for different redshift intervals.

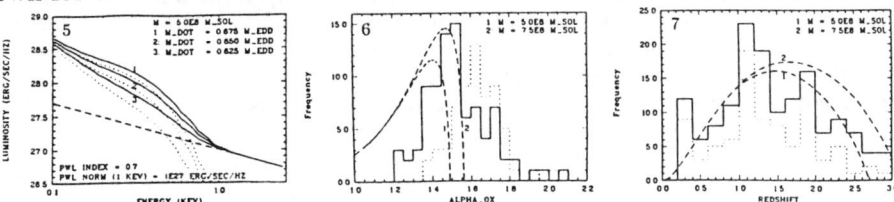

Fig. 5 — Accretion disk model spectra for different mass accretion rates and superposition with hard power law are plotted in the ROSAT PSPC sensitivity window. The set of spectral parameters is the same as in Figs. 3 and 4.

Fig. 6 — Distribution of α_{ox}. Dotted line: α_{ox} distribution determined from X-ray upper limits. Dashed lines: Model predictions of α_{ox} distribution for two different central masses. See text for other parameters not listed in the plot.

Fig. 7 — Number of quasars in the survey area as a function of redshift. Dotted line: Number of quasars detected in X-rays. Dashed lines: Model predictions of number of detected sources in each redshift range for two different central masses. See text for other model parameters not listed in the plot.

CAN AGN ALONE MAKE THE COSMIC X-RAY BACKGROUND?

DARRYL LEITER* and ELIHU BOLDT
Code 666, Laboratory for High Energy Astrophysics,
NASA/Goddard Space Flight Center, Greenbelt, MD 20771
* NRC Senior Resident Research Associate

ABSTRACT

Recent ROSAT X-ray observations of AGN have yielded important new information about the analytic structure of the AGN X-ray luminosity function and its evolution to redshift z=3 (Boyle et al. 1993). Using the luminosity evolution obtained within the cosmological context of $\Omega=0$ we find, as recently noted by Zdziarski, Zycki, & Krolik (1993), that AGN could readily make up the CXB (cosmic X-ray background). However, in this case we conclude that accounting for the CXB with accretion-powered AGN emission is incompatible with the observed mass function for present-epoch black hole galactic nuclei (both active and dormant). Although the luminosity evolution obtained with ROSAT for $\Omega=1$ is indeed compatible with this mass function such a scenario definitely falls short of accounting for all the CXB, even when considering unified AGN models. We suggest that in this case the X-radiation characteristic of highly compact PAG (precursor active galaxies) sources indicative of the initial black hole mass growth desired would provide a natural explanation of the residual CXB.

SUMMARY

Based on general energetics arguments Daly (1991) has emphasized that producing the CXB (cosmic X-ray background) via black hole accretion is clearly more tenable than doing so via other plausible mechanisms, at least for an underlying cosmology corresponding to $\Omega=1$. Following this lead, we have examined currently available information about the extragalactic X-ray flux under the assumption that essentially *all* this radiation is generated via black hole accretion. In particular, the extent to which canonical AGN can account for it is evaluated within this context. By combining the deep *ROSAT* X-ray selected AGN sample (median z=1.5) with the larger EMSS AGN sample (median z=0.3) Boyle et al. (1993) have succeeded in providing us with candidate solutions for the AGN X-ray luminosity function and its evolution needed for addressing this issue quantitatively. By considering 1) candidate solutions to the soft X-ray luminosity function of relatively unobscured AGN readily visible to *ROSAT*, 2) the more obscured related AGN implied by the unified model (Antonucci and Miller 1985; Awaki 1991) and 3) recently obtained Compton *GRO* broadband Seyfert X-ray spectra (Maisack et al. 1993), we are able to infer the total present-epoch black hole mass density emerging from the AGN phenomenon in previous epochs. We compare this value with that implied by the broadband X-ray flux associated with the entire extragalactic sky, foreground plus background. From this black hole integral mass constraint, as provided by the CXB, and the mass spectrum observed for

© 1994 American Institute of Physics

present-epoch AGN (Padovani, Burg and Edelson 1990) we determine which one of the two basic solutions for AGN cosmological evolution (i.e., corresponding to $\Omega=1$ or $\Omega=0$) is most tenable. The implications of this are then investigated in the context of a residual background of faint sources corresponding to PAG (precursor active galaxies) indicative of an earlier epoch (Leiter and Boldt 1982, 1992; Boldt and Leiter 1987) when all galactic nuclei presently with black holes were amply fueled and clearly active.

DISCUSSION

As emphasized by Soltan (1982), the total present-epoch black hole mass density arising from accretion can be determined directly from the *observed* total sky flux of all the resultant radiation, independent of the parameter values (Ω and H_o) characterizing the underlying Friedmann cosmology. Since the omnidirectional energy flux ($4\pi I$) of the CXB is observationally very well defined (Boldt 1987, 1992; Gruber 1992) it provides us with an especially good measure for this. In particular, the mass density associated with generating the CXB via accretion is given by

$$\rho(\text{CXB}) = (\varepsilon^{-1} - 1) \, \eta \, (1+\langle z \rangle) \, (4\pi I) / c^3, \tag{1}$$

where
$\varepsilon \equiv L(\text{bolometric}) / [c^2(dM/dt)]$,
$\eta \equiv L(\text{bolometric}) / L(\text{all X-rays})$, and
$\langle z \rangle \equiv$ the mean redshift of CXB photons
with respect to the observed energy flux.

Since the mean black hole seed mass (i.e., that prior to any growth due to accretion) is expected to be no larger than a Jeans mass ($\sim 10^6 M_o$, immediately after the recombination epoch), much smaller than that of a canonical AGN supermassive black hole, the total mass density of present-epoch black holes should be dominated by $\rho(\text{CXB})$, the density of accreted matter inferred by the CXB. This CXB implied black hole mass density is two orders of magnitude larger than that inferred from the Seyfert AGN mass function (Padovani, Burg &Edelson 1990). For comparison, the luminosity density (volume emissivity) corresponding to the Boyle et. al. (1993) X-ray luminosity function for AGN, extrapolated to z=4, can be used to evaluate ΔM, the associated mass per AGN accreted from z=4 to z=0. In general, ΔM scales with ε and η as follows:

$$\Delta M \propto (\varepsilon^{-1} - 1) \eta / n, \tag{2}$$

where n is the comoving number density of the AGN involved. We have calculated this for the two cases (corresponding to $\Omega=0$ and $\Omega=1$) considered by Boyle et al.(1993) already cited. In doing this we have taken the observed X-ray luminosity in the band (0.3-3.5keV) as 10% of the bolometric AGN luminosity, compatible with IR-UV estimates (Mushotzky, Done and Pounds 1993) but probably somewhat of an upper bound for sources with a "big bump" in the EUV. For Seyfert 1 galaxies we take the spectrum used by Boyle et al.(1993) (i.e., $\alpha=1.3$ at $E<1.5\text{keV}$, $\alpha=1$ for $E>1.5\text{keV}$) for energies < 3.5keV and the spectrum observed with *Ginga*

(Pounds, Nandra and Stewart 1992) at higher energies, with an exponential roll-off introduced to match the peak in (E_{S_E}) observed for NGC4151 with OSSE (Maisack et al. 1993). From this we calculate that η(Seyfert 1)=2.3 (i.e., a luminosity in broadband X-rays comparable to but somewhat less than that for all other bands combined). The accreted mass density ρ_A associated with the AGN phenomenon (z=4 to z=0) is then obtained as follows:

$$\rho_A = [n(\text{Seyfert})] \Delta M \qquad (3)$$

where n(Seyfert)=3.3[n(Seyfert 1)] and the unified model implies that the same value of ΔM applies to *all* types of Seyfert galaxies. The two solutions to the AGN luminosity function and its evolution considered here also imply two solutions for the mean initial black hole mass (M_i) at z=4. These solutions depend, of course, on the actual initial value of the Eddington ratio $\lambda \equiv L(\text{bolometric})/[1.3 \times 10^{38}(M/M_o) \text{ erg/s}]$ at z=4, still unspecified. The final *total* mass density ρ of the underlying black hole population is given by:

$$\rho = \rho_A + [n_S(M)_i]/f. \qquad (4)$$

The final mean black hole mass M_f (at z = 0) may be obtained as follows:

$$M_f = M_i + (\Delta M)f, \qquad (5)$$

where f *is* the fraction of galactic nuclei with black holes that are currently active. We interpret this fraction f as the probability of such a nucleus being in the "on" state (Cannizzo 1992) and indicative of the duty cycle for such activity in the past as well. The population of black holes underlying AGN then has a relatively large number density, $[n(\text{AGN})]/f$, possibly comparable to that of "normal" galaxies. The observed mean Seyfert AGN black hole mass to be compared with equation (5) is on the order of $10^7 M_o$ (Padovani, Burg and Edelson 1990). Mass curves for both Ω =0 and Ω =1, corresponding to Equations 2-5, are plotted in Figure 1. From the figure we see that the situation for $\Omega = 0$ is clearly unacceptable. However it is evident from this figure that scenarios for the CXB involving a unified model of accretion-fueled AGN have the potential of being consistent with the solution for the luminosity function and its evolution obtained by Boyle et al. (1993) within the context of Ω=1. However we find that this apparently acceptable solution definitely falls short of accounting for all the CXB, even when considering unified models for AGN. This difficulty occurs despite the fact that, when such a unified model is referred to the 2-10keV band, the *ROSAT* based luminosity density $[n \langle L \rangle)_0$ =1.7x10^{38}ergs s^{-1} Mpc^{-3}] is entirely consistent with that obtained from the correlation found between *IRAS* galaxies and HEAO-1 A2 observed CXB surface brightness fluctuations (Miyaji et al.1993).

Madau, Ghisellini and Fabian (1993) have succeeded in applying a unified AGN model to account for the CXB spectrum within the context of an Einstein - de Sitter cosmology (i.e., Ω =1) but they need strong continuous evolution of the comoving luminosity density up to z =5 to do it. This is clearly inconsistent with the ROSAT derived result that, for $\Omega =1$, the AGN luminosity evolves strongly *only* up to z=2 (Boyle et al. 1993). We find that a unified AGN model similar to one considered by Zdziarski, Zycki and Krolik (1993) for $\Omega =0$, but with the form of evolution needed for $\Omega =1$ taken up to z=4, can indeed be made to fit the broadband CXB spectrum within the context of an Einstein - de Sitter cosmology. Since this fit is found to require Seyfert spectra which exhibit a peak in EL_E at ~70 keV the effective redshift corresponding to the observed CXB peak in EI_E would then be at z(effective)=1.3. Although such a peak in Seyfert spectra may not be universal (e.g. IC4329a; Madejski et al. 1993), for this discussion we adopt the view that it is typical (Johnson et al. 1993). The Seyfert 2 spectra needed for this CXB fit correspond to self-absorption by $N_H = 10^{23} cm^{-2}$, exhibiting unit optical depth at ~3 keV (Morrison and McCammon 1983), enough to render these sources essentially inaccessible to the bandpass (<2keV) of *ROSAT* (as assumed). We find, however, that the Seyfert AGN normalization required for the unified model to adequately account for the CXB in this instance corresponds to an unacceptably large value [i.e., $(n \langle L \rangle)_0 = 2.8 \times 10^{38}$ ergs s^{-1} Mpc^{-3}(0.3-3.5keV)] for the present-epoch luminosity density. This not only exceeds the value obtained by Boyle et al. (1993) for the situation where $\Omega =1$ but is large enough to present us with a definite violation of the mass constraints previously described.

We have found that the difficulties described above can be easily resolved by noting that the underlying supermassive black holes which already exist at the onset of the canonical AGN phenomenon of supply-limited accretion must have undergone a previous growth phase where the accretion would be expected to be Eddington-limited. In this likely scenario (i.e., for $\Omega=1$), the residual CXB over and above the foreground of canonical AGN can be naturally explained by the characteristic X-ray emission from highly compact PAG (precursor active galaxy) sources associated with these numerous black holes, at redshifts just beyond the earliest AGN (Leiter and Boldt 1993).

REFERENCES

Antonucci, R. and Miller, J. 1985 ApJ, 297, 621
Awaki, H. 1991, PhD thesis, Nagoya University
Boldt, E., & Leiter, D., 1987 ApJ, 322, L1
Boldt, E. 1987, Phys. Rept., 146, 215
Boldt, E. 1992, in The X-ray Background, ed. X. Barcons & A. Fabian
 (Cambridge: Cambridge Univ. Press), p. 115
Boyle, B., et al. 1993 MNRAS, 260, 49
Cannizzo, J. 1992 ApJ, 385, 94
Daly, R. 1991 ApJ, 379, 37
Gruber, D. 1992, in The X-ray Background, ed. X. Barcons & A. Fabian
 (Cambridge: Cambridge Univ. Press), p. 229
Johnson, W. N., et al. 1993 Second Compton Science Symposium, University of Maryland
Leiter, D. & Boldt, E. 1982 ApJ, 260,1
Leiter, D. and Boldt, E. 1992 in Testing the AGN Paradigm, ed. S. Holt, S.
 Neff & M. Urry (New York: AIP) AIP 254, p 370
Leiter, D. & Boldt, E. 1993, in preparation

Madau, P., Ghisellini, G. & Fabian, A. 1993 ApJ, 410, L7
Madejski, G. et al. 1993 ApJ, submitted
Maisack, M., et al. 1993 ApJ, 407, L61
Miyaji, T., Lahav, O., Jahoda, K., & Boldt, E. 1993, in preparation
Morrison, R. & McCammon, D. 1983 ApJ, 270, 119
Mushotzky,R., Done, C. & Pounds, K. 1993 Ann Rev. Astron. Astrophys., 21, 717
Padovani, P., Burg, R. & Edelson, R. 1990 ApJ, 353, 438
Pounds, K., Nandra, K. & Stewart, G. 1992 in X-ray Emission from AGN and the CXB (eds W. Brinkmann and J. Trumper) MPE Report 235 p. 1
Soltan, A. 1982 MNRAS, 200, 115
Zdziarski, A., Zycki, P. & Krolik, J. 1993 ApJ Letters,414, L81

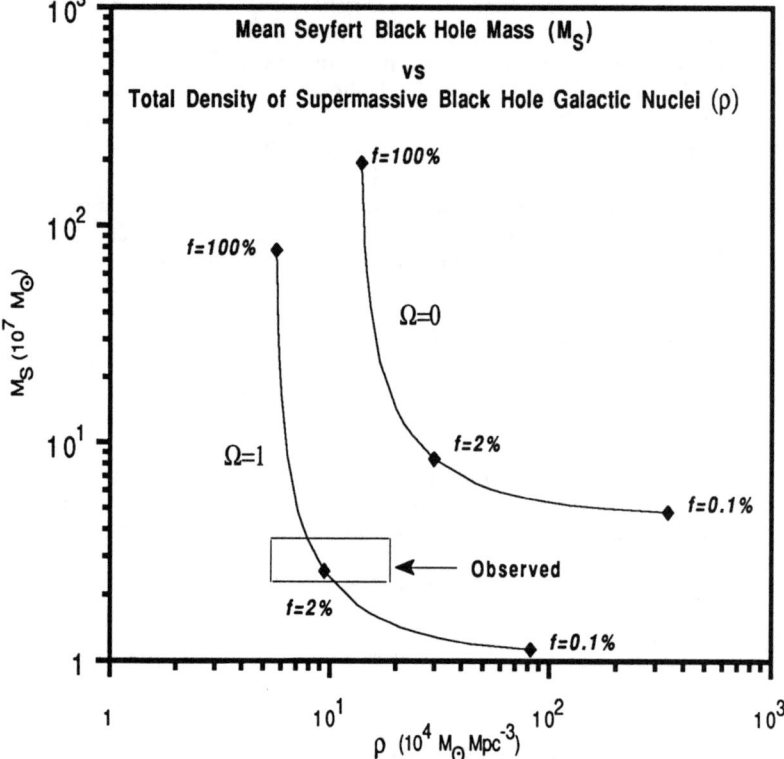

Figure 1. The lower bound on the mean present-epoch Seyfert AGN black hole mass and the lower bound on the total present-epoch black hole mass density (for both active and dormant black hole galactic nuclei) are shown for a range of values assumed for the underlying AGN duty cycle (f). The rectangle denotes the observed range of values defined by: a) the mean Seyfert black hole mass obtained from the observed Seyfert mass function and, b) the total black hole mass density estimated from the observed CXB (via Equation 1). The curves correspond to the mass growth during the luminosity evolution from $z = 4$ to the present for the two cases $\Omega = 0$ and $\Omega = 1$, taking $\lambda = 1$ at $z=4$. Points denoting the duty cycle for the AGN phase are shown for $f = 0.1\%$, 2% and 100%.

WORKSHOP PRESENTATIONS

ANALYSIS PROCEDURES FOR ROSAT PSPC OBSERVATIONS OF EXTENDED TARGETS

J. A. Mendenhall and D. N. Burrows
Dept. of Astronomy,The Pennsylvania State Univ.,University Park,PA 16802

S. L. Snowden
NASA/USRA, Goddard Space Flight Center, Greenbelt, Md. 20771

D. McCammon
Dept. of Physics,Univ. of Wisconsin,1150 University Ave.,Madison,WI 53706

ABSTRACT

Proper reduction of ROSAT observations of extended sources or the diffuse background requires techniques that are quite different from the standard point-source analysis. We describe in detail an appropriate set of procedures for this purpose for the X-ray Telescope, Position Sensitive Proportional Counter. We present the use of detector efficiency maps for proper flat-fielding of images. Sources of noncosmic background contamination and their modeling and subtraction are also discussed. IDL routines for implementing these data reduction procedures are being made available to the community through the ROSAT IDL library maintained at Goddard Space Flight Center.

1. Introduction

The analysis of PSPC observations of extended features is intrinsically more complex than that of point sources. In the latter case, providing the source is roughly centered in the field of view, event selection, vignetting correction, and exposure correction are straightforward and a reasonable estimate of the background can be made from an annulus surrounding the source. For extended sources, however, the background can vary significantly over the region of interest, requiring an independent assessment of its distribution over the detector. For studies of the diffuse background or features that might extend to the edges of the field of view or beyond, the integrated intensity of the background must be estimated from models of its various components. Furthermore, the telescope and detector quantum efficiencies vary over the field of view requiring detailed corrections.

2. Generation of Energy Dependent Exposure Maps

The proper efficiency-weighted (including vignetting, detector artifacts, and variations in the detector quantum efficiency) exposure correction for an observation of an object extending over more than a few arc minutes is obviously vital. The telescope effective area and detector quantum efficiency are functions of detector position, and therefore sky position. Furthermore, they are also all functions of energy which requires the exposure correction to be done on a band-by-band basis. Software for the creation of exposure maps has been written and incorporated into the ROSAT IDL library (Reichart 1993), the PROS analysis package (Worrall et al. 1992), and the EXSAS analysis package (Belloni 1992;

Zimmermann et al. 1992). This software uses the procedures discussed below.

The method and software developed for the creation of exposure images to correct pointed observations are straightforward. Maps of relative detector efficiency, called A_{eff} maps, have been created in detector coordinates in the pulse-height bands of Table 1, which explicitly incorporate all detector and telescope effects. These A_{eff} maps are used in conjunction with the pointed observation aspect and livetime information to cast efficiency-weighted exposure times in sky coordinates over the field covered by the observation, normalized to the on-axis effective areas.

The first step is the creation of a 3-dimensional binned aspect list which sums the exposure (livetime) from all intervals (typically 1 s) of the observation with the same Right Ascension and Declination offsets from the nominal pointing direction (in minimum steps of $\sim 15''$) and roll angle (in minimum steps of $\sim 15'$). This binning significantly reduces the computation time for the exposure map calculation and sacrifices no accuracy. Second, for each entry in the binned aspect list, the offsets and roll angles are used to rotate and position the A_{eff} maps on the sky. The exposure time from the list is multiplied by the detector map value, pixel by pixel, and added to the exposure image in sky coordinates.

The detector efficiency maps, A_{eff} maps, were created by using events from the ROSAT all-sky survey in detector coordinates to approximate a flat field. Point sources, bright extended objects, times of strong scattered solar X-ray and particle contamination, and times of short-term noncosmic background enhancements were excluded from the data set. Furthermore, an estimate of the residual particle background contribution to the data was subtracted in the creation of the A_{eff} maps. This allows a more accurate exposure correction to the background-subtracted data since the particle background counts are distributed differently from vignetted X-rays. It also explains the existence of negative-value pixels in regions of the maps with low statistics (e.g., under the window support ribs).

Creating the A_{eff} maps from such a pseudo flat field has an advantage over using the theoretical vignetting function in that it accurately reflects all detector and telescope nonuniformities. Specific examples of such nonuniformities are the shadowing by the wires and ribs of the window support structure, electronic "ghost" images (see below) in the R1L (and R1) band, and variations in the window thickness and therefore the detector quantum efficiency as a function of position. The A_{eff} maps are dependent on the X-ray spectrum and their creation for each pulse-height band reflects the average spectrum of the soft X-ray diffuse background (SXRB). This will create no problems in the lowest pulse-height bands where the vignetting is little changed over the energy range covered by the band. However, for the highest pulse-height band, if the spectrum of an extended object is much different from that of the SXRB, the vignetting correction will lose accuracy.

The A_{eff} maps for the two PSPCs are not the same. Besides detector-specific artifacts, there is a small shift in the position of the window support structures and the windows have slightly different thickness distributions. Since the main survey was done with the first PSPC (~ 180 days) and there were ~ 11 days of survey completion with the second PSPC, data exist to create detector maps for both PSPCs. However, the statistics of the survey completion are obviously worse than those of the main survey, and are inadequate for bands R3 through R7. For these bands, templates were created by shifting the detector maps of the first PSPC to correctly align the shadows of the window support wires and ribs with the second PSPC. The shifted detector maps were then normalized to the A_{eff} maps of the second PSPC over overlapping $5' \times 5'$ regions to give the correct telescope vignetting and detector quantum efficiency. The systematics of the detector artifacts in the R1 and R2 bands are sufficiently different between the two detectors to preclude using the same scheme for these bands. The detector maps for these bands were created in the same way as the maps for the first PSPC.

The maps were normalized to the average value of the central $10'$ diameter ($20'$ for the R1 and R1L maps) region of the PSPC where telescope vignetting is insignificant. The *average* shadowing by the window support wires is therefore not included in the exposure correction; however, it is included in the window transmission for modeling purposes. The spatial structure of the shadowing caused by the window support wires and the window support ribs is included in the exposure correction produced by the A_{eff} maps.

Figure 1 shows the second PSPC detector map for the R1 band. The effects of electronic ghost images are very obvious in the regularly spaced bright spots and somewhat less bright lines. The PSPC is an imaging proportional counter that makes use of induced charge on crossed cathode wires to obtain the position of accepted events. The two-dimensional position determination is done using the largest signals on the crossed cathodes, essentially interpolating the event position between the two nearest cathode wires in each direction. For very low pulse-height events, there is the possibility that only one cathode in one or both directions will have signals above the lower level discriminator of the analog electronics chain. In this case, the position determination degenerates to the center of the nearest cathode, yielding a line (if only one axis has a single nonzero cathode value) or a point (if both axes have only one nonzero cathode value each). Also visible in Figure 1 are slight waves in the electronic ghost-image lines. This detector artifact is due to the position correction algorithm. The algorithm corrects the event position based on the assumption that the X-ray was absorbed in the counter gas near the window. The bulging of the window support structure by the pressure of the counter gas bends the electric field lines in the electron drift region of the PSPC. This causes a displacement of the event position, an effect which has been calibrated and is included in the SASS event position-correction procedure. Since the low pulse-height events which contribute to the electronic ghost images have detected positions shifted to the wire positions, the correction is not the appropriate one. Electronic ghost images strongly affect only the R1 and R1L band detector maps, although the R2 band map also shows some irregularities. However, since the electronic ghost images are pulse height and not energy dependent, the R1 detector efficiency map created from the high-gain data is reasonably appropriate for correcting the R1L band data collected in the low-gain state (where no survey data exist to produce a flat field). This works because the R1L band at low gain includes the same pulse heights at its low end as the R1 band at high gain. The weighting by the source spectrum is of course slightly different, but this is a small effect. The R1 A_{eff} maps should be used for R1 band analysis for data collected during high-gain operation *and for R1L band analysis for data collected during low-gain operation*. We have created R1L A_{eff} maps for both detectors to be used in R1L band analysis *only* for data collected during high-gain operation.

Figure 2 shows the second PSPC detector map for the pulse-height channel range 42 to 131, the sum of bands R3, R4, R5, and R6. This is the map used by SASS to create the exposure map supplied in observation data sets. The events are distributed considerably more smoothly than in the R1 detector map, with the major features now due to the window support structure: the wagon-wheel shaped shadow from the ribs and the grid pattern from the coarse wire window-support mesh. Note that the shadows from the wires are properly straightened by the position correction algorithm, in contrast to the electronic ghost image lines. The effect of vignetting is also visible in the image with the central region having a higher efficiency than the outside.

3. Removal of Contamination

We have been able to identify five different components to the noncosmic background of PSPC observations. The first is particle-induced background (PB), caused by high-energy charged particles and gamma-rays that can penetrate the detector. The second is afterpulsing (AP), small pulse-height events that follow larger events in the detector and are believed to be caused by negative ion formation. The scattered solar X-ray background (SB) is produced by Thomson and fluorescent scattering of solar X-rays in Earth's atmosphere along the line of sight of the observation. Auroral X-rays and other sporadic events, most of which are probably also produced by low-energy charged particle interactions with the atmosphere or telescope, can produce high rates that last typically for a few minutes to a few hours, and are collectively referred to as "short-term enhancements" (STE's). Finally, there is a mysterious X-ray background that varies on a timescale of 12 days and appears most strongly in the R1 and R2 bands. These events are called "long-term enhancements" (LTE's), and evidence from the ROSAT all-sky survey (Snowden and Schmitt 1990; Voges et al. 1992; Snowden et al. 1993c) suggests that they affect observations at some level $\sim 90\%$ of the time. These backgrounds and the methods for their avoidance or modeling and subtraction are described individually below.

A. Particle Background

In most observations and in most energy ranges the particle background contribution is small: $< 6\%$ of the cosmic diffuse X-ray background rate. Allowing only data collected with the Master Veto (MV) rate < 170 counts s^{-1} eliminates times of severe PB contamination that cannot be modeled accurately. The remaining PB counts can easily be modeled using calibration data provided by Plucinsky et al. (1993), and subtracted from an observation. Regions near the edge of the detector affected by reduced PB vetoing efficiency should also be excluded from the data reduction.

Before subtraction, the modeled PB counts must be spatially distributed or "cast" into a background image in sky coordinates. The process is similar to the casting of efficiency-weighted exposure and the casting of the SB. Two new detector maps are required in which the pixel values represent the relative contribution of that part of the detector to the PB. One map, P_i, is for the internally-produced component which has a radial dependence (different for the two PSPCs), and one map, P_e, is for the externally produced component which is flat except where it is shadowed by the window support structure. These maps are available through the SASS, EXSAS, and PROS releases along with the A_{eff} maps of § 2, where P_i is generated from the analytic functions given by Plucinsky et al. (1993) and P_e is a devigneted version of the R3-R6 A_{eff} map. Both are normalized so that the sum of the pixel values over the entire detector equals 1. The $P_{i,e}$ detector maps are thus the probability distribution for the detection of a single PB event in any given pixel. The MV data from the event rate file and aspect and livetime data are used to model the number of PB counts in short time steps (1 − 2 s) and generate a binned aspect list with counts rather than the livetime of the exposure correction. Data from the aspect (attitude) file are then used to orient the P_i and P_e maps on the sky so the modeled counts can be transferred into sky pixels using the detector map probability distribution.

B. Afterpulses

A background component, variable in magnitude and related primarily to the particle background, became apparent in the summer of 1991. This component consists of very low pulse-height afterpulses (AP) which follow very shortly after a "precursor" event. Some of these afterpulses can be removed by rejecting all soft events occurring within 0.35 ms after another event. Unfortunately, except near a bright source, most of the precursor events are PB events which have been vetoed by the on-board electronics and are therefore not telemetered to the ground, which makes elimination of their afterpulses impossible.

The level of this additional contamination component depends on the date of the observation and insufficient data are available to determine its entire history. However, a reasonable estimate of its contribution can be made in most cases from analysis of the observation being analyzed. The first step is to select all events occurring within 0.35 ms after another event. The spectral shape of these afterpulses can be fit using a Polya function (Plucinsky et al., 1993),

$$Y(CH) = a_1 \times e^{\frac{-a_2 \times CH}{a_3}} \times \frac{CH^{a_2-1}}{a_3},$$

where $Y(CH)$ is the model value at channel CH, a_1 is the normalization constant, and a_2 and a_3 determine the shape of the Polya function. Note that this function is fit directly to the data without being folded through the detector response matrix. Next, the pulse-height spectrum of channels ≤ 25 (all events occurring > 0.35 ms after the preceding event) can be modeled using the model particle background from Plucinsky et al. (1993), the Polya function that fits the afterpulse spectrum, and a thermal spectrum with $T \sim 10^6$ K (which must first be folded through the detector response matrix, unlike the particle background and afterpulse spectra). The normalization of the PB spectrum is fixed by the model, while the amplitudes of the AP and thermal spectra are allowed to vary.

The AP counts are not distributed across the detector in the same pattern as PB events. We have prepared a detector map, P_{AP}, which will again be included with the SASS, PROS, and EXSAS distributions. It has large uncertainties, but is the best that can be done at this time. In most cases, the AP contamination will probably be small. The estimated number of AP counts can then be cast like the PB by using the P_{AP} map and assuming that the AP rates are proportional to the MV rate during the observation.

Scientific conclusions affected by any amplitude of the AP component within the limits allowed by the fitting procedure and the magnitude estimates of Plucinsky et al. (1993) should be treated with great caution.

C. Scattered Solar X-rays

The scattered solar X-ray background is highly time variable even in non-flaring conditions due to changes in viewing geometry during an observation (Snowden and Freyberg, 1993). For example, a single observation interval can start in satellite night, cross the terminator, and end in bright-Earth blockage. The SB count rate in that case goes from zero to (occasionally) more than 50 counts s^{-1}, with some contamination present at all times when the Sun-Earth-satellite angle is < 120°. This can be compared to typical uncontaminated sky rates of $4-15$ counts s^{-1}. This background constituent is significant for bands R1 through R5 and particularly affects bands R3 through R5 due to the strong O Kα line at ~ 0.54 keV.

Corrections for the SB have been discussed in detail in Snowden and Freyberg (1993). Again, a detector map and aspect information can be used to cast the modeled events into sky pixels. Since the SB originates as diffuse X-ray emission external to the XRT, the same set of detector flat-field maps used for creating exposure images, A_{eff}, can be used after normalization so that the integrated response equals 1.

D. Enhancements

Short-term enhancements are easily recognized in plots of count rate versus time (such as the background light curve plot produced by SASS) as irregular increases in the diffuse count rate lasting less than a few orbits. Since STE's probably have several sources (e.g., auroral X-rays, additional SB contamination from solar flares, enhanced charged particle precipitation, etc.), modeling of the spatial distribution across the counter is not feasible. While the number of excess counts can be determined from the count rate versus time plot, it is difficult to determine where those specific counts originated in the detector (and therefore where they appear on the sky). Thus it is usually safest to simply exclude the affected time intervals.

While auroral X-rays are a specific example of STE's, they deserve a special note. They can completely dominate a pointed observation, with count rates exceeding 200 counts s^{-1} in particularly bad geometries. Not surprisingly, the most sensitive geometries for auroral X-ray contamination are where the satellite is passing near the northern or southern particle belts and the observation is of a far northerly or southerly target, which requires a large zenith angle. In this geometry, the line of sight can pass through the particle belts at relatively low altitudes where the atmospheric density is large.

Long-term enhancements are readily identifiable in the uncleaned ROSAT all-sky survey data in the 1/4 keV band (Snowden et al. 1993c). They appear as stripes which circle the sky along lines of constant ecliptic longitude (cf. 1991 June 29 Science News cover photo or the 1992 ROSAT calendar). During the all-sky survey, ROSAT observed one great circle passing through the ecliptic poles each orbit. The great circle was rotated by $\sim 4'$ in ecliptic longitude each orbit. Since the PSPC field of view is $\sim 2°$, the part of the sky observed changed very little from one orbit to the next.

The LTE's are spectrally quite soft, appearing most strongly in the 1/4 keV band with an order of magnitude less contribution above 0.5 keV. Their strength usually varied by less than a factor of three around an orbit of the sky survey, while the look direction changed by 360°. Similar enhancements were observed with the HEAO A2 LED detectors (Garmire et al. 1992). The LTE's can be eliminated from the ROSAT sky survey data because the multiple coverage of all points on the sky over a period of at least two days allows the identification of time-varying background components. The contamination of a pointed observation by an LTE, however, can be difficult to detect because the timescale of LTE's is usually longer than a single pointed observation, and the variation with parameters such as satellite position or zenith angle is quite small. While the background light curve of a pointed observation may show some variation due to this component, particularly if the observation was spread over several days, there is no reason to expect that the minimum count rate of any pointed observation is unaffected by contamination.

While analysis of data from the ROSAT all-sky survey continues, we have as yet found no conclusive correlation of the enhancements with any parameter, including the satellite's day/night status, geographic and geomagnetic latitude, geomagnetic viewing angle, K_p index, and solar 10.7 cm flux. While the average severity of the contamination increases slightly with higher K_p value, LTE's occur at all values. This is consistent with similar findings for the HEAO LED survey (Garmire et al. 1992). The HEAO results suggested that the enhancements seen by that experiment were produced by charged particles, since the enhancements were absent from the middle layers of the proportional counter. On the other hand, the PSPC counts produced by LTE's are vignetted in the same way as the sky background (Snowden et al. 1993c), which suggests that they are produced by photons rather than by particles. It is therefore not clear whether the HEAO and ROSAT enhancements are produced by the same mechanism. Nevertheless, the fact that the time-variable component of the LTE's is vignetted in the ROSAT data simplifies the modeling and subtraction of the counts from pointed observations.

Any time variation of an LTE during a pointed observation can be removed by subtracting any excess over the minimum observed total rate (after removing all other sources of contamination). The excess counts for each pulse-height band are distributed using the same detector efficiency maps, A_{eff}, which are used to distribute the effective exposure and SB. As in the case of the SB modeling, the maps must first be normalized to one over the field.

This procedure minimizes the LTE contamination in a pointed observation, but many if not most observations are likely to include residual LTE contamination that cannot be determined (or eliminated) on the basis of the observation itself. For example, the 1/4 keV data from an MBM 12 observation, after all background subtraction, still show an intensity $\sim 41\%$ higher than the cleaned survey data (Snowden et al. 1993a). To allow a final correction of the zero level of pointed observations (to the background subtraction accuracy of the survey data reduction), cleaned all-sky survey data will be made available to observers covering a limited region in the directions of pointed observations.

4. Discussion

We have presented methods for modeling various components of non-cosmic X-ray background. We have also described the generation of energy dependent exposure maps, used for accurate flat fielding of images. Here we present a suggested method for the reduction of ROSAT PSPC observations towards diffuse targets.

For observations of the diffuse X-ray background or other extended targets, we begin our analysis with the examination of the SASS background count rate hardcopy (Figure 3 depicts a portion of the background count rate for an observation of the Draco nebula). The background count rate versus time plot provides an overall view of background contamination and an estimate of the difficulty of the data reduction. Periods of strong SB contamination will appear as sharply falling (rising) count rate enhancements at the start (end) of observation intervals. STE's appear as erratic and asymmetrical peaks anywhere in the observation intervals, while LTE's are seen as a gradual drift in the minimum count rate. The information on point sources identified in the field should be searched for strong or variable objects. If the contributions from individual sources are small compared to the total background count rate or the sources are constant, they can be ignored when modeling the temporal variation of the SB and LTE background components. If not, the affected regions should be excluded from the following procedures

Once initial inspections are complete, a time selection is applied to the accepted data to eliminate all periods where the Master Veto count rate averaged over $\sim$ 30 s is greater than 170 counts s^{-1} (Plucinsky et al., 1993). This criteria will eliminate periods when particle background events have not been accurately modeled. Afterpulse contamination is then partially accounted for through the elimination of all events occurring within 0.35 ms of a previous event. Times of obvious short-term enhancements are also eliminated at this point to make it easier to fit other background components.

Having applied the first selection criteria, light curves are generated from the accepted data. Scattered solar X-ray, charged particle, long term enhancement, and afterpulse backgrounds are then modeled according to procedures outlined above. Using the derived count rate versus time models for the SB, LTE, PB, and AP for the chosen bands and accepted time intervals, background counts are cast onto background maps in sky coordinates. Using livetimes, efficiency weighted exposure periods are cast onto exposure maps for the chosen bands in sky coordinates for the accepted time intervals. Next, events from the accepted time intervals are sorted into photon maps in sky coordinates for the chosen bands. All event, model background count, and exposure maps are then combined to form intensity images for the chosen bands (i.e., subtract all of the background maps from the event map on a pixel-by-pixel basis, and then divide by the exposure map). Point sources are then removed at this point in one of two ways. For statistical analysis, we remove point sources from the event, background, and exposure maps individually (before any spatial binning) by zeroing all pixels within a reasonable distance from the source location (e.g., within 1.5 times the FWHM of the spatial resolution at the radial position of the source). Alternatively, for making pictures of the field without gaps, we remove point sources from the intensity image after flat-fielding, filling in the resulting gaps with the average rate from an annulus surrounding the removed data.

Finally, in order to adjust the image to the correct sky rate, we compare the pointed observation average rates with the sky survey data from the same direction and subtract the difference from the intensity image produced previously.

References

Belloni, T. 1992, private communication
Burrows, D. N., & Mendenhall, J. A. 1991, Nature, 351, 629
Garmire, G. P., et al. 1992, ApJ, 399, 694
Plucinsky, P. P., Snowden, S. L., Briel, U. G, Hasinger, G.,
   & Pfeffermann, E. 1993, ApJS, submitted
Reichart, G. A. 1993, private communication
Snowden, S. L., & Freyberg, M. J. 1993, ApJ, 404, in press
Snowden, S. L., McCammon, D., & Verter, F. 1993a, ApJL, in press
Snowden, S. L., Mebold, U., Hirth, W., Herbstmeier, U., &
   Schmitt, J. H. M. M. 1991, Science, 252, 1529
Snowden, S. L., Pietsch, W., et al. 1993b, in preparation
Snowden, S. L., Plucinsky, P. P., Briel, U., Hasinger, G., &
   Pfeffermann, E. 1992, ApJ, 393, 819
Snowden, S. L., & Schmitt, J. H. M. M. 1990, Ap&SS, 171, 207
Snowden, S. L., et al. 1993c, in preparation
Voges, W., et al. 1992, in Proceedings of Satellite Symposium 3, Space
   Science with particular emphasis on High-Energy Astrophysics,
   ed. T. D. Guyenne & J. J. Hunt (Noordwijk:ESA Publications Division), 223
Worrall, D. M., et al. 1992, in Data Analysis in Astronomy IV,
   eds. V. Di Ges'u et al. (New York:Plenum), p.~145
Zimmermann, H. U., et al. 1992, in Data Analysis in Astronomy IV,
   eds. V. Di Ges'u et al. (New York:Plenum), p.~141

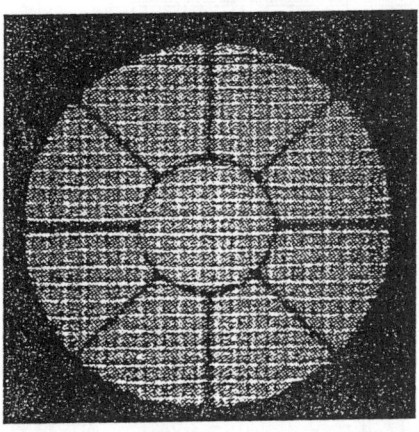

Figure 1: Grey scale image of the R1 band detector efficiency map for the second PSPC. Lighter shading represents higher efficiency.

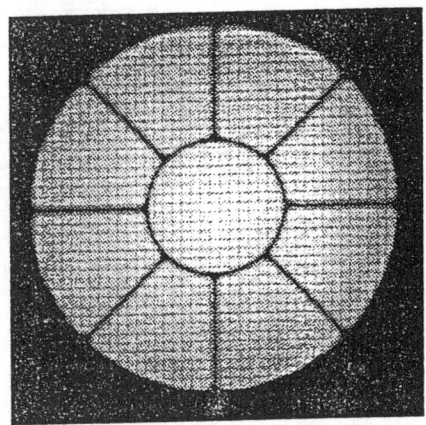

Figure 2: Same as Figure 1 except for the pulse-height (PI) channels 42 to 131 of the second PSPC. This is the detector efficiency map used by SASS to create the exposure map included in a data set.

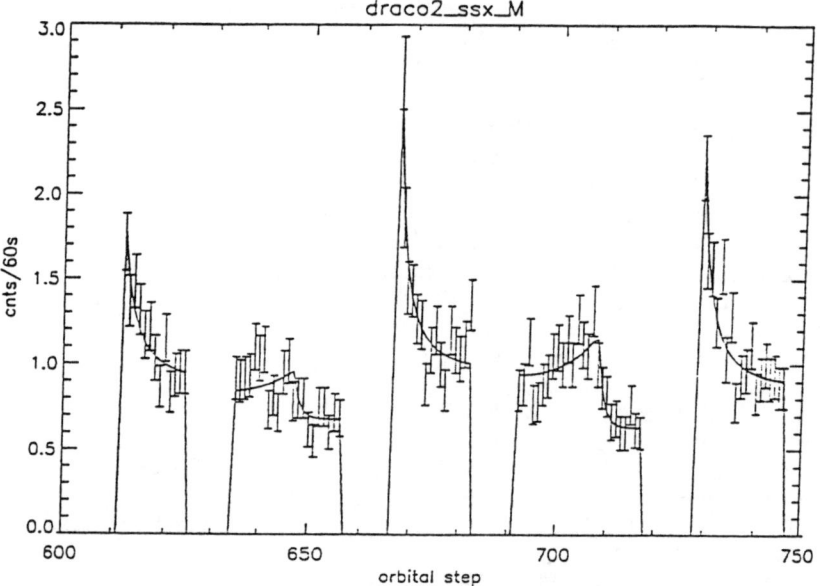

Figure 3: A portion of the background count rate light curve for an observation of the Draco nebula. Sharp increases at the end of obis are caused by scattered solar X-rays. Solid line represents fit to the data.

Table 1: Broad Energy Band Definitions

Band Name	PI Channels	SASS Channels	Energy[a]
ROSAT Bands			
R1	8 – 19	1[b] – 5	0.11 – 0.284
R1L[c]	11 – 19	3 – 5	0.11 – 0.284
R2	20 – 41	6 – 10	0.14 – 0.284
R3	42 – 51	11 – 12	0.20 – 0.83
R4	52 – 69	13 – 15	0.44 – 1.01
R5	70 – 90	16 – 18	0.56 – 1.21
R6	91 – 131	19 – 23	0.73 – 1.56
R7	132 – 201	24 – 30	1.05 – 2.04
SASS Bands			
A	11 – 41	3 – 10	0.12 – 0.284
B	52 – 201	13 – 30	0.51 – 2.01
C	52 – 90	13 – 18	0.47 – 1.29
D	91 – 201	19 – 30	0.76 – 2.02
Total	11 – 235	3 – 33	–

[a]10% of peak response, in keV

[b]SASS channel 1 also includes PI channel 7

[c]R1 band for the low gain state

QPTOOLS: TOOLS FOR CREATING AND MANIPULATING IRAF/QPOE DATA

M. A. Conroy

Smithsonian Astrophysical Observatory, Cambridge, MA 02138

Email ID
mo@cfa.harvard.edu

ABSTRACT

The QPOE data format was developed by IRAF to support photon event-list data. We are developing several tasks that allow users to generate QPOE files of their own description and to modify existing data files. For instance, users can create their own QPOE file by replacing columns in an existing file, or by importing *ASCII* lists of events.

1. Introduction

The Quick Position Ordered Event (QPOE) data format was developed by the Image Reduction and Analysis Facility (IRAF) to support photon event-list data. The filtering mechanisms that IRAF provides with this data structure make it a very powerful tool for data analysis. One of the key features of this data format is the automatic conversion from event-list format to *image* format. However, it is often desirable to view and manipulate the QPOE file in tabular form. Unlike the *table* data format developed by STScI, the QPOE format does not have an accompanying package of manipulation tasks. We have started work on developing a few of these basic utilities, modeled after existing IRAF *image* or TABLE tasks.

2. What is QPOE?

The IRAF/QPOE file structure was designed specifically to provide storage and access of event-lists (photon-lists) generated by photon-counting detectors, such as those common in High Energy Astrophysics. The representation of X-ray data in QPOE data structures is the heart of the PROS package. The interface provides several powerful capabilities:

- Dynamic conversion of event-list data to *image* arrays

- Rich set of built-in dynamic filtering capabilities, including run-time calculation of EXPOSURE time

- Support for World Coordinate System representations

- Support for dynamic spatial masking

- Support for application specific extensions

Since this data structure is so fundamental to ROSAT and *Einstein* analysis, we have been developing tools to extend the IRAF built-in capabilities. Since EUV is another project that uses the QPOE data format, they are developing QPOE tools as well.

3. Sample QPOE tasks

QPCREATE is a script that executes a sequence of IRAF tasks to produce a QPOE file from an input *ASCII* list.

```
<ASCII list>
<HEADER list>    ---->   FITS/BINTABLE    ---->   QPOE file
<GTI list>
```

The only requirement for the *ASCII* list is that it include 'x' and 'y' columns. Optionally, the user may generate a set of HEADER parameters by modifying the supplied template. The other optional file is an input list of 'good-time' intervals (GTI) that is used to define the correct exposure intervals.

QPCALC is a general task (modeled on the TABLES **tcalc** task) that performs algebraic operations on QPOE event-list attributes. This task evaluates an arbitrary expression that includes event-attribute names, constants, and operators, and either creates the specified event-list attribute in the QPOE file or overwrites an existing attribute. The usual arithmetic and logical operations are supported, as well as the most common arithmetic functions.

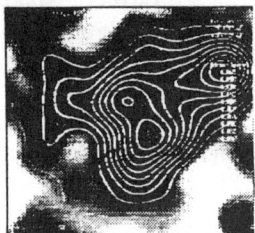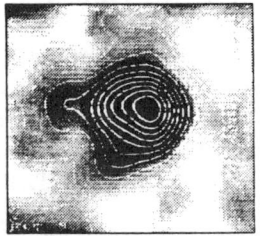

Figure 1. On the left is the original ROSAT observation of Cygnus A, combining 5 observing segments. On the right is the same observation, in which the segments have been separated, aligned to correct for aspect errors and re-combined.

QPLINTRAN is a task (modeled on IRAF **imlintran** task) that performs a general linear transformation on the X/Y axes of a a QPOE event-list. The task implements the following transformation equation, to transform (X1,Y1) into (X2,Y2):

```
X2 = Xout + (X1-Xin)*Xmag*cos(Angle) - (Y1-Yin)*Ymag*sin(Angle)
Y2 = Yout + (Y1-Yin)*Ymag*cos(Angle) + (X1-Xin)*Xmag*sin(Angle)

    where    (Xin, Yin)      - input reference pixel
             (Xout,Yout)     - output reference pixel
             (Xmag,Ymag)     - relative plate scale in each axis
             Angle           - the rotation angle
```

Two common examples of this transformation are 'shift' and 'rotate'. These two utilities are useful with ROSAT observations, to correctly align separate observing 'segments'. The **qpappend** task will then 'concatenate' the aligned segments.

```
xr> qpshift shockb.qp -13.0 10.0
xr> qpshift shockc.qp -10.0 5.0
xr> qpappend @shock.lst "" "cyga.qp" ""
```

QPGAPMAP is a special purpose task to generate corrected detector pixel coordinates from raw instrument coordinates. This process for the ROSAT/HRI instrument is referred to as 'de-gapping'. The detector coordinate system is the system used for all the HRI calibration maps. Once the detector coordinates are available in the QPOE file, it is possible to use them to 'look-up' the appropriate pixels of the calibration map.

QPPHASE is a task that calculates a period phase from the event-list times and produces a new event-attribute, 'phase'. IRAF allows user filtering on any event attribute, thus allowing 'on' and 'off' images of a pulsar to be separated. The 'phase' attribute is calculated as:

```
phase = time / period + 0.5 * (-1.0*pdot/period**2) * time**2
phi   = mod(phase,1)

      where   time    is the photon arrival time
                      (optimally barycenter-corrected)
              period  user input period in seconds
              pdot    is the user specified rate-of-change
                      in the period.

xr> qpphase crab.qp "" 33.4081E-3 4.22E-13 crabphase.qp
xr> xdisp crabphase.qp[phase=(0.5:0.6)]      'on'  PHASE
xr> xdisp crabphase.qp[phase=(0.7:0.8)]      'off' PHASE
```

4. Future Directions

The existing tools for manipulating QPOE files are a loose collection of tasks scattered throughout the IRAF layered packages such as PROS and EUV. As the existing support tools for QPOE mature and expand it might be reasonable to consider the following:

- Consolidate QPTOOLS into a single IRAF layered package
- Add a GUI interface to these tools
- Integrate the ST *table* file structure into QPOE
- Expand the user input options for QPCREATE
- Add more utilities to access QPLINTRAN and QPCALC for common tasks, e.g. QPFLIP

Acknowledgments

The PROS project is partially supported by NASA contracts to the ROSAT Science Data Center (NAS5-30934) and *Einstein* NAS5-30751.
PROS is available at no charge by contacting rsdc@cfa.harvard.edu, (6699::RSDC, rsdc@cfa).

TEMPORAL DATA SCREENING IN PROS

J. DePonte and M A. Conroy

Smithsonian Astrophysical Observatory, Cambridge MA 02138

Email ID

janet@cfa.harvard.edu, mo@cfa.harvard.edu

ABSTRACT

We present several tasks available in the PROS/IRAF environment to assist the user in screening and filtering event data. The technique employed translates all user filters into a common time-based filter. The time intervals are then used to filter events in the QPOE file. The rationalized data format (RDF) for ROSAT make both the unscreened photons and time-tagged ancillary data available to the user. Thus, with the temporal screening tools presented below the user has the ability to completely specify the criteria for acceptance or rejection of photons.

1. Introduction

ROSAT data distributed in the rationalized data format (RDF) provide the community with the complete (unscreened) events in the *Basic* FITS file (Corcoran et. al 1993). In addition to the primary event list there are several additional extensions. The status of instrument parameters are stored in the *temporal status interval (TSI)* FITS bintable extension. The limits of acceptable data quality for the screened events are stored in the *standard quality limits (STDQLM)* extension. The limits of acceptable data quality for the unscreened events are stored in the *all quality limits (ALLQLM)* extension. The start and stop times during which all observing conditions meet the *standard quality limits* are stored in the *standard good time intervals (STDGTI)* extension.

The RDF Basic FITS file is self-defining and supported in PROS by the QPOE data structure (Conroy *et. al* 1992). Ancillary data such as orbit, aspect, and housekeeping data are provided in separate FITS bintable files. The ancillary data is supported in IRAF by the TABLES data structure.

Several tasks have been developed in PROS/IRAF to implement the extended filtering capabilities that these data make available.

2. EVENT FILTERS derived from the *BASIC* FITS file

The *BASIC* FITS file provide user access to all parameters used by the standard processing system to screen photons. The standard (screened) event-list of standard processing or the complete (unscreened) event-list can be defined using these tasks. More importantly, the user may edit one or more of the attribute filters to create a time filter that defines an event-list with user-specified conditions.

MKHKSCR - This task creates an *ASCII* file with housekeeping limits in IRAF filter format. Many instrument parameters are recorded in the *TSI* extension, either as bit-encoded flags or binned data levels. The task utilizes a lookup table that matches quality parameters (either *ALLQLM* or *STDQLM*) with *TSI*s and provides the mapping between the two QPOE data structures. The output attribute filter is a list of temporal status identifiers with their quality limit range.

Example Data Quality Filter for 'Unscreened' HRI Events

```
hiback=1:100, hvlevel=12:16, viewgeom=1:5, aspstat=1:5,
asperr=1:14, hqual=0:0, saadind=1:3, saada=1:4, saadb=1:4,
temp1=1:8, temp2=1:8, temp3=1:8, logicals=%44X
```

HKFILTER - This task creates an *ASCII* file with time limits in IRAF filter format. The task reads the *TSI* attribute filter generated by MKHKSCR and applies it to the *TSI*s. The result is a list of time intervals during which all the limits expressed in the input screen are true.

Example Time Filter for 'Unscreened' Events

```
time=(2585242.00:2585666.00,2585668.00:2588182.00, ...
      2684948.00:2686271.00,2707077.00:2708791.00)
```

3. EVENT FILTERS derived from the *Ancillary* FITS Files

Users will often find that the quality of their data is dependent on *Ancillary* data not used by the standard processing system, and therefore not recorded in the *Basic* FITS file. Thus, we present tools that use *Ancillary* TABLE files to generate event-list filters.

Any TABLE with a time attribute column is a candidate file, though as a QPOE time filter only space-craft units are compatible with the times in the QPOE file. Some data in the *Ancillary* TABLES are raw data that are stored as levels in the *TSI* (i.e. high background in _evr.tab).Other TABLES columns are not available in the *TSI* (i.e. master veto rate in _evr.tab) for ROSAT.

TABFILTER - This task is a data quality selection task that allows the user to interactively display time dependent data stored in TABLES. The TABLE can be displayed and graphically edited to select data. The result is a time filter that can be written to the QPOE as a macro definition. Macro definitions are time filters that are stored in the QPOE header. The basis of this macro task is the EUV task DQSELECT.

TABFILTER Session

```
xp> tabfilter rp800020_evr.tab

 Loading table: rp800020_evr.tab

Command> list rp800020_evr.tab    ... lists table column names
Command> disp mv_aco              ... tex window display
Command> window                   ... manipulate graphics window
Command> edit mv_aco              ... choose min/max from graph
```

```
Command> gtnew                        ... init time filter
Command> gtgen                        ... generate time filter
                                          from selections
Command> filter rp800020.qp mvbk      ... add time filter to header
                                          as macro definition
Command> q                            ... exit

xp> imhead rp800020.qp long+          ... to see the macro def in hdr

xp> qpcopy 'rp800020.qp[time=(mvbk)]'
```

Tasks such as QPCOPY accept either *ASCII* filter files or QPOE filter macros to screen the events according to the time intervals expressed.

4. EVENT FILTERS in Use

Most users will find that a combination of QPOE-based and TABLE-based time filters will be needed to achieve optimal results. The following is an example using ROSAT HRI data:

- Run RFITS2PROS to make an 'unscreened' event QPOE file.

- Run MKHKSCR and build a 'standard' attribute filter; edit the file and remove the entry for HIBACK.

- Run HKFILTER to generate a filter of accepted times.

- Run TABFILTER on the event-rate file (_evr.tab)

  – Display and edit high background (BACKLEV) column
  – Generate a temporal filter for BACKLEV

- Run QPCOPY to filter the QPOE event-list through the time filters generated by HKFILTER and TABFILTER to create a new QPOE file with the specified screens.

Acknowledgments

This work is partially supported by NASA contracts to the ROSAT Science Data Center (NAS5-30934) and *Einstein* (NAS8-30751).

References

Conroy, M.A., Garcia, M.R., Mandel, E.G., Roll, J., Worrall, D.M. 1992, in Astronomical Data Analysis Software and Systems I, eds. D.M. Worrall et al. (Astronomical Society of the Pacific), 17

Corcoran, M.F., White, R., Conroy, M. 1993, in Astronomical Data Analysis Software and Systems II, eds. R.J. Hanisch et al. (Astronomical Society of the Pacific), 549

EVALUATION OF SOURCE COUNTS AND UPPER LIMITS IN CROWDED ROSAT PSPC FIELDS

V. Kashyap
5640, S. Ellis, AAC, U. Chicago, Chicago, IL 60637
kashyap@oddjob.uchicago.edu

G. Micela and S. Sciortino
Osservatorio Astronomico, Palazzo dei Normanni, 90134 Palermo, Italy

F.R. Harnden, Jr.
Center for Astrophysics, 60, Garden St., Cambridge, MA 02138

R. Rosner
Astron. & Astrophys. and Enrico Fermi Inst., U. Chicago, Chicago, IL 60637

ABSTRACT

We present a method for determining source counts, S/N, and upper limits at specified positions in a crowded ROSAT PSPC field using a combination of publicly available software packages. The algorithm is based on the so-called 'Local DETECT' method of source detection, and improves upon currently available software, to permit a meaningful comparison of non-detections with detected sources. We also present a recipe to obtain source counts and S/N in the case of point sources which overlap significantly.

1. INTRODUCTION

Standard detection software systems currently available for analysis of ROSAT data include LDETECT in IRAF/PROS, (L/M)DETECT and Maximum Likelihood (ML) in MIDAS/EXSAS. These are 'true' X-ray detection algorithms, in the sense that they search the X-ray image for possible sources. However, their usefulness is limited (cf.Micela et al.1993) in the case of crowded fields (where overlapping sources cannot be handled, and nearby strong sources may cause the algorithm to fail) or off-axis sources (where errors in the assumed model of the Point Spread Function (PSF) may dominate count determination). Further, there is no provision for easily estimating the upper limits for non-detections in these systems (except the ML method in EXSAS, results from which, however, are not directly comparable with results obtained from the other methods). We have therefore developed a recipe using IRAF and IDL (Interactive Data Language, created by Research Systems, Inc.) to supplement the PROS and EXSAS systems to overcome the above limitations. We emphasize that the recipe is designed to obtain good estimates of source counts, and *not* source positions.

We describe the algorithm used to determine source counts and upper limits for isolated sources in §2, a generalization of this algorithm applicable to overlapping sources in §3, and outline the recipe in §4. Finally, we discuss our enhancements, and illustrate their effects, in §5.

2. ISOLATED SOURCE

We use a method similar to that used in PROS LDETECT (derived from the *Einstein* Rev-1 data processing system, Harnden et al.1984) to compute counts from a point source. The counts in a detect cell centered on the source (C) are compared with the counts in a 'background' cell (T), also centered on the source,

$$C = \alpha S + B, \text{ and } T = \beta S + rB, \tag{1}$$

respectively, where α is the fraction of the source counts S falling within the detect cell, β is the fraction of S falling in the background cell, r is the ratio of the areas of the background to the detect cell, and B are the background counts in the detect cell. Adoption of a specific model for the shape of the PSF allows us to compute α and β for the given detect and background regions. Our procedure differs from that used in PROS and EXSAS in that we use circular detect and background cells to collect the counts, and we have adopted the most recently developed model of the PSF (Hasinger 1993). The former change allows us to easily correct for small overlaps of adjacent sources: If the cells are circular, these overlaps can be removed with a wedge-shaped cut originating from the center of the cells, leading to a trivial correction of α and β. The latter change allows us to obtain better estimates of α and β, and hence, better estimates of the source counts. The background and source counts,

$$B = \frac{(\alpha - \beta)C + \alpha Q}{\alpha r - \beta}, \text{ and } S = \frac{(r - 1)C - Q}{\alpha r - \beta}, \tag{2}$$

where $Q = T - C$ are the counts in the 'frame region' surrounding the detect cell, are obtained by solving Eqns (1), along with their associated errors,

$$\sigma_B = \sqrt{\frac{(\alpha - \beta)^2 \sigma_C^2 + \alpha^2 \sigma_Q^2}{(\alpha r - \beta)^2}}, \text{ and } \sigma_S = \sqrt{\frac{(r - 1)^2 \sigma_C^2 + \sigma_Q^2}{(\alpha r - \beta)^2}}. \tag{3}$$

The variances of C and Q are approximated as $\sigma_X^2 = (\sqrt{X}+1)^2$. The significance of a detection is evaluated as the ratio $S/N = S/\sigma_S$, and the threshold for detection at a specified S/N is obtained by computing S for which S/σ_S equals the threshold.

3. OVERLAPPING SOURCES

In cases where the overlap between suspected sources is significant, we use a generalization of the method described above to determine individual source counts and associated errors starting with a list of known (perhaps optical) positions of putative X-ray source positions. Since there is no standard X-ray detection system currently available that would find multiple sources in these cases, we assume that the X-ray source positions are coincident with optical positions, and center the detect cells on these positions.

If there are N sources in the region of interest with source strengths S_i, then the counts in a detect cell centered on source i (C_i) and the counts in a frame region surrounding all the detect cells (B),

$$C_i = \Sigma_j \alpha_{ij} S_j + \rho_i B, \text{ and } Q = \Sigma_j \beta_j S_j + B, \tag{4}$$

respectively, where ρ_i are the ratios of the areas of the detect cells to the area of the frame region (note that r used in §2 is the 'inverse' of the ρ_i), α_{ij} are the fraction of the counts from source j falling on the detect cell centered on source i, β_j are the fraction of the counts from source j falling in the frame region, and B are the background counts in the frame region. The detect and frame regions are in general irregular in shape, necessitating a numerical integration of the PSF over these regions. Eqns (4) constitute a set of $N+1$ equations in $N+1$ variables, and hence can be written as a matrix equation,

$$\mathbf{C} = \aleph \# \mathbf{S}, \tag{5}$$

where '#' denotes matrix multiplication, $\mathbf{C} = [C_1, C_2, \ldots C_N, Q\ ]^T$ and $\mathbf{S} = [S_1, S_2, \ldots S_N, B\ ]^T$ are column vectors (the superscript T denotes a matrix transpose). Note that $\aleph_{i \leq N, j \leq N} = \alpha_{ij}$, $\aleph_{i \leq N, N+1} = \rho_i$, $\aleph_{N+1, j \leq N} = \beta_j$ and $\aleph_{N+1, N+1} = 1$, from Eqns (4). Inverting the above equation, we obtain an estimate of the source strengths S_i, and the background in the frame region B. The variances of S_i and B are

$$\sigma_\mathbf{S}^2 = \mathbf{\Gamma}^2 \# \sigma_\mathbf{C}^2, \tag{6}$$

where $\mathbf{\Gamma}$ is the inverse of the matrix $\aleph$, and the squaring is carried out for each element individually. Note that in order to preserve the statistical independence of the C_i and Q, the detect regions and the frame region must not overlap each other.

4. RECIPE

We have developed our recipe using IDL, PROS and/or EXSAS to obtain estimates of source counts, S/N, and upper limits to sources observed with the *ROSAT* PSPC. The following steps summarize the recipe:

1. Obtain a source position list. This can be either a list from a pre-existing catalog, or an X-ray based list obtained using the detection algorithms in PROS and/or EXSAS.
2. At each of the source positions thus obtained, compute detect and background cell region descriptors. The sizes of the cells depend on the size of the PSF, which in turn depends on the chosen energy and the off-axis position of the source.
3. If some of the source positions overlap, remove the overlap by cutting a wedge shaped region out of the detect and background cells. Since we use circular cells, the correction introduced by this cut is mathematically trivial. This task requires the user to first find the overlaps (by viewing the detect and background cells on SAOimage, an image display program available with IRAF), and then to type in the extent of the wedge shaped cut, manually.
4. If the overlap is large, use the generalized version of the algorithm (see above), starting with a list of putative X-ray source positions.
5. Compute the counts in the detect and background cells (this can be accomplished with the IMCNTS task of PROS).
6. Compute the background counts, source counts, S/N, and if the source is not detected, a threshold for detection, using the algorithm outlined above.

5. DISCUSSION

We improve upon the methods used in PROS and EXSAS in that we allow arbitrary specification of the passband (note, however, that exposure maps can be computed in only a limited set of passbands); use circular detect and background cells with radii that vary across the field of view to match the variation in the FWHM of the PSF (note that the EXSAS ML method also does this); use the new model of the PSF (Hasinger 1993) to compute α and β, which are therefore expected to be more accurate; correct for small overlaps between adjacent point sources by cutting a pie shaped region out of both the detect and background cells, and correcting the values of α and β as required; obtain unique estimates of source counts even for point sources that overlap significantly; and provide a conceptually simple method of comparison between detected sources and non-detections.

To illustrate the changes introduced by our improvements, we show, as an example, the 3σ upper limits on source counts computed at optical catalog positions (Micela 1993) of Pleiades stars, using both the detect and background cell counts as determined in PROS, and the application of our recipe. If the cell size is held constant (Fig. 1), source counts necessary to obtain a detection increase as the off-axis angle is increased due to the decreasing value of the denominator in Eqns (3). Hence, matching the detect cell size to that of the PSF together with the adoption of a better PSF (Fig. 2), permits significantly lower estimates to be obtained: Our algorithm allows the detection of weaker sources at large off-axis angles.

Acknowledgments: We acknowledge support for this project from NASA, under NASA's ROSAT Grant (University of Chicago), NASA Grant NAG5-1794 (CfA), and from Agenzia Spaziale Italiana, Ministero della Università e della Ricerca Scientifica e Tecnologica and GNA-CNR (Palermo).

REFERENCES

Harnden, F.R.,Jr., Fabricant, D.G., Harris, D.E., Schwarz, J. 1984, SAO Special Report 393
Hasinger, G. 1993, *cf.ROSAT Status Report*, No.65
Micela, G. 1993, private communication
Micela, G., Sciortino, S., Kashyap, V., Harnden, F.R.,Jr., Rosner, R. 1993, in preparation

FIGURE CAPTIONS

1: Upper limits on source counts for detection at 3σ, at approximate optical positions of Pleiades stars, using the Gaussian PSF and square detect/background cells as used in PROS2.2. The off-axis position is with reference to the deep (50 ksec) *ROSAT* observation RP200068 (PI: R. Rosner). Different detect cell sizes are indicated by the following symbols: '.' - 30", '+' - 60", 'x' - 120", and '◇' - 240".

2: Better sensitivity (lower upper limits) can be achieved with a variable cell size. Symbols indicate the 3σ Upper Limits on Source Counts for sources assumed to be at optically cataloged positions, computed by matching the detect cell sizes to the size of the PSF (see text).

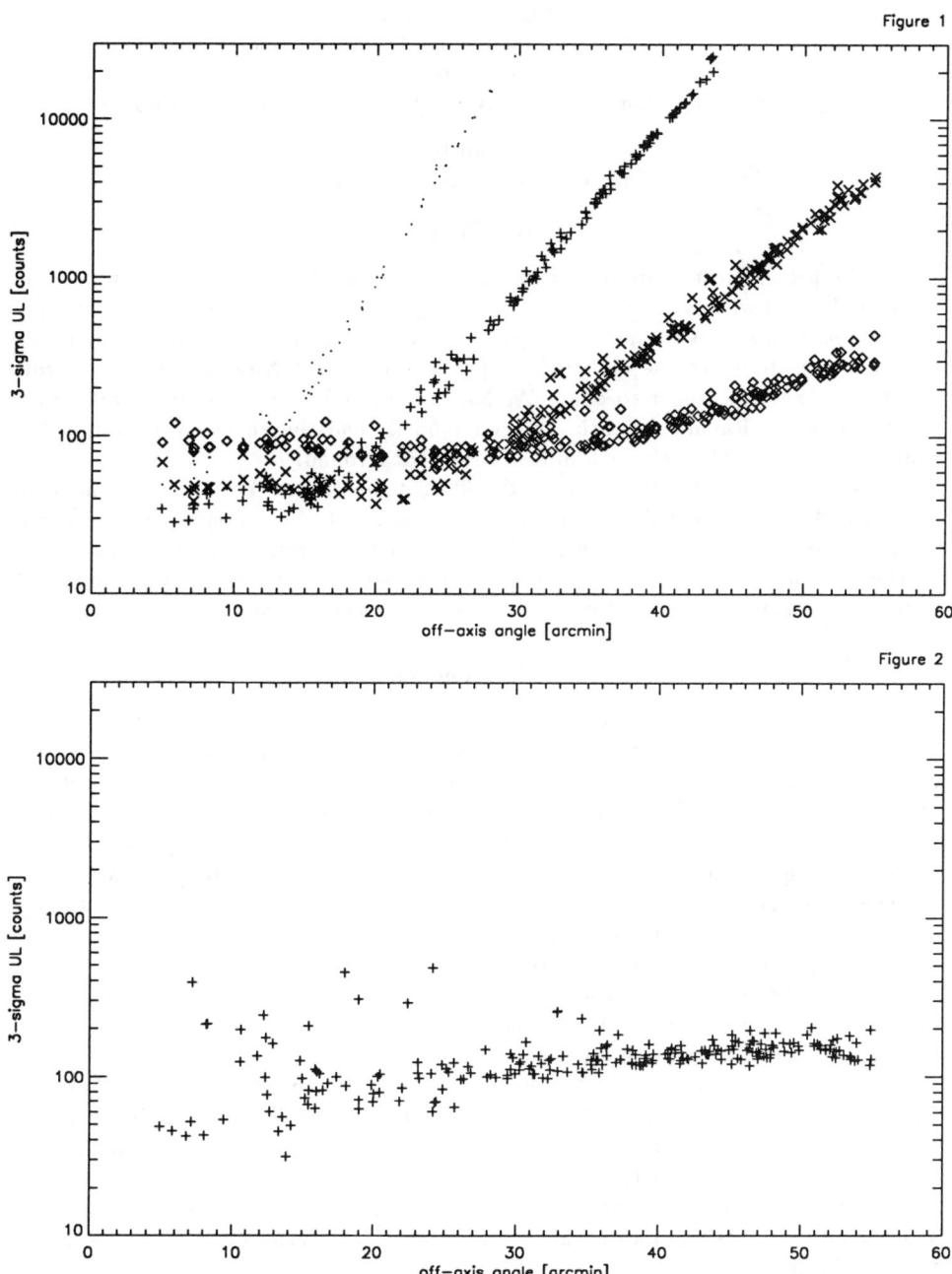

Figure 1

Figure 2

SIMULATED PSPC SPECTRAL FITS OF CORONAL X-RAY SOURCES

A. Maggio, S. Sciortino
Osservatorio Astronomico, Palazzo dei Normanni, I-90134 Palermo
F.R. Harnden
Harvard-Smithsonian Center for Astrophysics, Cambridge, MA 02138

Email ID
maggio@oapa.astropa.unipa.it

ABSTRACT

We present results from an extensive number of PSPC spectral fitting simulations, performed with the aim to ascertain the quality of the fit results as a function of photon counting statistics. In particular, we have explored the reliability of the results and their sensitivity to the assumed model, in the case of X-ray sources in thermal equilibrium, such as stellar coronae. We have considered both single-temperature and two-temperature Raymond-Smith emission models, and the effects of contamination from either low or relatively high diffuse X-ray background.

We show the relative errors on the fitted temperatures derived from the simulations, and discuss our ability to infer the presence of two components with different temperatures and emission measures in the observed spectrum, and to put constraints on them. Finally, we investigate on the chances that the presence of two components could be erroneously derived from the analysis of a single-temperature spectrum.

1. Introduction

Parametric fitting of low-resolution X-ray spectra is an example of a complex process, whose results depend on the assumed model, the instrument response, the contamination of the observed spectrum by sources unrelated with the target, the details of spectrum extraction and background subtraction, the photon energy binning scheme, as well as on the counting statistics and the method adopted for finding the best fit. As a consequence, questions such as how many parameters can be fitted, or how many counts are needed to get reliable spectral fitting results, can be reliably and quantitatively answered only by performing simulations.

For this reason, we have performed an extensive number of simulated spectral fits of coronal-type sources with two primary aims: (i) assessing the quality of the fitting results, and (ii) testing the correctness of source model inference, as a function of counting statistics and signal-to-noise ratio.

The source model spectra we have considered are single-temperature (1T) and two-temperature (2T) Raymond-Smith spectra (Raymond & Smith 1977, Raymond 1989) for an optically-thin plasma with cosmic abundances in thermal equilibrium. These are low-order approximations of the truly multi-thermal spectra expected from a magnetically-confined plasma, as observed in the solar corona, but they are most suitable for the investigation we want to carry out, since the spectral fitting results are more easily interpreted.

In all the simulated spectra we have included the ISM absorption due to a fixed hydrogen column density of 10^{19} cm^{-2}, thus focusing our attention on relatively nearby sources. We have also considered the influence of diffuse background contamination by modeling the latter as a power law with fixed photon index $\alpha = 1.5$, plus a monochromatic line component at ~ 0.5 keV representing the solar oxygen Kα emission scattered by the Earth atmosphere (Snowden & Freyberg 1993). We have assumed a fixed count rate ratio of 1.5 between the power law component and the line component, which is typical of observations heavily contaminated by geocoronal emission. In this respect we put ourselves in the worst situation.

The simulations were performed with the software XSPEC V8.34 in the following way: we produce independent synthetic spectra for the source plus background emission and the background only, where the source and the background have pre-selected count rates; then we χ^2 fit the background-subtracted spectrum discarding the first four channels, in the SASS 34-channel binning scheme. The latter choice follows a recommendation of the ROSAT Data Analysis Working Group.

2. Simulation Schemes

We have performed three different groups of simulations, each simulation consisting of at least 200 realizations. In the first group we have assumed a 1T model both as the input source model and the fitting model, with the purpose to evaluate the uncertainties on the fitted parameters. The second group comprises simulations where we have assumed a 2T model for the source and either a 1T or a 2T model for the fitting, with the aim (a) to estimate the uncertainties as before, (b) to test our ability to infer the presence of two components in the input spectrum, and (c) to check the sensitivity to high-temperature (> 1 keV) components. In the last group of simulations, we have assumed a 1T spectrum for the source and a 2T model for the fitting, in order to investigate the chances of erroneous inference of two components in a truly isothermal source spectrum.

We have fitted three parameters when a 1T model was assumed, namely the model temperature, T, the hydrogen column density, N_H, and the normalization, while in the case of a 2T fitting model, four free parameters have been considered, i.e. two temperatures and two normalizations, keeping N_H frozen at the input value.

We have performed simulations with a number of source net counts fixed between 150 and 2000, and considering three levels of background contamination: no background (ideal case), 1 cts/arcmin^2/sec (low), and 3.5 cts/arcmin^2/sec (high). Due to the limited space available, we will focus the following discussion on results obtained mainly in the low counting statistics regime, and on the model temperature determination. For more details we refer to Maggio et al. (1993).

3. Results

As a measure of the uncertainties on the fitted temperature, we have computed for each simulation in the first group the ratio $\Delta_{1\sigma}T/T$, equal to the inner 68% range of the fitted temperature values divided by the input T value. This number is about the expected 1σ relative uncertainty on the fitted temperature.

In Figure 1 we report this parameter for several sets of simulations, belonging to the first group, as a function of the number of net source counts. The uncertainties tend to decrease with increasing counting statistics, as expected, and their values depend on

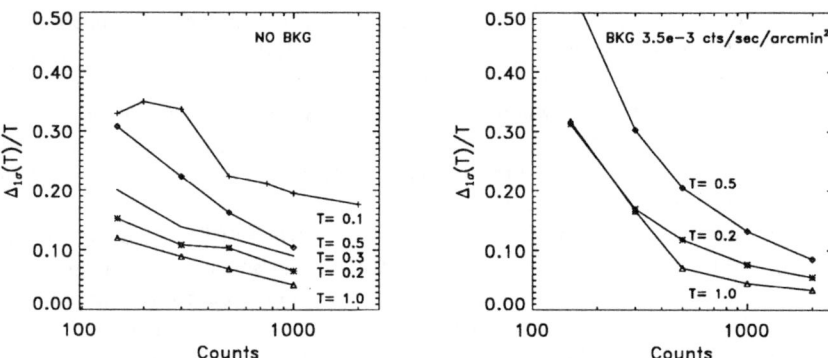

Figure 1: Relative uncertainties on fitted 1T model temperatures obtained from simulations of 1T spectra. Lines connect simulations with the same input temperature and different number of total net counts. (a) Results of simulations with no diffuse background contamination. (b) Results assuming an high background level.

the model temperature. The uncertainties are the lowest at 1 keV, and the highest at 0.1 keV, however the change is not monotonic: in particular, the uncertainties for model temperatures of 0.5 keV are relatively high. We have attributed this behavior to the presence of different features in the input spectrum (intense emission line blends) combined with the presence of the carbon edge in the PSPC effective area. If we include a relatively high background (Fig. 1b), the uncertainties rise steeply for low count rate sources but the increase is only few percent for a total number of net counts of 500 or greater.

In Figure 2 we have reported in a schematic way the results of two sets of simulations belonging to the second group. A total number of 300 net source counts was assumed. In the first set we have fixed the values of the two input temperatures at 0.2 and 0.5 keV, and we have varied the ratio of emission measures as reported in the abscissa. These simulations tell us that it is quite difficult to distinguish the presence of two components with so close temperatures, since the distributions of the fitted low- and high-temperature tend to overlap, even when no background contamination is present. The task appears easier to achieve if we consider the results of the second set of simulations (Fig. 2b), where we have assumed a low-temperature component at 0.2 keV, an high-temperature component at 1 keV, and we have also included an high level of background contamination. These simulations suggest that, except for cases with a very low ratio of emission measures, we should be able to correctly infer the presence of two components in the input spectrum even with few hundred counts and a relatively high background contamination.

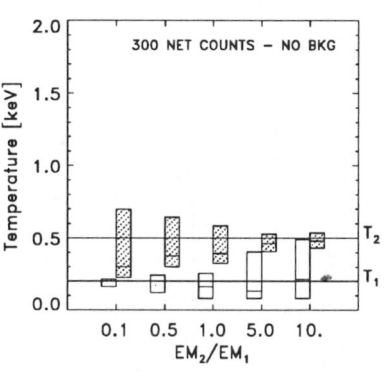
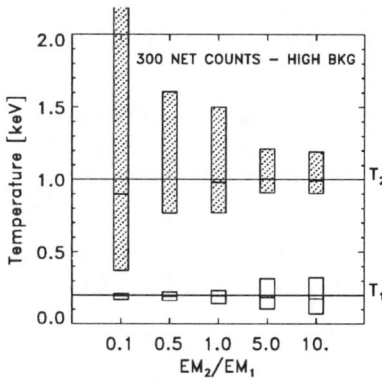

Figure 2: Boxplots of the results of two sets of simulations performed assuming a 2T model both for the source and for the fitting. The horizontal line drawn through each box is at the median of the distribution, the upper and lower edges comprise the middle 68% of the data. (a) The temperatures have been fixed at $T_1 = 0.2$ keV and $T_2 = 0.5$ keV, while the ratio of emission measures has been varied as reported in abscissa. No background contamination has been assumed. (b) As before, but $T_2 = 1$ keV, and a diffuse background flux of 3.5 cts/arcmin^2/sec has been assumed.

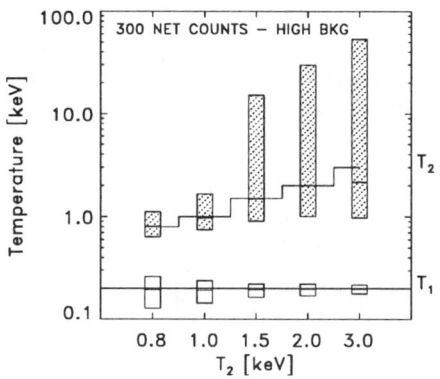

Figure 3: Same as in Fig. 2b, but fixing $EM_2/EM_1 = 1$ and varying T_2 as reported in abscissa.

In Figure 3 we show the results of simulations performed with the aim to test the sensitivity of the PSPC to high-temperature components in the input spectra. We have fixed the low-temperature at 0.2 keV, while the high-temperature assumes values from 0.8 keV up to 3 keV, with an emission measure ratio equal to 1. The number of sources net counts was again fixed at 300 (within statistics). We find that while the low-T component is always well constrained, we would be able to put only a lower limit to the value of the high-temperature. On the other hand, the presence of this second component seems to be correctly established.

Finally, in Figure 4 we show results of simulations which belong to the third group, where a 1T input spectrum has been fitted with a 2T model. We find either two components with very similar temperatures and comparable emission measure, or – for cold input spectra – two temperatures which may be well separated in few cases, with a trend of decreasing emission measure ratio, EM_2/EM_1, for increasing values of the high temperature. The chances to find a low-T around 0.2 keV, and an high-T > 0.8

keV with $EM_2/EM_1 > 0.5$, are restricted to 15 realizations out of 200 in the sole case of an input model temperature of 0.2 keV.

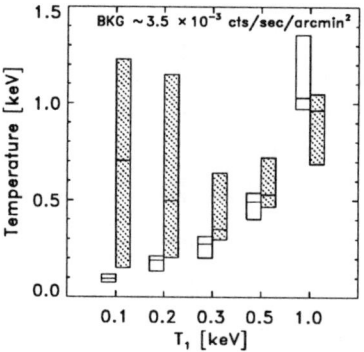 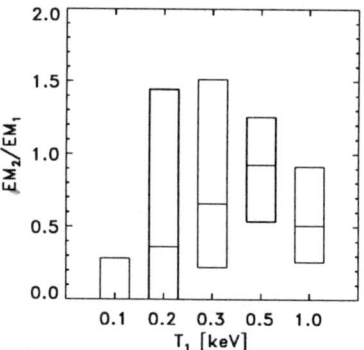

Figure 4: Boxplots of results of simulations performed assuming a 1T source spectrum and a 2T model for the fitting. A total of 300 net counts in the source spectrum and the indicated diffuse background flux has been adopted for all simulations. (a) Distributions of fitted temperatures. (b) Distributions of fitted emission measure ratio.

4. Conclusions

The simulations we have performed proved to be a powerful tool to get insight on the capabilities of PSPC spectral fitting in recovering the most interesting physical parameters for coronal-type X-ray sources. Based on the present results, we conclude that – *with few hundred counts in the input spectrum and a relatively high background contamination* –

- The temperature of an isothermal plasma can be well constrained (1σ uncertainty < 20%) if this temperature is around 0.2 keV or around 1 keV;

- The presence of two components in the source spectrum can be correctly inferred if the temperatures are well separated, and the ratio of emission measures, EM_2/EM_1, is greater than ~ 0.5;

- The presence of hot components (T > 1 keV) can be correctly recognized, but only a lower limit can be put on the high-temperature value;

- There are few chances of erroneous inference of two components with well separated temperatures and comparable emission measures from the analysis of a truly isothermal spectrum.

REFERENCES

Maggio, A., Sciortino, S., and Harnden, F.R. 1993, A&A, in preparation
Raymond, J.C., and Smith, B.W. 1977, ApJS, 35, 419
Raymond, J.C. 1989, private communication
Snowden, S.L., and Freyberg, M.J. 1993, ApJ, 404, 403

PREPARATION OF ROSAT PSPC DATA FOR DIFFUSE SPATIAL ANALYSIS

Jeffrey A. Mendenhall and David N. Burrows
Dept. of Astronomy,The Pennsylvania State Univ.,University Park,PA 16802

ABSTRACT

We demonstrate software used to prepare one of four ROSAT PSPC pointed observations towards the Draco nebula for diffuse analysis.

1. Introduction

ROSAT pointed observation data may be prepared for diffuse analysis through the removal of contamination and generation of energy dependent exposure maps. Sources of contamination include scattered solar X-rays, charged particles, short and long term enhancements and afterpulse events. The generation of energy dependent exposure maps are equally important in preparing data for diffuse analysis. These maps allow one to correct for the instrument induced count rate fluctuations across the field of view due to aspect variation, vignetting, rib/mesh obscuration and changing detector response over the ROSAT bandpass.

For this analysis we use one of four pointed observations towards the Draco molecular cloud obtained on September 29, 1991, and divide the data into three bands: C(PI 8-41), M(PI 52-90) and J(PI 91-201).

2. Removal of Contamination

All ROSAT PSPC observations contain some level of non-cosmic contamination. For point source analysis, subtraction of background measured in nearby regions effectively removes such contamination. However, background subtraction for large, diffuse sources such as supernova remnants and the soft X-ray background is not easy and must be accounted for my some other means.

Scattered solar X-rays (SSX) are perhaps the greatest influence on ROSAT diffuse studies and are present in almost every observation. Snowden and Freyberg (1993) have shown that fluorescent scattering off molecular nitrogen $K\alpha$ and atomic oxygen $K\alpha$ in the Earth's atmosphere dominate above 0.4 keV (M-band) while Thompson scattering is of primary importance below 0.25 keV (C-band). We have modeled the SSX intensity as a function of time for both C and M bands. This procedure fits a constant SXRB component plus a modeled scattered X-ray component to each observing interval's light curve (Snowden and Freyberg, 1993). Figure 1 depicts the modeled SSX contamination for six M-band observation intervals of our Draco data. Clearly, intervals 1, 3 and 5 exhibit severe contamination. For our analysis, we have removed all data corresponding to SSX greater than 15% of the best fit constant SXRB models. For other times, we have binned the SSX into C, M, and J bands and cast them onto normalized efficiency maps to create SSX background images for later subtraction.

Although the ROSAT particle background rejection efficiency is excellent (> 98%), residual events for this source of contamination must be accounted for. We have modeled the particle background event rate as a function of ROSAT master veto rate using the technique described by Plucinsky et al. (1993) (Figure

2). Once modeled, events are cast onto normalized efficiency maps and divided into C, M, and J particle background images for later subtraction. We find less than 2% contamination by charged particle events throughout this observation for C and M bands and less than 5% for the J band (Table 1).

Short and long term enhancements are the least understood of all contaminating elements in ROSAT analysis. These effects manifest themselves as short (< 1 day) gradual increases in accepted event rates or longer (> 1 day) constant or variable enhancements and appearing as stripes on the ROSAT all-sky survey. We have attempted to identify short term enhancements by searching for large (30%) increases in average event rate levels. Additionally, we have searched for long term enhancements by comparing event rates against the cleaned all-sky survey average rates in this direction. We find no evidence for short term enhancements for this observation. However, LTEs are present in both C and M bands. We have attempted to account for these events by first identifying the observational minima as free of LTE and casting relative differences between other levels onto C or M band normalized efficiency maps to create LTE background images for later subtraction.

Finally, afterpulse events have only recently been identified as a serious source of contamination for some C-band studies. Plucinsky et al. (1993) have found these spurious events occurring less than 0.35 milliseconds after precursor events and mimic photons below 0.25 keV. The source of these events are unclear but have been attributed to charged particle events. We have attempted to remove such events by filtering our data set to eliminate all events occurring within the afterpulse interval. We have eliminated 776 events from this observation. Unfortunately, afterpulse events may also follow rejected events which are not telemetered to the ground and so may be a source of unaccounted contamination.

3. Generation of Energy Dependent Exposure Maps

The exposure map of a ROSAT PSPC pointed observation provides the effective exposure time of each map pixel. The exposure map is critical for spatial analysis (particularly diffuse objects) because it allows one to correct for the instrument induced count rate variations across the field of view. The heart of an exposure map is the instrument map. In addition to window support obscuration and vignetting, the PSPC instrument map must account for apparent spatial variations produced by the photon induced charge cloud response to the anode/cathode grid. This response is dependent on wire spacing, voltage and photon energy. These effects are especially critical in the C-band map where the effects of small charge packets and the lower level discriminator manifest themselves as "ghost images".

An exposure map may be generated by casting the instrument map onto the sky, including the effects of aspect drift and wobble during the observation. For this demonstration, energy dependent exposure maps, based on uncontaminated observation intervals and deadtimes, have been generated for C, M and J bands.

4. Data Cleaning

The generation of "clean" data begins with the removal of observing times identified as heavily contaminated by scattered solar X-rays, particle background and enhancements. We find 22% of our C-band data and 60% of our M-band data unacceptable for analysis based on SSX and particle background modeling. Photon maps are next created according to pulse invariant bin filtering. Photon, scattered solar X-ray, internal and external particle background and long term enhancement maps are then flattened by the corresponding exposure maps. This process concludes with the removal of point sources and subtraction of SSX, PB, and LTE backgrounds from C, M, and J photon maps and smoothing.

5. Discussion

Proper modeling and subtraction of contamination is essential in the correct interpretation of ROSAT data obtained towards extended or diffuse targets. We find scattered solar X-rays are the dominant source of contamination with little effect by charged particles. Additionally, short and long term enhancements are difficult to predict but may also seriously affect one's data.

Software for these procedures have been developed at MPE and PSU and are being incorporated into the software database at GSFC and SAO (Snowden et al. 1993).

References

Burrows, D. N., & Mendenhall, J. A. 1991, Nature, 351, 629
Plucinsky, P. P., Snowden, S. L., Briel, U. G, Hasinger, G., & Pfeffermann, E. 1993, ApJS, submitted
Snowden, S. L., & Freyberg, M. J. 1993, ApJ, 404, in press
Snowden, S. L., et al. 1993, in preparation

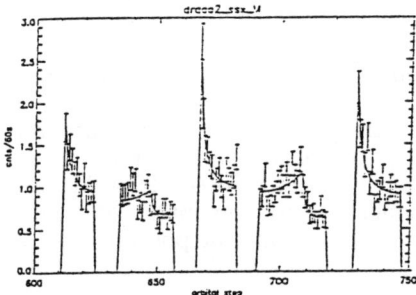

Figure 1: Modeled M-band scattered solar X-ray contamination for six observation intervals.

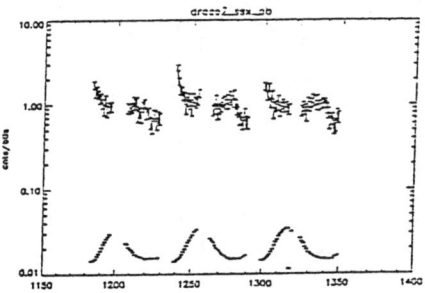

Figure 2: Modeled M-band particle background contamination. Upper data represents observation light curves. Lower data represents modeled particle background light curves.

Table 1: Levels of Contamination in Original Data

Band	Original Data	Scattered	Internal PB	Externl PB	Enhancements	Afterpulses
C	141265	4495	568	123	8736	776
M	22551	4687	306	141	3048	0

Table 2: Levels of Contamination in Final Data

Band	Final Data	Scattered	Internal PB	Externl PB	Enhancements	Afterpulses
C	101394	2206	359	79	6738	?
M	4517	68	71	33	990	0

CORRECTING FOR ASPECT SOLUTION ERRORS IN *ROSAT* HRI OBSERVATIONS OF COMPACT SOURCES

Jon A. Morse
Space Telescope Science Institute, Baltimore, MD 21218

Email ID
morsey@stsci.edu

ABSTRACT

Elongations in *ROSAT* HRI images of compact sources that are produced by errors in the aspect solution often inhibit the identification and description of extended X-ray emission associated with sources such as pulsars and active galactic nuclei. A simple method for reducing the effect of aspect solution errors is discussed and then applied to observations of a known soft X-ray point source and an active galaxy.

1. Introduction

X-ray point sources observed with the *ROSAT* High Resolution Imager (HRI; see Zombeck et al. 1990) often appear elongated over scales of $\sim 5'' - 10''$ from the image core. This elongation has been attributed to errors in the attitude correction as the satellite is wobbled during the observations, and affects sources with both soft and hard X-ray spectra (David et al. 1992). The orientation of the elongation is apparently uncorrelated with the wobble direction of the satellite and is not well understood. This effect has hampered efforts to identify and describe possible extended X-ray emission around sources such as pulsars and active galactic nuclei (AGN) within $\sim 10''$ of the central source.

In this contribution, I briefly describe a method for mapping out and correcting errors in the aspect solution in observations of compact sources. More complete discussions of this procedure appear elsewhere (Morse 1994; Morse, Wilson, & Elvis 1994). Investigators are cautioned not to over-interpret small-scale elongations in HRI observations without first addressing the possible effects of aspect errors.

2. Method

Targets observed with the HRI are usually dithered back and forth across the detector by wobbling the satellite, primarily to smooth out quantum efficiency variations across the detector. The wobble period is $\sim 300 - 400$ seconds and the wobble amplitude is of order several arcminutes. Errors in the aspect solution probably derive from inaccuracies in following bright stars in the 1' pixels of the XRT star tracker as the satellite is wobbled. (A second star tracker for the XRT malfunctioned early in the mission.) Note that point sources do not always appear elongated in HRI observations; sometimes the tracking is fine and point sources appear perfectly round.

To map out errors in the aspect solution in an observation of a compact source, one would ideally like to monitor the source centroid over very short time intervals and apply corrections to the original aspect solution as a function of time. Such a procedure has been used by Schmitt *et al.* (1994) to produce round point sources. However, many sources have count rates $\lesssim 1$ count s^{-1} and will not accrue sufficient counts to achieve accurate image centroids in short intervals. Because the source moves ~ 2 pixels ($= 1''$) per second across the detector during the wobble, even sub-images containing photons recorded over time intervals as short as 30 seconds may exhibit elongations.

We can overcome this obstacle by assuming that all photons *from the source* received at a given phase of the wobble are mapped into celestial coordinates using the same attitude correction. We can create sub-images from specific locations on the detector through which the source moved during the wobble and monitor the offsets in RA and Dec as a function of wobble phase. Sufficient counts for accurate centroiding can be obtained in the separate sub-images because we include photons recorded over several different wobble periods; the light-curve for each spatial bin contains periodic peaks as the source swept through the bin. However, by choosing small spatial bins on the detector, each sub-image is sampling an effectively short time interval.

Several steps for mapping out and correcting aspect errors. which can be performed in the IRAF/PROS software environment, follow:

- Create separate images from each orbit (OBI) to verify that there are no large relative systematic orbit-to-orbit shifts. If so, shift each OBI to a common center and restack into a single image.
- Display the observation in *detector coordinates* and create sub-images from small regions on the detector through which the source was wobbled. For targets with count rates of ~ 0.35 counts s^{-1} and total integrations of $15-20$ kiloseconds, $\sim 20-30$ bins covering the extent of the wobble in detector coordinates could be chosen which will have adequate signal-to-noise ratios for accurate centroiding.
- Measure the centroids of the sub-images and plot the offsets from a common center in RA and Dec versus position on the detector to identify systematic trends as a function of phase in the wobble. Then shift the sub-images to a common position and restack.

Note that this procedure is strictly designed for compact sources; the total area on the detector covered by the bins must contain all photons from any extended emission near the unresolved central source. Photons detected outside this area are thrown away. For sources with very extended X-ray emission ($> 20'' - 30''$; e.g., supernova remnants, some active galaxies) aspect error corrections can only be made by dividing the data into small time intervals, perhaps co-adding sub-images from the same phase of the wobble. and shifting the sub-images to a common center.

3. Application to HZ 43

I used the above method in an attempt to rid a high signal-to-noise on-axis HRI observation of the soft X-ray point source HZ 43 of its characteristic elongation (Fig. 1). The high count rate for this source (~ 12 counts s^{-1}) allowed me to divide this observation into 181 sub-images, each containing photons from only a small region on the detector through which the source passed during the satellite's wobble (Fig. 2). By measuring the positions of the individual image centroids, I found clear evidence for systematic offsets from a common mean by up to $\sim \pm 3''$ in both RA and Dec as a function of phase in the satellite wobble (Fig. 3). Shifting the sub-images to a common center and then restacking

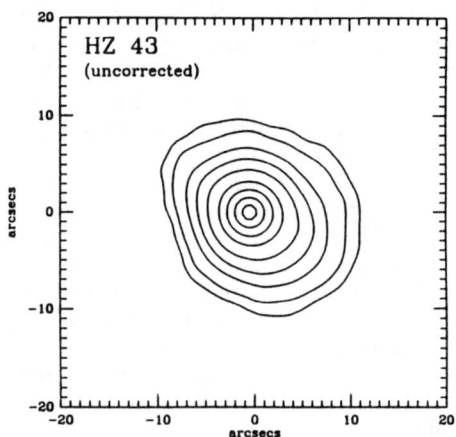
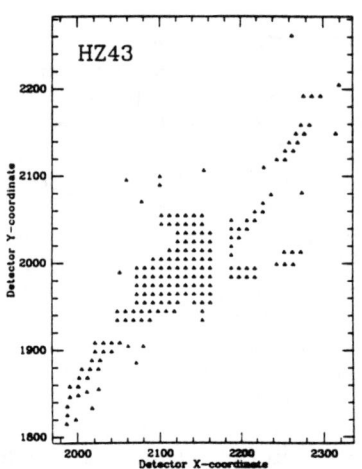

Fig. 1.— An HRI image of the soft X-ray point source HZ 43. Contours are set at 1%, 2%, 5%, 10%, 20%, 40%, 60%, 80% and 95% of the maximum intensity.

Fig. 2.— The layout of bins in detector coordinates from which HZ 43 sub-images were extracted, showing the movement of the source across the detector during the wobble.

them into a single image measurably improved the symmetry of the point spread function (Fig. 4). Some ellipticity persists in the corrected image of HZ 43, possibly because I did not use small enough bin sizes in detector coordinates to completely resolve out the errors in the aspect solution. This may be unavoidable for weak sources where larger bin sizes are required to obtain enough counts for accurate image centroiding.

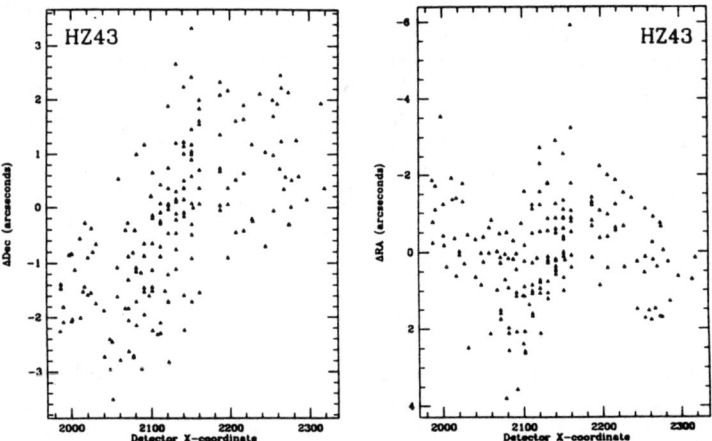

Fig. 3.— The offsets in RA and Dec versus the detector X-coordinate. Sub-images formed from the top of the wobble are at larger detector X-coordinate values. The offsets in RA and Dec as a function of position along the wobble show strong systematic trends.

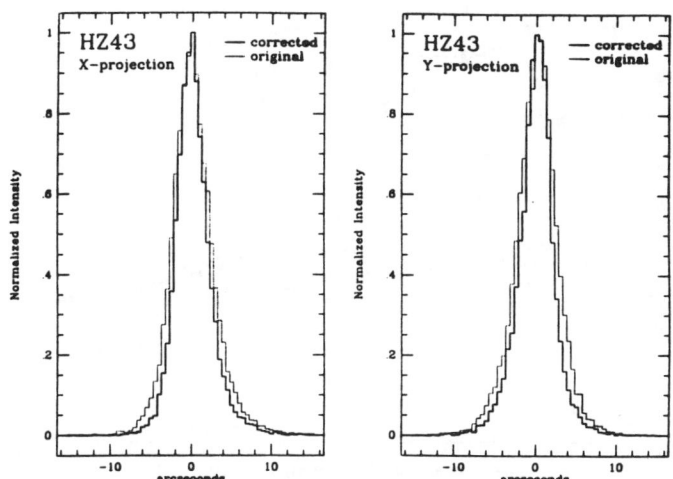

Fig. 4.— Intensity cuts across the original and corrected HRI images of HZ 43. The peaks are narrower and more symmetric in the corrected image but some asymmetries persist in the wings at flux levels below 20% of the peak.

4. Application to NGC 3516

The nearby Seyfert 1 galaxy NGC 3516 shows extended emission in the optical and radio wavebands (e.g., Miyaji, Wilson, & Pérez-Fournon 1992) over scales of $\sim 2'' - 10''$ from the active nucleus. Isolating co-spatial extended X-ray emission from the circum-nuclear environment can place significant constraints on the physical processes operating in these regions (e.g., see Wilson, this volume). Fig. 5 displays a contour plot of the NGC 3516 HRI image which shows an ellipticity along a PA of $\sim 40°$, roughly aligned with the 'Z'-shaped

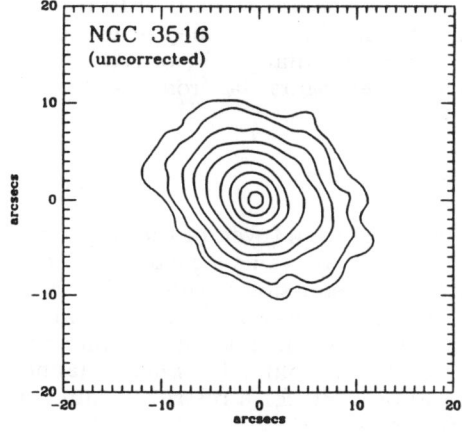

Fig. 5.— Raw HRI image of NGC 3516. Contours are set to the same levels as in Fig. 1.

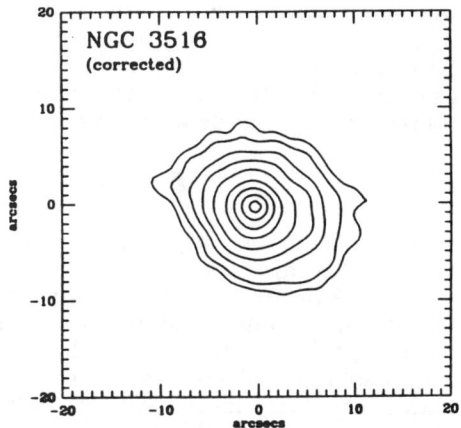

Fig. 6.— Corrected image of NGC 3516. Contours are set to the same levels as in Fig. 1.

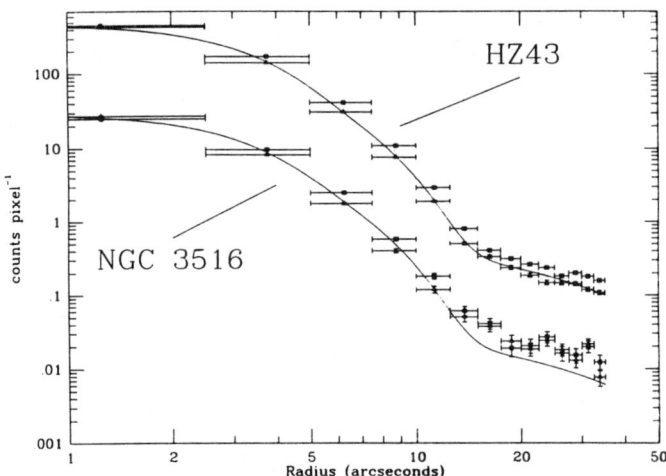

Fig. 7.— A comparison of azimuthally averaged radial brightness profiles for NGC 3516 and HZ 43. The open squares are from the uncorrected images and the solid triangles from the corrected ones. The solid line is a match to the corrected HZ 43 profile using the 2 Gaussian plus exponential parameterization of David et al. (1992). This fit has also been scaled and compared to the NGC 3516 profile.

extended narrow-lined optical emission. To determine whether this elongation results from errors in the aspect solution, I assumed that the central peak was from an unresolved point source and used the procedure described above to map out the image centroid offsets in RA and Dec. These offsets were found to lie along the direction of the elongation, and removing them yielded a more axisymmetric image (Fig. 6). Aspect errors appear to be responsible for much of the ellipticity in the original image. Comparing the azimuthally averaged radial brightness profiles of the original and corrected images of NGC 3516 and HZ 43, we cannot distinguish with certainty the X-ray morphology of NGC 3516 from a point source within $\sim 10''$ of the central source (Fig. 7). There may be evidence for extended emission beyond a radius of $10''$ in the radial profile of NGC 3516 as compared to HZ 43, but it is difficult to interpret regions far from the central source for the reasons given at the end of §2.

5. To Wobble or Not to Wobble?

Aspect errors vary from observation to observation (even of the same target) and for this reason are very difficult to correct. It seems logical that if the *scientific objectives* are to identify extended X-ray emission within $\sim 10''$ of a bright central source, the HRI observation should not be wobbled. It would be better in this case to be somewhat uncertain of the count rate rather than the aspect solution. There is some concern that by not wobbling, the micro-channel plates may decay more swiftly at certain locations, but for a restricted number of compact sources, this should not be a problem.

References

David, L. P., Harnden, F. R., Jr., Kearns, K. E., & Zombeck, M. V. 1992, The *ROSAT* High Resolution Imager (HRI) (Technical Rep., US *ROSAT* Science Data Center/SAO)
Miyaji, T., Wilson, A. S., & Pérez-Fournon, I. 1992, ApJ, 385, 137
Morse, J. A. 1994, PASP, submitted
Morse, J. A., Wilson, A. S., & Elvis, M. 1994, in preparation
Schmitt, J. H. M. M., Güdel, M., & Predehl, P. 1994, A&A, in press
Zombeck, M. V., Conroy, M., Harnden, F. R., Roy, A., Braeuninger, H., Burkert, W., Hasinger, G., & Predehl, P. 1990, in Proc. SPIE Conference on EUV, X-Ray and Gamma-Ray Instrumentation for Astronomy, ed. O. H. W. Siegmund & H. S. Hudson (SPIE Proc., 1344), 267

This research has been supported by NASA grants NAGW-2689 and NAGW-3268 to A. Wilson at STScI.

IMAGE DECONVOLUTIONS OF *ROSAT* HRI OBSERVATIONS: CHOOSING A POINT SPREAD FUNCTION

Jon A. Morse
Space Telescope Science Institute, Baltimore, MD 21218

Email ID
morsey@stsci.edu

ABSTRACT

Issues concerning the choice of a point spread function for image deconvolutions of *ROSAT* HRI observations are discussed.

1. Introduction

As use of the *ROSAT* High Resolution Imager (HRI) sharply increases in 1994, there will certainly be situations where investigators will want to apply image deconvolution techniques to HRI images to increase the effective spatial resolution. In this contribution, I briefly discuss some of the issues to bear in mind when choosing a point spread function (PSF) for the image deconvolutions.

Most of this discussion was presented by Larry David (SAO) during his talk at the symposium, and a detailed investigation of the in-flight HRI PSF can be found in the HRI Calibration Report (available via anonymous ftp on *legacy* at GSFC; see also David *et al.* 1992) that accompanies the December 1993 release of PROS.

2. Choosing a PSF

The biggest concern is how errors in the aspect solution will affect the science data and the chosen PSF. Aspect errors are presently not well understood (e.g., see Morse, this volume) and could lead to significant distortions of the deconvolved data, depending on what the important scale lengths are. Aspect errors manifest themselves primarily over scales of $\lesssim 10''$; if the important features in the science data are found on this or smaller scales, some care needs to be taken in choosing an appropriate PSF.

The following items summarize how to choose a PSF, in order of priority:
- A known point source in the same image within $\sim 6'$ of the target is the best choice for a PSF (assuming the target was observed on-axis). The $6'$ range is needed because the PSF begins to blow up at larger off-axis angles (David *et al.* 1992). This choice of PSF is best because aspect errors will affect both the target and the point source in the same way, and will be cancelled out in the deconvolution.
- In the absence of a point source within the same image, use an HRI point source observation available in the PROS calibration database observed at roughly the same off-axis angle. Keep in mind that aspect errors may affect the point source observation differently than the science data. If possible, one should try to correct for aspect errors in both the point source and the science data observations before deconvolving.
- Use an axisymmetric model PSF if no other PSF appears suitable. David *et*

al. (1992) provide a functional form for the azimuthally averaged PSF based on fits to several HRI point source observations.

As an example, consider the HRI image of NGC 3516 (see Fig. 5 of Morse, this volume), a nearby Seyfert 1 galaxy that shows extended emission in the optical and radio wavebands (e.g., Miyaji, Wilson, & Pérez-Fournon 1992) over scales of $\sim 2'' - 10''$ from the active nucleus. The raw HRI image shows an elongation along a PA of $\sim 40°$, roughly aligned with the 'Z'-shaped extended narrow-lined optical emission. Fig. 1 shows the result of deconvolving this image with 50 iterations of the Lucy-Richardson algorithm using an axisymmetric model PSF based on the azimuthally averaged radial brightness profile given by David et al. (1992). On the other hand, Fig. 2 shows a similar deconvolution but using an on-axis HRI observation of the soft X-ray point source HZ 43 as the PSF. Both images used to obtain the result in Fig. 2 were first corrected for aspect errors. It is obvious that very different scientific conclusions might be drawn from Figs. 1 and 2. An analysis shows that most of the ellipticity in the original image of NGC 3516 probably derives from aspect errors and not real extended X-ray emission.

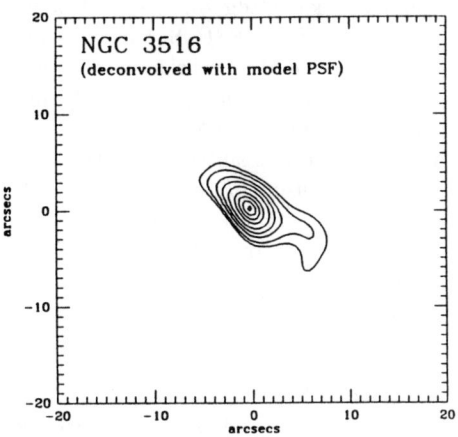

Fig. 1.— The deconvolved HRI image of NGC 3516 using an axisymmetric model PSF based on the azimuthally averaged radial profile given by David et al. (1992). Contours are set at 1%, 2%, 5%, 10%, 20%, 40%, 60%, 80% and 95% of the maximum intensity.

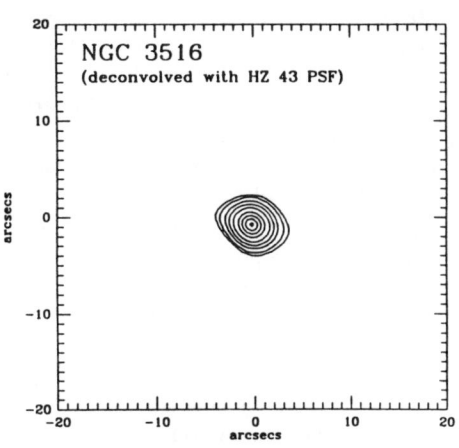

Fig. 2.— The deconvolved HRI image of NGC 3516 using an on-axis HRI observation of the soft X-ray point source HZ 43 as the PSF. Both observations were corrected for aspect errors prior to the deconvolution. Contours are set to the same levels as in Fig. 1.

References

David, L. P., Harnden, F. R., Jr., Kearns, K. E., & Zombeck, M. V. 1992, The *ROSAT* High Resolution Imager (HRI) (Technical Rep., US *ROSAT* Science Data Center/SAO)

Miyaji, T., Wilson, A. S., & Pérez-Fournon, I. 1992, ApJ, 385, 137

This research has been supported by NASA grants NAGW-2689 and NAGW-3268 to A. Wilson at STScI.

DETECTING X–RAY SOURCES WITH THE WAVELET TRANSFORM

Piero Rosati[1,2]
[1]Dept. of Physics – University of Rome *La Sapienza*, Italy.
Richard Burg[2,4], Riccardo Giacconi[2,3,4]
[2]Department of Physics & Astronomy, Johns Hopkins Univ., Baltimore (MD).
[3]European Southern Observatory, Garching, Germany.
[4]Space Telescope Science Institute, Baltimore (MD).

Email ID
rosati@stsci.edu

ABSTRACT

We present here preliminary results of a detection algorithm based on the *Wavelet Transform*. This technique is an alternative to the standard techniques based on Maximum–likelihood sliding box methods. It consists of an orthonormal decomposition of a signal into both space and scale through a multiscale filtering process. The position, net counts and morphological parameters of the sources can be automatically estimated, once a source model is assumed, without any knowledge of the local background. The detection itself does not require knowledge of the PSF. Simulations are required to assess the significance of the detection. The method works equally well for both point–like and extended sources. We have tested this technique on both simulations and ROSAT data. The wavelet analysis of the Lockman field, the deepest ROSAT-PSPC observation, shows that this algorithm is also well suited for particularly crowded fields. We compare our results with those obtained using other techniques. Our algorithm detects several extended sources (i.e. with characteristic scale significantly greater than the point response function of the telescope/detector system) in this field; two of these have recently been optically confirmed as clusters.

1. THE WAVELET TECHNIQUE

The wavelet analysis consists of unfolding a field into both *space* and *scale* (and possibly direction) by convolving it with a family of functions, called *wavelets*, which are generated by dilations and translations (and possibly rotations) of a unique *analyzing wavelet*. This is a real or complex valued function, $g_a(x,y)$, with specific smoothness and compactness properties whose fourier transform, $\hat{g}(\vec{\omega})$, satisfies the admissibility condition $C_g = \int_{R^2} |\hat{g}(\vec{\omega})|^2 \frac{d\vec{\omega}}{|\vec{\omega}|^2} < \infty$, which implies that $g_a(x,y)$ has zero mean.

Mathematically, the *Wavelet Transform* (WT) of a function $f(\vec{x}) \in L^2(\mathbf{R}^2)$ is the L^2–inner product (i.e. the convolution integral) between f and the wavelet family $\{g_{a,\vec{x}'}\}$:

$$\text{WT}: f(x,y) \in \mathbf{R}^2 \Longrightarrow w(x',y',a) \in \mathbf{H} = \mathbf{R}^2 \times \mathbf{R}^+,$$

where a is the *wavelet scale*. This gives the wavelet coefficients:

$$w(x',y',a) = \langle g_{a,\vec{x}'}|f\rangle = a^{-1}\int_{R^2} f(x,y) g^*\left(\frac{x-x'}{a}, \frac{y-y'}{a}\right) dx\, dy$$

(see, for instance, Grossmann et al. 1988, Daubechies 1988).
The WT has been used in astronomy to study, for example, low/high density structures

in galaxy catalogues (Slezak et al., 1993) and subclustering in optical clusters (Escalera et al., 1992). Commonly used analysing wavelets are the Marr and the Morlet wavelets:

$$g(\frac{r}{a}) = (2 - \frac{r^2}{a^2})e^{-\frac{r^2}{2a^2}} \qquad g(\frac{r}{a}) = \frac{2}{a^2}[e^{-\frac{r^2}{a^2}} - \frac{1}{2}e^{-\frac{r^2}{2a^2}}]$$

2. DISCUSSION

The detection algorithm works as follows. The WT of the data image is evaluated for a set of scales $\{a_1, a_2, \ldots\}$. The background counts and low frequency noise are averaged to zero by this operation. It is difficult to determine the statistical distribution of the WT of the white noise (especially for large scales). However, one can show that the variance of the WT of the white noise decays as $1/a^2$, the proportionality constant depending on the integral power spectrum of the mother wavelet and the white noise variance. This information together with off–line simulations (Fig.1), provides a natural detection threshold with an associated confidence level for detected features. No artificial parameters (such as detection/background box) are needed and the PSF has not been used at this stage.

Once a model for the source profile is assumed and its WT calculated (analytically or numerically), a fitting procedure in the *wavelet space* allows an estimate of the characteristic parameters of the source without having to estimate the the local background in the data. The final estimated parameters are those corresponding to the scale which maximizes the S/N of a given source in the wavelet space. This is shown in Fig.2: the contribution of the signal has a maximum when the wavelet scale matches the characteristic size of the source, whereas background fluctuations dominate at smaller scales ($1/a$ behaviour). The output parameters include the position of the source, its characteristic scale (e.g. FWHM for a gaussian, core radius and β-index for a King profile) and its net counts. The χ^2 of the fit can be used to discriminate between different source models.

We have applied the WT technique on a very deep PSPC image of the Lockman Hole in the H band (0.4 – 2.4 Kev). More than 100 sources are detected with a 90% c.l. in the inner 20' radius of the field. A comparison of the X–ray counts estimated with the WT technique and the multi–ML analysis (Hasinger et al. 1993) is shown in Fig.3.

Using the information of the telescope/detector system PSF, we can detect *extended sources* as those with characteristic scale significantly greater of the PSF width. From this analysis, two sources with $FWHM \simeq 1'$ are identified (id# 34 and 41 in Fig.4). These are two of several cluster candidates which have been studied in a deep optical follow up by groups at MPE, CalTech, JHU/STSci and Bologna. In particular, # 41 has been confirmed and its redshift measured ($z \simeq 0.34$).

3. REFERENCES

Daubechies I. 1990, IEEE Trans. Inf. Theory, 36, 961
Escalera E., Slezak E., Mazure A. 1992, A&A, 264, 379
Grossmann A., Kronland-Martinet R., Morlet J. 1988, in Wavelets,Time-Frequency
  Methods and Phase Space, ed J.M.Combes et. al (Berlin, Springer-Verlag), 1
Hasinger et al. 1993, A&A, 275, 1
Slezak E., de Lapparent V., Bijaoui A. 1993, ApJ, 409, 517

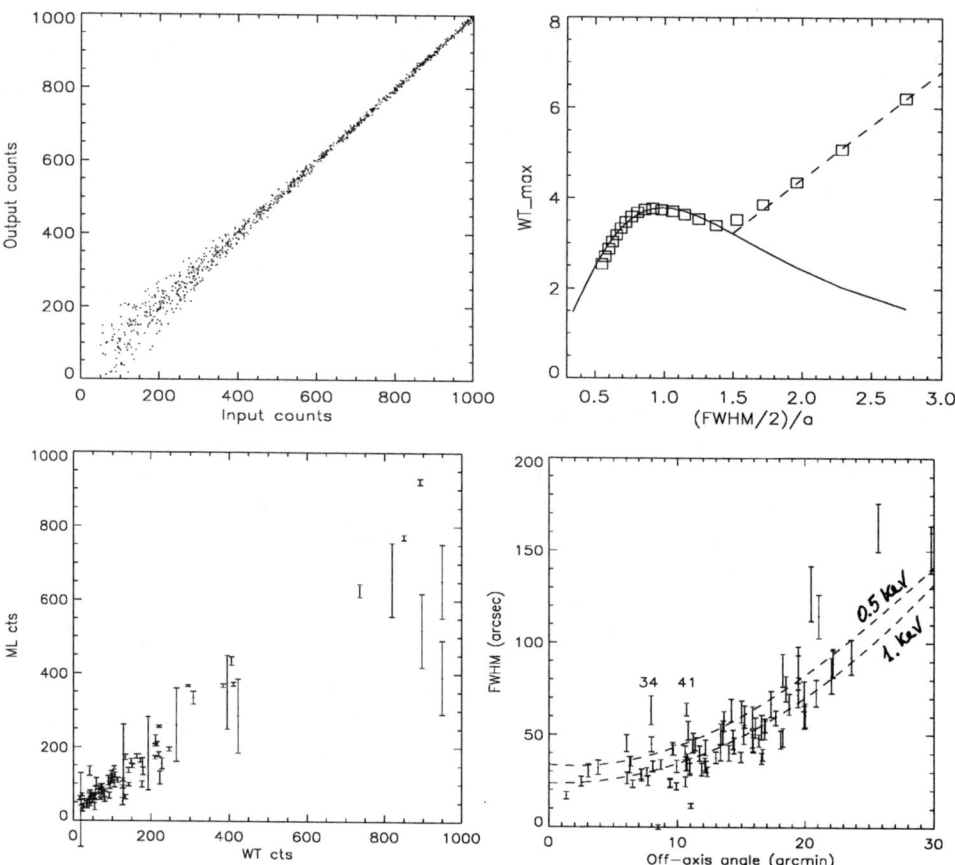

*Fig.*1 Comparison of input and output counts from 1000 simulated sources detected with the WT detection algorithm.

*Fig.*2 Maximum of the WT vs (a =wavelet scale)$^{-1}$ for a source with a given FWHM. The solid line is the expected curve for a gaussian source, the dashed line is the ($1/a$) expected behaviour of background fluctuations. The squares are the data.

*Fig.*3 Comparison of estimated counts of detected features in the Lockman field using the WT analysis and the multi-source Maximum Likelihood (ML) analysis (Hasinger et al. 1993). Only the ML counts errors are shown for clarity.

*Fig.*4 Estimated FWHM of the detected sources with the WT analysis. The PSPC FWHM is overplotted. Two sources with characteristic size significantly greater than the PSF are the optically confirmed clusters. The sources detected at $\theta > 20'$ where the PSF is highly distorted are not significant and should be treated with particularly large wavelet scale.

An IDL-Based ROSAT Data Analysis Package

Frederick M. Walter
ESS Dept, SUNY, Stony Brook NY 11794-2100

Email ID
fwalter@astro.sunysb.edu

ABSTRACT

I announce the availability of, and briefly describe, a set of IDL procedures for the examination and analysis of ROSAT PSPC and HRI data.

1. INTRODUCTION

The two major software packages available for examining and analyzing ROSAT data, IRAF/PROS and MIDAS/EXSAS, meet the needs of most users. However, all software has limitations, and good observers are often never completely satisfied by the job done by someone else's software. There is no single best way to analyze data.

Do we need more integrated packages for analyzing ROSAT data? My inclination was to say no, but I found myself doing just that once ROSAT data tapes began arriving at my door. I have been using IDL for over a decade, and so found it convenient to use IDL to examine the data. My needs, to read the data in, plot the images, extract sources, and performing timing and spectral analyses, are similar to those of most users. Eventually I put together series of IDL procedures which form a fairly complete ROSAT XRT data examination and analysis package. As the code was written with partial support from ROSAT funding, I have decided to make it available to the community.

Of course this software is customized to my peculiar analysis needs, and is no more appropriate to the general user than any other package. What can I offer that IRAF/PROS or EXSAS does not? Mainly, I offer software which runs in the IDL environment. Users of IDL need not be reminded of the power of the languge for data examination and for the display of data, nor do they need be reminded of the ease with which procedures can be modified to suit, or written from scratch. I offer this software both as a reasonably complete package for doing simple examination and analysis of ROSAT data, and as a starting point for more detailed tasks. This code complements the procedures available from the ROSAT IDL library at the HEASARC.

2. PHILOSOPHY

The procedures all use a set of variables stored in common. The data, either from the QPOE (.FITS) file or the images, are read into common variables. In this way the data, once read in, are accessible to all procedures, as well as for interactive use.

A main program is used to define the common blocks, so that they will be available interactively, though it is not necessary to do this. A driver can be used to read in the data and populate the common variables. By default, the QPOE file is read in, as are the broad-band image and exposure map (or the HRI image, if appropriate). One can read in more, or less, by setting the appropriate keywords. The events are stored in a structure variable.

In general, the user is not prompted for inputs; they are passed via keywords, with defaults pre-assigned. On-line help is available to remind the user what keywords are available. A more detailed LATEX documentation file exists.

The system currently runs under IDL 3.0 and 3.1, on both VMS and UNIX systems.

3. WHAT IS INCLUDED

The software includes code to:
- Read in and display FITS images.
- Overplot coordinate grids, X-ray sources, SIMBAD identifications, HST GSC entries, or arbitrary coordinates, on the image.
- Read in the QPOE file.
- Select photons from the QPOE file by X,Y position on the sky or in the detector, by arrival time, or by PH or PI bin.
- Merge QPOE files from different observations.
- Construct images from selected photons.
- Merge images from different observations.
- Extract source and background photons. These are placed into structure variables with the same structure as the full events list.
- Make pulse height spectra in either 256 bins or the 34 SASS bins. Optionally, a .PHA file readable by XSPEC can be produced.
- Perform timing analyses on the source counts. One can bin the data and plot light curves, plot the distribution of photon arrival intervals, perform K-S tests, and do FFT period-finding analyses.
- Undertake source detection, using code based on the CAL EINSTEIN system (under development).

I am continually adding to the set of procedures as I think of new things to do, and will consider suggestions to augment the package.

4. AVAILABILITY OF THE SOFTWARE

The software is available for the asking though a guest account on our VMS system at Stony Brook. It will eventually be made available via anonymous FTP. Complete documentation exists as a LATEX document. E-mail the author at the address above for specifics.

POSTERS

Stars

ROSAT HRI OBSERVATIONS OF T TAURI STAR PAIRS IN TAURUS-AURIGA

F. Damiani, G. Micela, S. Sciortino
Istituto ed Osservatorio Astronomico di Palermo, 90134 Palermo, ITALY

F.R. Harnden, Jr.
Harvard-Smithsonian Center for Astrophysics,
60 Garden Street, Cambridge, MA 02138, USA

Email ID
damiani@oapa.astropa.unipa.it

ABSTRACT

We present results from three ROSAT HRI fields in the Taurus-Auriga region, pointed towards unresolved pairs of T Tauri stars detected with the *Einstein* IPC. With the new observations we spatially resolve these stellar pairs, with separations of ∼20", and detect 7 out of 8 observed stars. Five other T Tauri stars are also detected in the HRI images.

These data demonstrate that usually both components in a binary T Tauri system have comparable X-ray luminosities. We do not detect X-ray flares from any star. Variability on longer timescales (∼15 years) may be present, but not exceed a factor of 2-3. The HRI data confirm the anticorrelation between L_x and stellar rotation period, found from IPC data, indicating a magnetic origin for the X-ray emission.

1. Introduction

Star-forming regions contain often many dozens of X-ray active stars (Herbig & Bell 1988), often occurring in groups so close together that they cannot be resolved with the *Einstein* IPC or the ROSAT PSPC, but are still within the spatial resolution capabilities of the ROSAT HRI. In their systematic study of X-ray emission from pre-main-sequence (PMS) stars in Tau-Aur, using IPC data, Damiani et al. (1993a) find that about half of the detected X-ray sources correspond to unresolved star pairs. This affects our knowledge of the true X-ray detection frequency, and of the correlation of X-ray emission with other properties of these stars.

2. X-ray Observations and Data Analysis

Our HRI images have been taken in Sept. 1992, with exposure times of 7237, 4729, and 5150 seconds, for the fields of the PMS star pairs DH/DI Tau, GI/GK Tau, and XZ/HL Tau, respectively. Each of the two former pairs is resolved in two distinct X-ray sources; in the latter pair, only XZ Tau is detected. The pair GH/V807 Tau is detected serendipitously, as an elongated X-ray source at 12.8 arcmin off-axis, indicating that both stars emit X-rays at a comparable level.

The recorded X-ray image of the pair GI/GK Tau is shown in Figure 1a. The slight overlap of the X-ray point-spread functions (PSF) of the two stars allows a good determination of the relative X-ray count rates. The discrepancy between the optical and HRI positions is always smaller than 5", within the expected range for the HRI.

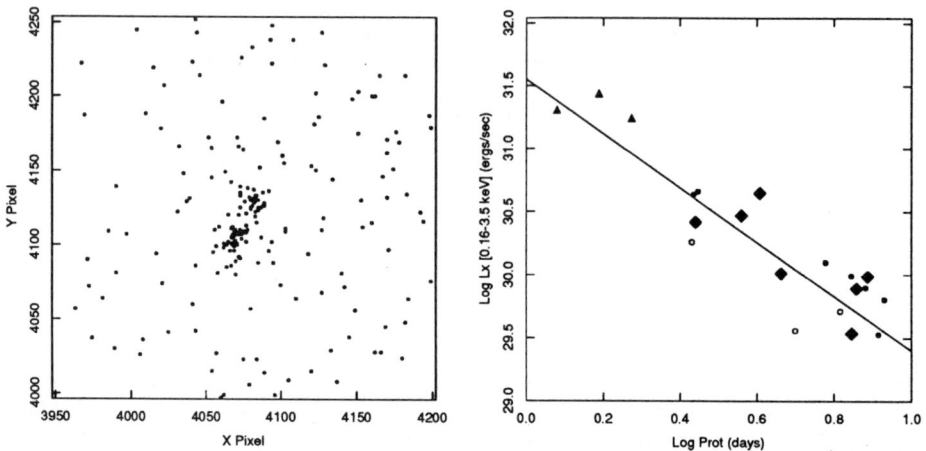

Figure 1: (*a*, left) ROSAT HRI image of the pair GI Tau (north) and GK Tau (south). (*b*, right) L_x vs. P_{rot} for the program stars (diamonds). Also shown are *Einstein* and ROSAT data for other T Tauri stars in Tau-Aur from Damiani et al. (1993a) (dots), and from Damiani et al. (1993b) (triangles).

3.1 The X-ray Detection Statistics of T Tauri Stars in Taurus-Auriga

In their study of PMS stars in Tau-Aur, Damiani et al. (1993a) detect in X-rays a fraction of 'classical' T Tauri stars which may range from 40% to 57%, corresponding respectively to the cases of only one or both detected stars in each pair. Therefore, the sampling of the L_x median value was not certain.

In our HRI sample, the detection fraction of PMS stars in pairs (7/8) is higher than its conservative estimate (4/8). Since these stars are not peculiar objects within their class, this suggests that in T Tauri pairs generally both stars emit X-rays at comparable level, at least when they have similar spectral types, as in our case. Therefore, the X-ray detection frequency of T Tauri stars should be higher than its conservative estimate based on low-resolution X-ray data, approaching its upper bound of 57%.

3.2 X-ray Variability

Strong variability of the X-ray emission has been reported for many PMS stars (e.g. Feigelson & DeCampli 1981, Montmerle et al. 1983, Walter & Kuhi 1984). In our sample, however, a Kolmogorov-Smirnov test does not detect for any X-ray source variability at more than 99% confidence level. This suggests, at least in the Tau-Aur region, a lower X-ray flaring frequency for T Tauri stars than previously suggested (Gahm 1980, Montmerle et al. 1983).

To study long-term variability, we have compared the ROSAT HRI data with the IPC data on the same stars. To compute (0.16-3.5 keV) X-ray luminosities we assume a single-temperature Raymond spectrum with $T = 1.5 \cdot 10^7$ K (as derived from IPC spectra of other T Tauri stars), deriving the absorption N_H from A_V. The HRI and IPC data (taken from Damiani et al. 1993a) for a given star are generally consistent within 1-2σ. Therefore, if variability is present, its amplitude does not exceed factors of 2-3.

3.3 X-ray Activity and Stellar Rotation

Using the derived HRI X-ray luminosities, we have studied the correlation between X-ray luminosity and (photometric) stellar rotation period P_{rot}. This is known for four of the observed stars in pairs (DH Tau, DI Tau, GI Tau, GK Tau), and for three of the serendipitous detections (V826 Tau, V827 Tau, V830 Tau).

A plot of L_x vs. P_{rot} is shown in Figure 1b. Dots are IPC data from Damiani et al. (1993a), triangles are (IPC and ROSAT PSPC) data from Damiani et al. (1993b), and diamonds are the new HRI data; the empty dots indicate IPC data for unresolved pairs, now representing only a minor fraction. The solid line is the best fit on the IPC data by Damiani et al. (1993a). Clearly, the new HRI data support further the correlation already found, as well as the picture (Bouvier 1990, Damiani et al. 1993a) of X-ray emission in T Tauri stars arising from magnetic dynamo activity.

It can also be seen from Figure 1b that some scatter around the average correlation remains, even for stars resolved with the HRI. If not a spurious effect, depending on the assumed X-ray spectrum, it might be due to X-ray variability, or extra X-ray emission from unknown companions.

4. Conclusions

New ROSAT HRI observations in the Taurus-Auriga region have allowed to resolve for the first time the X-ray emission arising from close (15-20 arcsecs) pairs of T Tauri stars. Our main results are:

(*i*) All X-ray sources associated with T Tauri stars and formerly detected with the *Einstein* IPC are again detected in our ROSAT HRI images. In all, we detect 12 distinct X-ray sources identified with T Tauri stars.

(*ii*) Seven out of eight stars in pairs are detected with the HRI. This suggests that previous conservative estimates underpredict the fraction of T Tauri stars with $\log L_x \geq 29.5$-30 (ergs/sec), which should rather be close to 60%.

(*iii*) No strong flarelike X-ray variability is seen in our HRI sources. Depending on assumed X-ray source spectrum and detectors cross-calibration, long-term variations are possible, with amplitudes below factors of 2-3.

(*iv*) The new HRI data confirm the anticorrelation between X-ray luminosity and P_{rot}, known from *Einstein* data, pointing to a magnetic origin for X-ray emission of PMS stars.

REFERENCES

Bouvier,J. 1990, AJ, 99, 946.
Damiani,F., Micela,G., Sciortino,S., Harnden,F.R. 1993a, ApJ, submitted.
Damiani,F., Micela,G., Sciortino,S., Favata,F. 1993b, proceedings of Eighth Cool Stars Workshop, Athens, in press.
Feigelson,E.D., DeCampli,W.M. 1981, ApJLett, 243, L89.
Gahm,G.F. 1980, ApJLett, 242, L163.
Herbig,G.H., Robbin-Bell,K. 1988, Lick Obs. Bull. n. 1111.
Montmerle,T., Koch-Miramond,L., Falgarone,E., Grindlay,J.E. 1983, ApJ, 269, 182.
Walter,F.M., Kuhi,L.V. 1984, ApJ, 284, 194.

AN X-RAY SPECTRAL STUDY OF ALGOL BINARY SYSTEMS

Stephen A. Drake
Code 668, NASA Goddard Space Flight Center, Greenbelt, MD 20771

Email ID
drake@lheavx.gsfc.nasa.gov

Nicholas E. White, Alan P. Smale, and Lorella Angelini
Code 668, NASA Goddard Space Flight Center, Greenbelt, MD 20771

Francis E. Marshall
Code 666, NASA Goddard Space Flight Center, Greenbelt, MD 20771

Steven H. Pravdo
168-222, Jet Propulsion Laboratory, Pasadena, CA 91109

ABSTRACT

The X-ray spectra of β Per (Algol), obtained by the *Einstein* Solid State Spectrometer (SSS), the Broad-Band X-ray Telescope (BBXRT), and the *ROSAT* PSPC, at a variety of epochs and orbital phases, are discussed, as well as new PSPC spectra of several other Algol systems. The spectra of β Per show evidence for anomalous, presumably circumstellar absorption at certain orbital phases, while the PSPC spectra of all 4 systems exhibit a soft thermal component with $T \sim 0.1 - 0.2$ keV.

1. Introduction

Algol binaries that are cool enough to contain a star with a convective envelope are an important class of coronally active objects. The prototype Algol (β Per) has a quiescent X-ray luminosity of $\sim 5 \times 10^{30}$ erg/s and exhibits X-ray flares lasting several hours in which the luminosity can increase by up to a factor of ten (White et al. 1986). The *Einstein* SSS spectra of Algol are well fit by a two-component thermal plasma model with temperatures of ~ 0.6 and $\sim 2 - 3$ keV that is similar to that inferred from the SSS spectra of RS CVn binaries (White et al. 1980, Swank et al. 1981). This similarity suggests that the X-rays are produced by a corona surrounding the K star component in this system, rather than, for example, shock heated gas from the collision of the gas stream with the B8 star primary component.

An X-ray survey of 9 Algols was made by White and Marshall (1983) using the *Einstein* IPC; they concluded that these systems show the same correlation of X-ray luminosity with rotation velocity that 'normal' coronal stars exhibit, but that, compared to the RS CVn systems, the X-ray luminosity of an Algol of the same orbital period is lower. White and Marshall suggested that, for RS CVn systems, the magnetic loop interactions between the two late-type components provide a larger magnetic volume to contain the X-ray emitting plasma than in Algol binaries, where such magnetic loop interactions should not occur.

We present here a preliminary analysis of recent PSPC observations of the prototype β Per (for which there is a very high-quality spectrum totalling

$\sim 52,000$ counts), and of 3 other Algols (U Cep, RZ Cas, and TW Dra). We also discuss high-resolution BBXRT and SSS spectra of β Per that cover a rather harder energy range ($\sim 0.4 - 10$ keV) than the PSPC ($\sim 0.2 - 2.5$ keV).

2. Discussion

The most striking new result is that β Per shows evidence at certain orbital phases ($\Phi = 0.2, 0.6, 0.9$) for an absorption column of $n_H \sim 5 - 8 \times 10^{20}$ cm^2 that is much greater than the expected interstellar column ($n_H \ll 10^{19}$ cm^2). (See Table 1). This conclusion is based on 2 of the 3 SSS spectra and on the BBXRT spectrum, all of which show this effect. There is no evidence for such anomalously high absorption in the SSS spectrum nor in the PSPC spectrum that were both obtained at phase $\Phi = 0.06$, i.e., near primary minimum when the K star is in front of the B star. Richards (1993) has discussed optical spectroscopy of Algol which show evidence for the existence of extensive intrastellar and circumstellar plasma in this system. The 'extra' absorption seen in some of the X-ray spectra of Algol may be another manifestation of this phenomenon.

Table 1. Best-Fit Parameters from X-Ray Modeling of β Per

Date (Year.Day)	Orbital Phase	χ_r^2	n_H (10^{20} cm^{-2})	kT (keV)	$\log EM$ (cm^{-3})
1979.041	0.89	1.13	4.9(+4.1, −3.4)	1.96(+0.26, −0.20)	53.18(+0.05, −0.05)
				0.75(+0.06, +0.05)	52.83(+0.14, −0.11)
1979.042	0.06	1.59	0.0(+2.4, −0.0)	1.90(+0.22, −0.18)	53.33(+0.04, −0.05)
				0.58(+0.08, −0.14)	52.71(+0.15, −0.23)
1979.221	0.67	1.13	8.3(+1.2, −1.2)	2.58(+0.18, −0.16)	53.54(+0.02, −0.02)
				0.64(+0.02, −0.02)	53.11(+0.04, −0.04)
1990.342	0.20	1.20	5.8(+2.1, −1.9)	2.39(+0.16, −0.15)	53.37(+0.03, −0.03)
				0.68(+0.02, −0.02)	52.94(+0.06, −0.06)
1992.030	0.06	0.84	0.2(+0.7, −0.2)	2.6(+5.9, −1.0)	53.21(+0.11, −0.08)
				0.67(+0.06, −0.09)	53.10(+0.09, −0.19)
				0.14(+0.03, −0.04)	52.61(+0.38, −0.16)

Note: The 1979 spectra were obtained with the SSS and have 86 degrees of freedom, the 1990 spectrum was obtained with BBXRT (260 d.o.f.), and the 1992 spectrum was obtained with the PSPC (11 d.o.f.) The errors quoted for all spectra represent the 90% confidence limits.

After properly taking into account the absorption, all 3 SSS spectra as well as the BBXRT spectrum of Algol are well-fit by a 2-component coronal plasma model in which one (the dominant) component has a temperature of $\sim 2 - 3$ keV, and the other component is at ~ 0.7 keV, as found by White et al. (1980). However, the PSPC spectrum of Algol is **not** well fit by such a 2-component model, but requires either a thermal plus power-law component model or a 3-component thermal plasma (3T) model, in which 2 of the components are the same as those identified from the SSS and BBXRT spectra, and the third component is very soft ($T \sim 0.15$ keV). Since the former class of models provide significantly poorer fits to the high-resolution spectra, we prefer the 3T model, although it clearly overconstrains the data, given the number of truly independent channels in a PSPC spectrum. The softest component has a much smaller emission measure ($\sim 25\%$) than either of the others and would not be detectable in either the SSS or the BBXRT spectra due to their harder energy responses.

The PSPC spectra of the 3 other Algol systems also require the presence of a very soft component: in TW Dra and RZ Cas, its temperature is similar to that found in Algol itself (~ 0.15 keV), while in U Cep the best-fit temperature of this component is very cool, ~ 0.07 keV. The PSPC spectrum of TW Dra requires only 2 thermal components (at 0.17 and 0.8 keV) for a good fit, while the spectra of U Cep and RZ Cas require an additional high-temperature component ($T \sim 3-4$ keV) similar to that previously inferred for Algol.

The plasma parameters inferred from the PSPC spectra of these Algol systems are fairly similar to those inferred from the analysis of All-Sky Survey PSPC spectra of RS CVn systems by Dempsey et al. (1993): they found that most RS CVn spectra were well-fit by 2-thermal components, with typical temperatures of 0.17 and 1.4 keV, and with the high-temperature component having the dominant emission measure ($EM_{low}/EM_{high} \sim 25\%$). The Algols analyzed here appear to have somewhat softer spectra than the RS CVn stars analyzed by Dempsey et al., due to the greater contribution of the low-T component to the total emission measure, but this needs to be confirmed using a larger sample of Algols with PSPC spectra. [Any interpretation of the Algol PSPC spectra faces the complication that it is now known that the PSPC response matrix is poorly determined for the softest energy channels (cf. Napiwotzki et al. 1993), and thus the inferred parameters of the soft plasma component are actually more uncertain than the formal fit errors would indicate]. If further analysis confirms this difference between Algols and other types of active stars, it would be tempting to associate this unusually strong, soft component with a manifestation of the mass transfer phenomena characteristic of these semi-detached binary systems, such as high-temperature accretion regions. However, there are difficulties with this hypothesis, as Wade et al. 1993 have recently discussed, based on their PSPC spectra of 4 other Algol systems. Further theorizing should probably be deferred until a more complete sample of Algol PSPC spectra has been studied and an improved PSPC response matrix is available. In the meantime, the recent *EUVE* observation of Algol that has been made might help clarify this issue, assuming that the total absorption column during the observation was not so high as to absorb out most of the EUV photons!

3. References

Dempsey, R.C., Linsky, J.L., Schmitt, J.H.M.M., and Fleming, T.A. 1993, ApJ, 413, 333
Napiwotzki, R. et al. 1993, A&A, 278, 478
Richards, M.T. 1993, ApJS, 86, 255
Swank, J.H., White, N.E., Holt, S.S., and Becker, R.H. 1981, ApJ, 246, 214
Wade, R.A., Stringfellow, G., and Polidan, R.S. 1993, in preparation
White, N.E., and Marshall, F.E., 1983, ApJ, 268, L117
White, N.E., et al. 1980, ApJ, 239, L69
White, N.E., et al. 1986, ApJ, 301, 262

Orion Trapezium - A view into the astrophysics of young stars with the ROSAT PSPC

Sven Geier

dpt. of astronomy, University of Maryland, College Park, MD 20742

Heinrich J. Wendker

Hamburger Sternwarte, Gojenbergsweg 112, 21029 Hamburg, Germany

Email ID

sgeier@astro.umd.edu — st2g101@staix1.hs.uni-hamburg.de

ABSTRACT

A ROSAT - PSPC image, a 10 ks exposure centered on the Orion Trapezium region, has been taken. Apparently all of the diffuse X-ray component, which was suspected to surround the Trapezium based on an EINSTEIN image, could be resolved into point sources. Due to the high surface density of bright sources (about 1.4 per square arcmin) we developed a special methodology in order to derive source parameters. We obtained a list of 172 sources inside the so-called inner ring of the ROSAT FOV, to a count rate of about 2.5×10^{-3}. The principal findings are: (1) The strongest source in the field is (expectedly) θ^1 Ori C. Since its unexpectedly hard spectrum can not be explained by interstellar extinction, it is necessary to postulate circumstellar absorption, possibly originating from a special Helium shell inside the stellar wind as predicted by Kudritzki et al. 1992. (2) The majority of the sources can be identified with pre-main sequence stars of the Orion Id/Ic cluster. (3) The physical features of the bright x-ray sources for which spectral information could be derived show a strongly correlated bimodal distribution, which is equally strongly correlated to the distribution in rotational periods (where this information is available). One can conclude, that one group needs an additional amount of circumstellar absorbing material and has a higher plasma temperature. We propose, that these stars belong to the subgroup d of the Orion OB1 association.

1. The spectrum of θ^1 Ori C

The high count rate of θ^1 Ori C enabled us to do a detailed study of the spectral features of this star. Although the bright x-ray source in the center of the Trapzium cluster represents a blend of θ^1 Ori C and E, which are only ~15" from each other, the x-ray flux of this double source is strongly dominated by the ~O5 star θ^1 Ori C, as indicated by the Einstein HRI observations (see Ku et al. 1981). This result was later reproduced by modelling the observed spatial flux distribution with point sources, assigning less then roughly ~ 10% of the Flux to a possible second source.

Figure I shows the result of a fit of a plasma spectrum (after Raymond and Smith, 1977) to the observed data. Besides the good agreement of data and model (confirming the assumption, that a one-source model is sufficient to fit the spectrum) the steep cutoff at low energies is obvious. This cutoff is properly reproduced by the model spectrum, but requires an absorbing column density of $\sim 12 \times 10^{21} \mathrm{cm}^{-2}$, nearly an order of magnitude more than can be possibly explained by interstellar material.

Values for the interstellar column density can be obtained by means of

- 21cm absorption (e.g. Muller, 1958) → $1.9 \times 10^{21} \text{cm}^{-2}$
- 21cm emission (e.g. Stark et al. 1992) → $2.0 \times 10^{21} \text{cm}^{-2}$
- Lα absorption (e.g. Savage et al. 1972) → $1.1 \times 10^{21} \text{cm}^{-2}$
- Color Excess: The value of $E_{(B-V)}$ for θ^1 Ori C (e.g. Bohlin and Savage, 1981) transforms directly into $\sim 1.9 \times 10^{21} \text{cm}^{-2}$ This value, even though in good agreement with the otherwise determined column densities, is at least questionable, since the ratio of neutral extinction to reddening $R = A_V/E_{(B-V)}$ in Orion deviates by about a factor of 2 from the galactic average of $R = \sim 3$.

Figure II shows the low energy (i.e. extinction sensitive) part of the spectrum — dots with error bars — together with the theoretically expected spectra for values of $N_H = 11.0$, 12.0 and 13.0 $\times 10^{21} \text{cm}^{-2}$ respectively. It is obvious from this plot, that the observed column density has to be a factor of 5 to 10 larger than the interstellar value. This can be explained by circumstellar extinction (see #3). Unlike the sources in section 3, the strong wind of θ^1 Ori C forbids a dust shell around this star. We therefore conjecture, that the extinction has to take place *in the wind itself*, e.g. due to a shell of partially neutral Helium, as proposed Kudritzki et al. (loc. cit.).

2. The list of optically identified point sources

After a complete reprocessing of the x-ray data, most of the point sources could be identified as pre-main sequence stars. 132 of the 172 point sources located in the so-called inner ring (a circular area of ~ 40' in diameter) are listed in the work of Parenago (1954, loc. cit.).

111 of the 172 sources are known variables, i.e. are listed in the *catalog of variable stars* (1982). 111 of the 150 sources inside the area covered in the proper motion survey of Johnes & Walker (1988) can be identified with optical sources in this survey. 28 of the sources coincide in position with emission line stars listed in the catalog of Herbig and Bell (1988).

The detection of x ray emission from main sequence stars of spectral types B5 to A3 (as reported by Caillault & Zoonematkermani, 1989) could *not* be confirmed.

3. The two groups of sources

A fit of a plasma spectrum, convolved with galactic absorption, yielded two important parameters of the 95 brightest sources for which a spectral analysis was possible: the plasma temperature and the absorbing column density.

Figure III shows a histogram of the hydrogen column densities N_H Two groups of sources can easily be distinguished: one with a value of $N_H \lesssim 1. \times 10^{21} \text{cm}^{-2}$ (the prominent peak to the left) and a group of sources with N_H values around $(12.\pm 2.) \times 10^{21} \text{cm}^{-2}$ In contrast to θ^1 Ori C however, these sources (mid- and late type pre main sequence stars) are not expected to have a violent wind. Any proposed circumstellar extinction may thus be the result of an excess of dust around these newly formed stars. Therefore the obscuring column density as observed from the x-ray spectrum of a stellar source may be an indicator of its evolutionary state. This idea is supported by the fact that, of the 26 sources with excess absorption (i. e. $N_H > 1. \times 10^{21} cm^{-2}$, for which an estimate on the probability of membership in the subgroup d of the OB association was available (Johnes, Walker loc. cit), 24 had been assigned a probability of 97% or more, one 70% and one 42%. All sources in our sample with membership probabilities below 33% show values of $N_H \leq 0.5 \times 10^{21} cm^{-2}$.

Figure IV shows the distribution of the fitted plasma temperatures. The bimodal distribution and the strong correlation with the absorbing column density are obvious. This graph however has to be viewed with caution, since the absorption *mechanism* for the more heavily obscured sources is different from the assumed model — the resulting temperatures *may* therefore be incorrect. Nevertheless the numerical value of the result (be it a real temperature or the reaction of the fitting algorithm to the additional absorption) is obviously a good indicator of membership in one of the groups.

Additional support for the assumption that this distribution is an effect of vestigial dust shells around the pre main sequence stars in the trapezium region (and therefore an effect of stellar evolution) comes from the bimodal distribution of rotational periods of the 10 sources for which periods were available (Attridge & Herbst, 1992). Since it is generally accepted that post T Tau stars undergo a "rapid rotator" phase as they contract onto the main sequence (e.g. Marcy et al. 1985 or especially Butler et al. 1987), one would expect a correlation of the rotational periods and the observed column densities, if the latter are an indicator of age. This correlation can be found for 8 of the 10 sources. The 9th one (V 1123 Ori) may be a misidentification (the difference of optical and x ray position is with 10 arcsec nearly three times as high as the average of 3.7 arcsec in this group). The tenth star, Il2203, however may be in the phase of transition from one group to the other: it has the lowest N_H value of the high N_H group, and the shortest rotational period of *all* the stars for which data was available.

Attridge, J. M., Herbst, W.; ApJL, **398**, L61
Bohlin, R. C., Savage, B. D.; ApJ,**249**,109
Butler, R. P., Cohen, R. D., Duncan, D. K., Marcy, G. W.; ApJL, **319**, L19
Caillault, J.-P., Zoonematkermani, S.; ApJL, **338**, L57
Herbig, G. H., Bell, K. R.; Lick obs. bull. #1111
Johnes, B. F., Walker, M. F.; AJ, **95**, 1755
Ku, W. H.-M., Righini-Cohen, G., Simon, M.; Science, **215**, 61, 1981
Kudritzki, R.-P.; preprint, 1992
Marcy, G. W.; Duncan, D., K.; Cohen, R. D.; ApJ, **288**, 259
Muller, C. A.; in Bracewell (ed.): IAU symp. #9, 360
Parenago, P.P.; Transactions of the astr. inst. Shternberg., Vol 25, 1954
Raymond, J. C., Smith, B. W.; ApJS, **35**, 419, 1977
Savage, B. D., Jenkins, E. B; ApJ, **172**, 49, 1972
Stark, A. A., Gammie, C. F., Wilson, R. W., Bally, J., Linke, R. A.; ApJS, **79**, 77, 1992

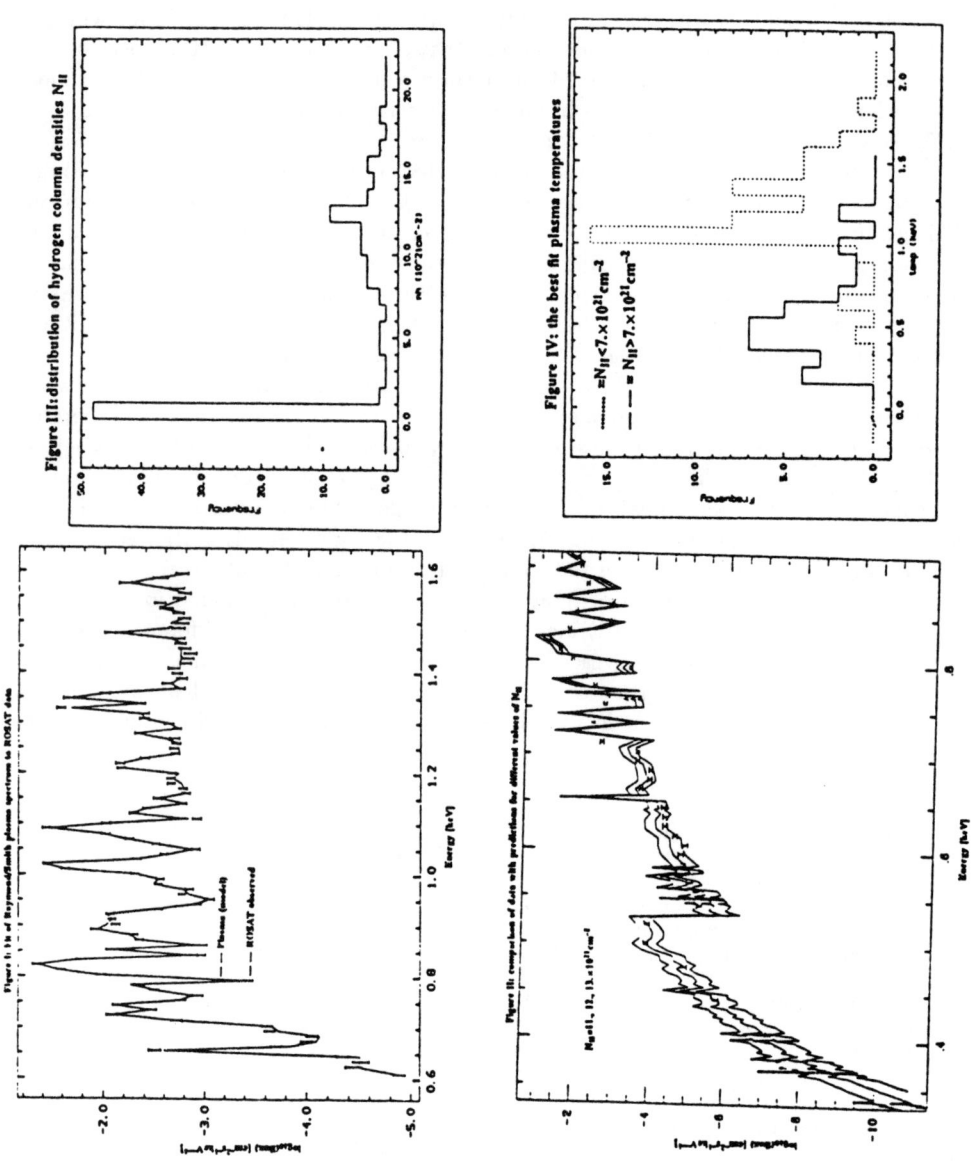

AN OBSERVATIONAL TEST TO DETERMINE THE PRESENCE OF INTRINSIC X-RAY ABSORPTION

Wayne L. Waldron
Applied Research Corporation
8201 Corporate Drive, Landover, MD 20785

ABSTRACT

A key factor in the analysis of X-ray data is determining the true nature of the X-ray absorption, i.e., the relative strengths of the intrinsic or "local" and ISM absorption components. In O stars, this local component is in addition to the fixed ISM absorption and refers to the X-ray attenuation caused by the massive, highly ionized stellar wind. By comparing IPC and PSPC hardness ratios, we show how this correlation can be used to verify the presence of the local absorption component. We conclude that the observed X-ray characteristics inherent in O stars are strongly influenced by the local absorption component. Although only demonstrated for O stars, this test should also be applicable to any X-ray source possessing IPC and PSPC data.

1. INTRODUCTION

To carry out X-ray spectral analyses, we need to know the X-ray source characteristics (thermal or non-thermal) and the amount of X-ray absorption encountered between the source and the observer. In general, the absorption consists of two terms, an ISM component and a local component. The local component is in addition to the fixed ISM absorption and refers to the X-ray attenuation caused by the massive, highly ionized stellar winds in O stars. Several studies (e.g., Waldron 1984) have shown that the neglect of local absorption can result in underestimating the strength and spectral characteristics of the intrinsic X-ray source. Furthermore, by determining the amount of local absorption present, constraints can be placed on the location of the X-ray source relative to the stellar reference frame. Determining the location of the X-ray source is crucial to understanding the mechanism responsible for the X-ray emission (e.g., see discussion by Cassinelli 1985).

Lacking observational evidence for the existence of the local component, most spectral studies to date have neglected this component. Although, over the past few years, evidence supporting local absorption has increased (MacFarlane et al. 1994; Corcoran et al. 1993; Waldron 1991), it would be very useful if a simple observational test could be performed before carrying out detailed spectral analyses. By comparing IPC and PSPC hardness ratio (HR) data, we illustrate how these data can be used as an effective diagnostic tool to ascertain the presence of a local absorption component.

2. DISCUSSION

Since we know the spectral responses of the IPC and PSPC instruments, we can predict the expected relationship between the IPC and PSPC HR data for a given spectral configuration (thermal or non-thermal) and absorption medium. The advantage of using HR data is that they are independent of distance and source emission measure. Using ISM absorption cross sections, a large range in Raymond-Smith spectra (log T_x = 6 - 7.5) and column densities (log N_H = 20 - 22), we calculate the predicted relationship between IPC and PSPC HR data. The results are shown in Figure 1 along with observed IPC and PSPC HR data for 18 OB stars. If the observed X-ray characteristics are controlled by ISM absorption, then all observed points should fall somewhere within the contours. Only B main sequence stars (PSPC HR < 0) follow the ISM model contours. This is expected since these stars have weak stellar winds. The O stars (PSPC HR > 0) do not fall within the contours. Therefore, for O stars, we conclude that the local absorption component must be the primary source of absorption, i.e., the ISM component provides only a minor contribution to the total X-ray absorption. One could argue that the group of O stars in the region of PSPC HR ≈ 0 to 0.2 appears consistent with the ISM model, requiring a large T_x and small ISM column density ($\approx 10^{20}$), but these stars have observed ISM column densities of order 10^{21}.

The reason for this behavior can best be understood by comparing ISM and stellar wind X-ray absorption cross sections (see Figure 2). Although a typical wind cross section is less than the ISM, we point out that wind column densities are usually 10-100 times larger than the observed ISM column densities. Therefore, when the wind absorption dominates the ISM absorption, the X-ray absorption between ≈ 0.3 - 0.6 keV becomes essentially constant, causing an apparent excess in X-ray emission in this energy range when one tries to fit a spectrum with ISM cross sections. Hence, the combined effects of ISM and stellar wind absorption, along with the PSPC higher sensitivity to softer X-rays and higher energy resolution is responsible for the observed IPC-PSPC HR behavior shown in Figure 1. Similar conclusions are obtained for bremsstrahlung and power law input spectra.

This work was supported in part by NASA contract NAS5-31220.

Cassinelli, J. P. 1985, "The Origin of Nonradiative Heating/Momentum in Hot Stars,"
    eds., A. B. Underhill & A. G. Michalitsianos, (NASA CP-2358), p. 2
Corcoran, M. F., et al. 1993, ApJ, 412, 792

MacFarlane, J. J., Waldron, W. L., Corcoran, M. F., Wolff, M. J., Wang, P., & Cassinelli, J. P. 1993, ApJ, in press
Waldron, W. L. 1984, ApJ, 282, 256
Waldron, W. L. 1991, ApJ, 382, 603

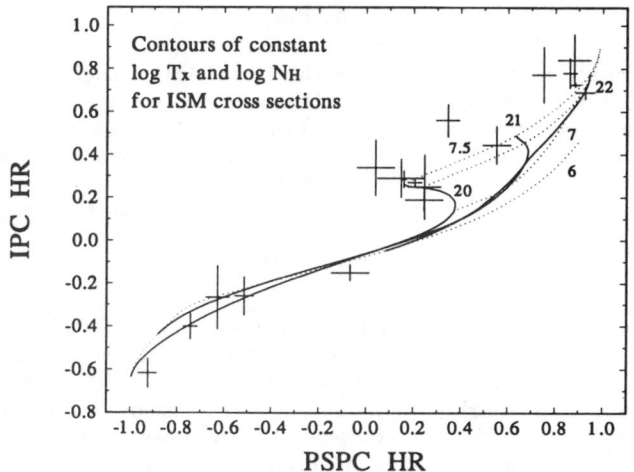

Figure 1. The predicted relationship of IPC and PSPC HR data for an ISM absorption model. The solid lines correspond to contours of constant log N_H and the dotted lines are contours of constant log T_x. The observed HRs for 18 OB stars are shown.

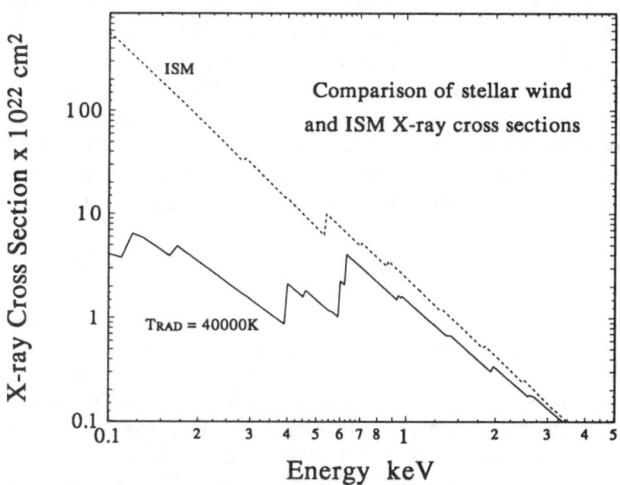

Figure 2. A comparison between ISM and stellar wind cross sections. T_{RAD} is the radiation temperature used. The resulting wind cross section is typical for most O stars.

Star Formation in Orion (the Constellation)

Frederick M. Walter
ESS Dept, SUNY, Stony Brook NY 11794-2100

Email ID
fwalter@astro.sunysb.edu

ABSTRACT

Star formation is pronounced in the Ori OB1D association at the present time, but within the past 5-10 Myr the entire Ori OB1 association formed, generating hundreds of O and B stars throughout the constellation of Orion. I have been using imaging X-ray data to identify and map out the low mass stars associated with this star formation episode. I describe my observing program using the ROSAT PSPC, and present initial results from the program.

1. INTRODUCTION

Low mass stars abound in OB associations. This fact is not well known, mainly because the low mass stars are some 10-15 magnitudes fainter at V than the OB supergiants which dominate the associations. Aside from the occasional classical T Tauri star, marked by its strong Hα emission, continuum variability, and/or near-infrared continuum excesses, few low mass stars have been identified in OB associations. These associations tend to be fairly loose, so it is not easy to select low mass association members by starcounts. One needs to use kinematic criteria (radial velocities and proper motions) or age-related criteria (high surface lithium abundances; extreme chromospheric and coronal flux levels) to confirm membership.

X-ray observations provide an efficient means of surveying associations for low mass members. Low mass pre-main sequence stars are highly active; on their convective tracks they are fully convective. X-ray luminosities of these stars range from a few $\times 10^{29}$ erg s^{-1} to about 10^{31} erg s^{-1} with coronal temperatures $\sim 10^7$K. These are well suited to the sensitivity of the imaging soft X-ray instruments on the EINSTEIN and ROSAT observatories.

The Orion complex is dominated by the OB supergiants of the belt and Trapezium, but is more extensive (see Blaauw 1991). Star formation has proceeded for over 10^7 years, generally from northwest to southeast, where the Orion Nebula and L1641 clouds are actively forming stars today.

2. THE CONSTELLATION IN X-RAYS

The ROSAT calendar survey image of Orion shows that the constellation is recognizable in X-rays. The hot stars of the belt and the Orion nebula stand out. But near the ecliptic the survey does not go deep enough to see the typical late type pre-main sequence star at a distance of nearly 500 pc. I decided that Orion was a prime target for determining the spatial distribution of low mass stars in OB associations, and for studying the history of star formation in associations. Orion OB1 is one of the nearest rich OB associations, and is easier

Figure 1. The hard band image of the belt of Orion. This plot covers roughly 5^h10^m to 5^h46^m in RA and $-6°$ to $+4°$ DEC (J2000), an area which includes much of the Ori OBIa and b subassociations, and the northern part of the c subassociation. δ Ori is the bright source closest to the center of the image; the belt runs from δ Ori to the lower left. The Orion Nebula is centered in the southernmost image. Each PSPC image subtends $2°$ on the sky.

to survey, given its greater distance, than the Scorpius OB2 complex (Walter *et al.* 1994).

My *ROSAT* survey consists of two parts: a linear swath from λ Ori in the north to the Trapezium in the south, and a more complete investigation of the belt. The N-S swath will let us investigate background star formation unassociated with identifiable subassociations. Our intent is to map out the spatial distribution and the history of low mass star formation in this complex. Is low mass star formation coeval and co-spatial with the massive stars? What is the initial mass function, and how does it compare to those of othe OB associations and T associations? Do low mass stars form near high mass stars, or as part of a general background?

My proposal resulted in 17 PSPC pointings, with a typical observing time of 7 ksec. I will add relevant archival data to the pool as it becomes public. The current hard-band image of the region centered on the belt of Orion, shown in the figure, reveals many X-ray sources. The brightest are the OB supergiants of the belt, foreground dMe stars, and a newly discovered CV system (Walter & Zoonematkermani 1994), but there are literally hundreds of other X-ray sources. The SASS analysis system found 636 discrete X-ray sources in these fields. Over 80% are significant at $>3\sigma$ confidence.

We have only begun to identify the optical counterparts. We find:

• 175 coincidences (to within 15 arcsec) with a star in the HST Guide star catalog.

• 92 coincidences (within 1 arcmin) of a stellar object in the SIMBAD database, including 15 O8-B7 stars, 17 B8-A2 stars, 6 other A stars, 40 F-M stars, and 14 nebular variables.

Many of the G and K stars appear to be pre-main sequence (PMS) stars; the M stars are foreground. There is one extragalactic identification in the SIMBAD database.

4. THE OPTICAL COUNTERPARTS

Over the past two observing seasons we have obtained optical spectra at Kitt Peak, using the GOLDCAM spectrograph on the 2.1m telescope, of stars in 56 ROSAT X-ray error circles, and have identified 37 low mass PMS stars. All show strong lithium λ6707Å absorption. Many show weak Hα emission, although some are likely to be classical T Tauri stars.

We note that only 15 of the 37 PMS X-ray sources have counterparts in the GSC; the rest are fainter than the completeness limit of the GSC. If we assume that most of the X-ray sources coincident with GSC stars are low mass PMS stars, then we expect that some 50% of the X-ray sources will be associated with low mass PMS stars.

REFERENCES

Blaauw, A. 1991, in *"The Physics of Star Formation and Early Stellar Evolution"*, edited by C.J. Lada and N.D. Kylafis (Kluwer, Dordrecht) p. 125.

Walter, F.M., Vrba, F.J., Mathieu, R.D., Brown, A., and Myers, P.C. 1994, AJ, **107**, in press.

Walter, F.M., & Zoonematkermani, S. 1994, this volume.

Cataclysmic Variables

Discovery of a New Cataclysmic Variable System

Frederick M. Walter & Saeid Zoonematkermani
ESS Dept, SUNY, Stony Brook NY 11794-2100

Email ID
fwalter@astro.sunysb.edu

ABSTRACT

We announce the serendipidous discovery of a bright new CV system.

1. INTRODUCTION

While tracking down low mass pre-main sequence stars near the Ori OB1a association, we obtained the optical spectrum shown in Figure 1. This is not the spectrum of a low mass pre-main sequence star; rather the blue continuum and the He I emission lines suggest that the star is a cataclysmic variable (CV) system. No previously known X-ray source or close binary system is known at this location. We announce the discovery of a new cataclysmic variable system, RJ051542+0104.7.

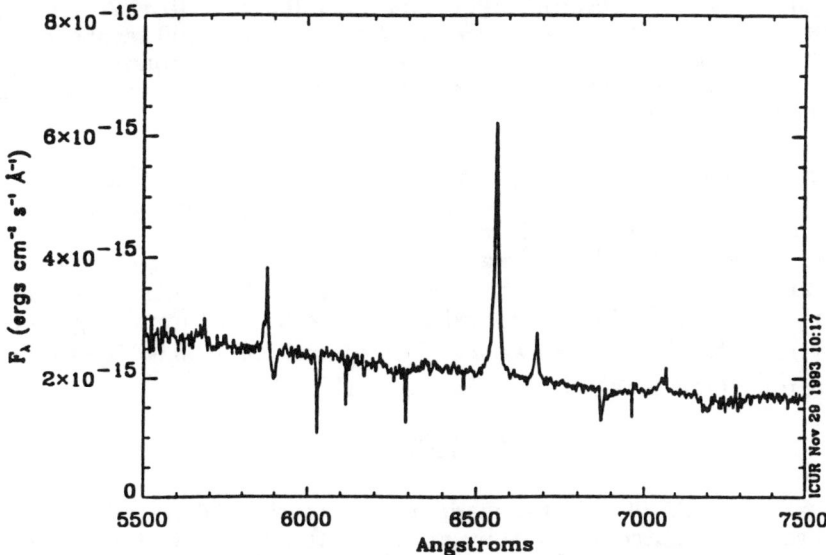

Figure 1. The optical spectrum, with a resolution of about 2Å. Note the blue continuum. The emission lines are He I and Hα. The narrow absorption features may be instrumental, but the broad Na D absorption feature is real.

The source was discovered in ROSAT PSPC pointing RP200930. The X-ray source position, from the SASS processing, is $5^h15^m42.2^s$ +1°4'42.5" (J2000), some 42 arcmin off-axis. The optical counterpart is the westernmost

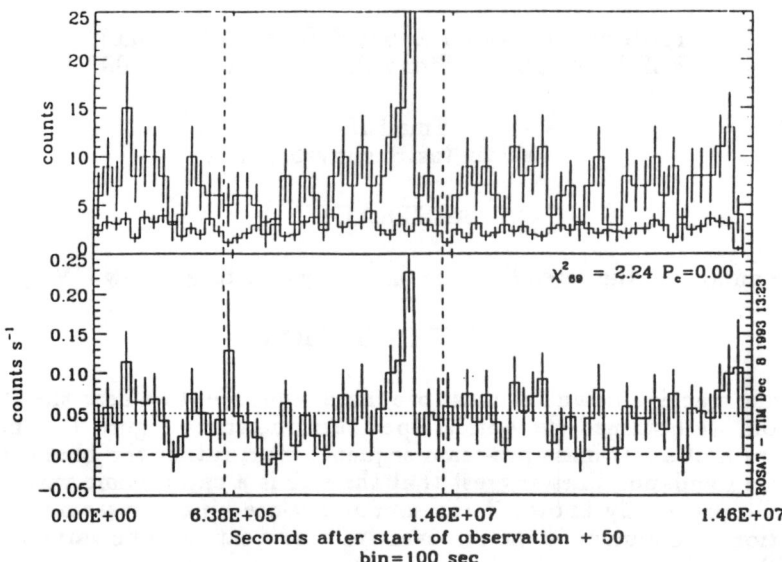

Figure 2. The "quiescent" light curve of RJ051542+0104.7. The upper plot shows the counts from the source and background in 100 second bins; the lower plot is the vignetting corrected, background-subtracted count rate. Dashed vertical lines mark the location of gaps in the data. There appears to be a small flare in the second interval.

and brighter of two stars at this position; it is not in the HST Guide Star Catalog. The position of the optical counterpart, by triangulating among 3 nearby stars, is within 10 arcsec of the X-ray position.

The X-ray source is fairly bright, with a mean vignetting-corrected count rate of 0.21 ± 0.006 counts s^{-1} in the PSPC. The optical star has a magnitude of about 15 on the POSS Sky Survey red print, as estimated from magnitudes of nearby stars in the Guide Star Catalog. The $\frac{f_x}{f_v}$ ratio of ~ 0.3 is high for normal stars, but reasonable for CVs (e.g., Stocke et al. 1983).

2. X-RAY VARIABILITY

Observation RP200930 was made at two epochs, 3.6 Ksec in September 1992 and another 6.1 Ksec on 27 February 1993. During the September 1992 observations, and again during the first orbit observed in February 1993, the source exhibited fairly steady "quiescent" emission at a rate of 0.12 counts s^{-1}, with one superposed flare (Fig. 2). Sometime during the 8.3 Ksec interval between observations, the character of the light curve changed dramatically (Fig. 3). The mean count rate doubled, but was anything but steady. Significant variations exist to timescales shorter than 2 seconds; on this scale the instantaneous count rate reaches 15 counts s^{-1}. (With the source so far off-axis, the wobble has no effect on the light curve.) We verified that this light curve was indeed intrinsic to the source, and was not a detector breakdown or other instrumental artifact,

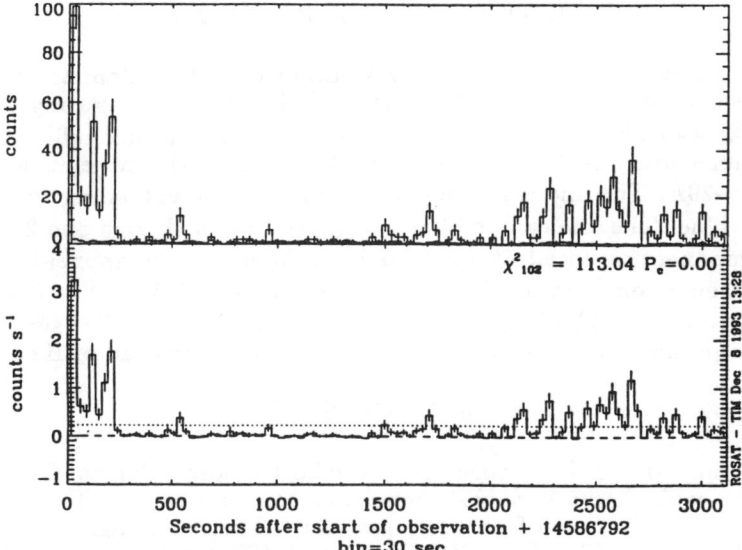

Figure 3. The erratic light curve seen during the last 3000 seconds of the observation, in 30 second bins. The source is highly variable on time scales down to 2 seconds (limited by photon statistics), with significant stretches of zero intensity.

by examining the background light curve (visible in Fig. 3) and the light curves of other sources in the field, none of which exhibited anything remarkable.

3. THE X-RAY SPECTRUM

We extracted the source counts within a 3.75 arcmin radius. The background region was an annulus between 3.75 and 11.25 arcmin from the source, excluding a sector containing another source. We binned the background-subtracted PI data into the 34 SASS channels, excluding channels 1-4 and 30-34 from the spectral fits. We fit the 1029 net source counts using XSPEC, finding acceptable fits for a blackbody spectrum with kT=60 eV and $n_H=4\times10^{20}$cm^{-2} ($\chi^2_{25}=27.5$) or a powerlaw spectrum with $\alpha=6.2$ and $n_H=6\times10^{20}$cm^{-2} ($\chi^2_{25}=20.0$). We cannot get an acceptable fit for a thermal plasma.

Cataclysmic variables exhibit two-component spectra. Although the data do not require it, we have fit 2-component spectra consisting of a black body plus either a powerlaw or a thermal bremsstrahlung. These yield modest reductions in χ^2. The PSPC is insensitive to the kT$\sim$10 keV hard component common in CVs.

There are no significant differences between the spectra obtained in the "quiescent" mode and in the the rapidly-varying mode.

4. THE OPTICAL SPECTRUM

The optical spectrum (Fig. 1) was obtained on 1993 January 1 using the 2.1m telescope at KPNO with the GOLDCAM camera and grating 47 (resolution ~ 2 Å; $\lambda\lambda 5500$-7600Å). The data were reduced and flux-calibrated in the usual manner using IRAF. The strongest line is Hα, with an equivalent width $W_\lambda(H\alpha) = -29$Å. The line is asymmetric, with emission extending some 40Å towards the blue. The line can be fit as two Gaussians, with $\sigma = 6$ and 20Å and the broad component centered 7Å to the blue of the narrow component. The other prominent emission lines are He I $\lambda 5876$, 6678, and 7065Å ($W_\lambda = 5.5$, 4.3, and 2.5Å, respectively). The brighter two of these lines have profiles similar to Hα. There is a prominent broad Na D absorption feature redward of He I $\lambda 5876$.

5. DISCUSSION

The optical and X-ray spectra are similar to those of known CVs (Cordova 1993). The rapid X-ray variability requires a compact object. The blue optical continuum is likely that of an accretion disk. The X-ray light curve is more amazing than some CVs, (e.g., Patterson et al. 1981) but it is not unique in its rapid variability.

The only possibly stellar feature is the Na D absorption. It is too broad to be interstellar. There is no evidence of TiO absorption bands, as would be expected from a luminous late-type companion.

By the time this paper appears, we expect to have obtained optical and near-IR colors, as well as an optical spectrum in the blue. Clearly, further spectroscopic and photometric observations of this object are warranted, as they may add to our understanding of the nature of the mass-transfer processes in close binary systems.

ACKNOWLEDGEMENTS

We thank M. Watson for beneficial discussions.

REFERENCES

Córdova, F. 1993. preprint, to appear in "*X-Ray Binaries*", eds W.H.G. Lewin, J. van Paradijs, and E.P.J. van den Heuvel (Cambridge:Cambridge)
Patterson, J., Williams, G., and Hiltner, W.A. 1981. ApJ **245**, 618.
Stocke, J.T., et al. 1983. ApJ **273**, 458.

ROSAT OBSERVATIONS OF 7 CATACLYSMIC VARIABLES

Paula Szkody
Department of Astronomy, University of Washington, Seattle, WA 98195

ABSTRACT

ROSAT observations are presented and discussed for the cataclysmic variables YY Dra, AH Eri, S193, Nova Her 91, TT Boo, HV Vir and SS UMi. The first 3 observations were done to search for spin periods of the white dwarf. Nova Her 91 was observed 19 months after outburst to determine the timescale for remnant envelope ejection in a fast nova. The last 3 systems are parts of an ongoing program of determination of accretion rates of dwarf novae with extreme outburst amplitudes and of CVs with orbital periods within the period gap.

1. YY Dra, AH Eri, S193

YY Dra is a 16th mag dwarf nova with an orbital period of 3.96 hr and a distance of 155 pc (Mateo, Szkody and Garnavich 1991). Patterson et al. (1982) found it to be a strong 2-10 keV source from Ariel and HEAO surveys. High speed optical photometry showed the presence of 3 periods: 275 and 550 sec in the U band, and 265 sec in the B band (Patterson et al. 1992). This presence of an optical periodicity at a shorter timescale than the orbital period is the typical signature of a DQ Her type magnetic system, where the magnetic field of the white dwarf is not large enough to lock the spin of the white dwarf to the orbit. ROSAT observations were done to ascertain the correct spin period of the white dwarf.

The ROSAT observation took place on Nov. 29, 1991 for 4245 sec, and revealed a source at a count rate of 0.82±0.01 c/s. The spectrum was best fit with a thermal bremsstrahlung model with kT = 3.8 keV and log N_H = 19.8. This spectrum and the above distance gives a 0.1-2.4 keV flux of 7.8×10^{-12} ergs/cm^2/s and a luminosity of 2.5×10^{31} ergs/s.

A periodicity was present at 265 sec. The presence of this period, together with the optical and spectroscopic periods, implies a model where the white dwarf spin period is 530 sec. The X-ray and B periodicities at 265 sec are the emission from 2 active poles, while the U period at 275/550 sec is the reprocessing of emission lines and Balmer continuum light in a structure moving with the binary period of 4 hr. Further details can be found in Patterson and Szkody (1993).

S193 is a 12-14th mag CV with no known orbital period, while BVR photometry shows an intermittent periodicity at 18.7 min (Garnavich, Szkody and Goldader 1988). ROSAT PSPC observations on May 11, 12 1992, covering 9059 sec, detected S193 with a count rate of 0.027±0.002 c/s. A thermal bremsstrahlung fit to the spectrum gives kT= 1 keV and log N_H = 20.82. The 0.6-2 keV flux was 2.9×10^{-13} ergs/cm^2/s and the luminosity (for a distance of 100 pc) is 3.5×10^{29} ergs/s. Simultaneous optical data showed a period at 16.5±2 min, but no consistent x-ray period was present throughout the data. Further details are in Szkody et al. (1993).

AH Eri is an 18.5 mag high latitude dwarf nova of unknown orbital period that shows a period of 42±2 min in optical photometry (Szkody et al. 1989). ROSAT PSPC observations on March 8,9 1992 for a total of 5747 sec showed a weak source (0.0046±0.0010 c/s) in the 0.6 - 1.6 keV channels. There were not enough counts to determine a spectrum. Optical observations on March 7,8 showed the usual period at 42 min, while the power spectra of the X-ray data shows a peak at 20.65 min in the combined data set, but not in each of the individual days. The correspondence of this period to that of the optical suggests that AH Eri may have 2 active X-ray poles (like YY Dra), but further data are needed to confirm this possibility.

2. Nova Her 91

Nova Her 91 = V838 Her was a very fast, very high amplitude nova with optical maximum near March 20, 1991. Optical data obtained in 1992 showed eclipses and provided an orbital period of 7.14 hrs (Leibowitz et al. 1992), while IUE data revealed an O,Ne,Mg nova with E(B-V) = 0.6 and d = 3.4 kpc (Starrfield et al. 1992). Nova Her was the first nova to be observed in X-rays at outburst (Lloyd et al 1992), which revealed a fairly hard source (10 keV with $\log N_H = 21.53$) at a count rate of 0.16 c/s.

ROSAT PSPC observations took place from October 14-18, 1992 (25 ksec effective time), when the nova was almost to its pre-outburst brightness. It was marginally detected ($7.8 \pm 2.9 \times 10^{-4}$ c/s in channels 0.6-1.3 keV; Szkody and Hoard 1993). This value for the count rate, together with a limit extracted from the archive from data obtained on March 24,25 1992, implies that the white dwarf in this fast nova must have ejected its envelope during or soon after its outburst. This is in contrast to slower novae such as GQ Mus, QU Vul, PW Vul and V1974 Cyg, which had high x-ray fluxes for 1-9 yrs after outburst.

3. TT Boo, HV Vir, SS UMi

TT Boo and **HV Vir** are extreme outburst amplitude (7 and 7.5 mag respectively) dwarf novae with orbital periods below the period gap. **SS UMi (PG1551+719)** is a lower amplitude system (5 mag) with a suggested period in the period gap. ROSAT PSPC observations are being conducted for a group of these kinds of systems in order to compare their accretion rates (inferred from the X-ray luminosity) to those of normal amplitude dwarf novae and to those outside of the period gap. The preliminary results available for the above 3 systems at quiescence show count rates ranging from 0.009±0.001 for the faintest system (TT Boo) to 0.045±0.003 for the brightest system (SS UMi). The ratio of the x-ray to optical fluxes indicate normal (rather than reduced) accretion is occurring in each case. Further incoming observations should refine this picture.

This work was supported by NASA grants NAG5-1927 and NAGW 3158 to the University of Washington.

4. References

Garnavich, P., Szkody, P., and Goldader, J. 1988, BAAS, 20, 1020
Leibowitz, E., Mendelson, H., Mashal, E., Prialnik, D. and Seitter, W. 1992, ApJ, 385, L49
Lloyd, H. et al. 1992, Nature, 356, 222
Mateo, M., Szkody, P. and Garnavich, P. 1991, ApJ, 370, 370
Patterson, J. and Szkody, P. 1993, PASP, 105, 1116
Patterson, J. et al. 1982, BAAS, 14, 618

Patterson, J. et al. 1992, ApJ, 392, 233
Starrfield, S. et al. 1992, ApJ, 391, L71
Szkody, P. and Hoard, D. 1993, ApJ, submitted
Szkody, P., Garnavich, P., Castelaz, M. and Makino, F. 1993, PASP, submitted
Szkody, P., Howell, S., Mateo, M. and Kreidl, T. 1989, PASP, 101, 899

BAYESIAN TIMING ANALYSIS OF THREE CATACLYSMIC BINARY SYSTEMS

P. E. Freeman, T. H. Metcalf, and D. Q. Lamb
Dept. of Astronomy and Astrophysics, University of Chicago, IL, 60637
E-mail: peterf@oddjob.uchicago.edu

ABSTRACT

The identification of periodic signals from cataclysmic binary systems is crucial for classifying them and interpreting their behavior. We use the Bayesian timing analysis technique of Gregory & Loredo[1] to attempt to detect signals in the ROSAT PSPC data of three such systems. This technique, conceptually similar to, but more sensitive then, epoch folding[2], involves directly comparing a model with no count rate variability with one describing the count rate as a stepwise distribution with m bins per period of length P, and phase ϕ (relative to the start of observation). This model comparison allows the computation of the odds favoring the more complex model. If the odds strongly support variability, one then estimates the period of variation by integrating the likelihood function over all values of m and ϕ for each value of P.

With this technique, we attempt to find the signal which Wood et al.[3] interpret as that of the orbital period, P_{orb}, of the He white dwarf PG1346+082. We also attempt to find the signals at the spin period, P_{spin}, and beat period, P_{beat}, of the DQ Her-type source 1H0542-407 and the possibly DQ Her-type source V426 Oph. Detection of a spin and beat period in V426 Oph would indicate that the binary system is not synchronized, aiding the DQ Her classification. We find that the odds do not favor a variable rate model for PG1346+082. The odds do strongly favor variable rate models for the latter two sources, but we cannot conclusively determine periods of variation, because of insufficient data and large gaps in the observations.

DETERMINING THE ODDS AND THE MOST PROBABLE PERIOD

We adopt the Bayesian approach to statistical inference in this work. A strength of this approach is its ability to directly compare two specified models, allowing the calculation of odds favoring one model over the other. If no a priori information favoring either model exists, then one affects the comparison by computing the ratio of model global likelihoods; the global likelihood is the integration of the likelihood function used to fit to the data over the allowed range of each model parameter.

In this work, we test for the presence of periodic signals in unbinned ROSAT PSPC time-of-arrival data containing N photons detected during a total livetime T, by comparing a model which assumes a constant rate of photon arrival with a model which assumes a variable rate with period P and phase ϕ, with each period divided into m stepwise bins. If P and ϕ are not known beforehand, the calculation of the global likelihoods yields the following expression for the odds favoring variability:[1]

$$O_{m>1} = \sum_{m=2}^{m_{hi}} \frac{1}{2\pi(m_{hi}-1)\log\frac{P_{hi}}{P_{lo}}} \binom{N}{N+m-1} \int_{P_{lo}}^{P_{hi}} \frac{dP}{P} \int_0^{2\pi} d\phi \frac{S(P,\phi)m^N}{W_m(P,\phi)}. \quad (1)$$

The factors 2π, $m_{hi}-1$, and $\log\frac{P_{hi}}{P_{lo}}$ represent the allowed ranges of the parameters of ϕ, m, and P, respectively, as chosen before the fit. Such expressions are called Ockham factors: they penalize the use of extra parameters to describe the data. The factor $S = \prod_{j=1}^{m} \tau_j m T^{-1}$ corrects for gaps in the data; τ_j is the livetime in the jth stepwise bin. The multiplicity, W_m, is the number of ways one can create a set of n_j values by distributing N events into m bins:

$$W_m(P,\phi) = \frac{N!}{n_1!\, n_2!\, \ldots\, n_m!}. \qquad (2)$$

This multiplicity is a sufficient statistic, containing all information about period and phase in the problem.

If the odds support variability ($O_{m>1} \gg 1$), one can then determine the probability that the variation has period P:

$$p(P) = \sum_{m=2}^{m_{max}} \frac{O_{m1}}{\sum_{j=2}^{m_{max}} O_{j1}} \int d\phi \frac{C}{P} \frac{1}{W_m(P,\phi)}, \qquad (3)$$

where C is the integral of W_m over the chosen values of P and ϕ.

RESULTS

PG1346+082. This source is part of an interacting binary white dwarf system, where two He white dwarfs with dissimilar masses exchange material. It is a photometric variable whose bright states show optical variability on time scales 1470-1490 s,[3,10] thought to be the orbital period P_{orb}. During other states this signal disappears or is replaced with a number of equal amplitude signals.

We find no evidence for variability in this source ($O_{m>1} \sim 10^{-2}$). Without a concurrent optical observation, we cannot say whether this lack of a signal is due to a lack of X-ray variability in general or whether the observation was made during an inopportune photometric state.

1H0542-407. Analysis performed by Touhy et al.[4] and Buckley & Touhy[5] on optical and EXOSAT data identify 1H0542-407 as a DQ Her-type source with orbital period ≈ 5.7 hr. Their use of the epoch folding technique led to the discovery of a strong signal at ≈ 1920 s in the EXOSAT Low Energy instrument, and a strong signal at ≈ 2100 s in the Medium Energy instrument. One can interpret the former as the spin period, P_{spin}, of the accretor, and the latter as the beat period P_{beat} ($P_{beat}^{-1} = P_{spin}^{-1} - P_{orb}^{-1}$).

The ROSAT PSPC data show that 1H0542-407 is variable ($O_{m>1} \sim 10^{82}$). Because of the low count rate, and lengthy gaps in the data, we do not find conclusive evidence for signals at either the expected P_{spin} or P_{beat}, though we find weak evidence for these along with sideband signals at 2350 and 2600 s (Figure 1). We hypothesize that many of the signals are artifacts of the ≈ 2500 s length of the individual observation intervals.

V426 Ophiuchi. A Fourier analysis of EXOSAT data by Szkody[6], which suggests $P_{spin} \approx 1$ hr, led to the identification of V426 Oph as a possible DQ Her-type source. More recent work[7] on optical and Ginga data suggests 28 and 56 minute periodicities, though no evidence for a beat period is reported. Hessman[8] suggests classifying V426 Oph as a Z Cam-type dwarf nova, and he and others[9] find no significant evidence for periodic signals.

The ROSAT PSPC data strongly support variability ($O_{m>1} \sim 10^{196}$), but the strongest signals are seen with much longer periods than the expected P_{spin}

296 Bayesian Timing Analysis

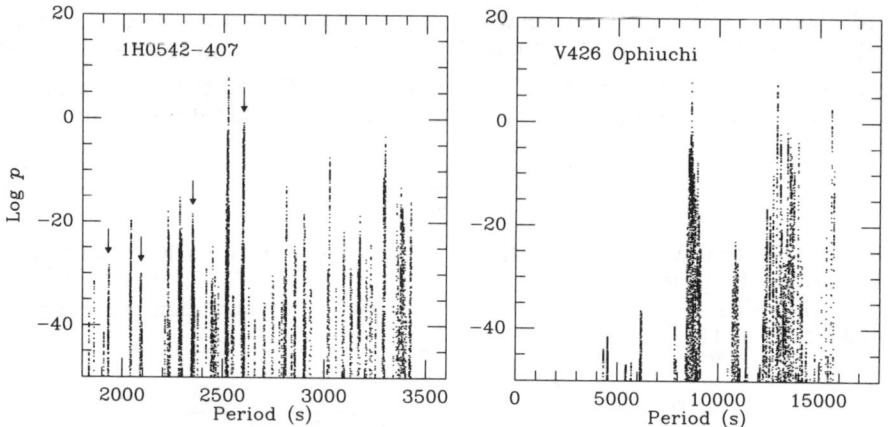

Figure 1. Left: the probability that the data support source variability, as a function of the period, for 1H0542-407. The arrows represent, from left to right, the expected P_{spin} and P_{beat} (following Buckley & Tuohy), and two possible sideband periods. Right: the same, for V426 Ophiuchi. The apparent signals are manifestations of the orbital signal.

(Figure 1). We find that these are manifestations of an orbital signal ($P_{orb} \approx 6.8$ hr): because of the large amount of dead time in the data, they cannot constrain the number of periods necessary to describe the orbital signal. We do find weak evidence for signals at 3447 (57.5 min) and 3928 s (65.5 min), which one can interpret as P_{spin} and P_{beat}. More constraining data are needed to determine whether these are indeed non-orbital signals from V426 Oph.

ACKNOWLEDGEMENTS AND REFERENCES

PEF acknowledges support of NASA Traineeship NGT-50778. We also acknowledge ROSAT Guest Observing Grant NAG 5-1646, and NAGW-830.

1. Gregory, P. C., and Loredo, T. J. 1992, ApJ, 398, 146
2. Leahy, D. A., et al. 1983, ApJ, 266, 160
3. Wood, M. A., et al. 1987, ApJ, 313, 757
4. Tuohy, I. R., et al. 1986, ApJ, 311, 275
5. Buckley, D. A. H., and Tuohy, I. R. 1989, ApJ, 344, 376
6. Szkody, P. 1986, ApJ, 301, L29
7. Szkody, P., Kii, T., and Osaki, Y. 1990, AJ, 100, 546
8. Hessman, F. V. 1988, A&A Supp, 72, 515
9. Hellier, C., et al. 1990, MNRAS, 242, P32
10. Provencal, J. L., et al. 1991, in White Dwarfs, eds. G. Vauclair and E. Sion (Dordrecht: Kluwer), p. 449

Diffuse Gas

ROSAT PSPC OBSERVATIONS OF THE ERIDANUS SOFT X-RAY ENHANCEMENT

Zhiyu Guo, David N. Burrows
Dept. of Astronomy,The Pennsylvania State Univ.,University Park,PA 16802

Wilton T. Sanders
Dept. of Physics,Univ. of Wisconsin,1150 University Ave.,Madison,WI 53706

ABSTRACT

We present our preliminary results of seven ROSAT PSPC pointed observations toward the Eridanus soft X-ray enhancement (EXE).

1. Introduction

The EXE is a prominent feature in the soft X-ray sky with a large angular size ($\sim 25° \times 35°$). It is of particular research interest because a wealth of astrophysical phenomena has been observed in this region, such as H_α filametary loops (Sivan 1974), HI expanding shell (Heiles and Habing 1974, Heiles 1976) and enhanced soft X-ray emission (Williamson et al 1974, Naranan et al 1976). Reynolds and Ogden (1979) concluded that UV photons from stars in the Orion OB association are the source of the ionization. The most recent and extensive studies of the EXE have been carried out by Burrows et al (1993) and Snowden et al (1993) using HEAO-1 A2 low energy detector data and ROSAT sky survey data, respectively. The authors have revealed interesting structures of the EXE, e.g. the absorption of the (galactic) northern half of $\frac{1}{4}$ keV emission from the region $b > -30°$, an X-ray "hook" and many shadows of the intervening clouds, and concluded that the EXE is a stellar wind bubble possibly reheated by a supernova explosion and overlaps an unrelated soft X-ray enhancement presumed to originate in the galactic halo.

In the following, we shall present preliminary results of seven ROSAT PSPC pointed observations toward the EXE as well as analysis procedures.

2. Data and Analysis

We name seven ROSAT PSPC pointed observations as erid2 to erid8. Erid2, erid3 and erid4 are pointed across a prominent IR filament that absorbs the $\frac{1}{4}$ keV band emission to obtain the column density and therefore dust/gas ratio if combined with IRAS maps. Erid5, erid6, erid7 and erid8 are pointed to the northern edge of the $\frac{1}{4}$ keV enhancement to obtain spectral variations across this boundary. Typical exposure time is 10,000 seconds.

We remove contamination in the data using standard procedures of diffuse source analysis presented by Snowden et al (1994). Contamination include scattered solar X-ray (SSX), particle background (PB), after pulse events (AP), short term enhancement (STE) and long term enhancement (LTE). Spectra are created using IDL routines with energy dependent exposure maps. PB and AP components are calculated using models given by Plucinsky et al (1993) . The contribution of AP is found much larger than that of PB, eg. for erid3, the contributions of AP and PB are 5 % and 0.6 % respectively (figure 1). The fits to spectra after subtracting the above contamination still have large reduced χ^2 which indicate the existence of residual contamination.

3. Discussion

We have created mosaic images for three pointed observations (erid2-4) across a prominent IR filament (from $(l,b)=(202°,-50°)$ to $(199°, -44°)$) and for four pointed observations (erid5-8) across the northern half of the EXE (plate 6). We have subtracted contamination of SSX,PB,AP,LTE and STE. LTE is most prominent in erid8 observation (the contribution is about 50 % of the observed flux) and is subtracted using the sky survey rates in this region (ROSAT all sky survey data courtesy of MPE Garching, Germany). These images have shown anticorrelations between X-ray emission and IR 100 μm emission at different levels. The X-ray images across the IR filament clearly show the dust/gas absorption to $\frac{1}{4}$ keV emission while transparent to $\frac{3}{4}$ keV emission. From the mosaic images of observations across the northern half of the EXE, we can see that $\frac{1}{4}$ keV emission decreases toward the Orion OB association while $\frac{3}{4}$ keV emission does not vary as much which means the $\frac{1}{4}$ keV absorber increases toward the Orion OB association. Spectral fits results will finally give us temperature and column density variations in these regions. Currently, spectral fits have not been completed because of residual contamination. We hope to solve the problem in the near future.

REFERENCES

Burrows,D.N.,Singh,K.P.,Nousek,J.A.,Garmire,G.P.,& Good,J. 1993,ApJ,406,92
Heiles,C. 1976,ApJ,208,L137
Heiles,C.,& Habing,H.J. 1974,A&AS,14,1
Naranan,S.,Shulman,S.,Friedman,H., & Fritz,G. 1976,ApJ,208,718
Plucinsky,P.P.,Snowden,S.L.,Briel,U.G.,Hasinger,G.,Pfeffermann,E. 1993,ApJ in press
Reynolds,R.J.,& Ogden,P.M. 1979,ApJ,229,942
Sivan,J.P. 1974,A&AS,16,163
Snowden,S.L.,Burrows,D.N.,& Sanders,W.T. 1993,submitted to ApJ
Snowden,S.L.,McCammon D.,Burrows,D.N.,Mendenhall,J.A. 1994,submitted to ApJ
Williamson,F.O.,Sanders,W.T.,Kraushaar,W.L.,McCammon,D.,Borken,R., & Bunner,A.N. 1974,ApJ,193,L133

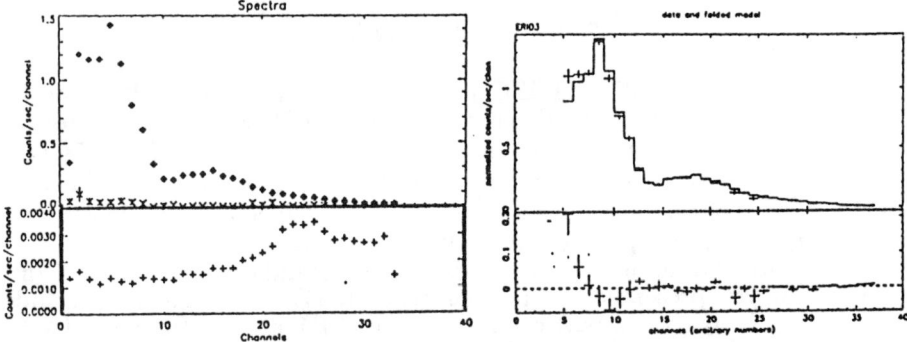

Figure 1. The left plot shows the source spectrum (diamonds), after pulse spectrum (asterisks) and particle background spectrum (pluses) of erid3. The right plot shows the XSPEC model spectral fit to the source spectrum after subtracting AP and PB contributions. The first and second PHA channels are not included. Large residuals are seen in the third and fourth PHA channels.

EXPLORING THE INTERSTELLAR MEDIUM WITH X-RAY SHADOWS

Q. Daniel Wang
CASA/JILA, University of Colorado, Boulder, CO 80303-0389
wqd@vela.colorado.edu

ABSTRACT

ROSAT observations of X-ray shadows cast by molecular clouds provide a new way to study properties of X-ray absorbing gas and X-ray emitting diffuse medium. We present our preliminary results on the R CrA cloud. Using six PSPC observations, we have mapped the gas column density distribution of the cloud. This X-ray measured distribution reveals features that are directly relevant to star forming activities in the cloud and offers a unique reference for calibrating various traditional methods used in studying molecular gas. Furthermore, we have found that the foreground X-ray emission of the cloud, harder than 10^6 K thermal plasma, accounts for a considerable fraction ($\sim 20\%$) of the line-of-sight 3/4 keV enhancement towards the general direction of the Galactic center. We conclude that the enhancement most likely represents a nearby hot superbubble with a characteristic temperature of 2.4×10^6 K and a pressure of 1.0×10^{-11} dyn cm^{-2}.

1. Introduction

We have developed a two-dimensional χ^2-fit algorithm for using X-ray shadow images in multi-energy bands to measure the gas column density distribution of a cloud as well as the foreground and background intensities relative to the cloud (Wang & Walter 1993). Such a direct measurement of the gas distribution is essential for determining accurately these intensities and is particularly useful for calibrating various traditional methods used in measuring gas distributions and masses of molecular clouds (Combes 1991 and references therein). Whereas the traditional methods rely on observing gas tracers such as CO and dust, we directly measure the X-ray absorption of hydrogen, helium, and trace species whose absorption cross sections are known to be insensitive to the physical and chemical states of the gas (Morrison & McCammon 1983).

2. Gas Distribution and X-ray Intensities

We have applied the algorithm to the ROSAT PSPC observations of the nearby molecular cloud R CrA (Rossano 1978; Wang & Walter 1993). The cloud is located in a region ($l \approx 0°; b \approx -18°$, $d \approx 130$ pc; Marraco & Rydgren 1981) relatively free from confusing interstellar features. Against the enhanced soft X-ray background in the general direction of the Galactic center (Garmire et al. 1992), the cloud appears as a deep shadow over the entire ROSAT energy range (e.g., Fig. 1). After discrete sources, various non-cosmic backgrounds, and instrument effects were removed, six individual PSPC observations were merged to form the maps that cover the cloud in the three PSPC broad bands: 0.1 - 0.4 keV (refered here as the 0.25 keV band), 0.5 - 0.9 keV (0.75 keV band), and 0.9 - 2 keV (1.5 keV band). Using these maps in our χ^2-fit, we have obtained

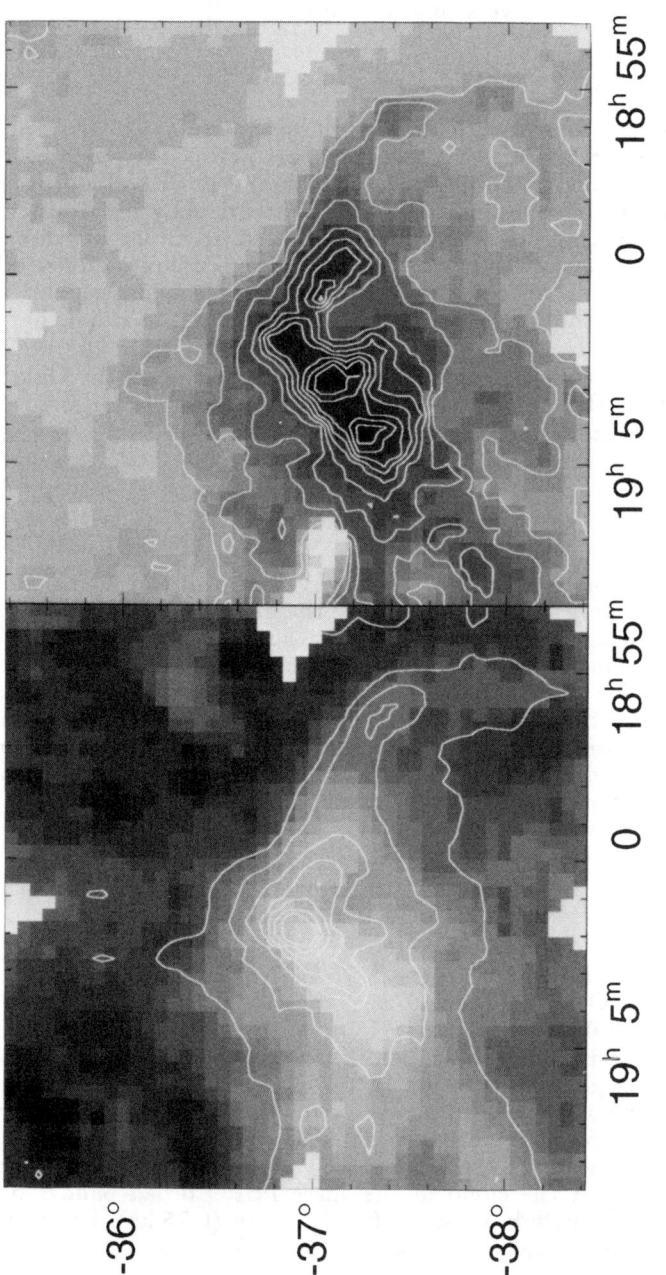

Fig. 1 (right panel) - Gray-scale map of diffuse X-ray intensity in the region of the R CrA molecular cloud. The map was constructed in the PSPC channel band 0.5 - 0.9 keV and was smoothed with a 3×3 median filter. The lower and upper gray-scale limits are 0.0 and 5.6×10^{-4} ct s^{-1} arcmin^{-2} with darker shading corresponding to higher flux. Bins that are completely blank represent regions where data have not been taken. The overlaid contours display the IRAS 100 μm intensity (a large-scale Galactic latitude-dependent background has been removed) at 4, 10, 20, 30, 50, 100, and 200 MJy sr^{-1}. Fig. 2 (left panel) - Gas column density distribution of the R CrA molecular cloud measured with ROSAT PSPC X-ray observations. The contour levels are 0.5, 1, 2, 3, 4, 5, 6, 8, and 9×10^{21} cm^{-2}.

the gas column distribution of the cloud (Fig. 2) as well as the foreground and background diffuse X-ray intensities, which are: 3.9 and 44 (0.25 keV), 0.67 and 4.5 (0.75 keV), and 0.057 and 2.6 (1.5 keV) in units of 10^{-4} ct s^{-1} arcmin^{-2}.

3. Properties of the X-ray Absorbing Gas

Using the X-ray measured gas distribution, together with observations of the cloud in other wavelength bands, we have examined gas properties of the R CrA cloud (Wang & Walter 1993). We have found that the core of the gas distribution coincides in position with the region of greatest visual extinction (Rossano 1978) and highest volume density indicated by the OH and H$_2$CO maps of the cloud (Vaile & Taylor 1982), but is away from the peak of both the dust and CO emissions (Fig. 1; Loren 1979), which are greatly enhanced in the western part of the cloud containing stars that have formed recently. As is evident in Fig. 2, the core consists of two lobes. The NW lobe, containing a cluster of dusty young stellar objects (Harju 1993), may represent an ongoing star forming region. The SE lobe, where gas volume density is highest but no stellar objects have been found, is probably the site of future star formation. The column density conversion $N(^{13}\text{CO})/N(\text{H})$ estimated in the eastern part of the cloud is approximately 2×10^{-6}, which is about twice higher than the value normally assumed. We have further found that the ratio of gas column density to 100 μm dust emission correlates strongly with local dust temperature, characterized by the ratio of IRAS 100 μm to 60 μm intensities, in regions with 100 μm intensities greater than 15 MJy sr^{-1} (Fig. 3).

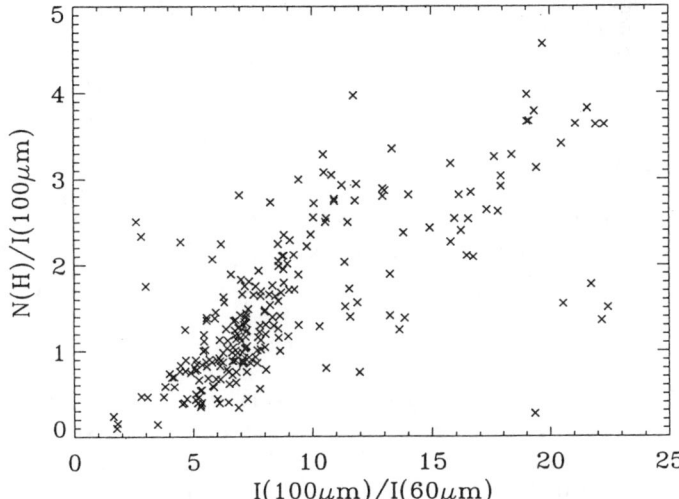

Fig. 3 Correlation between the ratio of gas column density to IRAS 100 μm intensity and the ratio of IRAS 100 μm to 60 μm intensities in the molecular cloud R CrA. The data points were collected from regions where the IRAS 100 μm intensity is greater than 15 MJy sr^{-1} (Fig. 1).

4. Properties of the X-ray Emitting Medium

It is now clear that the sun is embedded in a bubble that extends for perhaps 100 pc and is filled with X-ray emitting gas. The temperature of the gas, if characterized by a thermal plasma model (Raymond 1991), is about 10^6 K (McCammon & Sanders 1990; Snowden et al. 1993). However, the extent of any hot medium outside this hot Local Bubble had been difficult to determine until recently. With X-ray shadows of interstellar clouds, we can now explore the medium within and beyond the Local Bubble (see also Burrows & Mendenhall 1991, Snowden et al. 1993).

The R CrA cloud is well-placed for determining the nature of the enhanced 0.75 keV-band radiation towards the general direction of the Galactic center (Garmire et al. 1992 and references therein). The enhancement could be emission from a Galactic wind emanating from the Galactic Bulge (Garmire et al. 1992). However, since it is roughly contained within the northern boundary of radio Loop I (Berkhuijsen 1971), the enhancement may well represent a nearby hot superbubble (or several supernova remnants) produced by the Scorpio-Centaurus OB association. Our results on R CrA support this latter explanation. The foreground X-ray emission of the cloud most likely includes a considerable part of the enhancement since the differential foreground fluxes obtained in the three bands (§2) indicate a much harder spectrum than a 10^6 K thermal plasma. Therefore, it appears that the enhancement is indeed associated with LoopI that has an estimated distance comparable to the cloud. In fact, the cometary shape of the cloud, pointing towards the center of the circle fit to Loop I, can be interpreted naturally as the result of the interaction of the cloud with Loop I.

Assuming that R CrA is located with Loop I, we have analyzed X-ray spectral data collected from on-cloud and off-cloud regions to study the physical properties of hot gas in the direction of the cloud. The off-cloud spectral data from the NW region of Fig. 2 was first used to extract the spectrum of LoopI (Fig. 4). Specifically, we subtracted from the spectral data a combination of the Local Bubble contribution and the general (rather uniform) sky background, which were estimated approximately from the on-cloud spectral data (Fig. 5) at PSPC channel energies ≤ 0.4 keV and with a PSPC spectrum off the Galactic center region (Wang & McCray 1993) at > 0.4 keV. A satisfactory fit to the Loop I spectrum with a thermal model led to the estimates of the emission measure, temperature, and foreground absorption as $2.5^{+0.5}_{-0.4} \times 10^{-2}$ cm^{-6} pc, $2.4^{+0.3}_{-0.1} \times 10^6$ K, and $2.1^{+0.2}_{-0.5} \times 10^{20}$ cm^{-2} (with the error bars at the 90% confidence level). The best-fit temperature and absorption were then used to characterize the spectral shape of the Loop I contribution to the foreground X-ray spectrum of the cloud. This contribution, together with two additional components, were included in our model fit to the on-cloud spectral data (Fig. 5) that were collected in regions with $N_H \geq 5 \times 10^{21}$ cm^{-2} (Fig. 2). The spectral shape of the Local Bubble component was characterized by a thermal plasma of 10^6 K and an assumed absorption of 5×10^{18} cm^{-2}. The background component (significant only at energies ≥ 1 keV) was approximated by a power law of an energy slope 3.1 (a fit to the off-cloud spectral data). The best-fit mean absorption of the background component is $8.4^{+1.0}_{-0.9} \times 10^{21}$ cm^{-2}. The emission measures of the Local Bubble and the Loop I foreground contribution are $4.1^{+0.2}_{-0.3} \times 10^{-3}$ cm^{-6} pc and $4.8^{+0.7}_{-0.7} \times 10^{-3}$ cm^{-6} pc. Thus, the foreground accounts for about 20% of

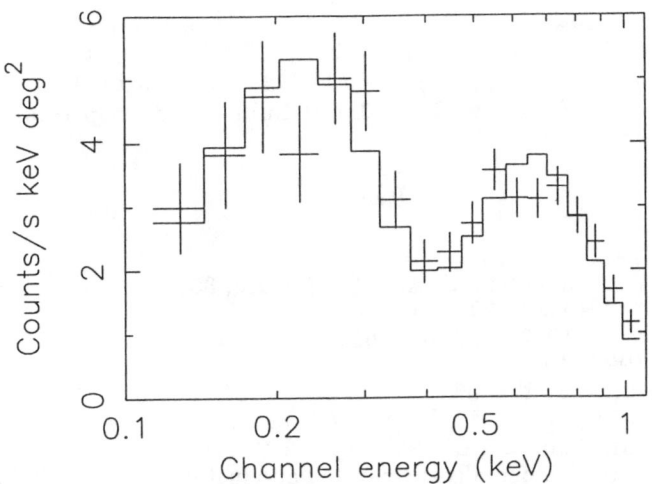

Fig. 4 - X-ray spectrum of Loop I in the region of the R CrA cloud. The spectral data were extracted from the off-cloud spectral data with subtractions of the contributions from the Local Bubble and the general sky background. The histogram represents the best fit of the data with a thermal plasma model (Raymond 1991).

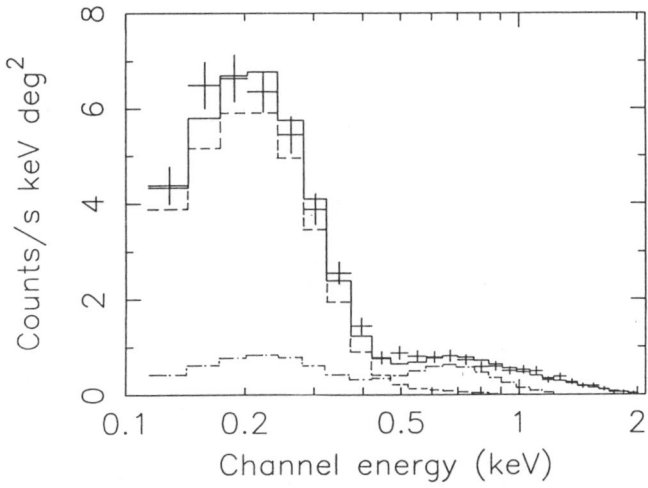

Fig. 5 - X-ray spectrum in the core region ($N_H \geq 5 \times 10^{20}$ cm^{-2}) of the R CrA cloud. The histogram represents the best fit of the spectrum with the three-component model (see text). The Local Bubble component (dashed histogram) and the Loop I foreground contribution (dash-dot histogram) were further plotted separately in the figure.

the total emission measure of Loop I, confirming that R CrA is within, but on the near side of, the Loop I superbubble. The hot gas pressure in Loop I ($p/k \approx 7.6 \times 10^4$ cm^{-3} K) appears much higher than that in the Local Bubble ($p/k \approx 1.3 \times 10^4$ cm^{-3} K) if both the Local Bubble and Loop I have a similar ling-of-sight size of 100 pc.

5. References

Berkhuijsen, E. 1971, A&A, 14, 359
Burrows, D. N. & Mendenhall, J. A. 1991, Nature, 351, 629
Combes, F. 1991, ARA&A, 29, 195
Garmire, G. P. et al. 1992, ApJ, 399, 694
Harju, J. et al. 1993, A&A, in press
Loren, R. B. 1979, ApJ, 227, 832
Marraco, H. G., & Rydgren, A. E. 1981, AJ, 86. 62
Morrison, R., & McCammon, D. 1983, ApJ, 270, 119
Raymond, J. C. 1991, (informally distributed and installed in XSPEC) an update
    to Raymond, J. C, & Smith, B. W. 1977, ApJS, 35, 419
Rossano, G. S. 1978, AJ, 83, 234
Snowden, S. L., McCammon, D., Verter, F. 1993, ApJL, 409, 21
Vaile, R. A., & Taylor, K. N. R. 1982, PASA, 4, 443
Wang, Q. D., & McCray, R. 1993, ApJL, 409, 37
Wang, Q. D., & Walter, F. 1993, ApJL, submitted

Supernova Remnants

X-RAY EMISSION FROM THE VELA SNR SHOCK REGION: SPECTRAL FITTING WITH A NON-EQUILIBRIUM IONIZATION MODEL

F. Bocchino, A. Maggio, S. Sciortino
Istituto e Osservatorio Astronomico di Palermo
Piazza del Parlamento 1, 90134 Palermo, ITALY

Email ID
bocchino@oapa.astropa.unipa.it

ABSTRACT

We report on the 5' scale spectral analysis of the X-ray emission from a region near the edge of the Vela SNR with a Non-Equilibrium Ionization (NEI) model. We have found significant variations of temperature, density, ionization time and interstellar absorption. We have identified an overdense region with higher density and lower temperature than the surrounding medium. That can be interpreted as an ISM cloudlet recently shocked by the blast wave, not yet in thermal equilibrium. Our independent determination of the Vela SNR distance is in agreement with the most recent results based on ROSAT All-Sky Survey data. Our analysis indicates the occurrence of fast electron-ion energy equipartition behind the shock.

1. The X-ray Observation

In order to perform spatially resolved spectral analysis, we have resampled a ~ 4 ksec PSPC observation of the Vela SNR North-East region in 25×25 spatial bins with an effective resolution of 5' (Fig. 1a). To avoid contamination from spurious events and electronic ghost images, we have discarded the photons at the end of OBI #1 (5.5% out of the data) and neglected SASS channels 1 and 2 in the spectral analysis.

2. The Non-Equilibrium Ionization (NEI) emission model

We numerically solved the system of NEI differential equations for each of the 12 most abundant elements (He, C, N, O, Ne, Mg, Si, S, Ar, Ca, Fe, Ni) to compute the population fraction, with two main assumptions: (i) the convective term in the diffusion equation of the population fraction has been neglected. This is a good assumption in the Sedov uniform expansion phase, but it could break down in case of mass transfer between cloudlets and the environment. (ii) The electron temperature, T_e, and density, n_e, remain constant during the ionization process. This assumption is good within the shock region since the Sedov model predicts slow variations in regions of 1-2 parsecs behind the shock front of a Vela-like SNR (Cui & Cox 1992).

We have used the Raymond-Smith code (Raymond & Smith, 1977; Raymond, 1989) for computing the grid of NEI X-ray spectra, which have been finally imported in XSPEC tabular form.

3. Results

NEI fits were performed in each of the 273 spatial bins with more than 500 counts using the temperature T, the ionization time τ, the normalization factor and the

X-ray Emission from the Vela SNR Shock Region

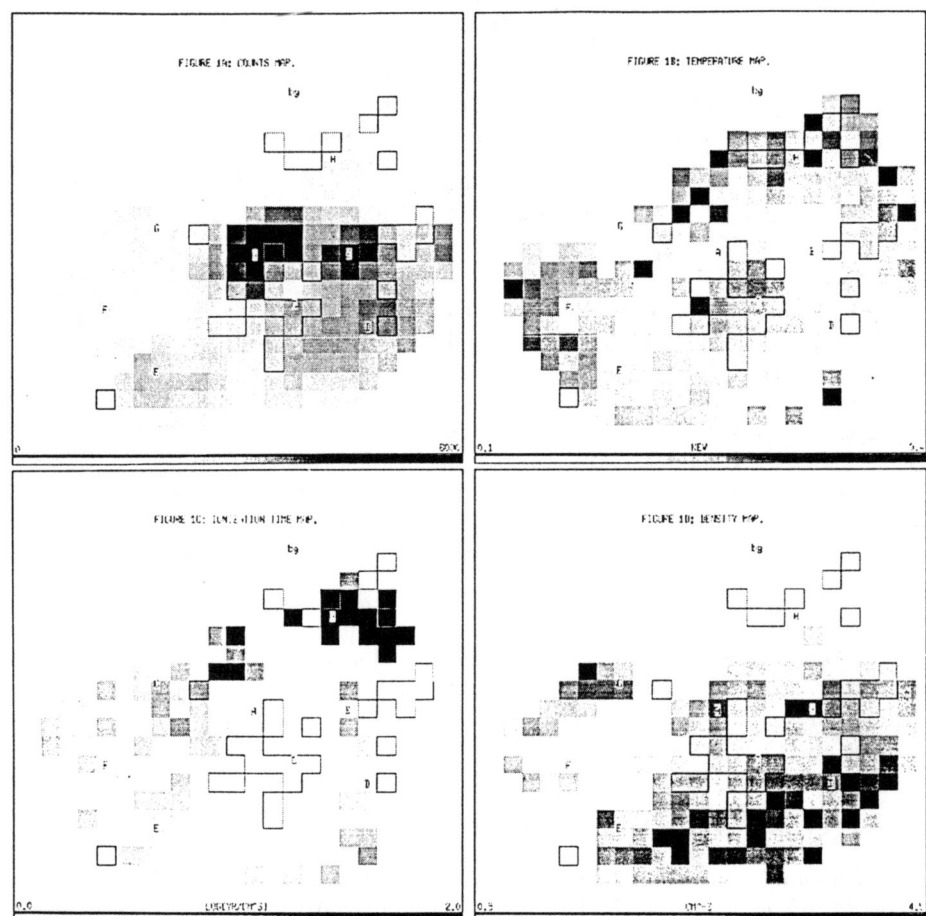

Figure 1: Rebinned X-ray image (a) and maps of NEI model parameters (b-d).

hydrogen column density as free parameters. The model results are acceptable, at the 95% confidence level, in 88% of the total investigated bins.

Best-fit NEI model parameters were arranged in maps with a resolution of 5' (Fig. 1b-1d). Overimposed letters identify zones discussed in the text. Contours highlight bins in which the NEI model fit is not acceptable. The density map has been computed assuming that the emitting plasma is confined in a thin shell ($\Delta r/r = 1/12$), according to the standard SNR shell model.

4. Discussion: the main shock

Since zones F and H show higher temperature and lower density than the other zones, we tentatively associate the emission observed in those hot zones with neutral ISM material shocked by the propagating main blast wave, and the colder zones (A, B,

D and E) with secondary shocks propagating in higher density ISM cloudlets.

According to this interpretation we derived the following main shock parameters. The **electron temperature** of the shocked ISM is $\sim 3 \times 10^6$ K (average value in regions F and H). The **ion post-shock density** is 1.0 ± 0.5 cm^{-3}; this implies a preshock density of ~ 0.3 cm^{-3}, in agreement with the previous Kahn et al. (1983) estimate of $0.15 - 0.65$ cm^{-3}. Assuming a fast electron-ion equilibrium, we found a **shock speed**, $v_{shock} = 390^{+90}_{-40}$ km sec^{-1}. The **post-shock pressure** is $2 - 5 \times 10^{-10}$ dyne cm^{-2}. The latter two results are in agreement with the Sedov expansion phase hypothesis (Cui & Cox, 1992). Applying the Sedov analysis and assuming an age of 11500 years (from radio observations of the Vela pulsar, Reichley, Downs and Morris, 1970), we estimated the initial blast energy, $E_0 = 2^{+5}_{-1} \times 10^{50}$ erg, and the shell radius, $r_{sh} = 12^{+10}_{-4}$ pc. Since the Vela SNR apparent radius in RASS observations is $\sim 7°$ (Aschenbach, 1992), we derive a SNR distance of $\sim 200^{+160}_{-70}$ parsec, well in agreement with the Oberlack et al. (1993) recent independent estimates.

To address the electron-ion equilibrium topic we tried to constrain the ratio $\eta = T_e/T_i$. If we assume a blast energy upper limit of 3×10^{51} erg (from SN explosion models), we can derive an upper limit for v_{shock} and T_i, and therefore we estimate $0.5 \leq \eta \leq 1$, supporting the hypothesis of a fast reaching of the electron-ion equilibrium.

5. Discussion: ISM cloudlets and secondary shock

Zones A and B might be regarded as belonging to an ISM cloudlet with a contrast density ~ 2 and an extension perpendicular to the line of sight of 1.5 pc. They have an electron temperature of 1.7×10^6 K corresponding to a secondary shock speed $v_{cl} = 300 \pm 40$ km sec^{-1}, and an ion density of 2-4 cm^{-3}. Using the best estimates of v_{cl} and v_{shock}, and the model of McKee and Cowie (1975), we derive an expected pressure enhancement of a factor ~ 1.5. In the pressure map, we observe an enhancement of ~ 2, so our data are in overall agreement with the McKee and Cowie model, considering statistical uncertainties. Assuming a contrast density of 2, Bedogni & Woodward (1990) derived $v_{cl}/v_{shock} = 0.84$, while the secondary shock is still propagating inside the cloud. Since the observed value is 0.8 ± 0.2, our data are in agreement with such a computation.

A rather flat temperature distribution is visible inside zone A and B, while one could expect a rather non-omogeneus distribution due to subsequent cloud evolution. Flat profiles can be explained if the emerging cloud shock is seen face on. The low observed ionization time supports such a geometry.

REFERENCES

Aschenbach, B., 1992, in "Space Science with Particular Emphasis on High Energy Astrophysics, J. Trümper ed. (MPE Preprint n. 227), 28
Bedogni, R., & Woodward, P. R., 1990. A&A, 231, 481
Cui, W., & Cox, D. P., 1992, in Winsconsin Astrophysics, preprint n. 428
Kahn, S. M., Gorenstein, P., Harnden, F. R. Jr., & Seward, F. D., 1985, ApJ, 299, 821
McKee, C. F., & Cowie, L. L., 1975, ApJ, 195, 715
Oberlack, U., Diehl, R., Montmerle, T., Prantzos, N., & von Ballmoos, P., 1993, Integral Workshop (1993), submitted
Raymond, J. C., & Smith, B. W., 1977, ApJS, 35, 419
Raymond, J. C., 1989, private communication
Reichley, P. E., Downs, G. S., & Morris, G. A., 1970, ApJ, 159, L35

A HIGH-RESOLUTION X-RAY STUDY OF THE SN1006 SUPERNOVA REMNANT

P. Frank Winkler
Department of Physics, Middlebury College, Middlebury, VT 05753
Email: winkler@middlebury.edu

Knox S. Long
Space Telescope Science Institute, Baltimore, MD 21218, Email: long@stsci.edu

ABSTRACT

ROSAT HRI images of two fields which together cover the northern half of SN1006 are presented. Comparison with the radio image of Moffett, Goss & Reynolds (1993) shows virtually identical morphology in the northeast, where the shock front is strongest in both bands. New optical images show the full extent of non-radiative, optical filaments in the northwestern portion of the shell. These filaments, which mark the present position of the shock front, are remarkably well correlated with the post-shock X-ray emission in this region.

1. Introduction

The supernova of 1006 A.D. is probably the brightest stellar event in recorded history (Stephenson, Clark & Crawford 1977). Originally identified as an X-ray source by Winkler & Laird (1976), the SN1006 remnant has been observed with most major X-ray instruments of the past two decades. Despite the absence of X-ray lines, Hamilton, Sarazin & Szymkowiak (1986) have argued that its soft X-ray emission is thermal and arises mainly from the reverse shock. ROSAT's soft X-ray sensitivity provides a superior probe for the structure of SN1006, which is at high galactic latitude ($+14.6°$) with low absorption.

2. Observations and Results

To date, we have obtained two pointings with exposure times of 8.0 and 12.8 ksec of the northeast and northwest portions of the SNR, both carried out 10-11 February 1991. The field centers were selected to achieve optimum resolution along the greatest length of the shock front.

To support the X-ray observations, we have obtained new optical images of SN1006 with the 0.6 m Schmidt telescope at CTIO, a 25 Å interference filter centered at the wavelength of Hα, and a Thomson 1024×1024 CCD. This combination gives a field of 31', well matched to the size of the remnant, at a scale of $1.83''$/pixel. A series of 16 dis-registered exposures with a total exposure time of 110 min was combined into a mosaic according to the procedure outlined in Winkler, Olinger & Westerbeke (1993).

The two ROSAT images were combined and smoothed with a $10''$ Gaussian to give the image shown in Figure 1. Figure 2 shows the 1517 MHz image of SN1006 obtained from the VLA in 1992 by Moffett, Goss & Reynolds (1993). It is evident that the X-ray and optical images have very similar morphology in the northeast region where the emitting shell is brightest. The ridges of emission in the two images, which suggest a deformed spherical shell, are correlated in extreme detail. Our Hα mosaic covers the entire remnant, but filaments are seen only along an arc to the northwest, much as previously reported by van den Bergh (1976) and Long, Blair & van den Bergh (1988). Figure 3 shows the northwest section of the Hα image, with the smoothed X-ray image shown as superposed contours.

3. Discussion

The new X-ray image shows a well-defined shock front extending around all of the remnant shell for which we have data. In the northwest, where the *Einstein* IPC images of Pye, *et al.* (1981) show a broad, indistinct ridge of soft emission, the ROSAT image indicates a sharp shell. This lends support to the hypothesis that the X-rays result from shock heating, despite the featureless *Einstein* SSS spectrum (Becker, *et al.* 1980).

In the northwest portion of the shell, where optical emission is seen, the X-ray shell follows the Hα filaments in exquisite detail. Balmer-dominated, or non-radiative, optical filaments result when neutral H atoms drift across the shock front to find themselves in a hot, ionized environment. There they can undergo charge exchange with fast-moving protons, leading to broad Balmer lines. No neutral H survives long in the hot, post-shock environment, so the Balmer filaments appear only in the *immediate* neighborhood of the shock. This is why they are so thin and wispy in appearance.

Figure 4 shows radial profiles of the emission in all three bands, taken in a 13° sector where the optical filaments are well-defined. The peak of the X-ray emission lies inside the Balmer filaments by approximately 5″, or 0.05 pc at a distance of 2 kpc for the remnant, while the radio peak lies further within, at a distance of 40″ (0.39 pc) behind the optical filaments. The small displacement between optical and X-ray emission suggests that, at least in the northwest, the contribution to the X-ray emission from the primary shock is much larger than suggested by the analysis of Hamilton, *et al.* (1986). The proper motion of the optical filaments has been measured at 0.30 arcsec yr^{-1} by Long, *et al.* (1988). The optical/X-ray displacement indicates that the heating and ionization timescales are short enough for X-ray emission to be well developed within 15 years after the shock passage.

Azimuthally, the X-ray and radio emission appear anti-correlated with the Hα filaments; the strongest X-ray and radio emissions stem from regions of the shell where no optical filaments are observed. Pre-ionization by X-rays emerging from the shock may explain the absence of Balmer filaments in these regions.

We are grateful to Greg Hanson for assistance in obtaining and reducing the optical data, and to Steve Reynolds for providing the VLA image shown in Figure 2. This work has been supported by NASA grant NAG5-1668, and also by NSF grant AST-9114935.

Becker, R.H., Szymkowiak, A.E., Boldt, E.A., Holt, S.S., & Serlemitsos, P.J. 1980, ApJ, 240, L33
Hamilton, A. J. S., Sarazin, C. L., and Szymkowiak, A. E., 1986, ApJ, 300, 698
Long, K.S., Blair, W.P., & van den Bergh, S. 1988, ApJ 333, 749
Moffett, D.A., Goss, W.M., & Reynolds, S.P. 1993, AJ, 106, 1566
Pye, J.P., Pounds, K.A., Rolf, D.P., Seward, F.D., Smith, A., & Willingale, R. 1981, MNRAS, 194, 569
Stephenson, F.R., Clark, D.H., & Crawford, D.F. 1977, MNRAS, 180, 567
van den Bergh, S. 1976, ApJ, 208, L17
Winkler, P.F., and Laird, F.N. 1976, ApJ, 204, L111
Winkler, P.F., Olinger, T.M., & Westerbeke, S.A. 1993, ApJ, 405, 608

314 A High-Resolution X-ray Study of the SN1006

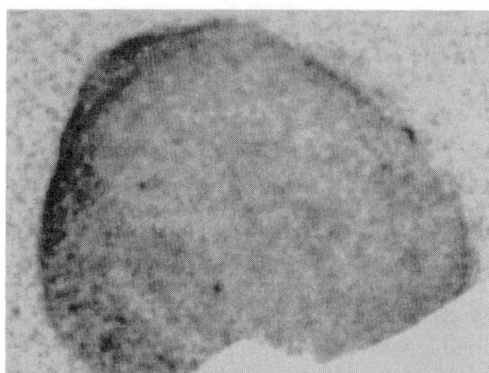

Figure 1. The combined ROSAT HRI image of SN1006, smoothed to a resolution of 10″. The two HRI pointings were centered just inside the shell in the NE and NW.

Figure 2. Grey-scale representation of the 1517 MHz VLA image, from Moffett, Goss, & Reynolds (1993). The diameter of the shell is 30′, or 17 pc at a distance of 2 kpc. The sharp linear feature crossing the radio shell in the east is a background radio galaxy.

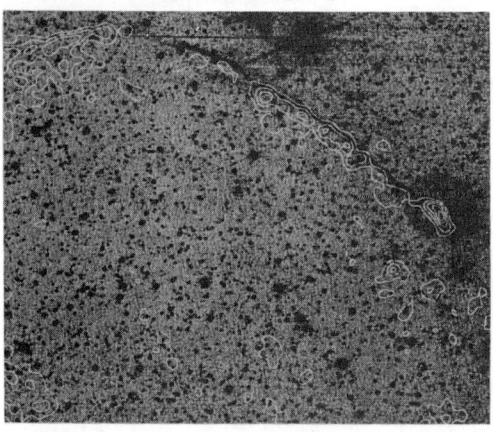

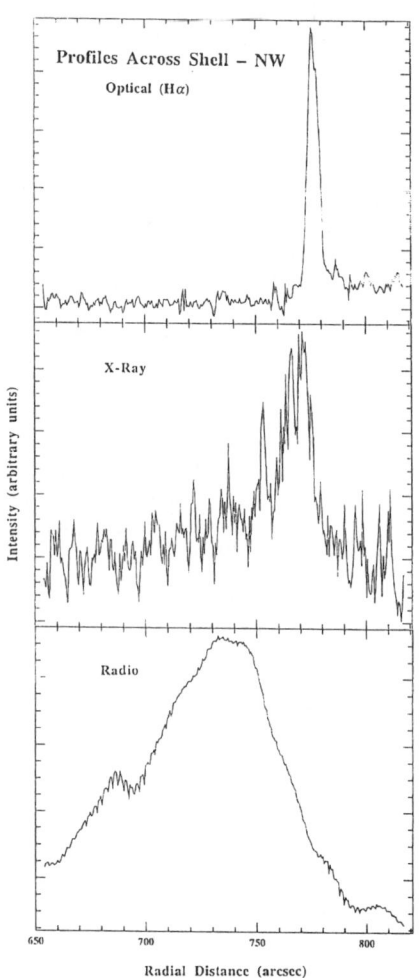

Figure 4. (above) Radial profiles taken in a 13° sector in the NW portion of SN1006. The optical profile is obtained from the Hα image, with a matched continuum image subtracted to remove most stars; X-ray and radio profiles are obtained directly from the HRI and VLA images. The X-ray peak occurs about 5″ inside the optical filaments.

Figure 3 (left). Hα image of the NW portion of SN1006, with the smoothed ROSAT HRI image represented in superposed contours. The X-ray shock front follows the optical filaments in detail, with the X-ray peak occurring about 5″ inside the Hα filaments.

3C400.2 - A SNR with a Centrally Condensed X-ray Morphology

Jon M. Saken, Knox S. Long
Space Telescope Science Institute, 3700 San Martin Dr., Baltimore MD, 21218
saken@stsci.edu, long@stsci.edu

William P. Blair
Dept. of Physics and Astronomy, Johns Hopkins University, Baltimore MD, 21218
wpb@pha.jhu.edu

P. Frank Winkler, and Brian M. DeChristopher
Dept. of Physics, Middlebury College, Middlebury VT, 05753
winkler@middlebury.edu, briand@middlebury.edu

ABSTRACT

We have used the ROSAT PSPC to examine the X-ray emission from 3C400.2, a supernova remnant which is a member of a class of remnants with centrally condensed X-ray and limb-brightened radio morphologies. The X-ray emission from this remnant is most likely due to a hot thermally emitting plasma rather than synchrotron emission. We compare the ROSAT observations to previous Einstein observations as well as radio and optical images of the SNR and discuss several explanations for its unusual X-ray morphology, including the evaporation of small clouds in a clumpy ISM and the interaction of multiple SNRs.

1. Introduction

In general, SNRs can be divided into two morphological classes, shell-like and center-filled. In SNRs with similar X-ray and radio morphologies, the shell-like remnants have X-ray emission characterized by a hot thermal plasma, while the center-filled remnants show X-ray emission due to synchrotron radiation. However a substantial number of radio, shell-like remnants appear centrally peaked at X-ray wavelengths. Other examples of this class of SNRs we are studying include HB3 and HB9. The X-ray emission mechanism and the reason for the dual morphology are unclear, although a few have been confirmed to be thermal sources.

3C400.2 has been well observed at radio, optical, and X-ray wavelengths. At radio wavelengths 3C400.2 has an elliptical, shell-like morphology, extending to the NW, with an angular size of 33'x28' (Dubner et al. 1994). Its optical emission consists of an incomplete shell of diffuse filaments only 16' (Winkler et al. 1993), poorly correlated with the radio. The brightest optical emission does not correspond with the brightest radio emission, and is somewhat offset from the center of the radio remnant.

The Einstein IPC X-ray observations reported by Long et al. (1991) show a smooth surface brightness with a peak well inside the radio shell and no evidence of an enhancement near the limb. They conclude that a Sedov model could not produce the observed X-ray morphology since more material is needed in the interior to produce the X-ray peak. Both power-law and thin-plasma models gave acceptable fits to the Einstein data, although Long et al. thought a thermal model more likely.

2. ROSAT Observations and Analysis

A total of three offset PSPC observations were obtained with ROSAT in order to search for any additional extended X-ray emission from 3C400.2. No emission was observed beyond the radio shell. The ROSAT observations do show that the X-ray emitting gas completely fills the radio shell (Fig. 1) as suggested by the Einstein observations. Here we discuss results from the central pointing which contains most, if not all, of the emission. This pointing consists of a series of observations taken on Oct. 4-10, 1992 for a total of 3570 sec. The total background-subtracted count rate from the SNR is 1.7 cts sec^{-1}.

A distance of 6 kpc was adopted. This is consistent with previous distance estimates and permits easy comparison with the Einstein measurements of Long et al. (1991). The global spectrum was extracted and fits were performed in XSPEC using both Raymond-Smith and power-law emission models. Power-law models yielded substantially worse fits ($\chi_r^2 > 5$) and can probably be ruled out, while the Raymond-Smith fits all yielded $\chi_r^2 \sim 1$. Background regions at a variety of off-axis angles were checked; no significant variations in the fits were found.

The results based on the best fit parameters and a comparison to the Einstein analysis are summarized in the table below. A filled spherical shell with $\Theta = 28'$ and $d = 49$ pc was assumed.

	Einstein	ROSAT
F (cts s^{-1})	0.7	1.7
N_H (cm^{-2})	7.7×10^{21}	7.1×10^{21}
T_X (K)	4.4×10^6	3.4×10^6
L_X (ergs s^{-1})[a]	8.4×10^{34}	6.5×10^{34}
EM (cm^{-3})	3.5×10^{58}	6.5×10^{58}
n_e (cm^{-3})	0.15	0.19
$m_X (M_\odot)$	260	330

[a] 1-4 keV for Einstein and 0.4-2.4 keV for ROSAT.

There is some evidence that the gas temperature is not uniform throughout the remnant. Separate fits were made to the extracted spectra from the X-ray bright central region and the low surface brightness SE region of the remnant. These were defined as a 6.1' × 7.5' elliptical region centered at RA 19:38:18, Dec. 17:18:44, and a 10.3' × 7.5' rectangular region centered at RA 19:38:44, Dec. 17:06:25 (J2000), respectively, with best fit values of $N_H = 3.3 \times 10^{21}$ cm^{-2}, $kT = 0.71$ keV and $N_H = 4.5 \times 10^{21}$ cm^{-2}, $kT = 0.42$ keV. Since N_H probably does not vary significantly between the two regions, we fixed it at an average value for the fits of 4×10^{21} cm^{-2} and obtained temperatures of 0.67 keV and 0.37 keV. Thus our preliminary analysis suggests the higher surface brightness regions of the SNR may have a higher temperature.

4. Discussion

The reason for the SNR's centrally-peaked X-ray morphology is poorly understood. Long et al. (1991) attempted to explain the X-ray emission from 3C400.2 (and

W28) in terms of cloudlet evaporation in the interior of the SNR. The relatively featureless appearance of 3C400.2 with modest temperature variations may lend support to this picture.

On the other hand, Dubner et al. (1994) suggest that the X-ray morphology of this remnant could also be the result of the interaction of two SNRs. This situation could arise, for instance, from multiple SNe within an OB association or from certain binary stars. This model could account for the two-lobed radio structure of 3C400.2. The possibility of a higher temperature in the region of the smaller ring may support this model. However, if a common SN energy and ISM density is assumed for the two shells, a Sedov model predicts that the the smaller shell should be four times hotter than the larger shell. This is not seen in the ROSAT data.

References

Dubner, G. M., Giacani, E. B., Goss, W. M., & Winkler, P. F. 1994, AJ, accepted
Long, K. S., Blair, W. P., White, R. L., & Matsui, Y. 1991, ApJ, 373, 567
Winkler, P. F., Olinger, T. M., & Westerkbeke, S. A. 1993, ApJ, 405, 608

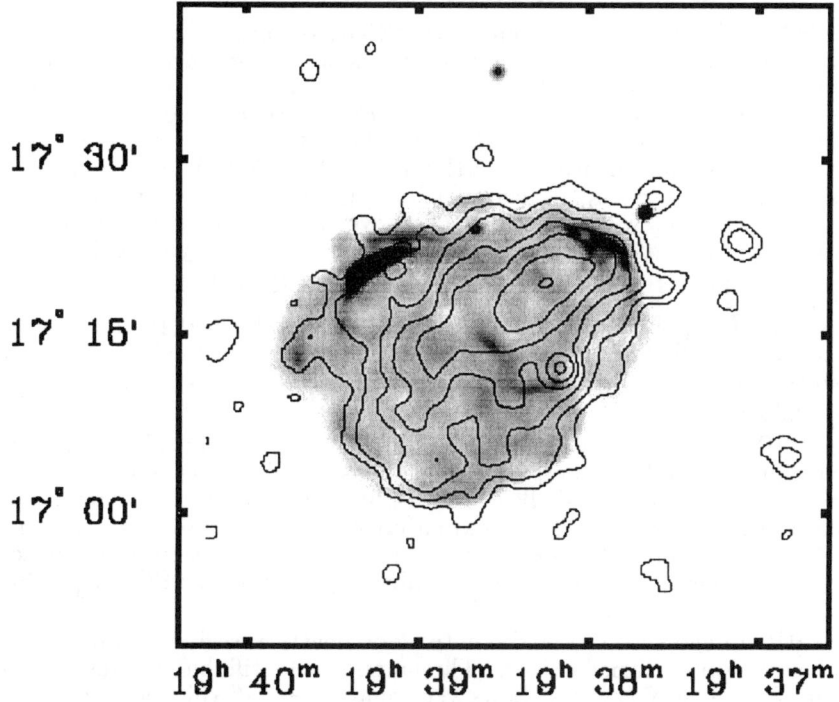

Fig. 1: Radio image from Dubner et al. (1994) with ROSAT PSPC contours of 3C400.2 overlaid. Contours levels are at $(1.2, 1.8, 2.6, 3.9, 5.8, 8.6, 12.8) \times 10^{-3}$ cts s^{-1} arcmin^{-2}; radio image linearly scaled from 0-1.5 $\times$ 10^{-2} Jy/beam.

ROSAT OBSERVATIONS OF IC 443

J.-H. Rho, R. Petre
Code 666, NASA/Goddard Space Flight Center
J.J. Hester
Dept. of Physics & Astronomy, Arizona State University
[1]Astronomy Department, University of Maryland

ABSTRACT

We present preliminary results of our spatial and spectral studies of a mosaiced ROSAT PSPC image of the IC 443 region, and an image of the X-ray bright northeastern region observed using the HRI. Although IC 443 has shell-like optical and radio morphologies, the X-ray emission appears amorphous and peaks somewhat inside the brightest part of the radio/optical shell. The PSPC observations reveal a second shell-like structure with $\sim 1.1°$ diameter which overlaps IC 443. The region by region spectra of IC 443 were well fit by a two-temperature Raymond-Smith model with temperatures of (0.1 - 0.25) keV and (0.4 -0.7) keV. The column density variations correlate with the brightness temperature of the CO from a molecular cloud in the line of sight. Spectral fitting of a band of low X-ray brightness in the NE of IC 443 reveals it to be a shadow of the second shell, indicating this structure to be a foreground object, probably another SNR.

1. INTRODUCTION

IC 443 is classified as a "composite" SNR, displaying shell-like optical and radio morphologies and showing anomalous structures such as its eastern arm filament (Fesen 1984) and different curvatures of the southwestern optical filaments. The X-ray emission, however, appears smoother and peaks somewhat inside the brightest part of the radio/optical shell. IC 443 (G189.1 +3.0) is in the Gem OB 1 association near the HII region S249. A thin molecular cloud stretches north-south across the center of the remnant. The remnant shows evidence of interactions with this cloud (DeNoyer and Frerking 1981), showing broad molecular lines and shocked H I gas (Giovanelli and Haynes 1979).

2. IMAGING AND SPECTRAL OBSERVATIONS

A series of 5 pointings using the PSPC was performed to map the IC 443 complex. In addition, a 8.2 ks HRI exposure was performed on the bright northeastern region. The PSPC count rate was 22.0 ± 0.1 cts s^{-1}; the HRI rate was 4.04 ± 0.03 cts s^{-1}.

The PSPC mosaic of the remnant is shown in Fig. 1 (color plate). The image shows high surface brightness in the northeast, peaking inside the brightest part of the radio and optical shells. The bright region is bifurcated by a band of low X-ray brightness running approximately east-west. Many new structures invisible to the Einstein IPC (Petre et al. 1988) in the western and southern portions of the image are revealed by the higher sensitivity PSPC image. Soft X-rays (E<0.5 keV) were not detected.

The HRI image shows that the entire northeastern region is composed of clumps of emission of various sizes. The smallest identifiable clumps are 17" (0.16pc) in size. The X-ray gas shows very sharp ($\sim$ 17" thickness) filaments along the shock front in the NE rim (Dec. 23°50'), where the line of sight is tangent to the shock. These filaments are probably produced by either direct shock heating or evaporation of high density gas associated with the H I cloud to the northeast of the remnant. The presence of shocked HI along the filament indicates the presence of material with a range of densities at the shock front.

Shown in Fig. 2 (color plate) is a PSPC mosaic of the entire IC 443 region. A second shell-like structure, with considerably lower surface brightness than IC 443 proper is apparent. This structure has a $\sim 1.1°$ diameter and is located inside of the 1.5° partial radio shell suggested by Braun and Strom (1986) to be connected to and behind IC 443. The shell is highly convoluted, unlike most shell-like SNR. The band of low surface brightness in the northeast of IC 443 proper seems to be a continuation of the second shell.

A detailed examination of the spatial distribution of the temperature and column density within IC 443 proper has been performed by combining 3 sets of the PSPC spectra. The regions are divided based on the X-ray contours and molecular cloud intensity. The spectra are well fitted using two temperature models of an optically thin, collisionally ionized plasma. The abundance is allowed to vary. The column density has a range of $(0.5 - 1.2) \pm 0.1 \times 10^{22}$ cm^{-2}, and its distribution is largely consistent with the brightness temperature of the CO from the molecular cloud in the line of sight (Cornett et al. 1977). The low and high temperature component have temperature ranges of $(0.11 - 0.21) \pm 0.05$ keV and $(0.38 - 0.7) \pm 0.1$ keV, respectively. The western and southwestern parts are no hotter than the east, which contradicts a theory that the difference in X-ray brightness between east and west is due to density gradient in the preshock medium (Parkes et al. 1977).

3. DISCUSSION

Two distinct variations in X-ray brightness have not been understood on the basis of previous observations: a band of low X-ray brightness in the NE connected to the eastern arm, and the discontinuity between the eastern and western parts of the remnant. On the basis of the PSPC observations, the band of low X-ray brightness in the northeast can be seen to be where the newly-discovered second shell crosses IC 443. Spectral fitting of the low surface brightness band shows it to be very different from adjacent, brighter regions. The best fit model employing a single absorbing column requires the band to have a much lower N_H (0.46×10^{22} cm^{-2}) than the neighboring regions (0.7-0.8×10^{22} cm^{-2}) as well as different temperatures. A much better fit is obtained using two distinct thermal components, each with its own absorbing column. This result indicates that two objects are superposed, with the newly-discovered shell in the foreground (cf. Braun and Strom 1986). It also shows that the distance of 1.5 kpc for IC 443 obtained from optical spectroscopy of eastern arm (Fesen 1984) pertains to the new shell, and not IC 443 proper. IC 443 itself is somewhere behind this object, probably at a distance from us of 1.5-2.5 kpc. From the fact that it emits X-rays, the shell is almost undoubtedly a previously unknown SNR.

REFERENCES

Braun, R., and Strom, R.G 1986, A & A., 164, 193.
Cornett, R.H., Chin, G. and Knapp, G.R. 1977, A & A, 54, 889.
DeNoyer, L.K. and Frerking, M.A. 1981, Ap. J., 246, L137.
Fesen, R.A. 1984, Ap. J., 281, 658.
Giovanelli, R., and Haynes, M.P. 1979, Ap. J., 230, 404.
Parkes, G.E., Charles, P.A., Culhane, J.L., and Ives, J.C. 1977, M.N.R.A.S., 179, 55.
Petre, R., Szymkowiak, A.E., Seward, F.D. and Willingale, R. 1988, Ap. J., 335, 215.

THE DIFFUSE X-RAY EMISSION OF CTB 109
J.-H. Rho[1], R. Petre
Code 666, NASA/Goddard Space Flight Center
[1]Astronomy Department, University of Maryland

ABSTRACT

We report on the X-ray spatial structures and spectral mapping of the supernova remnant CTB 109 (G109.1-1.0), as found using a 34,000 s ROSAT PSPC observation. X-rays arise from three distinct components: a semicircular shell, the centrally located X-ray pulsar 1E2259+586, and a jet-like structure between the pulsar and the shell. Spectral analysis of CTB 109 shows that the jet-like structure is thermal, with no evidence of synchrotron radiation. The spectra of the shell are adequately fit by a single temperature thermal model. A simultaneous non-equilibrium fit using BBXRT and PSPC spectra suggests extreme departure from ionization equilibrium within the interior of the remnant. A clear column density gradient is observed, running from northeast to southwest. The fit values of N_H fall in the range $(0.77-1.02)\pm 0.1 \times 10^{22} \text{cm}^{-2}$. A spectral hardness ratio map shows that a CO arm extending across the remnant and eastern shell is harder and the jet-like structure is softer than elsewhere. The X-ray structure associated with the CO arm region is due not only to absorption, but to the interaction of the SNR with the molecular cloud as well.

1. INTRODUCTION

The X-ray and radio emission of CTB 109 are confined to a hemispherical shell centered upon the X-ray pulsar IE 2259+586. There is no X-ray or radio shell to the west of the pulsar. The remnant is located in a very complicated region of the ISM, with a giant molecular cloud to its west. CO mapping by Tatematsu et al. (1987, A&A, 184, 279) has shown that the absence of the western portion of the shell is due to the interaction between the remnant and the molecular cloud. IE 2259+586 is a very strong X-ray source, with no radio counterpart (Gregory and Fahlman 1983, IAU101, 437). It has a 6.98 s period and a small positive period derivative. The curled jet-like X-ray structure extending eastward from the pulsar to the shell has been proposed either to originate from the central binary system (Gregory and Fahlman 1983) or to be related to the molecular cloud (Tatematsu et al. 1987). The spectroscopic distances of the exciting stars of HII regions associated with the molecular cloud have been estimated to be 3.6-5.5 kpc, and we assume here 4.0 kpc as the distance of the SNR.

2. X-RAY IMAGE AND SPECTRA

CTB 109 was observed for 34,000 s using the PSPC on July 9-10, 1991. The data presented here have been extracted from the ROSAT archive. The smoothed PSPC X-ray image is shown in Fig. 1 (color plate). The X-ray emission originates mainly from three components : the pulsar, the shell, and a jet-like structure connecting the eastern shell and the pulsar. There is faint X-ray emission where the molecular cloud extends into the remnant (called the CO arm hereafter). The count rates from the diffuse emission and from 1E2259+586 are 8.13 ± 0.02 cts s^{-1} and 1.30 ± 0.004 cts s^{-1}, respectively. X-rays with E < 0.5 keV were detected only from the jet-like structure and 1E2259+586. Despite the substantial improvement in sensitivity of the ROSAT map over previous images (*Einstein* IPC and EXOSAT), there is still no detection of X-ray emission from west of the pulsar.

The average photon energy ("color") map is shown in Fig. 2 (color plate). This map is virtually identical to the PSPC spectral hardness ratio (SHR) map between channels 70 - 250 ($\sim$ 0.7 -2.2 keV) and 30 - 70 ($\sim$ 0.3-0.7 keV). An apparent gradient in the average photon energy per pixel runs from the southwest to the northeast. A second PSPC hardness map (Fig. 3 [color plate]) has been made using the ratio between channels 131-250 ($\sim$ 1.3 -2.2 keV) and 40-131 ($\sim$ 0.3-1.3 keV). This hardness map shows that the CO arm extending across the eastern shell and the pulsar are harder and the jet is softer than elsewhere. We interpret the structure in the low energy (0.7 keV) cut SHR map as resulting from differential absorption, and those in the high energy (1.3 keV) cut hardness map from temperature variations.

The pulsar, the jet-like structure and shell show very different spectra. The spectrum of the jet-like structure requires two temperatures: (0.13 - 0.14)±0.05 keV and (0.29 - 0.31)±0.05 keV. A power law model is rejected, indicating that there is no evidence for synchrotron radiation. The spectra of the shell are adequately fit by a single temperature Raymond-Smith model. The best-fit temperature varies across the remnant, with values between 0.24±0.05 keV and 0.33±0.05 keV. The southern portion of the shell also can be well fit by a two temperature model, with 0.18±0.03 keV and $0.56^{+0.5}_{-0.3}$ keV. The column density gradually decreases from the northeast to the southwest with a range of (0.77-1.02) ±0.1 × 10^{22}cm^{-2}, largely consistent with the low energy cut hardness map. The CO arm region has lower N_H than the nearby jet-like structure, indicating the lack of X-rays is not due to absorption alone, but due to the interaction with the molecular cloud: where the shock crosses the clouds, it becomes radiative and emits less X-rays. The preferred model for a composite spectrum of the diffuse emission is a two temperature model, with temperatures of 0.17±0.02 keV and 0.56±0.15 keV. The 0.2-2.2 keV luminosity of the diffuse emission is 3.2× 10^{37} ergs s^{-1}. Assuming the emission arises in a shell with a thickness of 20 percent of the remnant diameter, the density is 1.2 cm^{-3}, and X-ray emitting mass is 415 $M_\odot$. Based on Sedov model, the shock velocity V_s is about 760 km s^{-1} and the age is about 10^4 yrs.

3. DISCUSSION

A fit using a non-equilibrium ionization model to combined BBXRT and PSPC spectra for a portion of the southern shell yields a shock temperature of 5.6 × 10^7 K, an ionization parameter $\eta = n_o^2$ E of 10^{49} erg cm^{-6} for $T_e \neq T_i$. The ionization timescale of 200 - 300 cm^{-3} yr suggests a substantial departure from ionization equilibrium. The absence of detectable nonthermal emission in CTB109 shows that the influence of the central pulsar on the emission from the remnant itself is minimal. In particular, it suggests that the precessing jet model used to explain the connection between the nonthermal X-ray lobes in W50 and SS 433 (Gregory and Fahlman 1983) is not applicable here.

The X-ray emission spatially coincident with the arm-shaped CO ridge from the western giant molecular clouds shows some intriguing properties. While there is an an apparent anticorrelation with the total X-ray surface brightness and the region appears hard in the hard cut SHR map, it appears especially bright in the soft cut SHR map. This suggests that the CO arm is embedded within the remnant, but near the front face. As most of the X-rays emanate from behind the arm, the overall effect is that of a shadow, and only the hardest X-rays penetrate the material. The soft emission arises from newly shocked material at the front of the cloud. Higher resolution spectra of this region should reveal more clearly the two distinct emission components.

AN X-RAY AND OPTICAL STUDY OF A CLOUD-BLAST WAVE INTERACTION

N. A. Levenson and J. R. Graham
Department of Astronomy, University of California, Berkeley, CA 94720

J. J. Hester
Department of Physics and Astronomy, Arizona State University
Tempe, AZ 85287

J. C. Raymond
Harvard-Smithsonian Center for Astrophysics, 60 Garden St.
Cambridge, MA 02138

R. Petre
MC 666, NASA Goddard Space Flight Center, Greenbelt, MD 20771

ABSTRACT

We present ROSAT HRI and optical emission line images of an isolated cloud in the southeastern region of the Cygnus Loop supernova remnant. Although at first glance this appears to be a small cloud that the blast wave has engulfed and shredded, we believe that this is the early stage of interaction between the blast wave and a large cloud and that projection effects are responsible for the apparent similarity to engulfed cloud models.

1. INTRODUCTION

Supernova remnants (SNRs) provide opportunities to study the interaction of blast waves and the non-uniform surrounding medium. As a SNR expands, the blast wave may overrun small clouds and destroy them. Alternatively, the edge of a SNR may remain a coherent shock as it encounters large clouds.

One place to observe a cloud-blast wave interaction is a small ($2' \times 4'$) knot in the southeastern region of the Cygnus Loop, a "middle-aged" SNR. Fesen *et al.* (1992) studied this knot using optical line-imaging and proposed that the blast wave has completely overrun the cloud there. We present ROSAT-HRI and further optical emission-line images of the southeastern knot. These X-ray and optical data suggest that this is the early stage of interaction between the cloud and blast wave.

2. OPTICAL AND X-RAY IMAGING

In the vicinity of the cloud the blast wave shows up as a sharp edge in the ROSAT HRI data. The cloud is located at the apex of a large scale indentation of the X-ray shell, and corresponds to a region of well defined X-ray filaments. There is good correspondence between X-ray and optical emission. The optical filaments that run from the northern end of the cloud in a NE direction, and the Hα feature to the west of the main cloud, are the best examples of correlated X-ray and optical line emission. (See Fig. 1.) However, correlation between optical and X-ray emission is not universal. The main body of the cloud corresponds to a local minimum in the X-ray emission, and the X-ray emission associated with the Hα streamer located at the southern end of the cloud consists of a bright knot of emission located at the end of this filament.

The ROSAT observations further indicate that the northern and southern sections of the Balmer filament are not continuous segments of the same shock, and projection effects are tremendously important in understanding them. In the southern section, strong X-ray emission occurs immediately behind the Hα filament whereas in the north, the X-ray edge appears to be significantly behind the Balmer filament.

3. A SMALL, ENGULFED CLOUD?

(Fesen et al. 1992) interpreted the southeast knot as a small, isolated cloud that the blast wave has completely engulfed. If this is correct then the curved Balmer-line filaments correspond to the diffracted shock. The cusp-like feature directly ahead of the cloud is the Mach disk, and the streamers result from shredding of the cloud.

4. OR A MORE COMPLICATED SITUATION?

Optical imaging and ROSAT–HRI data show that the physical situation is more complicated than it initially appears to be and reveal that the "engulfed cloud" description is inadequate.

First, we view the interaction almost tangentially, where the column density of emitting material is greatest. Looking at a blast wave that has engulfed a spherical cloud, a tangential observer would see in projection the edge of the diffracted shock and blast wave in front of the Mach disk. In these observations, however, we see the blast wave *behind* the Mach disk.

Second, deep Hα images (Fig. 2) show that the cusp–like feature associated with the "triple-point" at the intersection of the diffracted shock and the Mach disk, is in fact an "X." Balmer–filaments require a significant preshock neutral H fraction, which would not be present behind an X-ray emitting shock. The interior portions of this "X" thus *cannot* occur behind the main shock; their appearance must in fact be due to the line of sight projection of two shocks propagating in different directions. (Similarly, the Balmer-dominated filament to the west of the main knot is not a reverse shock because little neutral hydrogen would have remained after the passage of the initial X-ray shock.)

Finally, the larger scale structure of X-ray emission near the cloud also supports the view that this region of the Cygnus Loop does not present a simple case of a small, engulfed cloud. The X-ray image of Ku et al. (1984) and the optical images of Fesen et al. show an indentation of the shell extending well beyond the southeastern cloud. A small cloud could not alter the expanding shell to this extent.

This work was supported in part by grants from NASA.

REFERENCES

Fesen, R. A., Kwitter, K. B., Downes, R. A. 1992, AJ, 104, 719
Ku, W. H.–M., Kahn, S. M., Pisarski, R., and Long, K. S. 1984, ApJ, 278, 615

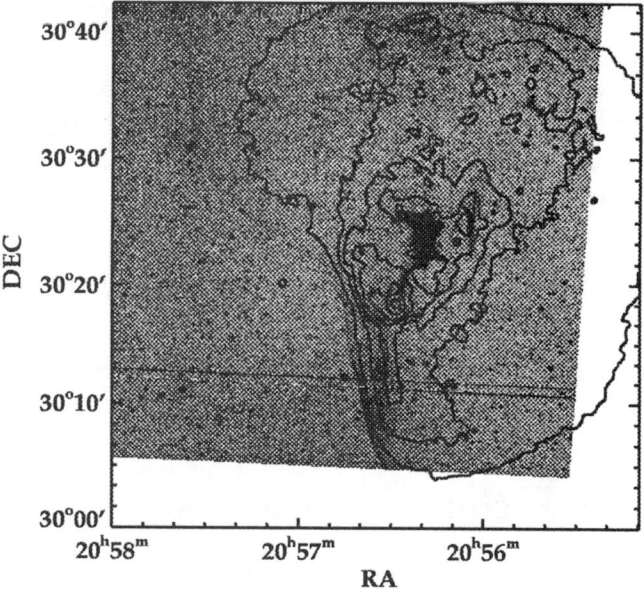

Fig. 1.—ROSAT-HRI contours overlaid on the Hα image of Fesen *et al.* (1992). This image traces faint Balmer-line filaments, which delineate the blast wave, and bright, radiative shocks driven into the cloud.

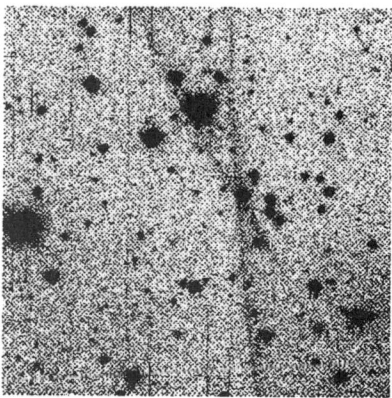

Fig. 2.—Deep Hα image from Lick Observatory showing the cusp point of the putative Mach disk. This image reveals that the morphology is due to a projection effect.

ROSAT OBSERVATIONS OF SNRs AS DISTANCE INDICATORS

N.E. Kassim, P. Hertz, S.D. Van Dyk, and K.W. Weiler
Code 7215, Naval Research Laboratory, Washington, DC 20375-5351

ABSTRACT

We explore the possibility that ROSAT observations of thermal X-ray emission from shell-type supernova remnants (SNRs) can be used to reasonably constrain the distances to both Galactic and extragalactic remnants. The only major assumption is that the initial kinetic energy (ε_o) of a supernova explosion is $\varepsilon_o = 10^{51}$ ergs. Fortunately, the derived distance is only weakly dependent on this assumption ($D \propto \varepsilon_o^{0.4}$). We evaluate this technique by applying it to SNRs with relatively well known distances, and our initial results indicate that the distances so obtained are relatively accurate. It therefore appears that the ROSAT all sky survey can be used to establish good quality distances to a large number of extended, shell-type SNRs. If confirmed by further tests, these data could greatly alleviate the long standing problem of unknown or very poorly established distance estimates to SNRs.

INTRODUCTION

SNRs are important to many areas of Galactic astrophysics because they are the major source of energy input to the interstellar medium (ISM), likely serve as the site of cosmic ray acceleration, and probably play a major role in triggering star formation. However, unlike HII regions, whose distances can be relatively well constrained by recombination line observations, Galactic SNRs have notoriously poorly determined distances. Of the currently known ~180 Galactic remnants only ~15% have distance determinations, and only a limited subset of these are likely to be very reliable (Green 1984). Most current means to establish distances to SNRs are either difficult and uncertain (HI absorption) or are too inexact to be useful (the "Σ-D" relation). Here we describe how soft X-ray observations, such as those provided by ROSAT, can be used to establish reliable distance estimates to a large number of known shell-type SNRs. We apply this Sedov procedure, developed in more detail for the case of G326.3-1.8 by Kassim, Hertz, and Weiler 1993 (hereafter KHW), to a few SNRs with well established distances in order to demonstrate its validity.

THEORY

KHW have reviewed the Sedov equations which describe the dynamics of shell-type SNR expansion during the adiabatic phase. By combining these equations one can solve for the "Sedov" distance D_s to the SNR as:

$$D_s = 9.03 \times 10^6 \, \varepsilon_o^{0.4} \, P(\Delta E, T)^{0.2} \, \theta^{-0.6} \, F_{xo}^{-0.2} \, T^{-0.4}. \quad (1)$$

Here ε_o is the initial energy of the SN explosion (thermal plus kinetic) in units of 10^{51} ergs, θ is the observed angular diameter of the SNR shell in arc-minutes, F_{xo} is the measured X-ray flux corrected for interstellar (and/or intergalactic) absorption in units of ergs sec^{-1} cm^{-2}, and T is the measured thermal temperature of the X-ray emitting gas in K. The analytical function $P(\Delta E, T)$ describes the power emitted by hot electrons in a low density plasma via free-free emission (Tucker and Koren 1971).

KHW have shown how equation (1) may be applied to ROSAT observations of shell-type SNRs for the case of G326.3-1.8. The ROSAT PSPC data were well fitted by a single Raymond-Smith thermal spectrum characterized by a temperature (T) and, after correction for interstellar absorption, an unabsorbed X-ray flux (F_{x0}) within the ROSAT 0.1-2.4 keV energy range. Since the source angular size θ could be directly measured for a resolved source, D_s could be determined if ε_o was assumed. KHW assumed the canonical value $\varepsilon_o=1$ and determined a distance of $D_s \sim 3.7$ kpc. This new, independent distance estimate is in agreement with and improved upon previous distance estimates of ~3-5 kpc based on much weaker arguments. Since D_s scales only weakly with ε_o ($D_s \propto \varepsilon_o^{0.4}$), this result implied that improved distance estimates could be established for the large number of extended shell-type SNRs which can be well fit by thermal X-ray spectra from ROSAT observations.

ASSUMPTIONS

The key assumptions of the Sedov analysis reflected by Equation (1) are: 1) the SNR shell is in the adiabatic expansion phase; 2) the measured X-ray temperature gives a reliable estimate of the SN shock velocity; and (3) $\varepsilon_o=1$ (i.e. 10^{51} ergs) is approximately correct for all SNe. Assumption (1) is the only one which can be checked self-consistently. KWH have shown how a calculation of the swept up mass can be used to ensure that a SNR has passed the free-expansion phase and entered the adiabatic expansion phase, and McKee (1982) has developed a test to show that the SNR has not yet entered the radiative phase.

Unfortunately neither assumptions (2) nor (3) can be firmly established on either theoretical or observational grounds. Calculations of Type I SNe by Nomoto, Thielmann, and Yokoi (1984) showed that ε_o only varies over the range 0.9-1.5, but there are no generally accepted ways of estimating the energy of Type II events. With the total energy released in the gravitational collapse of a star during a SN explosion of $\sim 10^{53}$ ergs, the assumed thermal plus kinetic energy ε_o of 10^{51} ergs is only ~1% of the total. Since the efficiency of transforming energy into thermal and kinetic forms is probably determined randomly by unique initial conditions, assumption (3) may be only roughly correct. Assumption (2) is related to complex issues involving the degree of thermalization of the kinetic energy in the shock, the extent of the equilibrium between ion and electron energies, and the equilibration of the electron velocity distribution and ionization non-equilibrium. Unfortunately these areas are directly related to poorly determined SN and strong shock theory, and major questions remain unanswered.

APPLICATION

Fortunately we may conduct an empirical test of the validity of the "Sedov" distance as expressed in equation (1). To do this we apply equation (1) to existing X-ray data from both the literature and the ROSAT archives to shell-type SNRs with thermal spectra for which distances are available from other sources. A comparison between the X-ray determined Sedov distance (D_s) and the independent distance estimate will indicate how reliably the method works. This relationship is summarize in Figure 1, which shows a good correlation for both Galactic and extragalactic SNRs. We are extending this test as more ROSAT SNR observations become available.

While the correlation shown in Figure 1 works well for most sources, it is noteworthy that it clearly fails in some cases. G18.9-1.1 is a good example of this. Aschenbach et al. (1991) were able to use an independent distance estimate of ~2 kpc from HI observations in order to calculate $\varepsilon_o=0.05$. The assumption of $\varepsilon_o=1$ therefore places the SNR nearly 4 times farther away, an error due to the unusually low explosion energy. It is obviously important to determine if there exist other observational diagnostics which may be used to warn against reliance on the X-ray determined distance for such "pathological" cases. (In addition to the ROSAT archives, references for the data in Fig. 1 are: Aschenbach et al. 1991; Gordon et al. 1993; Hwang & Markert 1994; Ku et al. 1984; Pfeffermann, Aschenbach & Predehl 1991; Saken et al. 1994; Seward & Harnden 1993.)

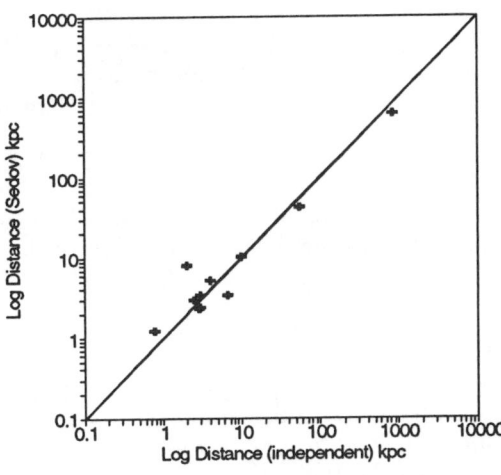

Fig. 1: "Sedov" X-ray distance vs. distance derived by other means for shell-type SNRs.

SUMMARY

We have conducted an empirical test of the ROSAT determined "Sedov" distances for SNRs based on an assumed initial kinetic energy of 10^{51} ergs on a sample of shell-type remnants with independent distance estimates. Our initial results indicate that the X-ray determined Sedov distance is a potentially powerful tool for improving distance estimates to the large number of shell-type SNRs with poorly determined distances which fall within the ROSAT all sky survey. While several key assumptions cannot be rigorously justified on theoretical grounds at this time, the method appears to offer a useful distance estimate in the absence of other data.

While the method appears to work well for most Galactic and extragalactic SNRs, it clearly fails in some cases. As we develop our database further, we hope to determine if there are observable characteristics which may be used to identify those pathological cases where the technique is inapplicable.

REFERENCES

Aschenbach, B., Brinkmann, W., Pfeffermann, E., Fürst, E., and Reich, W., 1991, Astron. Astrophys., 246, L32.
Gordon, S.M., Kirshner, R.P., Duric, N., and Long, K.S. 1993, Ap.J., 418, 743.
Green, D.A. 1984, M.N.R.A.S., 209, 449.
Hwang, U. and Markert, T.H. in these proceedings.
Kassim, N.E., Hertz, P., and Weiler, K.W. 1993, Ap.J. (in press).
Ku, W. H.-M., Kahn, S.M., Pisarski, R., and Long, K.S. 1984, Ap.J. 278,615.
McKee, C.F. 1982, *Supernovae: A Survey of Current Research*, ed. M.J. Rees and R.J. Stoneham (Dordrecht: D. Reidel), p 433.
Nomoto, K., Thielemann, F.K., Yokoi, K. 1984, Ap.J., 286, 644.
Pfeffermann, E., Aschenbach, B., and Predehl, P. 1991, Astron. Astrophys., 246, L28.
Saken, J.M., Long, K.S., Blair, W.P., Winkler, P.F., and DeChristopher, B.M. 1994, in these proceedings.
Seward, F. D. and Harnden, F.R. 1993 Ap.J. (in press).
Tucker, W.H. and Koren, M. 1971, Ap.J., 168, 283; also see erratum, Ap.J., 170, 62.

X-RAY AND RADIO EMISSION FROM W49B AND OTHER SUPERNOVA REMNANTS

John R. Dickel, Rosa Murphy, You-Hua Chu, and Daniel L. Goscha
Univ. of Illinois Astronomy Bldng., 1002 W. Green St. Urbana IL 61801
E-mail: johnd@sirius.astro.uiuc.edu

ABSTRACT

Comparison of x-ray and radio images of SNRs provides detailed information on the mechanisms responsible for the emission and on the evolution of the remnants.

1. Introduction

X-ray observations of supernova remnants (SNRs) delineate shock-heated gas while radio observations delineate the relativisitic particles and magnetic fields. The two forms of emission should be related, however, because the shocks that heat the thermal gas to x-ray emitting temperatures are also responsible for the acceleration of the relativistic electrons which produce the synchrotron radiation. Statistical studies indeed show a rough linear correlation between the x-ray and radio surface brightnesses of SNRs although individual remnants show large deviations from the mean relation (Berkhuijsen 1986; Dickel, Norton, and Gensheimer 1990). In addition, there can be very significant variations in the x-ray/radio brightness ratio across individual remnants.

These variations may have many causes, including:
1. Variable extinction which can affect the observed x-ray brightness
2. Different circumstellar densities which affect:
    a) shock speed - inversely dependent on density - which controls:
    temperature for x-ray brightness
    relativistic particle acceleration and magnetic field amplification for the synchrotron radiation
    b) x-ray emissivity - proportional to density^2
3. Strength and direction of ambient magnetic field for the radio synchrotron emission
4. Clumpiness of surroundings which affects the turbulent amplification for the synchrotron emission processes
5. Explosion energy -
    a) x-ray luminosity approximately proportional to E
    b) radio luminosity approximately proportional to $E^{7/4}$
6. Evolution -
    a) The initial x-ray emission arises in a shell between the forward shock and a reverse shock going back into the ejecta. Then, with time, the shock-heated ejected gas and engulfed clouds which have evaporated can expand to fill the entire volume of the SNR cavity. Eventually the gas in the center will cool and fade due to adiabatic work of expansion.
    b) The initial radio synchrotron emission is enhanced by turbulence in the shell. Next, the dense filaments responsible for the brightest radio emission will undergo thermally unstable compression and then evaporate in the hot surrounding gas so they will not be seen in the

interior of the SNR. The enhanced magnetic field in the outer part of the shell will also inhibit the movement of the radio emitting material back into the interior.
7. Presence of a compact central object which can power a filled volume of non-thermal emission at both x-ray and radio wavelengths.

Investigation of these various possibilities requires high resolution multi-wavelength comparisons of a large selection of SNRs, preferably at known distances. This project is being undertaken using both HRI and PSPC observations from *ROSAT* together with high-resolution radio observations. Carefully chosen Galactic and Large Magellanic Cloud SNRs of different ages in a variety of environments are being used in this study to separate the individual effects. Comparison of global properties as well as variations in the relative x-ray and radio brightnesses across individual SNRs allows determination of the detailed physics of the emission processes in both wavelength regimes plus assessment of the interaction with the surroundings as a remnant expands and evolves.

2. Results on W49B

The remnant W49B shows characteristics indicative of its evolutionary age. This SNR, at a distance of 10 kpc, has a diameter of 12 pc and is probably in the adolescent phase of its lifetime, just entering the Sedov blast-wave stage. Figure 1 shows a grey-scale image of a 33.7 ksec exposure with the *ROSAT* HRI and superimposed contours from a *VLA* radio image (courtesy of S. Reynolds and D. Moffett). The x-ray data have been convolved with a 15-arcsec Gaussian for improved signal-to-noise. Some x-ray emission is seen throughout the remnant but most of it comes from a very bright x-ray core in contrast to the radio emission from a complex shell with no central component. Only one small part of the shell, on the southeast corner, is prominent in x-rays. Full-resolution (6 arcsec) cuts across the bright central region of W49B show that the emission is extended and there is no point source above a 3-sigma level of 4 counts/beam. This result, combined with the thermal nature of the spectrum with a temperature of 2 keV deduced by Smith et al. (1985) from *EXOSAT* ME and GSPC observations, confirms the conclusion that the interior x-ray emission is entirely from hot shocked gas. Most of the gas in the shell heated by the forward shock has already cooled while the interior gas heated by the reverse shock and expanding back into the central region is still hot. The relatively soft response of the *ROSAT* HRI does pick up some parts of the cool shell.

3. Conclusion

There is faint x-ray emission from all parts of the SNR W49B but most of it is concentrated near the center of the remnant unlike the radio emission which arises in a shell near the perifery. This structure indicates that this SNR is in the adolescent phase of its lifetime.

4. Acknowledgements

This research was supported by NASA Grant NAG5-2018. We thank S. Reynolds and D. Moffett for the new radio image.

References

Berkhuijsen, E. M. 1986, A&A, 166, 257
Dickel, J. R., Norton, L., & Gensheimer, P. 1990, in High Resolution X-ray Spectroscopy of Cosmic Plasmas, ed. P. Gorenstein & M. Zombeck (Cambridge: Cambridge U. Press), 168
Smith, A., Jones, L. R., Peacock, A., & Pye, J. P. 1985, ApJ, 296, 469

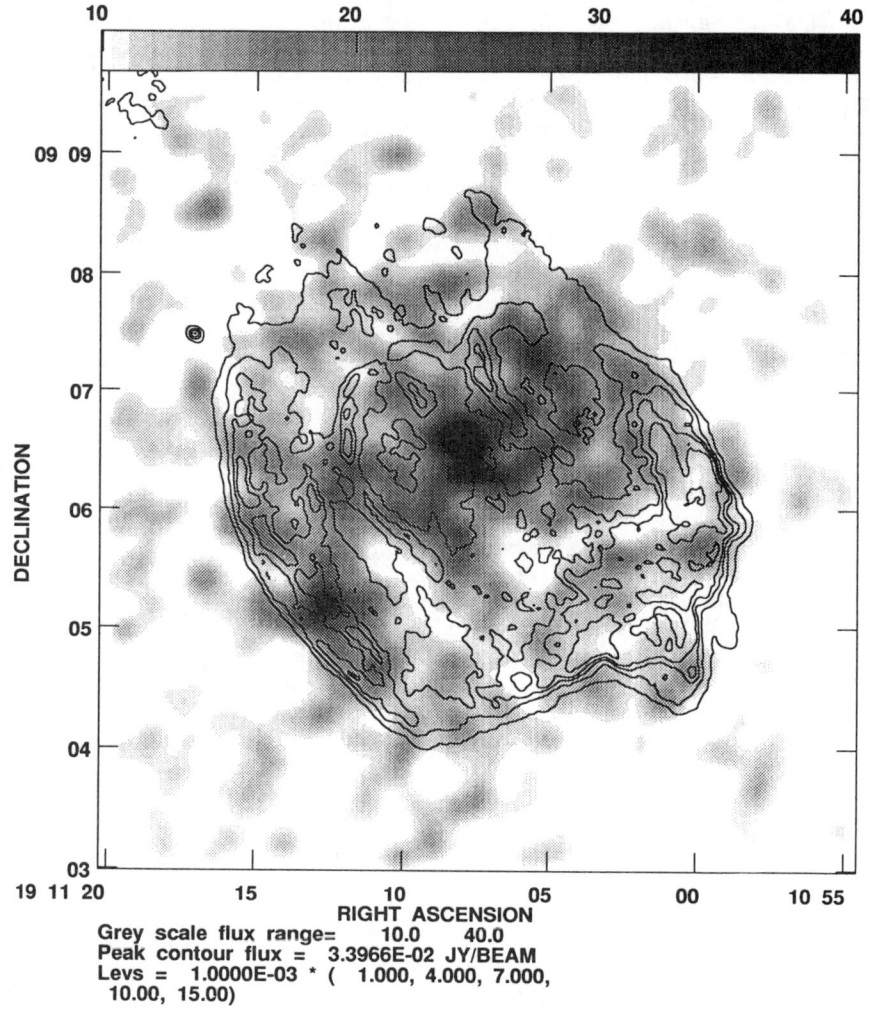

Figure 1. Grey-scale of the *ROSAT* HRI image of the supernova remnant W49B. The image has been convolved to a 15-arcsec point spread function. The units on the greyscale bar are counts/beam. The contours represent radio emission at a wavelength of 6.1 cm observed with the *VLA* by S. Reynolds and D. Moffett. The half-power beamwidth was 4 arcsec.

Normal Galaxies

STAR FORMATION TRIGGERED BY GALAXY INTERACTIONS – NGC 1792 / NGC 1808

Michael Dahlem
Space Telescope Science Institute, 3700 San Martin Dr., Baltimore MD 21218
dahlem@stsci.edu

ABSTRACT

Both the level and the location of star formation (SF) in galaxy disks can be influenced by external forces, like e.g. gravitational interactions. In the interacting galaxies NGC 1792 and NGC 1808 the highest level of SF is observed at those galactocentric radii, r, where gas can accrete most easily. In NGC 1808, a starburst occurs in a ring of massive HII-regions at its inner Lindblad resonance ($r \simeq 450$ pc), while in NGC 1792 the SF is concentrated in very luminous HII-regions at the galactic turnover of rotation ($r = 3.2$ kpc). Our observations suggest that, due to an interaction in the past, radial inward gas flows occured in both galaxies, which led to gas accretion at those radii and subsequently to the observed high-level SF.

1. INTRODUCTION

Statistical studies of large samples of interacting/non-interacting and barred/non-barred galaxies show an on average increased nuclear star formation (SF) activity of interacting and barred galaxies, as observed e.g. in the radio continuum and far-infrared. The nuclear emission of these galaxies is on average 5 times as strong as that of non-barred or field galaxies (e.g. Hummel et al. 1990 and references therein). Such enhanced activity is generally believed to be fed by gas accreting radially inward towards the centers of galaxies due a non-axisymmetric disturbance of the gravitational potential (e.g. Combes 1987). She found that gas moving inward will accrete near the inner Lindblad resonance (ILR), if present, on timescales of $\sim 10^8$ yr, typically. In the absence of a bar (and thus an ILR), gas can also accrete at the turnover radius of galactic rotation (Lesch et al. 1990). Building up a sufficiently high gas density at a galactic turnover of rotation, however, requires a longer time, up to 10^9 yr (B. Elmegreen, priv. comm.).

In order to further quantify statistical results and check model calculations as given above, selected pairs of galaxies have to be studied in detail. Here, our observations and results on NGC 1808/1792 are presented. Although relatively undisturbed at first sight, both the gaseous and stellar disks of NGC 1808 and NGC 1792 show slight disturbances, which hint at an interaction that happened some 10^8 yr ago, probably with a distant passage during the closest encounter (Dahlem et al. 1990; Dahlem 1992; Koribalski et al. 1993).

2. MULTI-FREQUENCY OBSERVATIONS

We observed NGC 1792 and NGC 1808 in several wavebands in order to investigate the gas kinematics, the distribution of SF regions, the level of the SF, and other related quantities. The results presented here were obtained using primarily Hα images, radio continuum maps obtained with the VLA, and two deep ROSAT HRI pointings. The latter have integration times of about 29 ksec (NGC 1808) and 26 ksec (NGC 1792). The data acquisition and reduction will be described in detail by Dahlem et al. (1994a,b; subm. to Ap. J.).

3. RESULTS AND DISCUSSION

In NGC 1808 a fast-rotating gas ring with a radius of about 450 pc has been detected in CO emission and HI absorption (Dahlem et al. 1990, Koribalski et al. 1993), in which a starburst is going on. In the ring, a number of radio continuum knots were found by Saikia et al. (1990), which these authors identify as young supernova remnants and compact HII-regions. This ring is located at the ILR of NGC 1808. High-resolution HI data provide evidence for inward radial gas motions in the outer disk, feeding gas into the central region (Koribalski & Dettmar 1993), which might have been caused by a bar or a gravitational interaction. Both processes are equally viable and possibly related to each other, though at present not distinguishable.

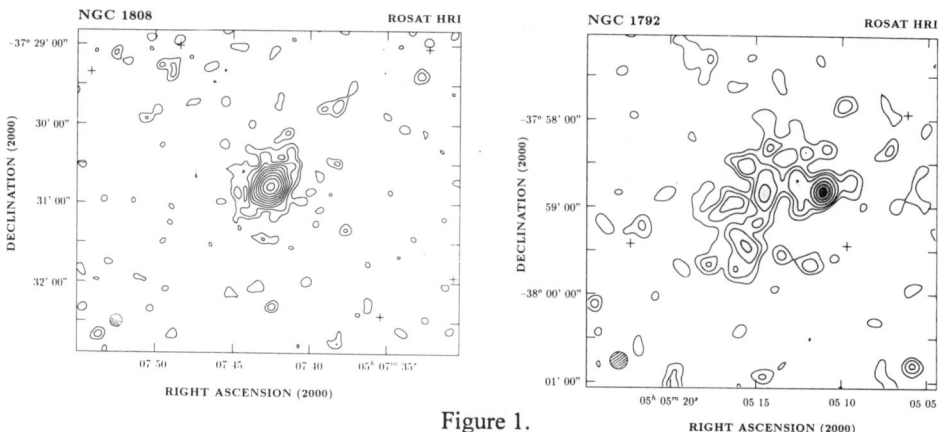

Figure 1.

We used ROSAT to investigate the soft X-ray emission of the starburst nucleus in NGC 1808. The emission distribution, as observed with the HRI, is shown in Fig. 1, left panel. The central emission complex is marginally resolved at a resolution of about 9". A formal deconvolution, assuming approximately gaussian shapes of both the emission distribution and the point-spread function, leads to an extent of the starburst nucleus of about 14"×9". This corresponds almost exactly to the extent of the gas ring and the distribution of the radio knots (see above). Thus, it is clear that the soft X-ray emission of NGC 1808 is dominated by the HII-regions and/or supernovae and their remnants in the starburst. The main emission source is probably hot gas, since a Raymond-Smith type spectrum fits the ROSAT PSPC data better than a power law (Junkes et al. 1994; in prep.). The integrated soft X-ray luminosity of NGC 1808 in the ROSAT band (0.1 to 2.4 keV) is about $L_x = 1.2 \times 10^{41}$ erg s^{-1} (Dahlem et al. 1994 subm. to Ap.J.).

Above the starburst conspicuous dust filaments extend into the halo up to z-distances of about 3 kpc. We found soft X-ray emission of hot gas associated with the outflow, north and NE of the centre. The integrated soft X-ray luminosity of the outflowing hot gas is $L_x \simeq 10^{39}$ erg s^{-1}.

The right panel in Fig. 1 shows our ROSAT HRI image of NGC 1792 (FWHM =

12"). A number of compact, mostly unresolved sources has been detected in the inner disk, which partly correlate with the locations of the prominent HII-regions. Spectral information from our PSPC data suggests that most of these sources are high-mass X-ray binaries. This is consistent with their high soft X-ray luminosities, which lie in the range of 5×10^{38} erg s^{-1} to 3×10^{39} erg s^{-1} (i.e. the Eddington luminosity of 4 to 22 $M_\odot$ stars).

For both galaxies the age of the on-going SF can be determined relatively well. In the case of NGC 1808 the presence of dust and neutral gas at z-distances of about 3 kpc yields a lower limit for the age of the starburst. Adopting a z-velocity of the outflowing matter of about 500 km s^{-1}, as measured in optical absorption lines, e.g. by Forbes et al. (1992) and Phillips (1993), the dynamical timescale for this material to reach z = 3 kpc is about 6.6×10^6 yr. The relatively small z-extent of the soft X-ray emission from the halo provides an upper limit of about 10^7 yr. In NGC 1792 we found a very good correlation of Hα and nonthermal radio continuum emission for most of the prominent HII-regions. This indicates that in these SF regions SNe have already gone off; the minimum age of the SF there is then on the order of 10^7 yr (the typical lifetime of O-stars of a few times 10^6 yr plus the dynamical time of the expanding cavities), cf. Leitherer et al. (1992). On the other hand, a number of HII-regions, primarily SW of the major axis, do not show up in radio continuum emission. Hence, the SF there (assuming the stellar mass function to be similar to that in the HII-regions emitting radio continuum) has to be younger, $t \leq 10^7$ yr. In NGC 1792, we are thus probably witnessing spatially propagating SF in its disk. Accordingly, the age of the currently on-going SF in both galaxies is similar, being $\sim 10^7$ yr. Taking into account the theoretical results by Combes (1987) on the timescales for the formation of GMCs from which stars can subsequently form at high rates, which is $\sim 10^8$ yr in case of weak perturbations, our results are consistent with a scenario implying that the currently observed high-level SF in NGC 1808 and NGC 1792 has been triggered by a gravitational interaction.

Acknowledgements: These results were obtained in collaboration with D. Bomans & J.-M. Will (Bonn Univ.), N. Junkes (Kiel Univ.), and G. Hartner (MPE Garching). This work was supported by NASA grant no. G002-72900.

4. REFERENCES

Combes F. 1987, Proc. NATO Conf. on "Galactic & Extragalactic Star Formation", ed. R. Pudritz
Dahlem M. et al. 1990, *Astr. Ap.*, **240**, 237
Dahlem M. 1992, *Astr. Ap.*, **264**, 483
Forbes D.A. et al. 1992, *M.N.R.A.S.*, **259**, 293
Hummel E. et al. 1990, *Astr. Ap.*, **236**, 333
Koribalski B. et al. 1993 *Astr. Ap.*, **268**, 14
Koribalski B. & Dettmar 1993, *ESO Messenger*, **71**, 37
Leitherer C. et al. 1992, *Ap. J.*, **401**, 596
Lesch H. et al. 1990, *M.N.R.A.S.*, **242**, 194
Phillips A.C. 1993, *Astron. J.*, **105**, 486
Saikia D.J. et al. 1990, *M.N.R.A.S.*, **245**, 397

A ROSAT Observation of the Spiral Galaxy NGC 6946

Eric M. Schlegel
MC 668, NASA Goddard Space Flight Center, Greenbelt, MD 20771

Email ID
eric@heasfs.gsfc.nasa.gov

ABSTRACT

I exceedingly briefly summarize two results of a 37ksec observation on NGC 6946.

1. Introduction

The X-ray study of nearby galaxies essentially started with the *Einstein* satellite (Fabbiano 1989, 1992). X-ray morphology studies using *Einstein* data were possible only for the nearest galaxies (e.g., M31, M33, M51, and the LMC). The mechanism for the X-ray emission for spiral galaxies is consistent with the summed contribution of individual sources, usually SNR, X-ray binaries, stars, and the hot ISM. The hot phase of the galaxy's ISM should appear as a diffuse thermal source. I report here on the *ROSAT* PSPC observation of the nearby, nearly face-on Scd galaxy NGC 6946. The observational details are presented in Schlegel (1993a).

2. Sources

The NGC 6946 image clearly shows approximately 6-10 sources (depending upon the chosen background level). Source S1 (the brightest) is likely a supernova remnant (Schlegel 1993b, Blair & Fesen 1993). The remaining sources need to be identified. Spectra of the individual sources, except for S1, can not be obtained due to their low count rates. The sources are, however, sufficiently bright to obtain count rates, so a differential and integrated number distribution can be constructed. The *Einstein* observation of NGC 6946 were of insufficient resolution to study these distributions.

Figure 1a (left) shows the differential number distribution (the number of sources per logarithmic bin versus the log of the X-ray luminosity) over the 0.5-2.0 keV band. Four logarithmic bins per decade were used to provide reasonable resolution. The observed distribution appears flat, but small-number statistics clearly dominate. From Trinchieri *et al.* (1990), the NGC 6946 result is comparable in sensitivity to the *Einstein* M33 distribution. The most striking difference is the complete lack of sources in the $10^{38.5-39.3}$ region in NGC 6946. A gap was also suggested by Trinchieri *et al.* in the M101 *Einstein* data, but the statistics were too poor to be certain of its presence. The lines in Figure 1a are not fits to the data but lines of slope -0.6 and -1.0. This range is expected if a universal power law distribution for X-ray binaries exists. The observed distribution is clearly flatter than a -1.0 power law, but is consistent with a slope of -0.6 within the uncertainties. Figure 1b (right) shows the integrated number distribution for the NGC 6946 sources (solid line) and the M31 *ROSAT* HRI sources (dotted line) (Primini *et al.* (1993)). The M31 distribution shows a turnover, at $\sim 10^{37}$ ergs s^{-1}. Most systematic, potential causes of the turnover can be eliminated (source confusion, locally high N_H, systematic under-estimates of L_X, etc.). The distributions for the 2 galaxies are quite similar in the small region of overlap. If source S1 did not exist, however, the two distributions would most likely be used to illustrate the overall similarity of the source distribution present in spiral galaxies.

3. The Overall Emission

The data were filtered in energy, creating an image for each of seven bands (defined approximately as in Snowden et al. 1994) and corresponding background images. The background images were averages of four sub-images obtained north, south, east, and west of the galaxy frame, offset by about 8' from the center of NGC 6946. Each band was smoothed by a Gaussian with a scale parameter appropriate to the point spread function at the mean energy of the band. Radial profiles were calculated for each frame. The radial profiles were used to verify the location and level of the background. The background images were then subtracted from the source images. The resulting images, presented as a mosaic and scaled to a common value, are shown in Figure 2.

While the image in Figure 2 includes the effective area of the PSPC, we can still conclude that the diffuse gas appears to exist at a temperature of about 0.8keV. Certainly, if the gas were emitting at a considerably lower temperature, the effective area is still sufficiently large that we would likely see some overall structure to the softest photons. Instead, we see what are likely source and background fluctuations (plus source 1 (Schlegel 1993a)). The images appear only slightly different when each frame is scaled optimally.

4. References

Blair, W. & Fesen, R. 1993, submitted to ApJ (Letters)
Fabbiano, G. 1988, ApJ, 333, 672
Fabbiano, G. 1989, ARAA, 27 87
Fabbiano, G. 1992, in *X-ray Binaries*, ed. W. Lewin, J. van Paradijs, & E. van den Heuvel, (Cambridge: Cambridge Univ. Pr.), 000
Primini, F., Forman, W., & Jones, C. 1993, ApJ, 410, 615
Schlegel, Eric M. 1993a, submitted to ApJ
Schlegel, Eric M. 1993b, submitted to ApJ (Letters)
Trinchieri, G., Fabbiano, G., & Romaine, S. 1990, ApJ, 356, 110

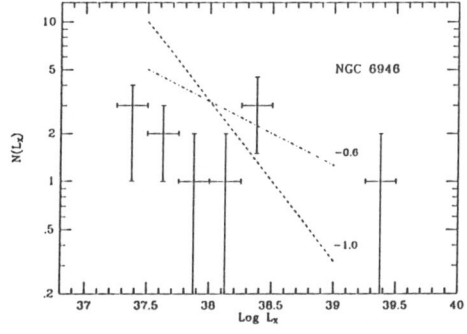
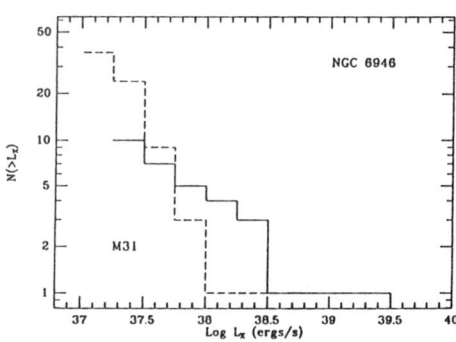

Figure 1.

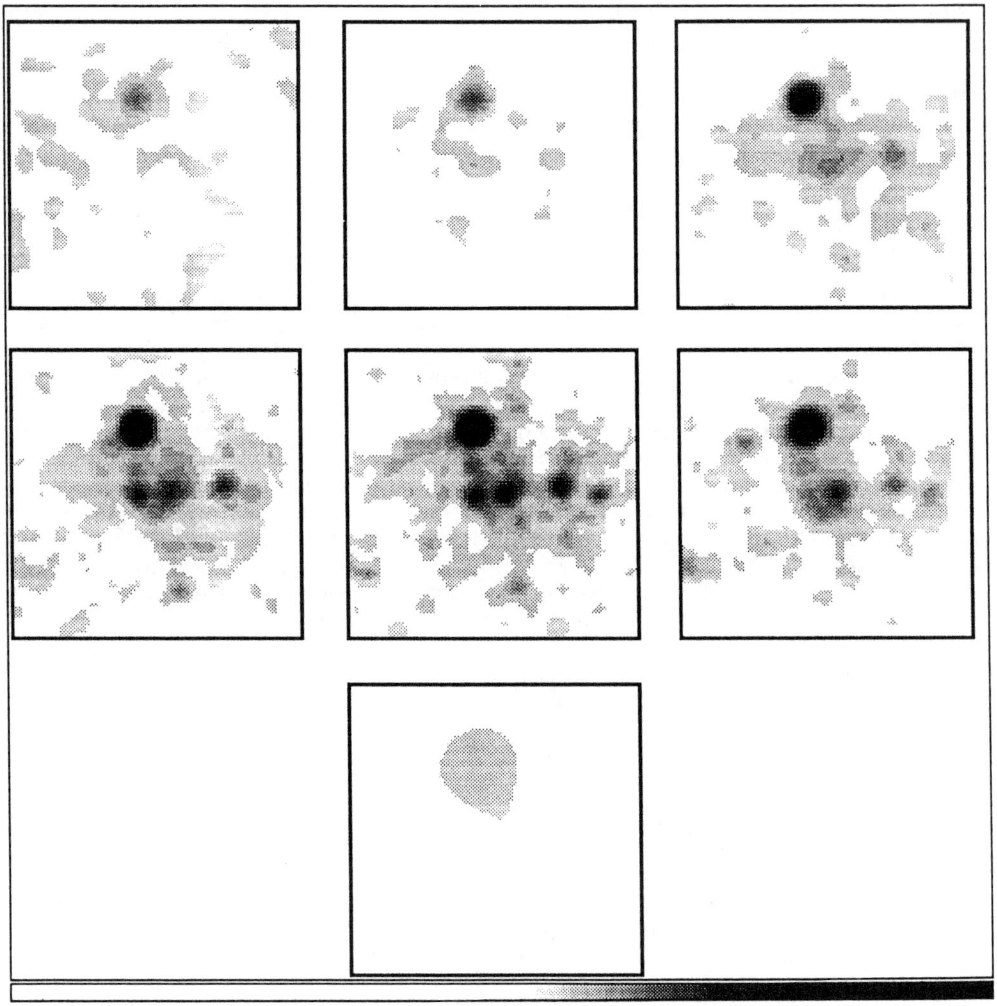

Figure 2.

A Deep X-ray Image of M33

Knox S. Long
Space Telescope Science Institute, 3700 San Martin Dr., Baltimore, MD 21218

Shawn M. Gordon
Center for Astrophysics, 60 Garden St., Cambridge, MA 02138

William P. Blair
Dept. of Physics and Astronomy, Johns Hopkins Univ., Baltimore, MD 21218

Philip A. Charles
Dept. of Astrophysics, Oxford Univ., Oxford OX1 3RH, United Kingdom

Email IDs
long@stsci.edu; sgordon@cfanewton.harvard.edu;
wpb@pha.jhu.edu;pac@oxvad.dnet.nasa.gov

ABSTRACT

A 50.4 ksec PSPC image of the nearby spiral galaxy M33 reveals 35 sources within 15 arcmin of the nucleus brighter than 1.3×10^{36} ergs s^{-1}. The bright sources that had been detected with Einstein in the same region are still visible. Several ROSAT sources are time variable, but the brightest source, the nuclear source, appears constant over the duration of our observations. The bulk of the sources in the galaxy are associated with Pop I tracers, and both the northern and the southern spiral arms of M33 are readily apparent in the image. There is additional diffuse (or unresolved source) emission throughout the inner portions of M33. There are 12 sources in the image that lie within 30 arcsec of optically-identified supernova remnants in M33. The spectra of these sources are soft compared to other sources of comparable brightness, and therefore it is likely that most of these sources are supernova remnants.

1. Introduction

M33 is one of the two nearest spiral galaxies outside the Milky Way. Because of its proximity (720 kpc; de Vaucouleurs 1978), relatively low inclination angle (57°, Searle 1971) and well-defined open spiral arms, M33 is an ideal galaxy for studying the X-ray properties of late-type spiral galaxies, and, as a result, both imaging proportional counter (IPC) and high resolution imager (HRI) studies of M33 were undertaken using the Einstein Observatory (Long et al. 1981; Markert et al. 1983; Trinchieri, Fabbiano, & Peres 1988). These studies revealed a dominant nuclear source, the brightest source in the Local Group with $L_x \sim 10^{39}$ ergs s^{-1}, and 14 other sources $L_x \gtrsim \sim 10^{37}$ ergs s^{-1} thought to be associated with M33. Most of the discrete sources are thought to be compact X-ray binaries, including one that is an eclipsing system with a 1.787 day period (Peres et al. 1989).

2. Observations

The ROSAT PSPC was pointed at M33 for a total of 50.4 ksec in the period

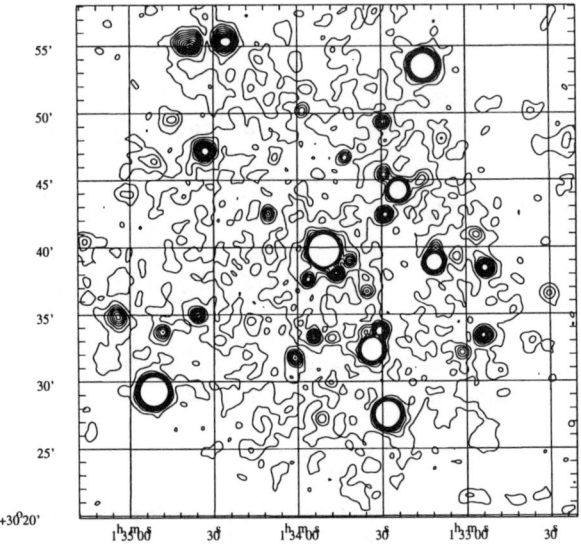

Figure 1: The ROSAT PSPC image of M33

between 1991 July and 1993 January. There were three blocks of observations. The first, which took place in 1991 July, was obtained during the so-called reduced pointing mode, when the nodding of the telescope which is carried out to smooth out the effects of the window support mesh on source counting rates was reduced. The other two observations in 1992 August and 1993 January were carried out under normal conditions.

A contour plot of the summed X-ray image of M33 is shown in Fig. 1. There are evidently a large number of point sources, dominated by the bright nuclear source that was observed with Einstein. From a comparison of the X-ray image and Hα images of M33, it is clear that many of the sources in M33 are associated with H II regions in the galaxy and that there is unresolved emission associated with both the southern and northern spiral arms of the galaxy.

3. Point Sources

The X-ray emission from M33 is complex, and therefore the identification of point sources is not straightforward. Based on the standard data products and an inspection of the images, we identified a total of 59 potential point sources. In order to evaluate the statistical significance of each source, we found the excess counts at each source position, and rejected any sources detected at less than 3 σ. A total of 48 sources survived, 35 of which lie within 15 arcmin of the nucleus of M33. The faintest source in our sample has a 0.1-2.4 keV luminosity of 1.3×10^{36} ergs s^{-1} assuming a power law spectrum $dF/dE \propto E^{-1}$ and log $N_H = 21$. The ROSAT luminosity function

N(>L) closely resembles the luminosity function obtained from the Einstein sources. The sources that were detected include all of the previously detected sources except X-9c, X-12, X-13, and X-15. Except for X-12, an HRI source, the other missing Einstein sources were all detected by Einstein at less than 3 σ.

The nuclear source contributes about 70% of the luminosity of the galaxy. The hard spectrum is similar to that observed with Einstein and is well fit in terms of a bremsstrahlung spectrum with kT $\geq$ 1.8 keV and log N_H $\sim$ 21.1. There are no obvious variations in the count rate of $\sim 0.7\,s^{-1}$ over the period of our observations.

4. Diffuse Emission

There is diffuse emission from the galaxy. The azimuthally averaged diffuse flux declines between 2 and 10 arcmin from the nucleus from $6 \times 10^{-4}\,s^{-1}\,arcmin^{-2}$ to about $10^{-4}\,s^{-1}\,arcmin^{-2}$. According to Trinchieri et al. (1988), the diffuse emission in M33 is softer than the emission from the bright point sources in M33. A comparison between the summed spectra of all of faint sources ($\leq$ 100 excess counts) and the spectrum extracted from the region between 4 and 10 arcmin (omitting all the regions with point sources) dramatically confirms their conclusion. For the diffuse X-ray emission, thermal bremsstrahlung fits yield log N_H of 20.6 and kT of 0.4 keV. For the sum of the faint sources, fits yield log N_H of 20.8 and kT of 1.1 keV. This argues strongly that the diffuse X-ray emission in M33 is not due to faint X-ray binaries.

5. Supernova Remnants

There are now 72 confirmed SNRs in M33 that have been identified by a combination of optical and radio techniques (Long et al. 1990; Gordon et al. 1993). Of the 35 sources within 15 arcmin of the nucleus of M33, 12 lie within 30 arcsec of known SNRs. Some, but not most, of these may be chance coincidences. If we assume that the 35 X-ray sources and the 72 SNRs are randomly distributed within 15 arcmin of the nucleus, then 3 chance coincidences would be expected. We have constructed hardness ratios (H+S)/(H-S) where H and S are net counts above and below 1.0 keV for the sources in M33. The sources that lie near SNRs are clearly softer than the the bulk of the sources. This argues strongly that the bulk of the X-ray sources that are spatially coincident with known SNRs are indeed SNRs. It also suggests that some of the other soft sources are likely to be SNRs as well.

This work is supported by NASA grant NAG 5-1539 to the STScI.

6. References

de Vaucouleurs, G. 1978, ApJ, 224, 710
Gordon, S. M., Kirshner, R. P., Duric, N., & Long, K. S. 1994, ApJ, 418, 743
Long, K. S., Blair, W. P., Kirshner, R. P., & Winkler, P. F. 1990, ApJS, 72, 61
Long, K. S., D'Odorico, S., Charles, P. A., & Dopita, M. A. 1981, ApJ, 246, L61
Markert, T. H., & Rallis, A. D. 1983, ApJ, 257, 571
Peres, G., Reale, F., Collura, A., Fabbiano, G. 1989, ApJ, 336, 140
Searle, L. 1971, ApJ, 168, 327
Trinchieri, G., Fabbiano, G., & Peres, G. 1988, ApJ, 325, 531

X-ray Emission from Giant H II Regions in M101

Rosa Murphy and You-Hua Chu
University of Illinois at Urbana-Champaign, Urbana, IL 61801

Email ID
rosanina@sirius.astro.uiuc.edu

ABSTRACT

We have examined the archival *ROSAT* Position Sensitive Proportional Counter and the High Resolution Imager images of the galaxy M101 to study the x-ray properties of its giant H II regions. All five giant H II regions, NGC 5447, 5455, 5461, 5462 and 5471, show x-ray emission. By observing their basic properties, such as size, hardness ratio and estimated luminosity, and by comparisons with better-known regions in the LMC, we suggest possible mechanisms for the sources of the x-ray emission.

1. Introduction

X-rays associated with the giant H II regions in M101 were first detected using the *Einstein Observatory*'s Imaging Proportional Counter (Trinchieri *et al.* 1990). The current work is part of an extended study of giant H II regions, which includes observations on the 30 Doradus region and other regions in Local Group galaxies. M101 is an excellent galaxy to study because it contains a number of giant H II regions, and is a face-on spiral so that these regions are very visible and extinction within M101 is minimized. We have adopted the figure of 6 Mpc for M101's distance.

2. Data and Analysis

M101 was observed with the *ROSAT* Position Sensitive Proportional Counter (PSPC), without the boron filter. The observation of M101 (ROSAT sequence number WP600108) was originally requested by Dr. S. Snowden. The PSPC exposure time was 34.5 ksec. A short exposure with the *ROSAT* High Resolution Imager (HRI) was also available. The image (WH600092) was originally requested by Dr. W. Pietsch and had an exposure time of 18.4 ksec. The PSPC data showed a diffuse central component and a number of individual sources distributed throughout the disk of M101. Five of the individual sources are coincident with giant H II regions; the others are either local or foreground sources.

We analyzed the data using PROS. Three of the giant H II regions appear extended in the PSPC image. These three are not detected in the HRI exposure; however, the emission of the other two regions appears to be a combination of an extended source with a strong point source. To get luminosity and spectral information, we analyzed the PSPC counts as a whole and in a series of energy bins. For each object we computed a "hardness ratio" equal to the difference divided by the sum of the (1 to 2.5) keV and (0.11 to 1) keV counts. We fit Raymond-Smith emission models to the spectra of objects with over 100 counts, with N_H as a free parameter. We estimated the source luminosity from the fitted model. We also fit Raymond-Smith models with N_H as a fixed parameter scaled from the values of A_V provided by Garnett (1993). Additionally, we obtained luminosity estimates from the observed count rates, using conversion factors given in Figure 10.3 of the ROSAT Mission Description. We assumed a temperature of 5×10^6 K and used the values of A_V provided by Garnett (1993). We also compared the x-ray emission from the giant H II regions of M101 to that of well-studied objects in the LMC: the 30 Doradus area (RP500131) and the N44 area (RP500093), as analyzed in the same manner.

3. Giant H II Regions and Comparison with LMC Sources

We assume that x-ray sources which appear coincident with giant H II regions are actually located within said regions, because the probability of coincidental superposition of five of the x-ray sources with the five giant H II regions is of order 10^{-8}. The x-ray characteristics of the regions are shown in Tables 1 and 2.

Table 1: Giant H II Regions in M101 - Observed Properties [1]

Object	RA (2000) hh mm ss	Dec (2000) dd mm ss	PSPC Counts	Hardness Ratio	X-ray Size	H α Size
NGC 5447	14 02 27	54 16 20	195 ± 21	0.21 ± 0.11	25″	55″× 16″
NGC 5455	14 03 00	54 14 19	60 ± 24	-0.81 ± 0.42	50″	27″
NGC 5461	14 03 39	54 18 55	136 ± 39	-0.22 ± 0.27	45″	66″× 25″
NGC 5462	14 03 53	54 21 49	150 ± 23	-0.41 ± 0.16	50″	90″× 34″
NGC 5471	14 04 29	54 23 40	88 ± 19	-0.79 ± 0.27	25″	25″

Table 2: Giant H II Regions in M101 - Derived Properties [2]

Object	N_H as free parameter			N_H as fixed parameter			L_x from counts	
	T	log N_H	L_x	T	log N_H	L_x	log N_H	L_x
NGC 5447	0.61	22.0	32.4	1.16	21.00	3.2	21.0	1.0
NGC 5455	–	–	–	0.28	20.89	0.74	21.0	0.3
NGC 5461	1.26	20.1	1.4	1.13	21.12	1.7	21.0	0.7
NGC 5462	1.00	20.0	1.1	0.92	20.81	1.2	21.0	0.8
NGC 5471	–	–	–	0.31	20.29	0.88	20.5	0.4

On the HRI image, NGC 5447 appears to have a point source embedded in diffuse emission. The hardness of the spectrum may be explained by a dominating energetic x-ray binary. Another explanation is that the highly energetic x-rays are produced by a young, hot supernova remnant (SNR) embedded in the H II region. There is too little information on NGC 5455 for us to speculate on the nature of its x-ray producing mechanism; we can only note the presence of extended soft x-ray emission. The x-ray emission of NGC 5461 and NGC 5462 may be due to a diffuse component or multiple sources, or both. The low number of counts makes the Raymond-Smith model fit uncertain. The emission below 0.5 keV may be intrinsic to the source or may contain contamination from the central diffuse x-ray disk of M101, which is very close to both objects. The hardness ratios of these objects are typical of those produced by a hot

---

[1] H α sizes from Israel et al. (1975); NGC 5471 from Chu and Kennicutt (1986)
[2] T in keV, L_x in 10^{38} erg s^{-1}

plasma. However, given the extended natures of these sources we can rule out the possibility that the x-rays from each region are produced by a single, hot object, such as an x-ray binary or young SNR. NGC 5471 has a HRI morphology similar to that of NGC 5447: a point source combined with diffuse emission.

The prototypical giant H II region 30 Doradus is rich in objects useful for comparison. (Chu et al. 1993b). Another useful source is N44, which features a SNR and an x-ray bright superbubble (Chu et al. 1993a). The hardness ratio and luminosity of NGC 5447 put it in the same range as an energetic SNR such as 30 Dor B or N158A, or an x-ray binary less energetic than LMC X-1. NGC 5455 is too faint for much analysis, and its hardness ratio falls well below that of the softest of the comparison objects. As with NGC 5455, NGC 5471's emission is very soft, and faint relative to the other regions. We suggest that these objects may each contain SNRs. NGC 5461 and NGC 5462 are both in the "medium-soft" range; NGC 5461 has a hardness similar to that of the N44 SNR, while NGC 5462 is comparable to the 30 Dor NE and N44 superbubbles. As both of these regions are extended in the PSPC image, it is likely that they contain a number of superbubbles and/or SNRs. The spatial resolution of the PSPC is too low to resolve individual SNRs, superbubbles, or x-ray binaries at a distance of 6 Mpc. The x-ray emission from the giant H II regions is likely to come from a collection of objects, similar to those of the 30 Doradus H II complex.

4. Conclusions

The giant H II regions NGC 5447 and NGC 5471 each contain a point source in addition to a diffuse component, while the other three regions, NGC 5455, NGC 5461, and NGC 5462, are dominated by diffuse emission. By comparing the sizes, hardness ratios, and estimated luminosities of M101's giant H II regions to those in which the x-ray emitting sources are known, we suggest possible mechanisms for the x-ray emission from these regions. The 30 Doradus giant H II region and the H II complex N44 in the LMC area are used for comparison. We suggest that the point sources in NGC 5447 and NGC 5471 are either luminous x-ray binaries or energetic SNRs. The diffuse emission in M101's giant H II regions could be due to the combination of the following three types of source: weaker x-ray binaries, isolated SNRs, and superbubbles with SNR activity near the shell wall. It is common that giant H II regions emit bright x-rays; however, for more distant objects it will be increasingly difficult to pinpoint the exact nature of the x-ray emissions.

5. References

Chu, Y-H., Mac Low, M.-M., Garcia-Seguro, G., Wakker, B., and Kennicutt, R.C. 1993a, ApJ, **414**, 213.
Chu, Y-H. et al. 1993b, in preparation.
Chu, Y.-H. and Kennicutt, R.C. 1986, ApJ, **311**, 85.
Garnett, D. 1993, personal communication.
ROSAT Mission Description 1991, NASA publication NRA 91-OSSA-3, Appendix F.
Israel, F.P., Goss, W.M and Allen, R.J. 1975, A & A, **40**, 421.
Trinchieri, G., Fabbiano, G. and Romaine, S. 1990, ApJ, **356**, 110.

ROSAT HRI OBSERVATIONS OF M33

Eric Schulman and Joel N. Bregman
Department of Astronomy, University of Michigan, Ann Arbor, MI 48109-1090
eric@astro.lsa.umich.edu, jbregman@asrto.lsa.umich.edu

ABSTRACT

Our 35 ksec ROSAT HRI observation of M33 reveals 37 X-ray sources stronger than about 2.3σ. Eight of the sources are coincident with supernova remnants, four are coincident with giant HII regions, and three are coincident with HI holes. M33 X-7 is a compact accreting eclipsing binary, similar to binary X-ray sources detected in the Galaxy. Our ROSAT data confirm the binary interpretation and allow us to measure the period to an accuracy of 0.001%. The nuclear source, M33 X-8, is not found to be variable in the ROSAT HRI observations, although it varied as much as 40% between *Einstein* HRI observations.

1. Observations and Data Reduction

We observed M33 for 35 ksec with the High-Resolution Imager (HRI) on ROSAT in 1992 Jan and 1992 Aug. The two observation sections have pointing centers which differ by about 5″ so the August observations were shifted before being merged with the January observations. Sources were found using *ldetect* as well as by determining the count rates at the coordinates of HI holes (Deul & den Hartog 1990) and supernova remnants (Long et al. 1990). We find 37 X-ray sources stronger than about 2.3σ, of which eight are coincident with supernova remnants, four are coincident with giant HII regions, and three are coincident with HI holes. Twelve of the sources were previously detected with *Einstein Observatory* observations. Two *Einstein* sources, M33 X-12 and M33 X-15 (according to the classification of Trinchieri, Fabbiano, & Peres 1988 and Markert & Rallis 1983), were not detected with ROSAT, despite the much higher sensitivity of the observations.

2. Variability Analysis

We studied variability within the ROSAT observations with three independent methods: the Cramer-Smirnov-Von Mises method and the Kolmogorov-Smirnov method (Eadie et al. 1971) which compare the cumulative distribution of photon arrival times with the distribution expected from a constant source, and a modified χ^2 test able to provide binning independent results (Collura et al. 1987). All three tests indicate that M33 X-7 is variable at the 99.999% confidence level. We find no convincing evidence for variability in the other 14 sources that are strong enough to perform the variability analysis upon.

3. An Eclipsing Binary X-Ray Source

M33 X-7 was interpreted to be an eclipsing binary by Peres et al. (1989) based on *Einstein* observations. We confirm this interpretation and are able, because of the 12-year gap between the *Einstein* and ROSAT observations, to determine a much more accurate period, which we find to be 1.78572 ± 0.00001 days. The period we determine for M33 X-7 is very close to that of Her X-1 (Tananbaum et al. 1972) and the low phase lasts about a quarter of the period as in Cen X-3 (Schreier et al. 1972). A more thorough analysis of this source is presented in Schulman et al. (1993).

4. The Nuclear Source

The nuclear source, M33 X-8, does not appear to exhibit X-ray variability. We find the 3σ amplitude upper limit for variability to be 6% on timescales of 8700 seconds, and 17% on timescales of 40 seconds. It is puzzling that the X-ray flux changed by less than 1% between 1992 Jan and 1992 Aug, since it decreased by 40% between 1979 Aug and 1980 Jan, and increased by 20% between 1980 Jan and 1980 Aug (Peres et al. 1989).

The origin of the X-ray emission from M33 X-8 has been a mystery for some time (Markert & Rallis 1983). Its X-ray luminosity of about 10^{39} erg s^{-1} is low for an AGN but quite high for a Galactic X-ray source, although a number of sources outside the Local Group have been found with comparable or larger X-ray luminosities (most recently, Collura et al. (1994) determined the X-ray luminosity of a source in M82 to be at least 5.0×10^{39} erg s^{-1}). The nucleus has no detected 6 cm emission, no hydrogen line emission observed, little or no forbidden line emission, and infrared colors quite unlike those of AGN. Yet the young stars in the nucleus make up only a small fraction of the young population of M33, so that the *a priori* probability of a binary in the nucleus with an X-ray luminosity ten times that of any other source in M33 is small.

We thank Alfonso Collura, Fabio Reale, and Giovanni Peres for their help with the variability analysis. This research was supported by NASA through grant NAGW-2135 and through NASA Graduate Student Researchers Program Fellowship NGT-50901.

References

Collura, A., Maggio, A., Sciortino, S., Serio, S., Vaiana, G. S., & Rosner, R. 1987, ApJ, 315, 340
Collura, A., Reale, F., Schulman, E., & Bregman, J. N. 1994, ApJ (Letters), in press
Deul, E. R., & den Hartog, R. H. 1990, A&A, 229, 362
Eadie, W. T., Drijard, D., James, F., Roos, M., & Sadoulet, B. 1971, Statistical Methods in Experimental Physics (Amsterdam: North-Holland)
Long, K. S., Blair, W. P., Kirshner, R. P, & Winkler, P. F. 1990, ApJS, 72, 61.
Markert, T. H., & Rallis, A. D. 1983, ApJ, 275, 571
Peres, G., Reale, F., Collura, A., & Fabbiano, G. 1989, ApJ, 336, 140
Schreier, E., Levinson, R., Gursky, H., Kellogg, E. M., Tananbaum, H., & Giacconi, R. 1972, ApJ, 172, L79
Schulman, E., Bregman, J. N., Collura, A., Reale, F., & Peres, G. 1993, ApJ, 418, L67
Tananbaum, H., Gursky, H., Kellogg, E. M., Levinson, R., Schreier, E., & Giacconi R. 1972, ApJ, 174, L143
Trinchieri, G., Fabbiano, G., & Peres, G., 1988, ApJ, 325, 531

Galaxies and Hot Gas

Observations of the Antenna-like Interacting Galaxy Pair Arp270: From X-rays to Radio Wavelengths

P. N. Appleton, C. Winrich,
Department of Physics and Astronomy,
Iowa State University, Ames, IA 50011
Internet address: pnapplet@iastate.edu, winrich@iastate.edu

G. Fabbiano
Center for Astrophysics, Smithsonian Astrophysical Observatory,
60 Garden Street, Cambridge, MA 02138
Internet address: pepi@cfa249.harvard.edu

and

P. M. Marcum
Washburn Observatory, Department of Astronomy,
University of Wisconsin, 475 Charter St., Madison WI 5305-1582
Internet address: marcum@madraf.astro.wisc.edu

ABSTRACT

Multi-wavelength observations have been made of the interacting galaxy system Arp 270 (=NGC3395/96). A comparison is made between recently obtained ROSAT PSPC observations and optical and near-infrared images of the system, as well as 21cm HI observations made at the VLA.

1. Introduction

There is now considerable evidence that galaxies with close companions are much more likely to show enhanced star formation rates than those which are isolated (e.g. Kennicutt 1990). It is likely that more than one process contributes to this effect, with mechanisms ranging from the funnelling of gas into galactic centers, to gravitational instabilities in massive central molecular disks or bar formation. However, despite many possibilities, a clear understanding of enhanced nuclear star formation is not yet forthcoming.

In the spirit of further exploration of interacting galaxies, we have made detailed observations of a number of interacting systems at many wavelengths, including recent ROSAT observations of Arp 270 (=NGC 3395/6).

2. Radio and Infrared and Optical Observations

We show in Figure 1 the integrated neutral hydrogen surface density map of the system made with the VLA. Huge HI plumes, reminiscent of the "Antenna" galaxies (NGC4038/9) (van der Hulst 1979) are observed eminating from the pair. The existence of the long plumes combined with their velocity structure suggests that strong tides are operating in the system and that the galaxies have been interacting for approximately 10^8 years. HI surface densities towards the centers of the galaxies are approximately 1.5×10^{21} at cm^{-2} over a VLA synthesized beam area of 30 x 30 arcsecs^2.

We also observed Arp 270 with the 2.1m KPNO telescope at the wavelength of the Hα line. Hα emission is found covering large parts of galaxies and filaments of emission are found between the galaxies and to the south of NGC 3396. K-band near-IR emission is also detected from the filaments from observations made at UKIRT. This suggests that the Hα filaments are associated with underlying star clusters. Making assumptions about the form of the initial mass function, we estimate the star formation rate to be 1.2 $M_\odot$/year (for stars of mass $0.1 < M < 100 M_\odot$) for both galaxies based on the Hα flux. Approximately 1.5×10^4 O5 stars would be required to produce the observed Hα flux.

3. ROSAT PSPC Observation

A 20,000s ROSAT PSPC observation was made of Arp 270 between June 2-4 1993. Figure 2 shows contours of X-ray emission. The two galaxies NGC 3395/6 are clearly detected at the center of the field along with a third unidentified point source (probably a background AGN). The two dwarf galaxies, IC 2608 and UGC 5927, which are seen in the HI map of Figure 1 to the south of Arp 270 are also detected with ROSAT.

Assuming a distance of 22 Mpc for Arp 270 (H_o=75 km s^{-1} Mpc^{-1}) we determine a total X-ray luminosity $L_x(0.1-2.4\text{keV}) = 8.65 \times 10^{39}$ ergs s^{-1}. This is approximately 5 times less than the X-ray luminosity of the "Antenna" as observed by the *Einstein Observatory* by Fabbiano and Trinchieri (1983) of 4.5×10^{40} ergs s^{-1} (after adjusting the latter for an assumed value of $H_o = 75$ km s^{-1} kpc^{-1}).

Spectral fitting of the X-ray emission from Arp 270 is consistent with both a soft and hard component being present. Approximately 10 percent of the flux comes from a soft component (kT = 0.36 keV) and 90 percent from a hard component (kT> 3 keV).

4. The Origin of the Soft Component?

We consider that there are at least four possible sources for the soft X-ray component in Arp 270. These are: a) emission from hot gas associated with massive stars formed in the star forming regions, b) emission associated with many supernova remnants in the disks of the two galaxies, c) hot coronal halo gas from possible nuclear superwind outflows, similar to M82 and, d) contributions from X-ray binary sources within the disks of the galaxies.

We consider a) and c) to be the most likely possibilities. The X-ray luminosity of the Carina nebula has recently been determined by ROSAT to be 2×10^{35} ergs/s (Corcoran, this conference). This nebula contains approximately 10 O and WR stars. The soft X-ray flux of Arp 270 corresponds to the X-ray flux of 4×10^3 Carina nebulae or roughly 4×10^4 O stars. This is similar to the estimated number of O stars responsible for the Hα emission (Section 2).

It is also plausible that the soft X-ray emission could originate in hot gas from supernova explosions. Model fitting to the X-ray spectra suggested a relatively low neutral hydrogen column density to the X-ray source. Since the VLA maps suggest large HI column densities in the disk, then c) rather than b) above is favored. Also the total number of supernova remnants needed in b) is excessively high, even allowing for the high rates of star formation determined in Section 2. Gas blown out of the disk in a manner similar to that seen in

M82 would seem a plausible explanation for the observed soft component, and would be less easily absorbed by the HI in the galaxy. If the volume occupied by the X-ray gas was similar to that of the Hα emission, then approximately 4 x $10^7 M_\odot$ of hot gas would be required. This mass of gas could be created in a few million years if the current rate of star formation was maintained (0.1-1 $M_\odot$ yr^{-1} for stars with $M > 10 M_\odot$).

This work was funded under the NASA-ROSAT grant NAG 5-2385.

5. References

Fabbiano, G. & Trichieri, G. 1983, ApJ (Letts), 266, L5
Kennicutt, R. 1990, in *Paired and Interacting Galaxies*, ed. J. W. Sulentic, W.C. Keel & C. M. Telesco, (NASA Pub. 3098), p269
van der Hulst, J. M. 1979, AA, 71, 131

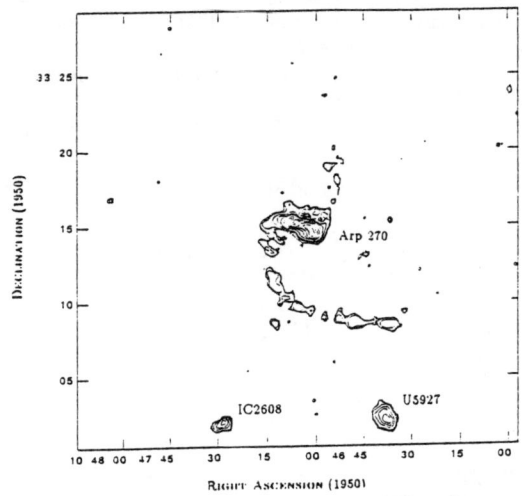

FIG 1. - Integrated HI map showing Arp 270 and two companions.

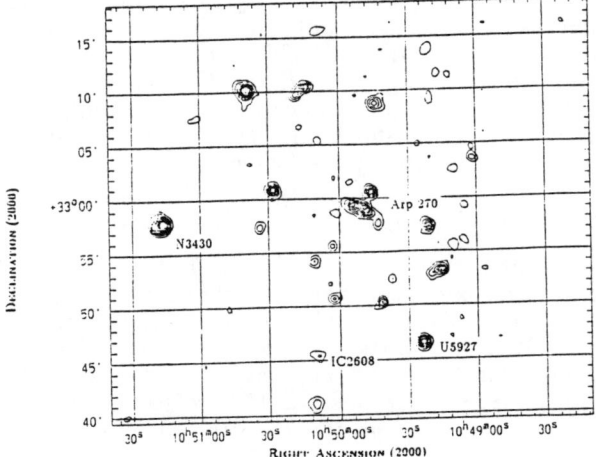

FIG 2. - The center portion of the PSPC field of view.

DARK MATTER IN THE ELLIPTICAL GALAXY NGC 1407

Raymond E. White III, Victor Andersen, & Chris Williamson
Department of Physics & Astronomy, University of Alabama

Email ID
white@merkin.astr.ua.edu

ABSTRACT

The elliptical galaxy NGC1407 is the brightest member of the Eridanus group of galaxies. A 22,000 sec ROSAT PSPC pointing centered on NGC 1407 shows that it is embedded in a flattened distribution of gas (with axial ratio 2:1) extending to at least 120 kpc (25') from the galaxy along the major axis (adopting the surface-brightness-fluctuation distance to NGC 1407 of 16.4 Mpc). The hot gas extends 11-22 times further than the optical effective radius of NGC1407, so it is likely that at least some of the gas belongs to the group rather than NGC1407 itself. Outside a central cooler region, indicative of a cooling flow, the gas is roughly isothermal (perhaps declining slowly outward) with a temperature of ~ 1 keV. The mass-to-light ratio is $M/L_B \approx 70$ within 70 kpc of the galaxy, which implies that $\sim 85\%$ of the enclosed mass is dark.

1. Introduction

The elliptical galaxy NGC1407 is the brightest member of the Eridanus group of galaxies and is centrally located in the subclump known as "Cloud A". Figure 1 shows the distribution on the sky of over 50 Eridanus group members, with NGC 1407 marked. NGC 1407 was chosen because its *Einstein* IPC detection was at a level typical for an individual, bright elliptical galaxy.

2. X-ray Properties

A 22,000 sec ROSAT PSPC pointing centered on NGC 1407 shows that it is embedded in a flattened distribution of gas (axial ratio 2:1) extending to at least 120 kpc (25') from the galaxy along the major axis (adopting the surface-brightness-fluctuation distance of 16.4 Mpc). The hot gas extends 11-22 times further than the optical effective radius of NGC1407, so it is likely that at least some of the gas belongs to the group rather than NGC1407 itself.

The azimuthally averaged X-ray surface brightness profile of the diffuse emission surrounding NGC 1407 is shown in Figure 2. A reasonable fit to the surface brightness profile is provided by the following two component model:

$$S_X(r) = 3.7 \times 10^{-3} \theta^{-2} + 5 \times 10^{-4} \theta^{-0.3} \text{cts s}^{-1} \text{ arcmin}^{-2} \quad (1)$$

where θ is the radius from the center of NGC 1407 in arcmin. The temperature distribution of this extended emission is shown in Figure 3.

Also marked just to the SW of NGC 1407 is the neighboring (at least in projection) elliptical NGC 1400, which was also detected in X-rays. Whether NGC 1400 is physically near NGC 1407 or even bound to the Eridanus group is unclear, since the peculiar velocity of NGC 1400 relative to the group mean exceeds 1000 km s^{-1}, which is much greater than the velocity dispersion of the group (~ 250 km s^{-1}).

3. Mass Distribution

Assuming the gas centered on NGC 1407 is in hydrostatic equilibrium, the distribution of total mass is given by:

$$M(r) = -\frac{kTr}{\mu G}(\Delta_r n + \Delta_r T), \qquad (2)$$

where $\Delta_r n$ and $\Delta_r T$ are logarithmic gradients (slopes) of the gas density and temperature profiles and μ is the mean mass per gas particle. In the outer parts, the gas density goes as $n \propto r^{-0.7}$ and the temperature distribution is consistent with being isothermal (although it may be declining slightly outward - more calibration assessment must be done...), so the total mass within 70 kpc (15′) is $1.8 \times 10^{12}\ M_\odot$. Neglecting the possibly interloping galaxy NGC 1400, the enclosed (blue) luminosity of NGC 1407 is $2.5 \times 10^{10} L_\odot$. Thus the mass-to-light ratio $M/L_B \approx 70$ and roughly 85% of the enclosed mass is dark.

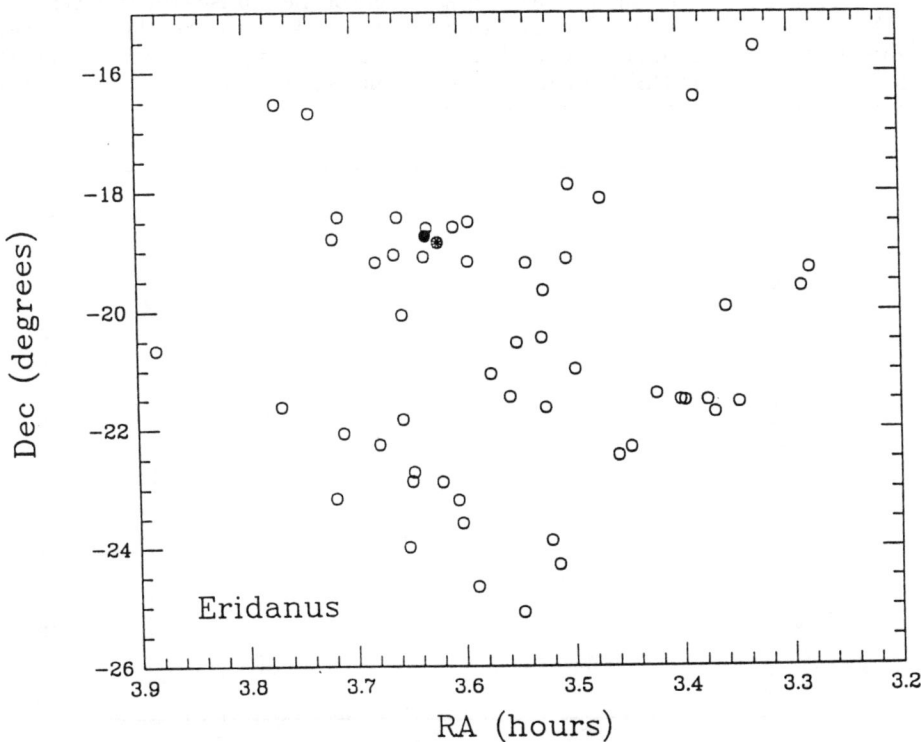

Figure 1 — Distribution of Eridanus Group galaxies on the sky. NGC 1407 is indicated by solid circle, while NGC 1400 is indicated by starred circle.

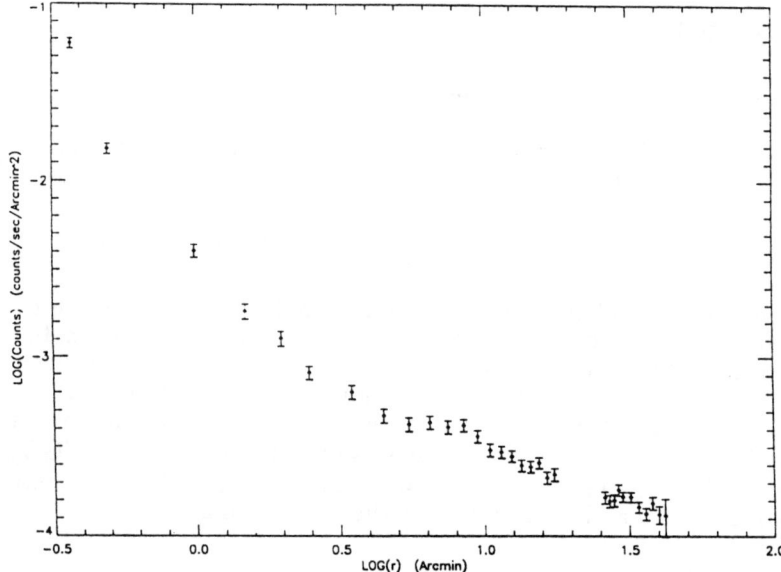

Figure 2 — X-ray surface brightness profile of the diffuse emission centered on NGC 1407. The adopted energy range is 0.5-2.4 keV and the latest energy-dependent exposure maps were used. Diffuse emission may be detected to nearly the full extent of the PSPC field, since we do not see the profile unambiguously flattening into background in the outer parts.

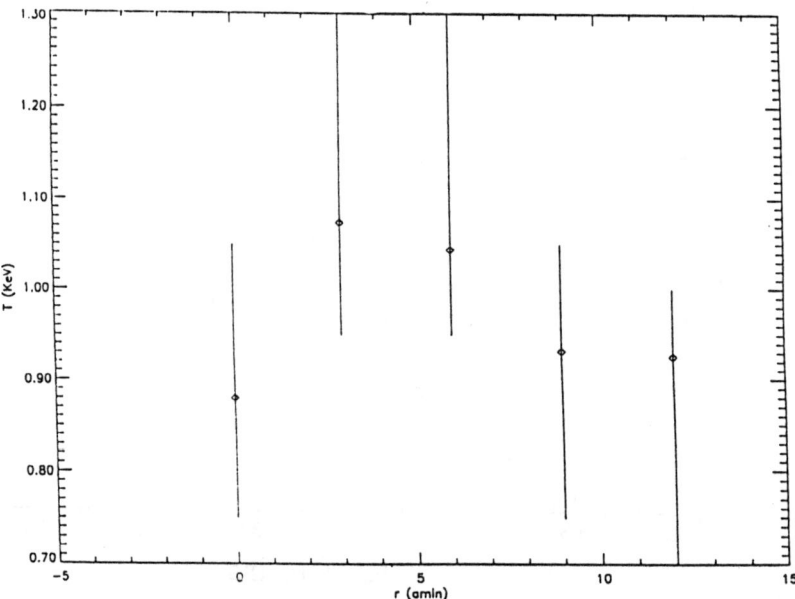

Figure 3 — Temperature distribution centered on NGC 1407.

PSPC OBSERVATIONS OF THE NGC 3607 GROUP OF GALAXIES

M. Loewenstein & R. Petre
NASA Goddard Space Flight Center, Code 666, Greenbelt, MD 20771

Email ID
loewenstein@lheavx.gsfc.nasa.gov

ABSTRACT

We present our preliminary analysis of a 24 ksec PSPC observation of the NGC 3607 group of galaxies. There is no evidence of extended group emission more than 50 kpc from the brightest galaxy in the group. The x-ray emission from the vicinity of NGC 3607 is asymmetric and is dominated by a 0.4 keV plasma, but shows evidence of an additional hard component. Possible origins for the spatial structure and the hard spectral component are discussed.

1. Motivation

It was well-established by analysis of x-ray observations made with the Einstein IPC that the most luminous early-type galaxies contain extended distributions of hot gas. Thermal emission from this gas dominates the x-ray emission in these galaxies, and the fact that they could retain such massive interstellar media provided indirect evidence for the presence of dark matter.

However, the IPC could not definitively determine the nature of the x-ray emission from low and moderate luminosity individual galaxies. In addition to x-ray emission from hot gas, a contribution from an ensemble of discrete (x-ray binary) sources is presumably present. Since the discrete source contribution should be roughly proportional to the optical luminosity, but the hot gas component has a steeper dependence, the x-ray emission should be dominated by discrete sources below some transition luminosity. Theoretical models predict that for the average observed SNIa rate and a universal ratio of dark to luminous matter of 9, the transition luminosity should be at about $3 \ 10^{10}$ in solar units. In addition, the observed x-ray luminosity is equal to the predicted contribution from x-ray binaries (assuming the same ratio of discrete x-ray to optical luminosity as in the bulge of M31) at approximately the same luminosity. As a result, we have observed three early-type galaxies (NGC 1380, NGC 4697, and NGC 3607) that have optical luminosities close to this value of interest with the ROSAT PSPC in order to determine, using both the spectral and spatial capabilities of the PSPC, whether the x-ray emission originates in hot gas, an ensemble of x-ray binaries, or a combination of these. Which of these alternatives is correct depends on the properties of the population of x-ray binaries, and on how the SNIa rate and dark matter content scale with galaxy luminosity.

We present preliminary results of our analysis of NGC 3607 here. This a rich field for study as it contains, in addition to NGC 3607 another low luminosity elliptical galaxy, NGC 3608, the AM Her system DP Leo, and numerous unidentified point sources ($\sim$ 30 with more than $5 \ 10^{-3}$ cts s^{-1}). In addition, NGC 3607 and NGC 3608 are members of a group of galaxies, providing an opportunity to search for diffuse intragroup gas. This field was observed for 23865 s in May/June 1993; we co-add data from a 8910 s, public, AO-1 observation of

DP Leo in some of our analysis.

2. Spatial Analysis

The 'hard band' image and contour map display a striking amount of asymmetry and inhomogeneity. These features show up even more prominently in the hardness ratio image and contour map (see figure below) where the extended emission seems to break up into secondary peaks.

One possible explanation is that we are seeing the effects of small numbers of bright discrete sources (x-ray binaries) that make a significant contribution to the x-ray flux. This is supported by spectral evidence for a hard component (see below).

In a comparison of 'soft' and 'hard' surface brightness profiles split into eastern and western halves, the hard emission appears more irregular beyond 2' (although this may be due in part to the lower source-to-background ratio in the soft band). However, the luminosity in the western excess (several 10^{39} erg s^{-1}) could only be explained by a large number of x-ray binaries – in apparent contradiction to the asymmetry in the emission.

Another possibility is that we are seeing an intragroup component (perhaps a coooling flow, given the inhomogeneity of the emission) that is centered westward of the center of NGC 3607. Indeed the x-ray emission is considerably more extended than the galactic light (half-light radii: $\sim$ 6 kpc in the optical, $\sim$ 15 kpc in x-rays) and extends out to $\sim$ 50 kpc. However, there is no evidence for the positive temperature gradient that might be expected if the more extended emission were in equilibrium in the deeper group potential well, and no evidence for intragroup gas beyond 50 kpc from the center of NGC 3607.

A third explanation might be an asymmetric superwind. NGC 3607 is a strong $H\alpha$ emitter; however, the optical emission line region seems to be confined to the inner 1.5 kpc of the galaxy.

3. Spectral Analysis

We have performed spectral fits to the $\sim$ 1700 counts within 5' of the center of NGC 3607. The spectrum is well fit by a thermal plasma model with $kT = 0.50$ (90% confidence limits: 0.45-0.57) keV and $Z = 0.15(0.11 - 0.25)$ solar plus a soft component (very steep power-law, very low temperature thermal plasma, or very low temperature blackbody) that accounts for $\sim$ 10% of the total flux. Such a soft component was detected by Kim, Fabbiano, & Trinchieri in very low luminosity ellipticals observed with the IPC. However, the magnitude of the component is on the order of the fluctuations in the soft background over the field, and so we are uncertain whether it is intrinsic to the galaxy.

The number of counts at energies greater than 0.9 keV is 5σ above a simulated $kT = 0.5$ keV, $Z = 0.1$ solar spectrum; and indeed, a fit of comparable goodness to that above is obtained if an additional hard component (modeled as 5 keV bremmstrahlung emission; any model with peak at energy greater the PSPC bandpass will do as well) is considered. The best fit model of this type has $kT = 0.41(0.35 - 0.51)$ keV and $Z = 0.29(> 0.14)$ solar for the primary component. We favor this model since it allows reasonable values of the metallicity. The hard component contributes $\sim$ 20 ($<$ 35)% of the 0.5-2.5 keV flux. It is natural to associate this hard component with the x-ray binaries that may account for some of the fluctuations in the x-ray image. (It cannot be an AGN since the hard component remains even if the central 1' is excluded from the spectral fits).

4. Summary

The x-ray emission from the S0 galaxy NGC 3607 cannot be characterized as a smooth, isothermal ball of gas. The emission is inhomogeneous on the scale of kiloparsecs and is asymmetric as well. X-ray binaries, a superwind, or an intragroup cooling flow are possible explanations for these irregularities, but none are completely satisfactory.

The most sensible fit to the x-ray spectrum has as its dominant component a ~ 0.4 keV thermal plasma of undetermined metallicity, with additional significant low and high energy components comprising $\sim 8\%$ and $\sim 17\%$ of the total flux, respectively. The low energy component may be an artifact of a non-uniform local background. The high energy component could be an ensemble of x-ray binaries and might be related in some way to the spatial structure.

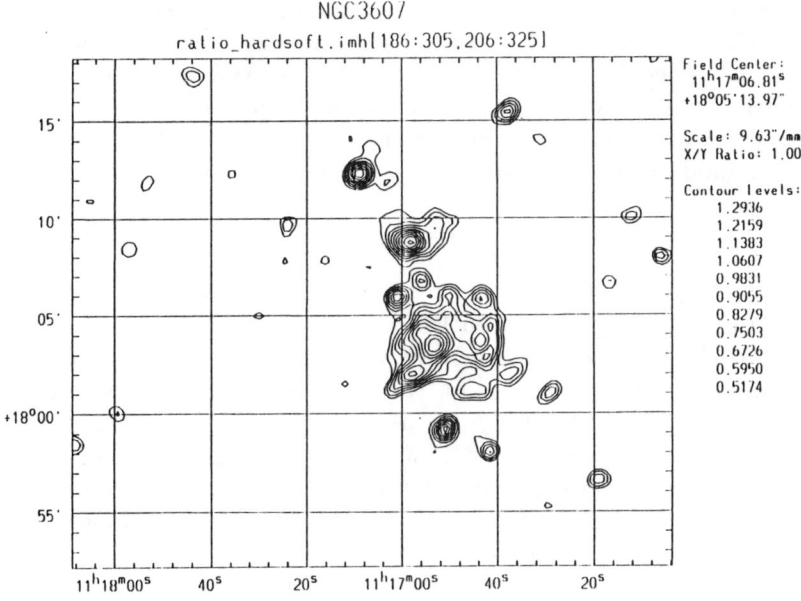

HARDNESS RATIO CONTOUR MAP

Contour map of hard band (0.5-2.0 keV) image divided by soft band (0.1-0.4 keV) image. The individual images were created using the energy-dependent exposure maps, and smoothed with a 22.5" gaussian. NGC 3608 is to the north and slightly east of NGC 3607. About 30 ksec of data are used, including both the AO3 NGC 3607 pointing and the AO1 DP Leo pointing, and excluding about 2 ksec of high background data. The hardness ratio in the background is about 0.2.

HOT ENTRAINED GAS IN THE JET OF NGC 4258 (M 106)

Gerald Cecil & Chris De Pree
Dept. Physics & Astronomy, UNC, Chapel Hill, NC 27599-3255
Andrew S. Wilson
STScI & Astronomy Dept., U. Maryland, College Park, MD 20742

ABSTRACT

This Seyfert galaxy has a helically twisted, bisymmetric nuclear jet that flows mostly in the gas disk. Our ROSAT results are: 1) The jet dominates the soft x-ray emission. After correcting pointing-errors of up to $\pm 6''$ in our 25 ksec HRI exposure, we find that the SE jet branch is spatially unresolved in width along much of its $2.'5$ length. The NW branch is more diffuse, and the extensive radio "plateaus" also emit x-rays. 2) The jet spectrum from the 6.5 ksec PSPC archival exposure is noisy, but is best fit as a Raymond-Smith plasma with $kT = 0.30$ keV, $\log N_H \approx 20.0$ cm^{-2}, and luminosity 1.6×10^{40} ergs s^{-1} between 0.1 and 2.4 keV. Gas at this temperature can arise from shock speeds of $350 - 500$ km s^{-1} (depending on preshock excitation), and may have been entrained as the jet scrapes along molecular clouds. The gaseous excitation and radial velocities from our published optical, emission-line spectra of the jet are consistent with these shock speeds if gas flows along helical paths. The PSPC spectral fit is poor above 0.6 keV. An additional hard component is required that peaks near the nucleus.

1. Introduction

The jet (Fig. 1) has been extensively studied in the radio continuum (e.g. van der Kruit, Oort, & Mathewson 1972) and optical emission lines (e.g. Martin et al. 1989.) Recent kinematic studies trace its interaction with large complexes of molecular clouds (e.g. Plante et al. 1991), and support the notion that it flows in the dense gas disk. An intriguing aspect of the jet is its organization into distinct helical strands when observed at $1-2''$ resolution (so far attained only at optical wavelengths), both spatially (Ford et al. 1986) and kinematically (Cecil, Wilson & Tully 1992, CWT hereafter). CWT used the Einstein image (Fabbiano, Kim, & Trinchieri 1992) to show that the SE jet is a resolved x-ray source, with $L_x \approx 2 \times 10^{40}$ ergs s^{-1}. Our full report will appear in ApJ (1994.)

2. X-Ray Properties of the Jet

To obtain the jet spectrum (Fig. 2d), we used the PSPC data from within the dark region of Fig. 2b. Whatever the absorption, simple free-free, exponential, and power-law forms could not fit both low and high energy bands of the PSPC. However, a Raymond-Smith plasma did provide a reasonable fit (Fig. 2d.) In our ApJ paper we exclude synchrotron and inverse Compton origins for the x-ray emission. Thus, we have found the tightest correlation seen to date between

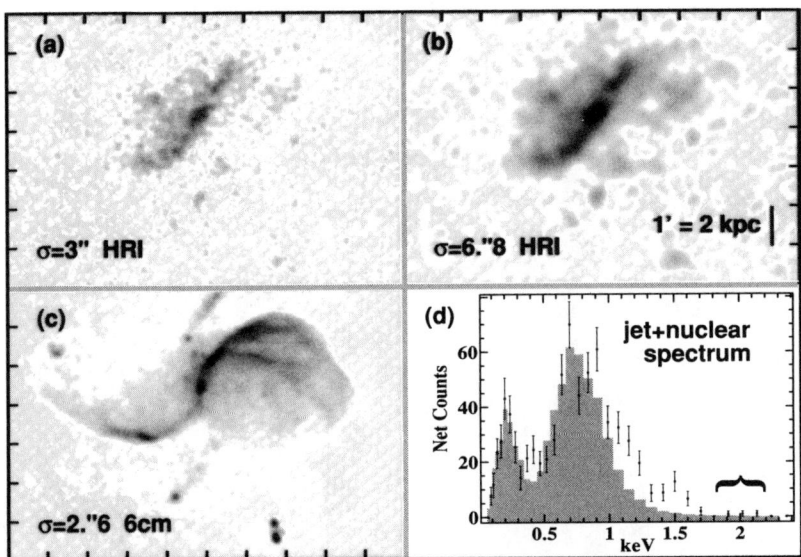

Fig. 2: X-ray and radio continuum emission from NGC 4258. N at top. X-ray images were binned into 3″ pixels. (d) The jet spectrum with best-fit Raymond-Smith plasma in gray. The brace marks bins that constrain the hard component.

thermal x-ray emission and a radio jet. The fit might improve with a shock model, or at least by allowing the temperature to change along the jet.

We assume that the hot gas is a shocked, cylindrical jet cocoon of elliptical cross-section (with major axis in the disk plane) and extent no larger than the bright emission in our HRI image. The 72° projection angle of the disk broadens our 19″ upper limit on the full x-ray width of the jet to 1′ (2 kpc.) Assuming a total extent of ≤10″ (≤350 pc) perpendicular to the dense disk, the shocked x-ray gas has volume $V_{65} \leq 10^{65}$ cm^3 and mean electron density $n_e \geq 0.006/\beta$ cm^{-3}, where $\beta = \sqrt{\eta V_{65}}$ and η is the filling factor. The gas mass and cooling time are $8\beta \times 10^5$ $M_\odot$ and 10β Myrs. Its pressure of $\geq 3/\beta \times 10^{-12}$ erg cm^{-3} may be similar to the $\approx 3 \times 10^{-11}$ erg cm^{-3} derived for relativistic electrons in energy equipartition at larger radii along the jet. The optical emission-line velocities and gaseous excitation (CWT) are suggestive of cooling by a hot, shocked gas.

We thank Mike Corcoran for retrieving the archival PSPC data, and Pat Crane for his VLA image. This work is supported by NASA and NSF grants to UNC.

Cecil, G., Wilson, A.S., & Tully, R.B. 1992, ApJ, 393, 134 (CWT)
Ford, H.C., et al. , 1986, ApJL, 311, L7
Kim, D.W., Fabbiano, G., & Trinchieri, G. 1990, ApJ, 390, 365
Martin, P., Roy, J.-P., Noreau, L., & Lo, K.Y. 1989, ApJ, 345, 707
Plante, R., Lo, K.Y., Roy, J.-R., Martin, P., & Noreau, L. 1991, ApJ, 381, 110
van der Kruit, P.C., Oort, J.H., & Mathewson, D.S. 1972, A&A, 21, 169

Clusters of Galaxies

AN X-RAY MOSAIC OF THE PERSEUS CLUSTER

Michael P. Kowalski

Code 7629.2, Naval Research Laboratory, Washington, DC 20375-5352

Email ID
kowalski@xip.nrl.navy.mil

ABSTRACT

Although cooling flows make the centers of rich clusters inherently interesting, most of a cluster's gas mass is found at large radii. Moreover, this gas may still be infalling and may not be in equilibrium. Therefore, a study of this extended gas through its X-ray emission may provide answers to such problems as the distribution of total mass within a cluster, the relationship between clusters and superclusters, and the interaction between the cluster gas and its member galaxies.

Four pointed observations were made of the Perseus cluster using the ROSAT PSPC. Each observation was located 50 arcmin (1.6 Mpc at $H_o=50$ km/s/Mpc) from the cluster center (NGC 1275), and the observations are spaced evenly in azimuth. We report here on the construction of a mosaic of images from these observations.

1. INTRODUCTION

Images of the central region around NGC 1275 (0.5-3 keV) obtained from the Einstein Observatory (c.f. Branduardi-Raymont et al. 1981, Fabian et al. 1981) have confirmed the existence of a point source coincident with the nucleus of NGC 1275 that accounts for a significant fraction of the emission. A region of enhanced emission 5 arcmin in radius surrounds NGC 1275, and this has been interpreted as a cooling flow. The cooling flow merges into the cluster extended thermal emission, which within about 17 arcmin of NGC 1275 (the useful limit of the IPC image), is elliptical in shape. With data from the scanning instrument SPARTAN-1, Snyder et al. (1989) were able to map the cluster X-ray emission (1-10 keV) out to 30 arcmin and quantify the elliptical structure out to this radius. They found evidence for isophotal shifting and twisting as well as varying ellipticity, all of which suggests the possibility of a merging subcluster. The ROSAT sky survey observation (Schwarz et al. 1992) shows a region of enhanced emission 15 arcmin to the east of NGC 1275 that is cooler than the extended emission, a result supported by the hardness-ratio of SPARTAN-1 scans. This was interpreted as evidence for a merging subcluster. The limit of cluster emission has not been determined yet. Ulmer et al. (1980) were able detect faint emission as far as 150 arcmin from the cluster center in data taken with the scanning HEAO-1 A-1 instrument, although source confusion may be present.

2. ANALYSIS

Images from our 4 ROSAT offset observations were created in the band 1.0-2.4 keV to avoid most sources of noncosmic background (Snowden et al. 1993). The solar-scattered background, after-pulsing, and long-term enhancements affect the countrate primarily at low energies and are negligible at energies above 1.0 keV. For each OBI of all 4 observations we examined the background light curve and found no evidence for any strong short-term enhancement. Therefore, only the particle background correction was applied to our images (Snowden et al. 1992, Plucinsky et al. 1993). To avoid regions at large radii where the particle background model is not valid we multiplied both the particle background and counts images by a mask with a radius of 55 arcmin.

Three short OBIs (3, 4, and 5) of the north field showed evidence of incorrect aspect solution, and therefore these data and times were eliminated from the analysis. The net live times are 31855 seconds for the west field, 15263 seconds for the east field, 35032 seconds for the south field, and 30659 seconds for the north field.

At the time of this analysis, only the SASS-provided mex.imh file was available to us for exposure information. We divided each observation's mex file by the net live time to obtain a "normalized" exposure map, and then divided this image into the counts map to produce a vignette-corrected counts map which also included the effects of detector nonuniformity and rib shadowing. Although the bandpass of the counts image and the exposure map are not identical, we expect the result of this on our broad-band image is small.

The corrected images were spatially registered to each other using the location of the strong cooling flow at the cluster center (NGC 1275), and the registration was checked using other point sources in overlap regions. The accuracy of registration is about 0.5 pixels (15x15 arcsec pixels). To check the validity of our vignette- and particle-background corrections, we compared the countrate between corrected fields in many registered regions. These regions were selected to be free from resolved strong point sources to avoid the possibility of a point source varying between the times of the different observations. In all such overlap regions tested, the diffuse emission countrate was found to be consistent at the 2-sigma level between different fields.

A counts mosaic was created by summing the 4 corrected images in register, and a countrate mosaic was obtained by dividing this file by the equivalent mosaic of the sum of live times. This was blocked to 30 arcsec pixels and smoothed with a gaussian which had a sigma of 2 pixels. Fig. 1 is a greyscale and contour image of the resulting mosaic. Eighteen logarithmic contours are displayed, separated by a factor of 1.5 and where the highest contour is a factor of 1.5 down from the peak map value.

3. DISCUSSION

In Fig. 1, point sources identified with cluster galaxies such as IC 310 as well as other galactic and extragalactic sources are clearly visible especially where the resolution is best at the field centers. Henry and Briel (1991) have found that the number of resolved point sources around A2256 is greater than one might expect by chance coincidence, and a similar quantitative examination is underway here.

Because the diffuse background has not been subtracted, the limit of diffuse cluster emission is still undefined (at this sensitivity) and may extend to the edge of the field and beyond. However, the countrate at a radius of 80 arcmin is still 3-5 times higher than at 100 arcmin which suggests significant cluster emission. There is good agreement between the inner regions of our mosaic and the Einstein IPC, SPARTAN-1, and ROSAT sky survey images. At radii greater than 30 arcmin, the general shape of the emission is still elliptical with more evidence for isophotal shifts and twists. The emission east of NGC 1275 is less extended than in other directions. This may be a real difference due to the passage of an infalling subcluster (see below), or may be partially caused by galactic absorption which increases to the east and north. (The galactic plane is located 13 degrees to the northeast.)

There is no strong visual evidence for any shock structure at radii greater than 30 arcmin from NGC 1275 as contours appear evenly spaced. Bunched contours might be expected in the case of a merger and have been found for A2256 (Briel et al. 1991). A moderately strong source lies near the edge of the north field about 1.5 degrees north of the cluster center, and this is coincident with another cluster or group of galaxies (Ulmer et al. 1993) of size 30-40 arcmin. It remains to be determined whether this group/cluster is falling into Perseus.

4. REFERENCES

Branduardi-Raymont, G., et al. 1981, ApJ, 248, 55
Briel, U.G., et al. 1991, AA, 246, L10
Fabian, A.C., et al. 1981, ApJ, 248, 47
Plucinsky, P.P., et al. ApJ, 418, in press
Schwarz, R.A., et al. 1992, AA, 256, L11
Snowden, S.L., et al. 1993, preprint
Snowden, S.L., et al. 1992, ApJ, 393, 819
Snyder, W.A., et al. 1990, ApJ, 365, 460
Ulmer, M.P., et al. 1993, private communication
Ulmer, M.P., et al. 1980, ApJ, 236, 58

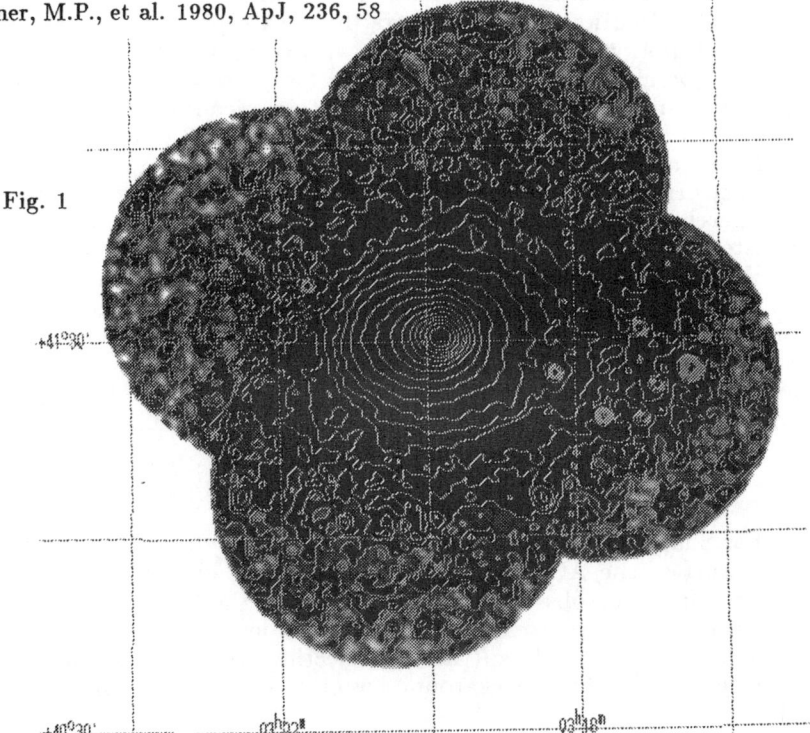

Fig. 1

ROSAT PSPC OBSERVATIONS OF THE SHAPLEY SUPERCLUSTER

Jeffrey O. Breen
Dept. of Astronomy, University of Virginia, Charlottesville, VA 22903
(job5g@Virginia.edu)

Somak Raychaudhury
Harvard-Smithsonian Center for Astrophysics, Cambridge, MA 02138
(somak@cfa.harvard.edu)

ABSTRACT

We present PSPC observations of two rich Abell clusters in the Shapley Supercluster, the richest known concentration of galaxies in the local Universe. We measure temperature and density profiles for the X-ray emitting intracluster gas in each cluster. We also highlight details of our data reduction process which we have found to be useful, such as the use of high resolution energy-dependent exposure maps generated by recently-released IDL routines.

1. Introduction

The Shapley Supercluster in northern Centaurus is the densest known baryonic mass concentration in the $z < 0.1$ Universe. This region has received considerable attention in recent years due to its proximity in the sky to the direction of motion of the Local Group with respect to the CMBR frame. The Shapley Supercluster is probably too far away ($z = 0.05$) to be the "Great Attractor", but it is clear that it has a very significant contribution to the local perturbation of the Hubble Flow (Raychaudhury 1989). Recent studies (Breen et al. 1994) have shown that clusters belonging to the Shapley Supercluster are on average brighter in the X-ray than a global luminosity function might suggest. The Supercluster also provides an interesting environment for studies of merging subclusters.

We present here ROSAT PSPC observations of two of the richest clusters in the Shapley Supercluster that have never been observed before with a position-sensitive detector. The cluster Shapley 8 (A3558) is part of a chain of four clusters in the center of the Supercluster that shows evidence of merging. The other cluster A3571 is one of the X-ray brightest clusters in the sky, and lies on the outskirts of the core of the Supercluster (about 40 h_{50}^{-1} Mpc from Shapley 8).

2. Observations

For Shapley 8, we used WP800076, a 30 ksec observation centered on the cluster, part of the AO-1 cycle, that we obtained from the data archive. We eliminated from the observation the times when the background count rate exceeded the mean background by $\geq 30\%$: this left us with data corresponding to an exposure of 26.9 ksec (thereby rejecting only 11% of the exposure but getting rid of 17% of the background counts). We generated high-resolution ex-

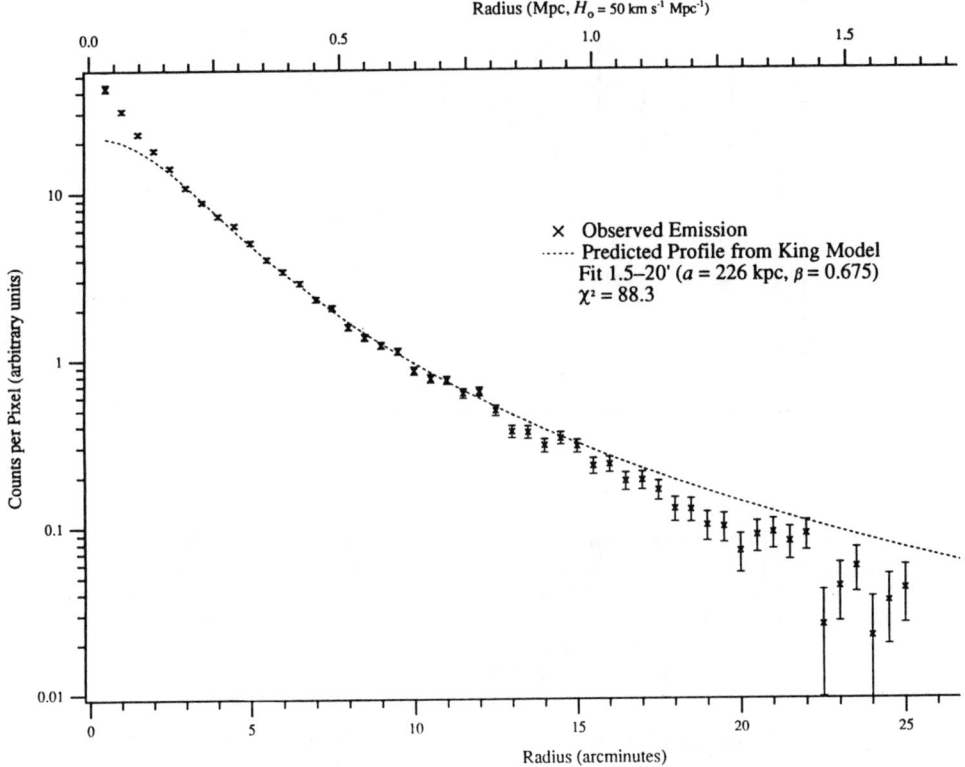

Figure 1. The Surface Brightness Profile for the cluster A3571 for the energy range 0.42–1.3 keV. A King model is a good fit to the profile out to 20' ($= 1.4\,h_{50}^{-1}$ Mpc) from the center of the cluster.

posure maps in each of the seven bands using the recently-released IDL routines (Make_Emap, part of the ROSAT IDL Library, due to Snowden, Mendenhall & Reichert), which were separately applied to the observations in the corresponding energy ranges. We calculated background values for empty regions in the field, and excluded point sources that do not belong to the cluster.

For A3571, we used RP800287, a 6 ksec exposure which is one of several clusters we observed in this region as part of the AO-3 cycle. The observation was reduced in the same way as above, except that it is not centered on the cluster, so certain azimuthal sectors were manually excluded from the analysis to avoid the effect of the grid. Figure 1 shows the radial profile of the hot gas in A3571, where we have used only the observation in the bands corresponding to 0.42–1.3 keV. We fit the profile to a King Model convolved with the PSPC point-spread function matched to the observation's off-axis angle (we used a calibration observation of AR Lac, which is at the same angle off-axis as the center of A3571). The profile shows an excellent fit out to about 20' from the center of the cluster.

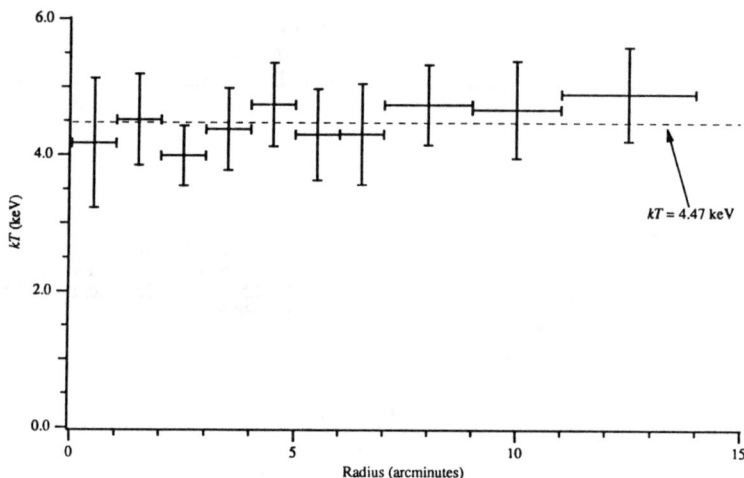

Figure 2. The radial dependence of the gas temperature of the cluster Shapley 8 (A3558). The central cD galaxy ($kT \sim 2.5$ keV) has been excluded from the figure. The remaining emission is binned in annuli and fit to a Raymond-Smith model described in the text. The X-ray emitting gas in Shapley 8 is very nearly isothermal at $kT = 4.5$ keV out to 15' ($= 1.2 h_{50}^{-1}$ Mpc) from the center of the cluster.

Figure 2 shows the temperature profile of the cluster Shapley 8, the central cluster of the Shapley Supercluster. We use XSPEC to fit a Raymond-Smith model to the spectrum in each bin, fixing the redshift of the cluster to the known value ($z = 0.048$) and the metal abundance to the cosmic value. Excluding the central cD galaxy, the temperature profile of the hot gas in the cluster turns out to be remarkably isothermal at 4.5 keV out to at least 15 arcmin from the cluster center. The net flux in the 0.5–4.5 keV range is measured to be 5.3×10^{44} erg s^{-1}. Repeating the same exercise for the central annuli corresponding to the cD galaxy, the central temperature turns out to be 2.5 keV.

We will use these observations to compare the x-ray luminosity, size parameters, gas density, temperature and gas mass of the clusters in this dense environment to galaxy clusters in the field. We have a detailed catalog of galaxy positions, redshifts and magnitudes of this region (Raychaudhury 1989), which will enable us to construct dynamical models of the clusters in the Supercluster. Using the optical and X-ray data together, we will study the efficiency of galaxy formation in rich environments.

We have extensively used the IRAF/PROS and IDL environments and the XSPEC package in our reduction of observations. This research has been supported by NASA grant NAG5-2145.

References

Breen, J., Raychaudhury, S., Forman, W., & Jones, C. 1994, *Ap.J.*, in press.
Raychaudhury, S. 1989, *Nature*, 342, 251.

A ROSAT PSPC OBSERVATION OF THE LENSING CLUSTER A1689

S. Daines, C. Jones, and W. Forman
Harvard-Smithsonian Center for Astrophysics
A. Tyson
Bell Laboratories

Email ID
sjd@cfa.harvard.edu

ABSTRACT

Using the surface brightness distribution from the ROSAT PSPC and the Ginga broad-beam temperature, we derive a range of acceptable potentials for the cluster A1689. The faint optical arcs observed in A1689 lie at radii which are too large to be produced by simple smooth potentials, consistent with the X-ray data and with plausible elongation along the line-of-sight. They thus reflect superposition or substructure.

Although the x-ray observations show no evidence of substructure, the cluster velocity dispersion is too large to be due to a single cluster with the gas temperature and mass measured through the X-ray observations. The galaxy radial velocity data then suggest the superposition of one or more subclusters directly along the line-of-sight to the primary A1689 cluster.

The cooling flow region in the cluster core is well resolved by the PSPC. Spectral fitting of the inner 1' shows emission from gas cooler than the broad-beam temperature, and consistent with a massive cooling flow of $\approx 500 M_\odot$ yr^{-1}, in agreement with the mass deposition rate found from a deprojection analysis of the imaging data.

1. Introduction

Arcs and arclets in the cores of clusters are highly magnified, gravitationally lensed images of faint background galaxies (Lynds & Petrosian 1989, Soucail et al. 1987, 1988). Tyson, Valdes, and Wenk (1990) detected many small, faint, weakly lensed arclets arising from galaxies behind the cluster A1689, as well as several fainter, greatly elongated arcs. In a smooth potential, the large arcs determine the mass within the critical radius of the cluster. The relatively recent capabilities of x-ray observations to measure cluster gravitational potentials and the studies of giant arcs and arclets to determine the mass distribution of the lensing cluster has led several investigators to compare the inferred cluster properties (Hammer 1991, Babul and Miralda-Escude 1994).

A1689 was observed with the ROSAT PSPC for 13,957s from 18 - 24 July 1992, giving 10253 source counts within a radius of 8.5 arcmin. Extrapolating from the ROSAT band by assuming a 7.8 keV thermal spectrum and column density $N_H = 1.79 \times 10^{20}$ cm^{-2} (Stark et al. 1992), we find a 2 - 10 keV flux of 1.70×10^{-11} erg cm^{-2} s^{-1} and a source-frame 2 - 10 keV luminosity of 2.85×10^{45} erg s^{-1}. Results of isophote fitting to the adaptive-smoothed image confirm that the X-ray image is close to circular, with ellipticity 0.1 and centered within 4" of the brightest cluster galaxy. The surface brightness profile from the PSPC combined with the Ginga X-ray spectrum, places improved constraints on the cluster mass distribution near the core and allows us to make a preliminary comparison with the position of the most distorted lensed arclets.

2. Constraining the Mass Distribution in A1689

We use the X-ray data to constrain the core mass by comparing broad-beam temperatures, predicted from the ROSAT surface brightness distribution (which determines the gas density) and three different parameterized gravitating mass distributions, with that observed. The X-ray emission in A1689 is strongly concentrated towards the cluster core, and the broad-beam Ginga temperature is therefore representative of the core gas temperature. We may then calculate also the surface mass density and compare to that implied by the largest of the observed lensed arc(let)s. The radial temperature solutions are derived by deconvolution of the surface brightness profile (Fabian *et al.* 1981) and solving for hydrostatic equilibrium in a variety of different shaped potentials.

We use three forms for the cluster mass distribution with radius, and model the cluster either as a sphere or as a spheroid prolate along the line-of-sight. Spherical potentials are:

A: a non-singular isothermal sphere (we start with a model with mass core radius of $1.1'$, the core radius of the X-ray gas).

B: a "softened isothermal" potential with $\rho(r) = \rho_0/(1 + (r/r_c)^2)$

C: a generalization of B, with $\rho(r) = \rho_0/(r/r_c)^\alpha(1 + (r/r_c))^{\gamma-\alpha}$ and $\gamma = 2$, $\alpha = 1$.

The largest axis ratios of X-ray isophotes (and hence potential) observed in clusters are $\lesssim 1.3$, corresponding to an axis ratio of $\lesssim 1.75$ for the gravitating matter. This ellipticity along the line-of-sight in A1689 would increase the mass surface density by a factor of $\lesssim 1.3$ for a given X-ray temperature.

A1689 shows many small, tangentially elongated arclets, and several larger, fainter arcs longer than $10''$. The three large arcs nearest to the brightest cluster galaxy have radial distances from it of $47''$, $43''$, and $46''$. A comparison of the positions of the "largest arclets" to the different mass models shows that the smooth potential seen by the X-ray gas contributes $\lesssim 50$ per cent of the critical density at the arc radius, independent of assumed shape. Plausible ellipticities along the line-of-sight increase this to $\lesssim 65 - 70$ per cent.

Although the x-ray observations alone show no evidence of cluster substructure, the cluster velocity dispersion is too large to be due to a single cluster with the gas temperature and mass measured through the x-ray observations. The galaxy radial velocity data then suggest the superposition of one or more subclusters directly along the line-of-sight to the primary A1689 cluster. The faint arcs observed in A1689 lie at radii $\approx 45''$ which are too large to be produced by the primary A1689 cluster if spherical, also requiring an enhancement in the mass surface density.

The lensing and X-ray data are then consistent with two scenarios. The simplest explanation for the positions of the large arcs closest to the cluster center is that they trace a critical radius at $\approx 45''$, with the surface mass density smooth and boosted by the superposition of an accurately aligned subcluster or extreme ellipticity (*i.e.* substructure along the line-of-sight). The large arcs at $45''$ could also reflect structure in the surface mass density on small scales rather than a near-circular critical radius, as is clearly the case with the arcs furthest from the cluster center at radius $\approx 75''$ which are spatially associated with bright galaxies. It is likely that substructure and ellipticity will occur

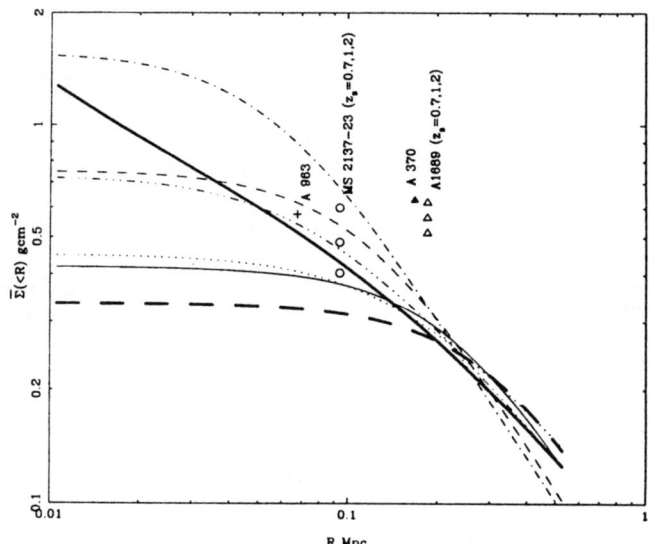

Figure 1 shows the mean surface mass density for each potential (normalized to a broad-beam temperature of 9 keV) and the radial positions of large arcs in A1689 and several other clusters, plotted at the critical density for each cluster redshift and background source redshift (if known).

preferentially in the same direction, aligned with large scale sheets or filaments which gradually empty into the main cluster. Alignment in filaments also greatly increases the probability of several superpositions and one accurate alignement.

If we assume that the range of "reasonable" potentials for A1689 are characteristic of other lensing clusters, we can compare these to the giant arcs observed in other clusters. In Figure 1, we have compared the scaled arc radii and critical densities for several clusters to our potential models for A1689. The arcs for A963 and MS2137 lie at small radii, and are consistent with the steeper potentials found for A1689. However, (like the faint arcs in A1689), the large arcs in A370 fall at radii which are too large to be due to a single cluster potential and must reflect superposed substructure.

This work is supported by the Smithsonian Astrophysical Observatory and NASA contracts and grants NAS8-39073 and NAG8-1536.

References

Babul, A. and Miralda-Escude, J. 1994, preprint
Fabian, A.C., Hu, E.M., Cowie, L.L., Grindlay, J., 1981, Ap. J., 248, 47
Hammer, F.,1991 Ap. J. 383, 66.
Lynds, R., Petrosian, V., 1989, Ap. J. 336, 1
Soucail, G., Fort, B., Mellier, Y., Picat, J.P., 1987, Ap. J., 172, L14
Soucail, G., Mellier, Y., Fort, B., Mathez, G., Cailloux, M., 1988, Ap. J., 119. L19
Stark, etal, 1992, Ap. J. (Suppl), 79, 77
Tyson, J.A., Valdes, F., Wenk, R.A., 1990, Ap. J., 349, L1

ROSAT OBSERVATION OF ABELL 1795: TEMPERATURE AND MASS PROFILE

H. Brunner, S. Weimer, H. Westphal, R. Staubert
Astronomy Institute, University of Tübingen, Germany

Email ID
brunner@ait.physik.uni-tuebingen.de, susanne@univers.sub.de

ABSTRACT

A 20 ksec ROSAT PSPC observation of the cluster of galaxies Abell 1795 was used to perform a spatially resolved, spectral model fit, permitting a non-parametric determination of the de-projected radial temperature and density distribution (Figs. a, b). We derive the radial distribution of gravitating, gas, and galaxy mass (Fig. c), the dark-to-visible mass fraction, and determine the cooling time (Fig. e) and cooling flow.

1. Observation and Data Preparation

We have observed Abell 1795 for 20 ksec with the ROSAT PSPC, collecting a total of $\sim$ 80,000 source counts (ROSAT observation WG700145P). Count rate spectra in the energy range from 0.2 to 2.1 keV were accumulated in 22 concentric rings, extending 11 minutes of arc from the cluster center. The ring radii were selected, such that $\sim$ 3000 source counts fall into each ring. This is sufficient for the determination of significant X-ray spectra in each ring. Areas contaminated by bright sources as well as areas close to, or outside of the inner ring of the PSPC support structure were excluded from our analysis.

2. Model Fitting

We fit the X-ray data to a spherically symmetric model where the de-projected temperatures and densities are defined at 9 radial distances and linear interpolation is performed between these distances. We assume local Raymond-Smith emission (Raymond and Smith 1977), using 0.5 × solar metal abundances. The local emission is integrated numerically over the line of sight, corrected for interstellar absorption ($N_\mathrm{H} = 1.14 \times 10^{20}$ cm^{-2}), and convolved with the ROSAT PSPC point spread function for the appropriate energy and off axis angle. The model flux predicted for each of our 22 concentric rings is folded through the response of the ROSAT PSPC detector and fit to the observed spectra. A Hubble constant of $H_0 = 50$ km/sec Mpc is used, which, at a redshift of $z = 0.062$, implies a scale of 108 kpc/arcmin.

3. Mass Determination

We calculate the radial distribution of the gravitating mass, using the well known formula:

$$M(r) = -\frac{kT_g r}{\mu m_p G}\left(\frac{d\ln n_e}{d\ln r} + \frac{d\ln T_g}{d\ln r}\right).$$

The gas mass distribution is determined by integrating over our best fit densities, using the appropriate element abundances. The radial distribution of the mass in galaxies is determined by fitting a King profile to the galaxy distribution, using galaxy counts by Baier (1979), and assuming a mean mass of 2×10^{11} $M_\odot$ per galaxy.

4. Cooling Flow

The cooling time t_c and cooling flow $\dot{M}$ are calculated following Sarazin (1986) and Fabian (1984):

$$t_c(r) = 8.5 \times 10^{10} \left[\frac{10^{-3} cm^{-3}}{n_p(r)}\right] \sqrt{\left(\frac{T}{10^8 K}\right)} \ yrs.; \quad \dot{M} = \frac{4\pi r^2 \rho(r)}{dt_c/dr}.$$

5. Results

We find a pronounced drop of the gas temperature at $r < 100$ kpc (~ 1'). The temperature at the cluster center is $T \sim 2.5$ [2 − 4.5] keV ([·]: 90 % confidence range). A maximum of $T \sim 7.5$ [5 − 11.5] keV is reached at $r \sim 110$ kpc and a marginally significant drop to $T \sim 4.5$ [2 − 5.5] keV is observed at larger radii (Fig. a).

From our best fit temperature and density distribution the total gravitating mass within $r < 1.2$ Mpc (11') is determined to be $\sim 7 \pm \times 10^{14}$ $M_\odot$, using the formula given above. Integrating the best fit densities yields a total gas mass inside $r < 1.2$ Mpc of 1.0×10^{14} $M_\odot$. The galaxy mass within $r < 1.2$ Mpc is found to be $1.2 \times 10^{13} M_\odot$. See Fig. c for the radial distribution of the gravitating (M_{grav}), gas (M_{gas}), and galaxy (M_{gal}) masses and Fig. d for the dark ($M_{grav} - M_{gas} - M_{gal}$) to visible ($M_{gas} + M_{gal}$) matter distribution. Note the relative over-abundance of the dark matter component at small radial distances $r < 200$ kpc (Fig. d), confirming similar results by Eyles et al. (1991) derived on the Perseus cluster and Briel et al. (1992) on the Coma cluster.

The cooling time is found to drop below the Hubble time at radial distances below ~ 110 kpc, agreeing with the observed drop in the gas temperature at a similar radius (see Figs. a and e). The derived cooling flow of ~ 400 $M_\odot/yr.$ is in full agreement with previous results (e.g. Edge and Stewart 1991).

References

Baier, F. W., 1979, Astron. Nachr., 300, 133
Briel, U. G., Henry, J. P., Böhringer, H., 1992, A&A, 259, L31
Edge, A. C. and Stewart, G. C., 1991, MNRAS, 252, 414
Eyles, C. J., Watt, M. P., Bertram, D. et al., 1991, ApJ, 376, 23
Fabian, A. C., 1984, Nature, 310, 733
Raymond, J. C. and Smith, B. W., 1977, ApJS, 35, 419
Sarazin, C. L., 1986, Rev. Mod. Phys., 58, 1

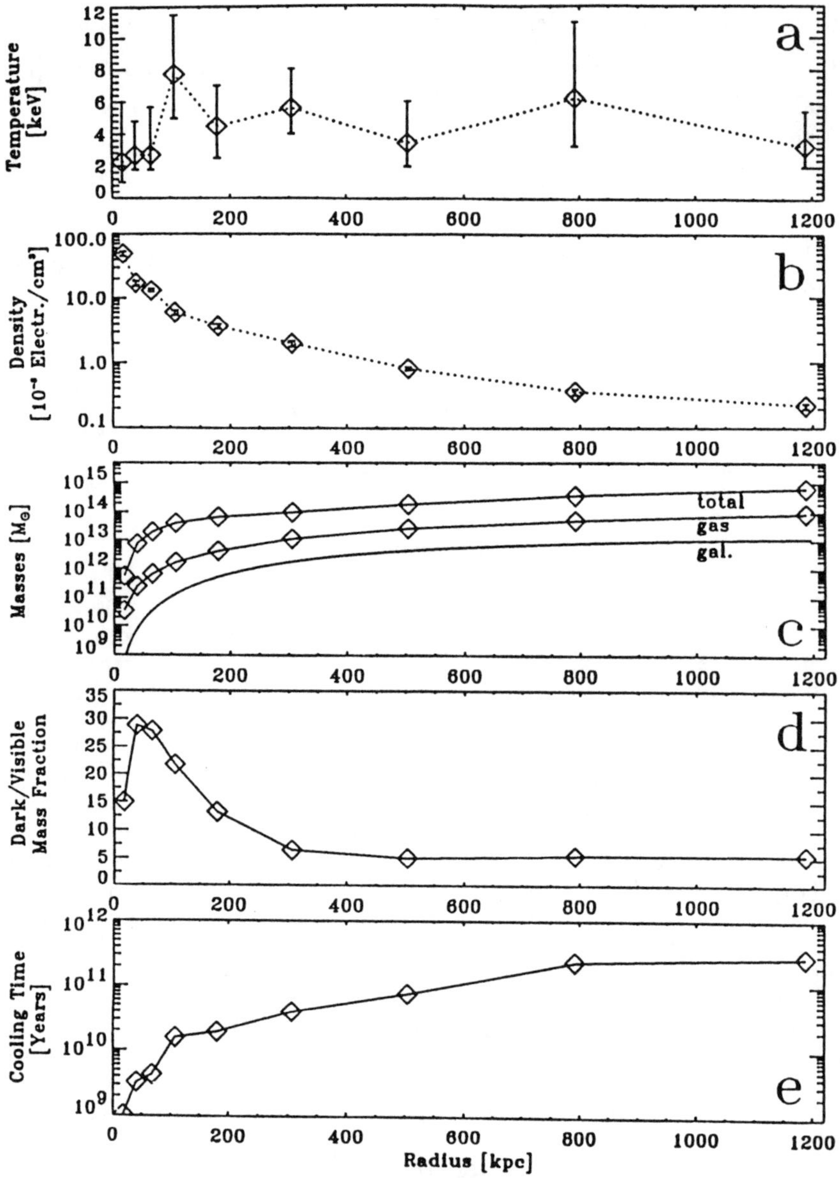

a, b: Best fit temperatures and electron densities at 9 radial distances. 90 % confidence errors ($\chi^2_{min} + 26.0$ for 18 free parameters) are given. **c:** Gravitating, gas, and galaxy masses within radius r, as determined from best fit temperatures and densities, and from galaxy counts. **d:** Dark-to-visible matter fraction for matter within radius r. Note the relative over-abundance of the dark matter component at $r < 200$ kpc. **e:** Cooling time as estimated from best fit temperatures and densities. Note that the cooling time drops below the Hubble time at roughly the same radial distance where a reduction in the gas temperature is observed (Fig. a).

ROSAT OBSERVATIONS OF DISTANT CLUSTERS OF GALAXIES

Megan Donahue
STSCI, 3700 San Martin Drive, Baltimore, MD 21218

Email ID
donahue@stsci.edu

John T. Stocke (UColo), Simon L. Morris (DAO), & Keith A. Arnaud (GSFC)

ABSTRACT

We present the results of ROSAT PSPC and HRI observations of two distant clusters, MS0735+74 and MS0451-03, selected from a sample of serendipitous X-ray sources. MS0735+74 at $z = 0.216$ is a cooling flow candidate because it has extended Hα emission in the central brightest galaxy. MS0451-03, at $z = 0.55$, is the most luminous cluster in the Extended Medium Sensitivity Survey with $L_x > 10^{45}\,\mathrm{erg\,s^{-1}}$. We have been able to place signficant limits on T_x, core radii and the β parameter, and cooling flow contributions. Our preliminary results show that "cooling flows" can be identified by their extended Hα emission. Distant clusters probably do not have more intrinsic absorption than nearby clusters. More tentatively, luminous clusters at high redshift may be cooler and may have somewhat lower M_{gas}/M_{total} (at 1 Mpc scales) than clusters of comparable luminosity at low redshift.

1. X-ray Clusters and Evolution of Large Scale Structure

We need to study distant clusters in the X-rays because they are just now forming and collapsing linearly. Therefore, their distribution and masses retain some "memory" of the initial density perturbations. By measuring temperature functions at different epochs, we can learn about the shape, size, and evolution of the spectrum of density perturbations at the cluster mass scale (Henry & Arnaud 1991). The most distant clusters are the most likely to show evidence of evolution in structure, temperature, and X-ray luminosity. We already know that the X-ray luminosity function (LF) of distant clusters differs from that of nearby clusters (Gioia et al. 1990a; Edge et al. 1990). Deep X-ray observations of individual clusters reveal how clusters evolve. Evolution of the X-ray LF may be caused by temperature evolution (or evolution of the $L_x - T_x$ relation), increases in soft X-ray absorption by cooler gas in the clusters, and/or increases in luminosity of rich clusters as small cluster merge to form large ones.

Studying the hot gas in cluster at all redshifts is relevant to questions about structure and cluster evolution. The hot gas responds to the underlying distribution of cluster mass if the gas remains in nearly hydrostatic equilibrium. Gas emission improves our view of cluster substructure and potential depth as the depth of the X-ray observation increases, in contrast to galaxy counts which are ultimately limited by the number of galaxies in the system. Since X-rays directly count the hot baryons in clusters, and hot baryons are the dominant luminous mass component of rich clusters, X-ray observations are crucial for measuring cluster M/L ratios. Cluster M/L may change systematically over time, particularly if delayed infall is a significant process as suggested by Kaiser (1991).

2. Temperatures, Column Densities, and Structure

We give here an update of on-going analyses of two distant clusters of galaxies. Both clusters are members of an X-ray selected sample of clusters of galaxies, a subset of the EMSS (Gioia et al. 1990b). All scales and 0.2-2.5 KeV luminosities are computed assuming $H_o = 50$ km/sec/Mpc and $q_o = 0.5$. In the following subsections, a King profile is assumed to be $I(r) \propto (1 + (r/r_c)^2)^{-3\beta+1/2}$, where r_c is the core radius. In isothermal models β corresponds to the ratio of kinetic energy in the galaxies to the thermal energy in the gas.

MS0735+74 ($z = 0.216$, 4.45 kpc/") was identified as a candidate "cooling flow" cluster by Donahue, Stocke, & Gioia (1992; DSG) because it had extended Hα emission and low-ionization level emission lines with [OII] 3727 $\gg$ [OIII] 5007. We confirm this identification with the ROSAT HRI observation of a central high surface-brightness peak in the central portion of the cluster.

T_x is $2.8^{+1.4}_{-0.8}$ KeV, significantly lower than expected for a cluster of its luminosity at low redshift. The cooling flow may contribute up to 20% of this emission. $N_H = (2.9 \pm 0.3) \times 10^{20}$ cm^{-2}, not significantly different from the expected Galactic absorption along that line of sight. The cluster luminosity within 2' of the center is 6.1×10^{44} erg sec^{-1}.

We fit radial surface profiles of the cluster to King models and to King models with a Gaussian central source. The fit improved significantly when we included the Gaussian central source, corresponding to the putative "cooling flow." The core radius is 230 ± 40 kpc, and $\beta = 0.80 \pm 0.25$, consistent with those quantities in clusters of low redshift (e.g. Jones & Forman 1984). The mass cooling rate is of order 125 $M_\odot$ yr^{-1}, given the central excess emission and temperature.

MS0451-03 ($z = 0.55$, 7.38 kpc/") is the most luminous cluster in the EMSS, and indeed, one of the most luminous clusters known. This cluster is hotter than Rosat can measure. Since the PSPC is sensitive to low surface-brightness emission, and has reasonable spatial resolution, we could fit a core radius of 360^{+80}_{-36} kpc and a β of $1.00^{+0.4}_{-0.3}$ to this cluster, as well as the soft X-ray absorption of $3.2 \pm 0.5 \times 10^{20}$ cm^{-2} and a lower limit for the cluster temperature of 3.5 KeV. An ASCA observation of this cluster will measure the temperature and iron abundance. The cluster luminosity within 100" is 1.56×10^{45} erg sec^{-1}. The mass cooling rate in this cluster is under 125$M_\odot$ yr^{-1}.

Neither of these clusters have a *strong* excess of soft X-ray absorption that may be associated with a cooling flow (White et al. 1991) or processes that preferentially occur in distant clusters as suggested by Wang & Stocke (1993).

3. Cluster Masses

We estimated the central density (Henry & Henriksen 1986), and derived the gas mass within 1 Mpc h_{50}^{-1} with parameters from the surface brightness fit, assuming that the cluster density $\rho = \rho_o(1 + (r/a)^2)^{-3/2}$. We estimated the total mass within that radius by assuming that the cluster is isothermal and in nearly hydrostatic equilibrium, and we get that:

$$M(r) = 10^{14} M_\odot \beta T(\text{KeV}) r(\text{Mpc}) \frac{(r/a)^2}{1 + (r/a)^2}. \quad (1)$$

We note that $M(r)$ as well as M_{gas} diverge with radius. These expressions cannot apply outside cluster radii of 1-3 Mpc where the gas and galaxies are not yet virialized and the X-ray surface brightness is low.

Table 1: Preliminary Estimates of Cluster Quantities within 1 Mpc

	MS0735+74	MS0451-03
Redshift	0.216	0.55
n_o (cm^{-3})$h_{50}^{1/2}$	4.3×10^{-3}	3.6×10^{-4}
$M_{gas} h_{50}^{-5/2}$ ($M_\odot$)	6.4×10^{13}	5.9×10^{13}
$M_{total} h_{50}^{-1}$ ($M_\odot$)	2.5×10^{14}	$> 5.0 \times 10^{14}$
$M_{gas}/M_{total} h_{50}^{-3/2}$	25%	< 12%

As can be seen from Table 1, we derive a somewhat smaller gas/total ratio for MS0451-03 than what has been observed in nearby clusters. Since our temperature is a lower limit, the mass ratio can only be smaller than what we list. The gas/total ratio for MS0735+74 is typical of clusters nearby (Edge & Stewart 1991).

4. Optical Nebulae of Distant Cooling Flows

An optical emission-line nebulae with $L_{H\alpha} = 7.3 \times 10^{40}$ erg s^{-1} lies in the central galaxy of MSS0735+74 (DSG). Our long-slit spectra show emission line ratios typical of class II "cooling flow" nebulae (Heckman et al. 1989) such as Perseus (N1275) or A2597 (N6166), with line FWHM of < 800 km/sec. If extreme UV photons from the cooling gas is responsible for powering the filaments, the maximum Hα to X-ray surface brightness ratio is 0.03 (Donahue & Voit 1991), consistent with what we observe after we degrade the Hα image. X-ray confirmation that this cluster indeed contains a cooling flow supports the contention of DSG that their detections of extended Hα in EMSS X-ray clusters are also cooling flows. If this contention is true, at least 40% of X-ray selected clusters contain cooling flows.

This contribution was poster 8.04.

- Donahue, M., Stocke, J. T. & Gioia, I. M. 1992, ApJ, 385, 49 (DSG).
- Donahue, M. & Voit, G. M. 1991, ApJ, 381, 361.
- Edge, A. C. et al. 1990, MNRAS, 245, 559.
- Edge, A. C. & Stewart, G. C. 1991, MNRAS, 252, 414.
- Gioia, I. M. et al. 1990a, ApJ, 356, L35.
- Gioia, I. M. et al. 1990b, ApJS, 72, 567.
- Heckman, T. M, Baum, S. A., van Breugel, W. J. M., & McCarthy, P. 1989 ApJ, 338, 48.
- Henry, J. P. & Henriksen, M. J. 1986, ApJ, 301, 689.
- Henry, J. P. & Arnaud, K. A. 1991, ApJ, 372, 410.
- Jones, C. & Forman, W. 1984, ApJ, 276, 38.
- Kaiser, N. 1991, ApJ, 383, 104.
- Wang, Q. D., & Stocke, J. T. 1993, ApJ, 408, 71.
- White et al. , 1991, MNRAS, 252, 72.

COMPLEX SPATIAL STRUCTURES IN SUNYAEV-ZEL'DOVICH DECREMENT CLUSTERS ABELL 665 AND CL0016+16

John P. Hughes and Mark Birkinshaw
Harvard-Smithsonian Center for Astrophysics
60 Garden Street, Cambridge, MA 02138

Email ID
jph@cfa.harvard.edu, mb1@cfa.harvard.edu

ABSTRACT

We report on our analysis of deep X-ray observations of Abell 665 and CL0016+16 obtained by the ROSAT position sensitive proportional counter (PSPC) as part of a project to characterize the density and temperature structure of clusters with well-measured Sunyaev-Zel'dovich decrements. The X-ray images of both clusters show dramatic departures from the standard single-component hydrostatic isothermal-β model.

1. Analysis

The PSPC observed A665 in April 1991 for 38641 s. Fig. 1a shows the 0.4–2.4 keV band X-ray map of the cluster after the data were background subtracted, exposure corrected, and adaptively smoothed. The structure of the cluster is complex. The main cluster component extends to a radius of 8'. Near its center, but offset from it by about 1', is a smaller and brighter, but clearly extended, second component. Numerous point sources, nearly all of which remain unidentified at this time, also appear in the field.

Maximum-likelihood fits to the raw two-dimensional image data were carried out. Two isothermal-β models, with parameter values $\beta = 0.85$, $R_c = 2.7'$ and $\beta = 1.24$, $R_c = 0.9'$, were clearly required, as was a previously unrecognized point source near the core of the second component. The residuals from this fit showed the presence of two additional surface brightness features each seemingly associated with a cluster component. These features were modeled as prolate spheroids of hot gas. In total five spatial components were required for a good fit. Spatially resolved PSPC X-ray spectroscopy showed that the 2 cluster components were consistent with having the same temperatures. Furthermore, it was possible to set a strong lower limit on the temperature of either component: $kT > 5$ keV, ruling out the presence of an active cooling flow.

The cluster CL0016+16 was observed by the PSPC in July 1992 for 43157 s. Fig. 1b shows a map of the cluster's 0.4–2.4 keV band X-ray emission after processing as above. The structure of this cluster is also complex. We performed elliptical isophotal analysis on the smoothed data. The ellipticity increases from 0 near the cluster center to a maximum of 0.24 at a radius of 80" before decreasing to 0 again at 120". Over this radial range the position angle of the major axis is roughly constant at 50°, but at 120" it changes abruptly by nearly 90°. From 120" to 200" the ellipticity is in the range 0.05 to 0.1. The derived positions of the ellipse centers vary with radius by less than ~5".

The distribution of galaxies (Dressler and Gunn 1992), as well as data on gravitationally lensed arcs in CL0016+16 (Ellis 1993), shows a bimodal spatial structure. We found that the galaxy distribution is consistent with 2 nearly

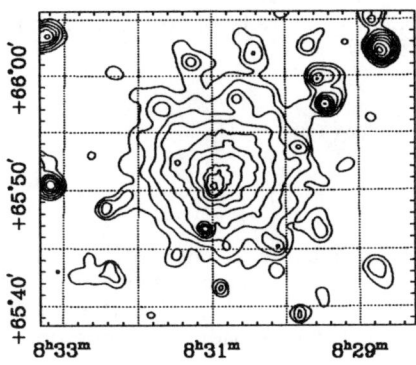 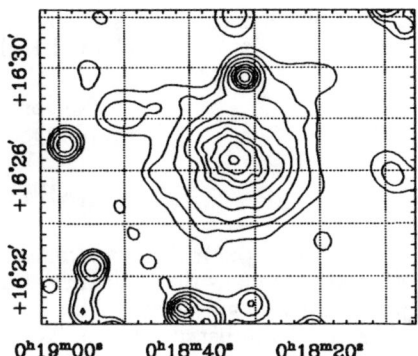

Fig. 1 (a) Left: ROSAT PSPC X-ray map of Abell 665. Contours are logarithmically spaced from 0.25% to 90% of the peak value. (b) Right: ROSAT PSPC X-ray map of CL0016+16. Contours are logarithmically spaced from 0.5% to 90% of the peak value.

equal mass groupings, separated by some 90″, and lying on either side of the bright central core of the X-ray image. When a similar two-component model for the gas distribution was fit to the X-ray data, the derived positions, separation, and relative intensities of the components were significantly different from the galaxian values. In particular the X-ray image appears to be dominated by a β-model component centered near the X-ray peak, between the 2 galaxy groupings.

2. Conclusions

A665 shows two subclusters in the process of merging. This is supported by the large β value (of 1.24) for the smaller, merging, component, which indicates a steep radial fall-off in its gas density. We believe the ellipsoidal clouds found in our analysis to be the remnants of the gas stripped off the outer atmosphere of the subcluster during a passage through the core of the main component. These observations agree well with numerical simulations of the merger of a dominant cluster and a small subcluster (Roettiger, Burns, & Loken 1993).

The optical data for CL0016+16 strongly support a scenario in which we are viewing this cluster as it undergoes a merger of two nearly equal components. At some level, this is supported by the strong ellipticity seen at $\sim 80″$ in the X-ray image. However, the presence of a single central X-ray surface brightness peak indicates that the gas distribution is more complex than just the superposition of two isothermal-β components: e.g., the core might contain an unresolved point source or even a cooling flow. Additional study of the gas and galaxy distributions using the commmon cluster gravitational potential as the starting point should clarify these issues.

3. References

Dressler, A., & Gunn, J. E. 1992, ApJS, 78, 1
Ellis, R. 1993, Proceedings of the 37th Yamada Conference, *Evolution of the Universe and its Observational Quest*, ed., K. Sato, in press.
Roettiger, K., Burns, J., & Loken, C. 1993, ApJ, 407, L53

EVOLUTION IN NEARBY CLUSTERS OF GALAXIES AND THE BUTCHER-OEMLER EFFECT

James A. Rose and Andrew Leonardi
Department of Physics and Astronomy, CB#3255, University of North Carolina
Chapel Hill, NC 27599

Nelson Caldwell
Whipple Observatory, SAO

ABSTRACT

Multi-fiber spectra of galaxies in the Coma cluster in conjunction with ROSAT PSPC images demonstrate that the SW region of Coma is undergoing a major evolutionary event. This event, which appears to involve the infall and merging of a substructure, is likely related to the evolutionary phenomena observed in clusters at high redshift.

1. Introduction

The Coma cluster is generally considered to be the ultimate example of a dynamically-relaxed nearby rich cluster of galaxies. However, recent X-ray and optical observations of Coma have strongly altered that simple picture. As is described below, the SW region of the cluster is the site of unusual activity which manifests the important evolutionary processes currently taking place there.

2. X-ray and Optical Observations in the Coma Cluster

Recently acquired ROSAT PSPC observations of Coma (Briel *et al.*1992 and White *et al.*1993) as well as Spacelab 2 observations (Watt *et al.*1992) have revealed the existence of a secondary peak in the X-ray distribution centered $\sim$45' SW of the cluster center. Moreover, multi-fiber spectra obtained with the Hydra positioner at the KPNO 4-m by Caldwell *et al.*(1993) have demonstrated that $\sim$1/3 of the early-type galaxies in the SW region of Coma show evidence for recent star formation. In contrast, virtually all of the spectra of galaxies in the central region of Coma are, as expected, dominated by the light of old stars. The situation is illustrated in Fig. 1 (which is taken from Caldwell *et al.*1993), where the positions of all galaxies observed with Hydra are overlaid on the ROSAT X-ray map of Coma obtained by Briel *et al.*(1992). Normal-spectrum galaxies (dominated by an old stellar population) are plotted as x's, abnormal-spectrum galaxies (showing evidence for recent star-formation and/or nuclear activity) are denoted by filled squares, and the positions of a few bright galaxies are marked with plus signs for reference. An important feature of Fig. 1 is that while the great majority of the abnormal-spectrum galaxies lie close to the secondary X-ray peak, on average they are $\sim$10' closer to the cluster center than the secondary peak. To illustrate what is meant by an "abnormal"-spectrum galaxy, in Fig. 2 we compare the "abnormal"-spectrum galaxy D94 with a composite of normal-spectrum galaxies. Note that while the Balmer absorption lines are enhanced in D94 (indicating the presence of A stars in its spectrum), there are no signs of emission lines. As is discussed in Caldwell *et* al.(1993), the abnormal Coma spectra are very similar to the "post-starburst" spectra found frequently in clusters at $z\sim0.3$-0.5 which exhibit the Butcher-Oemler effect (Butcher and Oemler 1978). In addition, long-slit spectra obtained with the MMT have shown that the "post-starburst" nature of the abnormal-spectrum Coma galaxies typically can be traced out to a distance of at least 6" from the nucleus of the galaxies, which corresponds to ~ 3 kpc. Hence the recent star formation activity has occurred over a sizable area of these galaxies.

3. A Merging Substructure in Coma?

The most straightforward explanation for the range of phenomena seen in the SW region of Coma is hierarchical growth of the cluster through an infalling gas-rich group. One would expect such merging events to be more common in the past and more concentrated towards the cluster cores. However, there may be serious problems with this explanation for Coma. First, the velocity dispersion of the abnormal spectrum galaxies, which is ~ 1300 km/sec, is at least as high as, and likely higher than, the rest of the cluster, and certainly inconsistent with the dispersion of ~ 500 km/sec expected for a local substructure. Second, on the deep ROSAT PSPC images the secondary X-ray peak appears to be diffuse and asymmetric, unlike that expected from an infalling gravitationally bound substructure. It therefore appears that the dynamical situation is even more interesting than originally suspected.

4. Connection with the Butcher-Oemler Effect

The evolutionary event occurring in the SW region of Coma has produced the same kind of "post-starburst" spectra in the early-type galaxies that are prevalent in star-forming galaxies in rich clusters at $z\sim0.3$-0.5. Hence the phenomenon seen in Coma may hold vital clues to the physical origin of the Butcher-Oemler effect. Since the Coma cluster is more than 10 times closer to us than the typical intermediate-z cluster, we have the opportunity here to study an evolutionary phenomenon at much higher S/N ratio and spatial resolution in both optical and X-ray than is possible at $z\sim0.3$-0.5.

5. Studies of Other Clusters

Deep ROSAT images of relatively nearby rich clusters of galaxies are revealing X-ray substructures in many of them, indicating that recent dynamical evolution has been a common phenomenon. Consequently, we are now expanding our program to other clusters than Coma. Recently, we obtained Argus multi-fiber spectra of galaxies in the cluster DC2048-52, and in October a deep PSPC image has been obtained of DC2048-52 with ROSAT.

Briel, U. G., Henry, J. P., & Bohringer, H. 1992, A&A, 259, L31
Butcher, H., & Oemler, A. 1978, ApJ, 219, 18
Caldwell, N., Rose, J. A., Sharples, R. M., Ellis, R. S., & Bower, R. G. 1993, AJ, 106, 473
Watt, M. P., Ponman, T. J., Bertram, D., Eyles, C. J., Skinner, G. K., & Willmore, A. R. 1992, MNRAS, 258, 738
White, S. D. M., Briel, U. G., and Henry, J. P. 1993, MNRAS, 261, p8

382 Evolution in Nearby Clusters of Galaxies

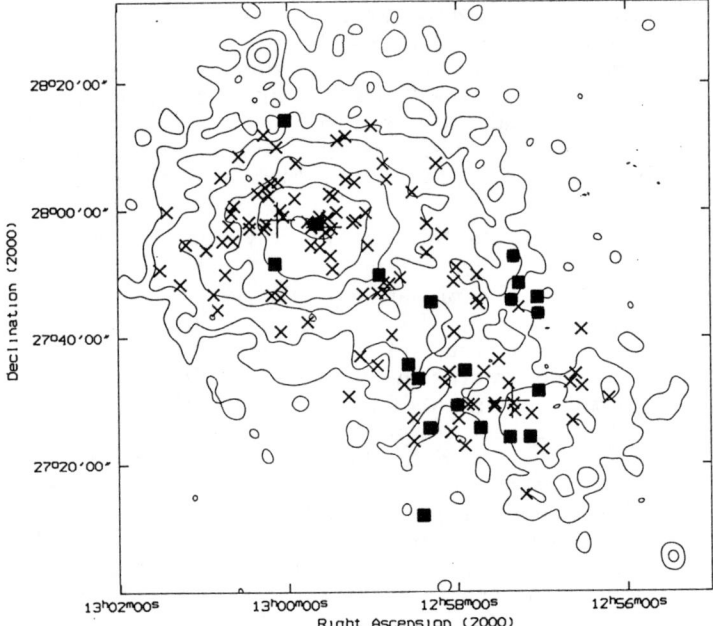

Fig. 1. The positions of all galaxies observed with Hydra are overlaid on the ROSAT X-ray map of Coma obtained by Briel et al. (1992). Normal-spectrum galaxies (dominated by an old stellar population) are plotted as x's, abnormal-spectrum (showing evidence for recent star-formation and/or nuclear activity) are denoted by filled squares, and the positions of a few bright galaxies are marked with plus signs for reference.

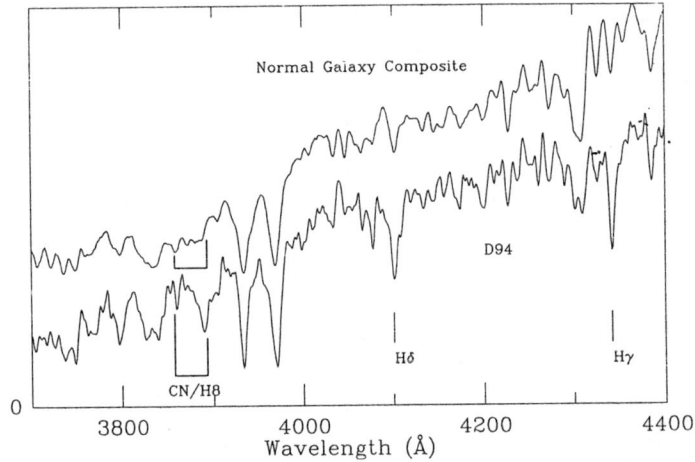

Fig. 2. The spectrum of a "post-starburst" galaxy in the SW region of Coma (D94) is compared to the composite spectrum of many normal-spectrum galaxies near the center of the cluster.

ROSAT OBSERVATIONS OF BRIGHT X-RAY CLUSTERS: ABELL 2142 AND ABELL 2199

Hassan Siddiqui, Gordon Stewart and Jackie Butcher
Department of Physics and Astronomy, University of Leicester, Leicester, UK

Alastair Edge and Roderick Johnstone
Institute of Astronomy, Cambridge, UK

ABSTRACT

We present detailed analyses of the X-ray data for Abell 2142 and Abell 2199. Standard deprojection techniques were applied to the data to obtain 3D temperature and density profiles. The projected PSPC temperature profile in conjunction with the integrated Ginga spectrum of the clusters provides constraints on the total cluster mass. With additional optical data, the minimum baryonic mass is inferred.

INTRODUCTION

Abell 2199 and A2142 are amongst the ten brightest clusters in the 0.1 - 10 keV band. A2199 is unique as it is the only cluster yet detected in the extreme UV (by the ROSAT WFC), partly due to the presence of a very low galactic column. A2142 is an exceptionally X-ray luminous cluster ($L_x \approx 4 \times 10^{45}$ ergs s^{-1}) classified optically as a binary. The results presented here are from observations using the ROSAT PSPC, with additional data from Ginga observations. Contour images of these clusters are shown in fig 1. The exposure times obtained were 10 and 15 ks, and total cluster count rates of 8.0 and 3.3 ct s^{-1} for A2199 and A2142 respectively. From the spectral data available, we obtained 2-d temperature and column density distributions, and hence inferred a value for the mass deposition rate of the cooling flow. Deconvolution of the surface brightness distributions (fig 2) yielded the 3-d density distribution and detailed cooling flow properties. This method also produced a 3-d temperature distribution, which after 'reprojection' was compared to the spectral data to constrain the range of gravitational potentials used. Finally, using optical data, a lower limit to the baryonic mass fraction was determined (fig 3).

PRINCIPAL RESULTS AND CONCLUSIONS

Baryonic content: Within a radius of 2 Mpc, the total gas mass for A2199 is $\approx 10^{14}$ M$_\odot$ and for A2142 is approximately 4×10^{14} M$_\odot$ within a radius of 2.5 Mpc. The total gravitational masses obtained using the assumptions of hydrostatic equilibrium and spherical symmetry are in good agreement with those inferred from the observed velocity dispersions. The total mass in the case of A2199 is $0.4 - 0.6 \times 10^{15}$ M$_\odot$ (< 2 Mpc) and $0.9 - 1.5 \times 10^{15}$ M$_\odot$ (< 3 Mpc) for A2142.

These results give a lower limit to the total baryonic fraction of 20% (A2199) and 35% (A2142) at 2 Mpc. This is much higher than that predicted by standard nucleosynthesis models assuming the inflationary paradigm of a $\Omega_0 = 1$ universe is correct (ie $\Omega_b = 0.06$, for $H_0 = 50$ km s^{-1} Mpc^{-1}), and typical of recent X-ray observations of clusters[1,2]. Baryon segregation in clusters cannot account for this descrepancy, as our data shows a gradual *increase*, rather than decrease, in the fraction of baryons at larger radii. Simulations of baryon infall during cluster formation have also produced much lower values than that observed[1]. Further work in reconciling theory with observations is required, unless the density of the universe is indeed less than the closure value. Alternatively, our standard interpretation of the element abundances is in error.

Presence of Cold Gas: A2199 and A2142 possess strong cooling flows of ≈ 250 and ≈ 650 M$_\odot$ yr^{-1} respectively. The spectrum of the central regions of A2142 and A2199 requires an intrinsic partially-absorbing column of $\sim 10^{20} - 10^{21}$ cm^{-2}. Studies of SSS data[3] also show excess absorption, although their results are marginally larger than ours. If this absorption is due to cold gas clouds, then up to 10^{12} M$_\odot$ of gas has been deposited in this form by the cooling flows.

Cluster Morphology: Both clusters show ellipticity (the axis ratio for A2199 is 1.3 and for A2142 is 1.6). In A2199 the major axis of the cluster gas is parallel to that of the central cD galaxy. This effect is found for a number of clusters[4] suggesting that the dynamics of the cluster as a whole are related to the evolution of the central galaxy. A more complex situation appears in A2142, as the line connecting the two brightest galaxies in the cluster, and the individual orientation of these galaxies all lie along the elongation axis of the cluster gas. This suggests that a merger is taking place along this line, between two subclusters of which these two galaxies were the most dominant members. It would be expected that, due to reheating and sustained bulk motion, the merger would disrupt a cooling flow, which would take a considerable amount of time ($\sim 10^9$ yr) to be re-established[5]. We must therefore be witnessing the merger at a relatively early stage.

References

1. White, S.D.M., Navarro, J.F., Evrard, A.E., and Frenk, C.S. 1993 *Nature* **366**, 429

2. Henry, J. P., Briel, U. G., and Nulsen, P. E. J. 1993 *Astron. Astrophys* **271**, 413

3. White, D. A., Fabian, A. C., Johnstone, R. M., Mushotzky, R. F., and Arnaud, K. A. 1991 *M.N.R.A.S.* **199**, 883

4. Rhee, G. F. R. N., and Latour, H. J. 1991 *Astron. Astrophys* **243**, 38

5. McGlynn, T. A., and Fabian, A. C., 1984 *M.N.R.A.S* **208** 709

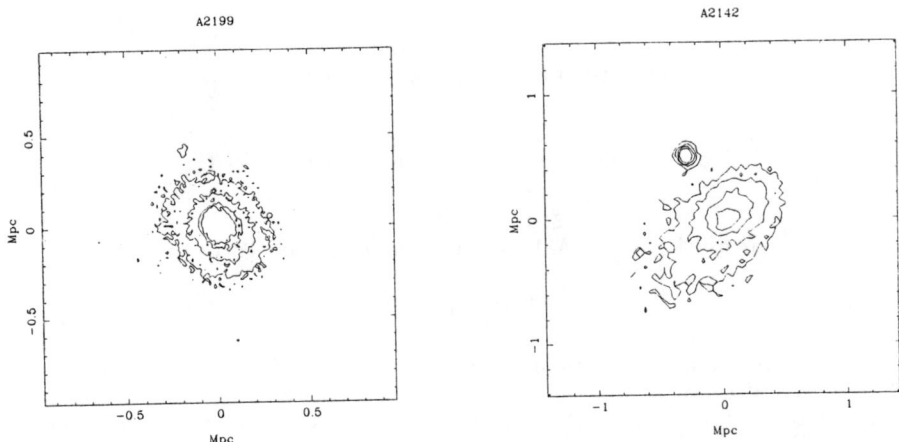

Figure 1: Contour images of A2199 and A2142

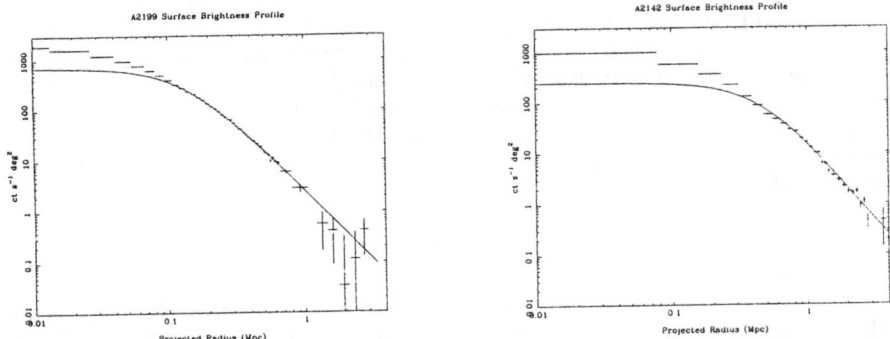

Figure 2: Surface brightness profiles obtained from the imaging data. The models shown are King fits to the data excluding the inner region of each cluster, which is affected by the cooling flow. The best-fitting core radii found for A2199 and A2142 are 126^{+69}_{-63} and 446^{+171}_{-140} kpc respectively; the corresponding β values are $0.62^{+0.07}_{-0.06}$ and $0.69^{+0.10}_{-0.07}$ respectively.

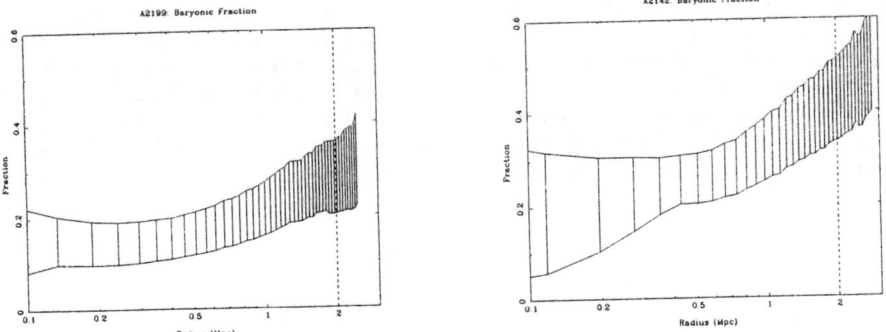

Figure 3: The fraction of baryonic matter within each cluster, estimated from the gas, gravitational and optical mass profiles. The dashed lines indicate the lower limits at 2 Mpc.

X-RAY LUMINOSITIES OF DISTANT RADIO-SELECTED CLUSTERS OF GALAXIES

Jennifer L. Sokoloski
Department of Physics, UC Berkeley

Ruth A. Daly
Department of Physics, Princeton University

& Simon J. Lilly
Department of Astronomy, University of Toronto

ABSTRACT

A study of distant clusters of galaxies with radio sources was undertaken to study the evolution of the intracluster medium with redshift, and to determine if this evolution is related to the evolution of the radio source population. We obtained ROSAT HRI data for four clusters of galaxies that contain radio sources at redshifts of about 0.5. The results of these observations are presented here. Two of the clusters were detected, and two have interestingly low upper bounds on their X-ray luminosities. The detected X-ray emission is unresolved, and in each case could originate from either the hot intracluster medium or the radio source. The fact that the two clusters that contain FRII radio sources have upper bounds on their X-ray luminosities, while the two clusters with compact or amorphous radio sources have detected X-ray luminosities supports the notion that the evolution of the environments of FRII radio sources with redshift may be linked to the evolution of the intracluster medium with redshift, as discussed by Wan et al. (1994a,b).

1. Introduction

The environments of powerful, classical double, Fanaroff-Riley class II (FRII) radio sources evolve with redshift. These sources are found in relatively low-density environments at low redshift, and in both high and low-density environments at redshifts of about 0.5 (Hill & Lilly 1991; Yates, Miller, & Peacock 1989). This evolution may be understood in terms of the evolution of the intracluster medium (ICM); FRII sources may be found in distant clusters because the high pressure ICM that is known to exist in low-redshift clusters may not be in place in more distant clusters. To investigate the relation between the evolution of the environments of FRII radio sources and the evolution of the intracluster medium we obtained ROSAT HRI data for 4 clusters of galaxies containing radio sources at redshifts of about 0.5. The results and implications of these observations are presented in brief here; a more detailed discussion is presented by Sokoloski, Daly, & Lilly (1994).

2. Data Analysis and Results

Three fields were observed with the ROSAT satellite's HRI camera, on 12/9/91, 12/10/91, and 1/8/91 respectively. X-ray fluxes and luminosities were determined for the radio galaxy 53W076 and the radio loud quasar 53W080, while upper limits on these quantities were obtained for the radio galaxies 1130+34 and 1245+34 (see Table 1). Fluxes and flux densities f_ν at an observed energy of 1 keV were calculated with the PROS software using a power law spectral model $f_\nu \propto \nu^{-\alpha_x}$ with $\alpha_x = 0.4$, and Galactic neutral hydrogen absorbing column densities were obtained from Heiles (1975). The specifics of the data analysis procedure are described by Sokoloski et al. (1994). Throughout the paper we use a Hubble constant $H_o = 100$ km s^{-1} Mpc^{-1} and de/acceleration parameter $q_o = 0$. The results of the data analysis are presented in Table 1.

Table 1: ROSAT HRI X-ray Results

Source	z	HRI Counts	S/N	f_ν (0.5-2.0 keV)	L_x (0.5-2.0 keV) 10^{43} erg s^{-1}
53W076	0.390	8.5 ± 3.2	2.7	2.9 ± 1.1	0.60 ± 0.22
53W080	0.546	47.9 ± 9.2	5.2	19.0 ± 3.6	8.2 ± 1.6
1130+34	0.512	10.3 ± 6.7	1.6	< 12.2	< 4.4
1245+34	0.409	< 0	n/a	< 2.7	< 2.5

The source names and redshifts are listed in columns 1 and 2. The total HRI counts are listed in column 3. The source counts were obtained from regions with radii $r \approx 50$ kpc and $r \approx 350$ kpc for 53W076 and 53W080 respectively, and regions with radius $r \approx 350$ and $r \approx 270$ kpc for 1130+34 and 1245+34 respectively. The signal to noise ratio is listed in column 4. The flux density integrated over the zero redshift 0.5 to 2.0 keV energy range is given in column 5 in units of 10^{-14} erg cm^{-2} s^{-1}. The limits quoted for 1130+34 and 1245+34 are 3σ upper limits computed withiin radii of $r \approx 350$ and $r \approx 270$ kpc respectively. The 0.5 to 2.0 keV luminosity integrated over the 0.5 to 2.0 keV energy range at the source has been computed and is listed in column 6 in units of 10^{43} erg s^{-1}.

3. Origin of the X-ray Emission

One of the main questions we wish to address in this study is the degree to which the intracluster medium is in place in high-redshift clusters. To this end, we compare our data to that from two types of low-redshift samples. First, the X-ray luminosities for the 4 clusters listed in table 1 are compared with the thermal emission detected from optically similar clusters at low redshift, using the EINSTEIN survey results of Abramopoulos & Ku (1983) and Jones & Forman (1984). Second, the relation between 2.0 keV X-ray luminosity density and 5.0 GHz radio core luminosity density for radio sources both in clusters and in the field found by Fabbiano et al. (1984) is used to determine whether the detected X-ray emission could be due to AGN processes alone. The results of these two comparisons are discussed in detail by Sokoloski et al. (1994). The result is that the detected X-ray emission could arise from either a hot

ICM or the radio source. The two upper limits place interesting constraints on the intracluster media in 1130+34 and 1245+34, and on the evolution of the intracluster media of clusters that contain FRII radio sources.

It is interesting that the two clusters that contain FRII radio sources, 1130+34 and 1245+34, have fairly low upper limits on their X-ray luminosities, while the clusters that contain radio sources with amorphous radio structure have detected X-ray emission. These results are mirrored by the larger high-redshift sample compiled by Wan et al. (1994a,b). Thus, it appears that the evolution of the environments of FRII radio sources with redshift is related to the evolution of the intracluster medium, though it is not clear at this stage whether this is due to the evolution of the intracluster medium in clusters with a given mass, or the evolution of the masses of the clusters. This will be determined by future studies.

It is a pleasure to thank Lauren Jones, Lin Wan, and Ed Groth for helpful discussions. This work was supported in part by the National Aeronautics & Space Administration, and the U. S. National Science Foundation.

REFERENCES

Abramopoulos, F., & Ku, W. H.M. 1983, ApJ, 271, 446.
Fabbiano, G., Miller, L., Trinchieri, G., Longair, M., & Elvis, M. 1984, ApJ, 276, 115.
Heiles, C. 1975, Astron Astrophys Suppl, 20, 37
Hill, G. J., & Lilly, S. J. 1991, ApJ, 367, 1.
Jones, C., & Forman, W. 1984, ApJ, 277, 38.
Sokoloski, J., Daly, R. A., & Lilly, S. J. 1994, ApJ, submitted
Wan, L., Daly, R. A., Jones, L., & Lilly, S. J. 1994a, ApJ, submitted
Wan, L., Daly, R. A., Jones, L., & Lilly, S. J. 1994b, these proceedings
Yates, M. G., Miller, L, & Peacock, J. A. 1989, MNRAS, 240, 129

HIGH RESOLUTION X-RAY AND OPTICAL IMAGING OF THE CENTRAL GALAXY IN THE CENTAURUS CLUSTER

W. B. Sparks, R. I. Jedrzejewski, F. Macchetto
Space Telescope Science Institute, 3700 San Martin Drive
Baltimore, MD 21218

Email ID
sparks@stsci.edu, rij@stsci.edu, duccio@stsci.edu

ABSTRACT

HST FOC and ROSAT HRI X-ray imaging is presented of the galaxy NGC4696. The X-ray image shows similar morphology to ground-based optical images of the emission-line gas and dust in this object — a one-armed spiral mostly to the South of the nucleus. A morphological correspondence between cool and hot gas phases is required in the "merger" interpretation of this and other similar systems. That interpretation has been proposed as a viable alternative to cooling-flows, without the large mass depositions of the cooling-flow scenario, and relying on quite different physical processes to explain the observations. A prediction of the merger model, in which cold accreted gas draws energy from pre-existing hot coronal gas via electron conduction, was that the X-ray and optical gaseous morphologies should be closely related. Here, we present a preliminary analysis of high resolution X-ray imaging data, recently obtained with the ROSAT satellite, as well as high resolution HST FOC data of the core regions of this galaxy.

1. Introduction

NGC4696 has been the focus of intensive study because of its key role in understanding the physics of interstellar gas and dust in early-type galaxies and clusters. This galaxy at the center of the Centaurus Cluster has been cited as a classical "cooling-flow" galaxy, in which the hot X-ray emitting coronal gas cools through thermal instability and becomes visible in the optical as a system of emission line filaments (Fabian et al. 1982).

On the other hand, a detailed analysis of the properties of the emission line filaments by Sparks, Macchetto & Golombek (1989) showed that they are dusty and that the dust is apparently normal. The surface brightness of the filaments is high compared to the cooling flow model and the emission is more localized to the galaxy centre. Kinematic peculiarities are also known to exist within the stars of the galaxy (Danziger & Focardi 1988). For these and other reasons, the alternative interpretation that the gas/dust system represents merger debris was proposed by Sparks, Macchetto & Golombek (1989) and independently from an analysis of the IRAS data by de Jong et al. (1990).

In this model, thermal interaction between the cool accreted gas and hot, pre-existing coronal gas both cools the hot gas (to mimic a cooling-flow) while heating the cool gas, giving optical and infra-red emission. This succeeds quantitatively in explaining the optical line-emission surface brightness and flux, the dust content and infra-red properties. The prediction made by Sparks et al.(1989), and pursued theoretically in Sparks (1992), is that the X-ray coronal gas must be related to the optical line emission gas quite closely.

In order to test whether the coronal gas has any 'knowledge' of the cooler optical gas, we obtained ROSAT HRI observations of NGC 4696.

2. Observations and Results

ROSAT HRI observations of duration 16.3 KSec were obtained 20–22 Jan 1993. In addition, an earlier effort to obtain these data resulted in the acquisition of 0.5 KSec 8–11 January 1992. There are Einstein HRI observations amounting to 9.4 KSec also available as archival material, which were obtained during 1979, day 218. Each of these data sets has been examined. Einstein ICP observations are described in detail in Matilsky, Jones and Forman (1985).

HST FOC images were taken of the core of NGC 4696 on 23–25 July 1991. 80 minutes of exposure time were obtained with the F372M filter (a medium-band filter centered on the O [II] 3727Å line), along with exposures in neighboring continuum regions and the ultraviolet.

3. Discussion

Preliminary inspection of the ROSAT data show (i) a one-armed spiral morphology similar to but more extensive than the optical one-armed spiral, seen in both HII gas and dust (Fig. 1a). (ii) the peak of the X-ray emission, accepting at face value for now the aspect solutions as provided for the two ROSAT datasets and the Einstein data, is located within the $H\alpha$ filament system, offset from the galaxy nucleus by about 6–10 arcsec (Fig. 1b).

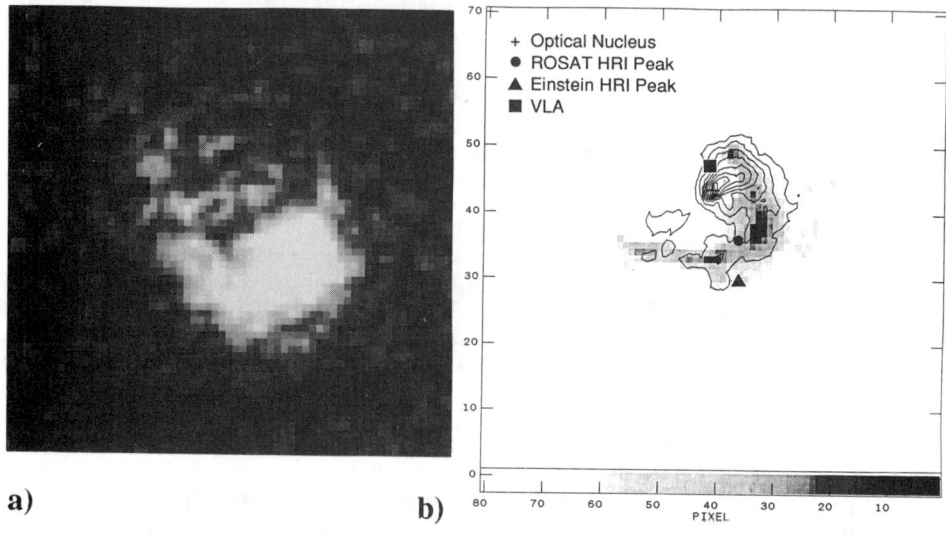

Fig. 1a) ROSAT HRI image of the central 1.5 × 1.5 arcmin of NGC4696. Fig. 1b) From Sparks, Macchetto & Golombek (1989), a contour map of the $H\alpha$ emission superposed on a grey-scale of the dust absorption. Added for comparison are the locations of the peaks of the X-ray emission in the ROSAT and Einstein HRI images, using default aspect solutions. The optical position was taken from HST observations, and also shown is the VLA position of the peak of the radio emission from O'Dea et al.(1993).

This suggests that the two phases are indeed related, which is a crucial result. Our ground-based work leads us to favour a merger origin for the cooler gas and dust, and so so this would imply that merging can affect the hot phase. The alternative, that the filaments are after all from the "cooling-flow" type process, would require further analysis to understand the optical and infra-red properties of the filaments.

The HST imaging data (Fig. 2) clearly shows the presence of more than one component in the nucleus. The separation of the two most obvious components is 0.26 arcsec, which is approximately 70 pc at an assumed distance of 59 Mpc. The core regions thus resemble those of M31 (Lauer et al. 1993) and Mrk 315 (Mackenty 1993), which have both been proposed as the products of recent accretion/merger events.

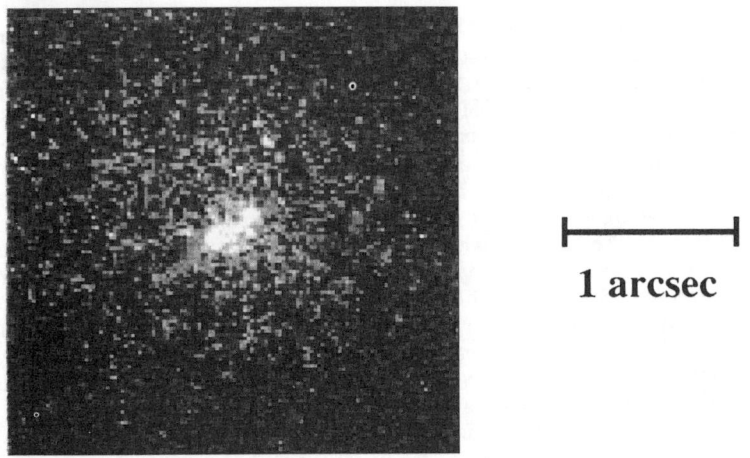

Fig. 2 HST FOC image of the core of NGC 4696 in the F372M filter, showing the multiple nucleus.

Future quantitative analysis of these data should enable us to make progress distinguishing between the models. Comparison with theoretical profiles for both thermal conduction fronts and cooling-flow condensations will be carried out, and extraction of the limited spectral data inherent in ROSAT HRI observations will be of great interest.

References

Danziger, J. and Focardi, P., 1988. In *Cooling Flows in Clusters and Galaxies*, ed A.C. Fabian.
de Jong *et al.*, 1990. *Astron. Astrophys.*, **232**, 317.
Fabian, A., Atherton, P., Taylor, K. & Nulsen, P., 1982. *M.N.R.A.S.*, **201**, 17P.
Lauer, T.R. *et al.*, 1993. *Astron. J.*, **106**, 1436.
Mackenty, J., 1993. In *STScI Newsletter*, Vol 10 No. 1.
Matilsky, T., Jones, C. & Forman, W., 1985. *Astrophys. J.*, **291**, 621.
O'Dea *et al.*, 1993. *Astrophys. J.*, submitted.
Sparks, W.B., Macchetto, F. & Golombek, D., 1989. *Astrophys. J.*, **345** 153.
Sparks, W.B., 1992. *Astrophys. J.*, **399**, 66.

THE EVOLUTION OF THE INTRACLUSTER MEDIUM IN CLUSTERS WITH EXTENDED RADIO SOURCES

Lin Wan, Ruth A. Daly
Department of Physics, Princeton University

Lauren V. Jones
Department of Astronomy, University of Alabama

& Simon J. Lilly
Department of Astronomy, University of Toronto

ABSTRACT

A sample of high-redshift X-ray clusters was compiled by combining our ROSAT observations of four distant clusters with results available in the published literature. This sample of 26 distant clusters is divided into two subsamples, those with and without FRII radio sources. It is found that the clusters with FRII radio sources tend to be less luminous X-ray sources than the high-redshift clusters that do not have FRII radio sources, though there are exceptions.

In addition to comparing the two high-redshift subsamples, each subsample is compared with a low-redshift sample of clusters with known optical and X-ray properties. The high-redshift clusters with FRII sources also tend to be underluminous in the X-ray relative to low-redshift clusters with similar optical properties (richness), and significantly underluminous in the X-ray if the expected X-ray emission from the radio sources is taken into account, though the highest power FRII sources appear to be exceptions. The high-redshift clusters without FRII radio sources show no tendency to be underluminous relative to optically similar low-redshift clusters.

This suggests that the evolution of the environments of FRII radio sources with redshift is closely linked with the evolution of the intracluster medium. It is interesting to note that the fraction of clusters in our distant sample that contain FRII radio sources and are underluminous X-ray sources is similar to the fraction of clusters missing from the high-redshift X-ray luminosity function relative to that for low-redshift clusters. If further investigations indicate that high-redshift clusters with and without FRII radio sources are of similar richness, this would imply that the evolution of the cluster X-ray luminosity function is more strongly linked with the evolution of the intracluster medium than with the evolution of cluster masses, and that the relation between cluster mass and X-ray luminosity established for low-redshift clusters evolves with redshift.

1. Introduction

There is a substantial amount of evidence that suggests the environment may play an important role in determining the radio properties of radio sources. At low redshift, the low-luminosity, "edge-darkened" FRI radio sources usually inhabit rich cluster environments (Longair & Seldner 1979; Prestage & Peacock 1988, hereafter PP88), while the powerful, "edge-brightened" FRII sources tend to lie in either small groups of sub-Abell richness or isolated fields (Longair & Seldner 1979; PP88). This difference between the environments of FRI and FRII sources is thought to be due to the fact that rich clusters contain a hot intracluster medium (ICM) whose high pressure severely disrupts the jets coming from the radio core and/or the related shocks, and thus prevents the formation of "classical-double" FRII sources, except for sources with very high beam power, such as Cygnus A (PP88).

Recent studies of the cluster environments of radio sources at $z \sim 0.5$ reveal that higher redshift FRII sources often inhabit much richer environments than they do at low redshift, while FRI sources inhabit similar environments at both high and low redshift (Hill & Lilly 1991, hereafter HL91; Yates, Miller, & Peacock 1989). If, indeed, the high-pressured ICM prevents FRII sources from forming in rich clusters at low redshift, then the change in environment of FRII sources from a redshift of zero to a redshift of 0.5 would indicate a change of the ICM. Namely, the ICM of some clusters at redshift of about 0.5 might be of lower pressure and thus sustain an FRII source. The lower pressure could result from a lower gas temperature, lower gas density, or both, and clusters with a low-pressured ICM would in turn have a low X-ray luminosity.

With this in mind, we compiled a high-redshift sample consisting of 26 X-ray clusters with and without radio sources by combining clusters available in the published literature with the four clusters for which we have ROSAT HRI data. The X-ray properties of high-redshift clusters with and without FRII radio sources are compared, and each high-redshift subsample is compared with the low-redshift cluster sample of Abramopoulos & Ku (1983). The results of these comparisons are presented in detail by Wan et al. (1994); the results are summarized here.

2. Results

The 26 clusters that comprise our high-redshift cluster sample are listed in Table 1. The sources are drawn primarily from the Henry et al. (1982, 1992) samples, and the Sokoloski, Daly, & Lilly (1994a,b) sample. Sources known to contain FRII radio sources are marked with an asterisk, and the 0.5 to 2.0 keV luminosities have been estimated assuming an X-ray spectral index of 0.4 (appropriate for thermal bremsstrahlung emission), a Hubble constant of 100 km s^{-1} Mpc^{-1}, and de/acceleration parameter $q_o = 0$. The comparison of the distribution of X-ray luminosities for the Henry et al. (1982, 1992) samples, which contain several FRII sources, with the low-redshift Abramopoulos & Ku (1983) sample shows that the distributions are similar, so no obvious selection bias is evident for the distant cluster sample. This suggests that the distant cluster sample may be representative, though it is not statistically complete.

Of the 26 clusters, 8 are known to contain FRII sources. The X-ray luminosities of the clusters containing FRII sources are low compared with those without FRII sources, except for 3C215 and 3C295. Note that 5 of the clusters with FRII sources have fairly low upper bounds on their X-ray luminosities.

3C 295 is a source with a radio power comparable to that of Cygnus A–the extremely powerful FRII source known to be in a rich cluster at low redshift. Such high radio power might enable 3C295 to survive as an FRII source even in a cluster with hot, dense ICM, as is the case for Cygnus A.

It is not yet clear whether the X-ray emission from 3C 215 is from the cluster around it or from the quasar itself. It was observed in an X-ray survey of quasars and was found to have a spectral index of $0^{+0.8}_{-0.2}$ (Wilkes & Elvis 1987), suggesting an AGN origin of the X-ray emission, though this spectral index is consistent with thermal bremsstrahlung emission from the ICM. As is the case for most of the FRII sources in this study, the X-ray detections and bounds are consistent with the X-ray emission expected from the AGN based on the Fabbiano et al. (1984) relation between radio and X-ray emission; this is discussed in more detail by Wan et al. (1994).

For the 8 clusters with FRII sources, the mean X-ray luminosity and standard deviation of the mean, including the 5 upper bounds as detections, is: $L_x = (5.6 \pm 1.6) \times 10^{43}$ erg s^{-1}. If 3C295 and 3C215 are excluded for the reasons discussed above, $L_x = (3.1 \pm 0.6) \times 10^{43}$ erg s^{-1}. For the 18 clusters without FRII sources, the average X-ray luminosity, including 1 upper bound as a detection is: $L_x = (6.7 \pm 1.3) \times 10^{43}$ erg s^{-1}. It is not clear at this stage whether this is due to evolution of the masses of the clusters or evolution of the ICM within clusters of a given mass.

It is a pleasure to thank Jeno Sokoloski for helpful discussions. This work was supported in part by the National Aeronautics & Space Administration, and the U. S. National Science Foundation.

Table 1: X-Ray Data

Source Name	$L_{x(0.5-2keV)}$ (10^{43}erg/s)	Redshift	Abell Class	Source Name	$L_{x(0.5-2keV)}$ (10^{43}erg/s)	Redshift	Abell Class
53WO76	0.6	0.39	0 or 1	A370	11	0.373	0
53WO80	8.23	0.546	>0	A908	5.6	0.390	0
PKS0116+08	6.04	0.594	0	A913	5.1	0.366	1
1130+34*	<4.4	0.512	>2	0302.5+1717	4.05	0.394	–
1245+34*	<2.1	0.409	1 or 2	0024+1654	3.2	0.39	2
3C215*	13.7	0.411	1 or 2	0302+1658	8.10	0.424	–
3C295*	11	0.46	1	1621.5+2640	7.35	0.426	–
3C19*	4.0	0.482	–	0147.8-3941	2.50	0.373	–
3C244.1*	<4.90	0.428	1	1512.4+3647	7.59	0.372	–
3C268.3*	<1.5	0.371	1	2053.7-0449	10.1	0.583	–
3C330*	<1.4	0.549	1	1333.3+1725	8.91	0.460	–
0016+1609	24.6	0.540	3	0303+17	3.3	0.450	–
0418.3-3844	2.24	0.350	–	A895	<1.5	0.36	1

REFERENCES

Abramopoulos, F., & Ku, W. H.-M. 1983, ApJ, 271, 446.
Fabbiano, G., Miller, L., Trinchieri, G., Longair, M., & Elvis, M. 1984, ApJ, 277, 115.
Henry, J. P., Soltan, A., Briel, U., & Gunn, J. E. 1982, ApJ, 262, 1.
Henry, J. P., Gioia, I. M., Maccacaro, T., Morris, S. L., Stocke, J. T., & Wolter, A. 1992, ApJ, 386, 408.
Hill, G. J., & Lilly, S. J. 1991, ApJ, 367, 1.
Longair, M. S., & Seldner, M. 1979, MNRAS, 189, 433.
Prestage, R. M., & Peacock, J. A. 1988, MNRAS, 230, 131.
Sokoloski, J., Daly, R. A., & Lilly, S. J. 1994, ApJ, submitted
Sokoloski, J., Daly, R. A., & Lilly, S. J. 1994, these proceedings
Wan, L., Daly, R. A., Jones, L. V., & Lilly, S. J. 1994, ApJ, submitted
Wilkes, B. J., & Elvis, M. 1987, ApJ, 323, 243.
Yates, M. G., Miller, L., & Peacock, J. A. 1989, MNRAS, 240, 129.

AGN

HI AND THE X-RAY SPECTRUM OF NGC 6251

M. Birkinshaw and D.M. Worrall
Harvard-Smithsonian Center for Astrophysics Cambridge, MA 02138-1596

Email ID
mb1@cfa.harvard.edu, dmw@cfa.harvard.edu

ABSTRACT

The ROSAT PSPC spectrum of NGC 6251 can be fitted with either a multiple-component model obscured by the gas column within our Galaxy or by a single power-law component obscured by this Galactic gas and an additional column $N_H = 1.3 \times 10^{21}$ cm^{-2}. We have used the VLA to search for the HI line in absorption towards the radio core of NGC 6251 to set limits on the total X-ray absorbing column. We find marginal evidence for a weak, broad, absorption feature associated with NGC 6251 which could provide the extra HI column suggested by the simpler X-ray spectral fit.

1. NGC 6251

The elliptical galaxy NGC 6251 is well known for its exceptional radio source (Perley, Bridle & Willis 1984) which contains a bright VLBI-scale component associated with the active nucleus of the galaxy, a prominent jet which extends several arcmin from this central component, and bright radio lobes. The galaxy is also known to be an X-ray source, although before our *ROSAT* PSPC observation it was unclear whether the X-ray emission is associated with the active nucleus, a galactic atmosphere, or the radio jet.

2. X-ray data

A detailed spatial analysis of the ROSAT data (Birkinshaw & Worrall 1993) found that the X-ray structure is strongly dominated by an unresolved source, but that this component is embedded in an X-ray halo which is likely to be associated with an extended galactic atmosphere. No detectable X-ray emission was found from the radio jet.

The PSPC spectrum of the unresolved emission from NGC 6251 can be fitted with three plausible models: a single steep power law with an absorption excess $\Delta N_H = (1.3 \pm 0.6) \times 10^{21}$ cm^{-2} above the expected Galactic HI column (of 5.6×10^{20} cm^{-2}); the superposition of a flat power-law component and a thermal emission component, with no excess N_H; or the superposition of hot and cool thermal emission components, with no excess N_H. The presence of extended X-ray emission suggests that the second and third of these options are more probable than the first: in Birkinshaw & Worrall (1993) the second possibility was chosen as the more likely since the hot component cannot be maintained in hydrostatic equilibrium in NGC 6251.

However, to rule out the first of these possibilities we must prove that there is no excess absorption on the line of sight towards NGC 6251. Since NGC 6251 is at Galactic latitude +31°, the 21-cm HI line should be a good indicator of the X-ray absorbing column.

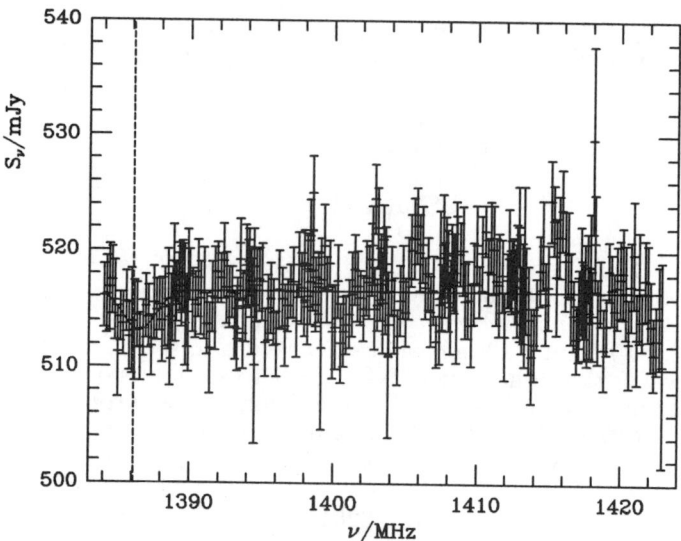

Figure 1. The flux density of the central component of NGC 6251 as a function of frequency. Eight separate 31-channel spectra are plotted: no baseline corrections have been made. The frequency of the HI line at the systemic velocity of NGC 6251 is indicated by the vertical dashed line. The curve shows a Gaussian fit to the most significant absorption feature in the HI data (see Sec. 4).

3. HI observations

We searched for gas that might cause the low-energy absorption in our PSPC spectrum of NGC 6251 by looking for neutral hydrogen in absorption against the bright central core of the radio source. Data were taken with the VLA in its A array, where the high resolution of the telescope (the FWHM of the synthesized beam is about 2 arcsec) ensures that the core source is well separated from jet emission. Eight spectra, each of 31 channels (with channel width 195 kHz), were taken for ~ 30 minutes each to cover a velocity range $v_{lsr} = -550$ to $+7910$ km s^{-1} with substantial overlap. The data should be sensitive to absorbing gas in NGC 6251, gas with anomalous velocities in our Galaxy, and intervening material.

The VLA data were reduced in the usual fashion, with flux density and spectral bandpass calibrations based on observations of 3C 286. The individual spectral channel maps provide estimates of the flux density of the core of NGC 6251 at each frequency. The resulting spectrum is shown in Fig. 1.

4. HI analysis

A detailed search for absorption features at any frequency in Fig. 1 leads to the 3σ limits on the peak optical depth of absorbing gas, τ_0, as a function of apparent line FWHM, $\Delta\nu$, given in Table 1.

Table 1. Optical depth limits

$\Delta\nu$/MHz	τ_0
0.25	< 0.041
0.5	< 0.027
1.0	< 0.019
2.0	< 0.015
4.0	< 0.011

If any absorbing gas is assumed to have a spin temperature T_S, then the corresponding limits to the HI column are

$$N_H = 4 \times 10^{22} \, \tau_0 \, (\Delta\nu/\text{MHz}) \, (T_S/100 \text{ K}) \text{ cm}^{-2} \, ,$$

so that no single feature with intrinsic velocity FWHM < 400 km s^{-1} can cause the $\Delta N_H = 1.3 \times 10^{21}$ cm^{-2} excess required by the power law model for the X-ray spectrum of NGC 6251 unless the gas has $T_S > 100$ K.

A careful search of the HI spectrum (Fig. 1) locates one feature with modest ($\sim 2\sigma$) significance. This feature has amplitude -3.5 ± 1.7 mJy and is centered at 1386.51 ± 0.95 MHz with intrinsic velocity FWHM = 700 ± 400 km s^{-1}. Its center is offset by only 100 ± 200 km s^{-1} from the systemic velocity of NGC 6251. The absorbing column associated with this feature is

$$N_H = (8 \pm 4) \times 10^{20} \, (T_S/100 \text{ K}) \text{ cm}^{-2} \, ,$$

which might be sufficient to cause the excess absorption in the fit of the PSPC spectrum to a simple power law component, and could be associated with the weak dust features reported by Nieto et al. (1983). This feature escaped detection by van Gorkom et al. (1989), probably because of its weakness and location at the edge of their spectral band.

6. Conclusions

No high-column absorbing gas has been detected in the line of sight towards the nucleus of NGC 6251, but the weak evidence for the presence of a faint, broad, absorption feature suggests that there might be sufficient gas in NGC 6251 to produce the observed PSPC spectrum from a simple power law. More sensitive HI data are needed to test the reality of the absorption feature.

This work was supported by NASA grants NAG5-1648 and NAG5-2312, and NASA contract NAS5-30934. The VLA is operated by the National Radio Astronomy Observatory under a cooperative agreement with the National Science Foundation.

References

Birkinshaw, M. & Worrall, D.M., 1993. ApJ, 412, 568.
Nieto, J.C., Coupinot, G., LeLièvre, G. & Madsen, C., 1983. MNRAS, 203, 39P.
Perley, R.A., Bridle, A.H. & Willis, A.G., 1984. ApJS, 54, 291.
van Gorkom, J.H., Knapp, G.R., Ekers, R.D., Ekers, D.D., Laing, R.A. & Polk, K.S., 1989. AJ, 97, 708.

ROSAT OBSERVATIONS OF THE BLAZARS PKS 1034-293 AND PKS 1335-127

L. Maraschi,
Dipartimento di Fisica, Università di Genova, Italy

A. Ciapi, G. Fossati,
Dipartimento di Fisica, Università di Milano, Italy

G. Tagliaferri
Osservatorio Astronomico di Brera, Milano, Italy

A. Treves
SISSA/ISAS, Trieste, Italy

Email ID
maraschi@astmiu.mi.astro.it

ABSTRACT

ROSAT PSPC observations of the blazars PKS 1034-293 and PKS 1335-127 are presented. In both cases the spectra are well fitted by a single absorbed power law. For PKS 1034-293 we obtain $F_{1keV} = 0.23 \pm 0.02$ μJy, $\Gamma = 1.23^{+0.54}_{-0.50}$, $n_H = 1.3^{+1.7}_{-1.2}$ (10^{20} cm^{-2}) (90 % uncertainty). The column density is significantly smaller than the galactic value derived from Stark *et al.* 1992. For PKS 1335-127 we find $F_{1keV} = 0.50 \pm 0.02$ μJy, $\Gamma = 2.29^{+0.55}_{-0.51}$, $n_H = 6.60^{+3.00}_{-2.23}$ (10^{20} cm^{-2}): in this case the column density derived from the fit is in good agreement with the recent determination by Jackson *et al.* 1993.

1. Introduction

BL Lac objects and Highly Polarized Quasars (HPQs) share the properties of a flat or inverted radio spectrum, high polarization in the radio and optical bands and violent variability at all frequencies. However they are distinguished by the strength of the emission lines, by the luminosity and redshift distribution and by the morphology of the extended radio emission (*e.g.* Ghisellini *et al.* 1993).

A further important difference between BL Lacs and HPQs may be the shape of the X-ray spectrum. *Einstein* and *EXOSAT* observations showed that HPQs generally have flatter spectra than BL Lac Objects (Worrall and Wilkes (1990), Sambruna *et al.* (1993)). The studied objects were however chosen by different observers with various criteria. With the scope of acquiring new information in a softer band, with a more sensitive instrument, and in the spirit of studying spectral properties starting from well defined samples, we initiated a program of X-ray observations with *ROSAT* of sources within the sample defined by Impey and Tapia (1988). This was selected on the basis of systematic optical polarization measurements of quasar-like objects in the Kühr *et al.* (1981) catalogue of radio-sources, stronger than 2 Jy at 5 GHz. Further criteria of selection for *ROSAT* observations were the optical brightness and the low Galactic hydrogen column.

Both sources have a measured redshift (0.312 for PKS 1034–293 and 0.539 for PKS 1335–127, Stickel et al. 1989, 1993) determined from the observation of emission lines with an equivalent width which indicates that both objects are HPQs.

Together with PKS 0537–441 (Treves et al. 1993) the two objects are the first observed within our *ROSAT* program.

2. Observations

The observations were held on January 2nd, 1992 and on February 1st, 1992 with duration of 3973 sec and 3615 sec, respectively for PKS 1034–293 and PKS 1335–127.

The spatial analysis was performed with the MIDAS/EXSAS dedicated software; after examination of the image, the source counts for PKS 1335–127 were derived integrating over a circular source area of 2.5' radius and subtracting the background evaluated on the surrounding corona of radii 2.5' – 5'.

In the case of PKS 1034–293 a source with an intensity $\sim$ 20% of the target object was present $\sim$ 5' away. Therefore, to avoid possible contamination we considered a source area of radius 2', and the background was evaluated on four circles of 4' radius close to the target and not affected by sources.

Standard corrections for vignetting, dead time and effective area were applied. No variability is apparent from inspection of the light curves. Spectra were produced from the integrated exposures with the energy channels rebinned to have bins statistically meaningful ($S/N \gtrsim 5$). This yielded 10 bins for PKS 1034–293 and 15 bins for PKS 1335–127.

The spectral analysis was performed with the XSPEC package, assuming the simple model of an absorbed power law (cross sections from Morrison and McCammon (1983) were used). The best fit was obtained through χ^2 minimization, considering (i) Γ as free parameter and n_H fixed at the Galactic value, (ii) Γ and n_H free.

The results are reported in the table. The uncertainties are 90 % confidence intervals, for one or two parameters of interest (case i) and ii) respectively).

Results from Rosat PSPC observations

source	n_H fixed			n_H free			
	n_H^{gal} (10^{20} cm^{-2})	Γ	χ_r^2	n_H^{fit} (10^{20} cm^{-2})	Γ	χ_r^2	F_{1keV} (μJy)
PKS 1034–293	4.74	$2.05^{+0.26}_{-0.32}$	2.70	$1.31^{+1.68}_{-1.23}$	$1.23^{+0.54}_{-0.50}$	0.99	0.23 ± 0.02
PKS 1335–127	6.02	$2.18^{+0.18}_{-0.19}$	0.95	$6.60^{+3.00}_{-2.23}$	$2.29^{+0.55}_{-0.51}$	1.01	0.50 ± 0.03

For PKS 1335–127 the column density derived from the fit is consistent with the Galactic one, which was accurately measured by Jackson et al. 1993.

For PKS 1335-127 the fitted power law slope is (2.3 ± 0.5), somewhat steeper than the average value of 1.5 ± 0.2 derived for HPQs by Worrall and Wilkes (1990) in the *Einstein* band.

For PKS 1034-293 the fit with fixed n_H is poor and the fit with free n_H yields a value significantly lower than the galactic one as derived from the catalogs of Stark et al. (1992) and Dickey and Lockman (1990). This may indicate the presence of a soft excess. Thus, if the galactic n_H is confirmed, altogether the spectrum appears concave with a soft component dominating at low energies.

Both sources were previously observed with *Einstein* (Worrall and Wilkes, 1990; Impey and Neugebauer, 1988) at an intensity level close to the present one.

3. Discussion

Observations of the two sources at other wavelengths are relatively scarce. Photometry was unsystematic, but still clearly indicating high variability. We have collected data available from the literature in order to construct the broad band energy distribution. In the case of PKS 1034-293 the X-ray flux is in excess and the X-ray spectrum is flatter than the extrapolation from lower frequencies. This suggests that X-rays derive from a different process than the radio to optical emission which is thought to derive from synchrotron emission. The possible soft excess indicated by the spectral fit may instead be interpreted as the high energy extension of the synchrotron emission. Thus, though formally similar to the X-ray spectra of Seyfert galaxies, both components of the X-ray spectrum of this object could have a non thermal origin.

The case of PKS 1335-127 is more ambiguous. Nevertheless, if the observed X-rays were the extension of the optical spectrum the object should have been very bright at the time of the X-ray observations and the optical spectrum substantially flatter than previously measured. Therefore also in this case it seems more likely that the observed X-rays belong to a separate spectral component. Simultaneous multifrequency observations are required to clarify this point.

References

Dickey J.M. & Lockman F.J., 1990, *Ann. Rev. Astron. Astrophys.*, **28**, 215
Ghisellini G et al. , 1993, *Ap. J.*, **407**, 65
Impey C.D. & Neugebauer G., 1988, *A. J.*, **95**, 307
Impey C.D. & Tapia S., 1988, *Ap. J.*, **333**, 666
Jackson N., Browne I.W.A. & Warwick R.S., 1993, *A&A*, **274**, 79
Kühr et al. , 1981, *A&AS*, **45**, 367
Morrison R. & McCammon D., 1983, *Ap. J.*, **270**, 119
Sambruna R. et al. , 1993, submitted to *Ap. J.*
Stark A.A. et al. , 1992, *Ap. J. Suppl.*, **79**, 77
Stickel M., Fried J.W. & Kühr H., 1989, *A&AS*, **80**, 103
Stickel M., Kühr H. & Fried J.W., 1993, *A&AS*, **97**, 483
Treves A. et al. , 1993, *Ap. J.*, **406**, 447
Worrall D.M. & Wilkes B.J., 1990, *Ap. J.*, **360**, 396

X-ray Variability of the Quasar 4C 39.25 and Bent Relativistic Jets

Yun Fei Zhang[1] and Alan P. Marscher
Department of Astronomy, Boston University

Email ID
zhang@orion.harvard.edu   marscher@buast0.bu.edu

ABSTRACT

We have observed the peculiar superluminal quasar 4C 39.25 with the ROSAT PSPC at two epochs. We examine both short-term (constant flux on intraday and several-day timescales) and long-term (30% increase over 2 yr) variability of the soft X-ray emission. The entire radio to submillimeter spectrum also rose by 30% over the same period. This supports the twisted relativistic jet model for 4C 39.25 (Marscher et al. 1991), in which the Doppler beaming factor increases monotonically as the velocity vector of the superluminally moving component (here identified as the source of the X-rays) continues to bend toward the line of sight.

1. Introduction

The quasar 4C 39.25 (0923+392; z=0.699) is an object classified as a "peculiar superluminal" radio source (Marcaide et al. 1989; Marscher et al. 1991). On parsec scales, the compact double structure (components $\underline{a}$ and $\underline{c}$) observed in the 1970's has changed to a multi-knot jet, with knot $\underline{b}$ moving superluminally at $3.5h^{-1}c$ ($q_0 = 0.5$ and $H_0 = 100h$) between $\underline{a}$ and $\underline{c}$, which remain stationary relative to each other. A fourth component detected at 1.3 and 0.7 cm on the western side of the jet could be the "core" of the source (Alberdi et al. 1993a,b). The radio spectrum of the source peaks at about 8 GHz with a power-law slope of -0.52 above 10 GHz, in contrast to the higher turnover frequency and flat $\sim$ 1 to 50 GHz spectrum typical of most compact, variable radio sources. The radio light curve of 4C 39.25 peaked in 1978, reached a minimum in 1984, and has been increasing monotonically since. The position angle of the magnetic field in the VLA core $\lesssim 0\rlap{.}''1$ component rotated from being parallel to the jet axis in 1984 (Kollgaard et al. 1990) to perpendicular to the jet axis in 1987 (Marscher et al. 1991).

The Einstein Observatory IPC detected X-ray emission from 4C 39.25 that was best fit by a single power-law spectrum $S_{keV} = 0.4(h\nu)_{keV}^{-0.37}$ μJy with absorbing column density $N_H = 1^{+9}_{-1} \times 10^{19}$ cm^{-2} (Wilkes & Elvis 1987). With a Galactic neutral hydrogen column density of 1.69×10^{20} cm^{-2} reported from 21 cm line emission (Stark et al. 1984), an extra soft X-ray spectral component was suggested to explain the deficit in column density derived from the X-ray measurements.

Using our ROSAT PSPC data, we examine both the short-term and long-term variability of the soft X-ray emission from 4C 39.25. We also discuss the bent jet model and X-ray radiation mechanism in light of the radio and X-ray observations.

---

[1] Also at Harvard-Smithsonian Center for Astrophysics

2. ROSAT Observations

Table 1 lists our ROSAT PSPC observations of 4C 39.25 and the resultant spectral model together with the earlier Einstein IPC spectral model (Wilkes & Elvis 1987). A single component power-law is used to fit the data. The uncertainties listed correspond to 90% confidence levels.

Table 1: Summary of X-ray Observations at Quasar 4C 39.25

Obs. Date	Net Counts	S(1keV) (μJy)	α	N_H (10^{20} cm^{-2})
1993 Apr 22	1356	$0.69^{+0.092}_{-0.092}$	$1.32^{+0.42}_{-0.44}$	$2.02^{+1.37}_{-1.04}$
1993 Apr 24	1589	$0.67^{+0.055}_{-0.050}$	$1.30^{+0.28}_{-0.24}$	$1.87^{+0.70}_{-0.61}$
1993 Apr 26	1793	$0.69^{+0.067}_{-0.061}$	$1.11^{+0.33}_{-0.30}$	$1.43^{+0.81}_{-0.71}$
1993 Apr 28	1905	$0.64^{+0.066}_{-0.067}$	$1.15^{+0.35}_{-0.32}$	$1.46^{+0.89}_{-0.73}$
1993 Apr 30	2082	$0.69^{+0.069}_{-0.070}$	$1.22^{+0.35}_{-0.32}$	$1.73^{+0.90}_{-0.75}$
1991 Apr 15–23	741	$0.52^{+0.091}_{-0.097}$	$1.17^{+0.61}_{-0.51}$	$1.57^{+1.59}_{-1.12}$
1979 Oct 19	1016	$0.4\ ^{+0.2}_{-0.2}$	$0.37^{+0.05}_{-0.02}$	$0.1\ ^{+0.9}_{-0.1}$

3. The Bent Jet Model and the X-ray Emission from 4C 39.25

The bent model proposed to explain 4C 39.25 (Marscher et al. 1991; Zhang et al. 1990; Alberdi et al. 1993a) consists of a slightly bent (by several degrees) jet with main axis oriented close to the line of sight and a compressed, shocked region moving relativistically along the jet. The base of the jet is misaligned relative to the line of sight, resulting in a weak flux density for the VLBI core.

Under such a scenario, selective enhancement by Doppler favoritism makes the sections of the jet pointing most closely toward the line of sight brighter than the remainder of the quiescent jet. This leads to the bright stationary components *a* and *c*. This apparent parsec-scale double structure remains stable as long as the jet geometry is unchanged. One possiblility for the bending mechanism is large-scale Kelvin-Helmholtz instabilities (Hardee 1990). A shock propagating down the jet is responsible for the superluminal component, *b*, whose apparent speed changes with the angle of the velocity vector. The variation in the Doppler factor of component *b* is the main cause of time variability in the brightness and polarization of 4C 39.25. The dominant polarization of the shock (moving component) corresponds to a magnetic field oriented parallel to the shock front, as observed with VLBI (Zhang et al. 1994). The bent jet model also predicts that the variation in the Doppler beaming factor of the shock should affect both the synchrotron and X-ray flux densities in the same way if the X-rays are synchrotron self-Compton (SSC) emission from the superluminal component.

With the SSC model parameters determined via radio observations, it is possible to predict the expected X-ray flux density of the shocked component (see Marscher, 1987 for formulae). The spectrum of component *b* in 4C 39.25 peaks with a flux density of 4.5 Jy at 8 GHz. The VLBI observations give an angular diameter of 0.54 mas and the superluminal motion indicates that $\delta \approx 5h^{-1}$ ($q_0 = 0$ and $H_0 = 100h$; Alberdi et al.

1993a). The SSC model then predicts a self-Compton flux density of about 0.6 μJy at 1 keV, very close to the observed value.

4. Discussion

We find no evidence of intraday soft X-ray variations from 4C 39.25, nor was there any significant variation on timescales of 2–8 days in late April 1993. This is consistent with the absence of intraday radio variations from the source (Quirrenbach et al. 1992), and agrees with the VLBI results that the main emission region is separated by several parsecs from the central engine of the source and has a size of about 1 pc. However, there was a ~ 30% increase in the X-ray flux density from April 1991 to April 1993. The radio flux density of the source increased by the same factor over the same two-year period. The one-to-one correspondence between the two wavebands agrees with the prediction made by the relativistic bent jet model, if the X-rays arise from the SSC process in the superluminal component. The predicted SSC X-ray flux density of 4C 39.25, based on the parameters measured by VLBI and single-dish observations, agrees with the value measured by ROSAT to within the accuracy of the calculation. In the ROSAT PSPC observations, we do not detect the soft X-ray excess reported by Wilkes & Elvis (1987) based on Einstein Observatory IPC observations. The spectral index of the X-ray emission is, however, considerably steeper than the Einstein value (and also than the high-frequency radio slope but similar to the infrared slope). Our best-fit column density N_H is consistent with the Galactic value (Stark et al. 1984) within the 90% confidence level.

ACKNOWLEDGMENTS: This material is based on research supported by NSF grant AST-9116525 and by NASA grants NAG5-1943 and NAG5-1637.

References

Alberdi, A., et al. 1993a, *Astrophys. J.*, **402**, 160
Alberdi, A., et al. 1993b, *Astron. Astrophys.*, **271**, 93
Hardee, P. E. 1990, in *Parsec-Scale Radio Jets*, ed. J. A., Zensus & T. J., Pearson (Cambridge: Cambridge University Press), 266
Kollgaard, R. I., Wardle, J. F. C., & Roberts, D. H. 1990, *Astron. J.*, **100**, 1057
Marcaide, J. M., Alberdi, A., Elósegui, P., Schalinski, C. J., Jackson, N., & Witzel, A. 1989, *Astron. Astrophys.*, **211**, L23
Marscher, A. P. 1987, in *Superluminal Radio Sources*, ed. Zensus, J.A. & Pearson, T.J. (Cambridge: Cambridge University Press), 280
Marscher, A. P., Zhang, Y. F., Shaffer, D. B., Aller, H. D., & Aller, M. F. 1991, *Astrophys. J.*, **371**, 491
Quirrenbach, A., et al. 1992, *Astron. Astrophys.* **258**, 279
Stark, A. A., Heiles, C., Bally, J., & Linke, R. 1984, Bell Labs, available on-line at U.S. ROSAT Science Data Center
Wilkes, B. J. & Elvis, M. 1987, *Astrophys. J.*, **323**, 243
Zhang, Y. F., et al. 1994, in preparation
Zhang, Y. F., Marscher, A. P., Shaffer, D. B., Marcaide, J. M., Alberdi, A., & Elósegui, P. 1990, in *Parsec-Scale Radio Jets*, ed. J.A. Zensus & T.J. Pearson (Cambridge: Cambridge University Press), 66

Excess X-ray Absorption toward Giga-Hertz Peaked Radio Sources

Yun Fei Zhang[1] and Alan P. Marscher
Department of Astronomy, Boston University

Email ID
zhang@orion.harvard.edu   marscher@buast0.bu.edu

ABSTRACT

We report excess X-ray absorption toward two sources with quite different redshifts: the $z = 2.07$ quasar PKS 0528+134 and the relatively nearby ($z = 0.077$) radio galaxy OQ 208 (1404+286). Both sources exhibit features common to Giga-Hertz Peaked spectrum (GPS) radio sources. Excess X-ray absorption has been found toward other high-z sources (many also GPS sources) and interpreted as either intrinsic to the host galaxy/cluster or absorption by an intervening galaxy or cloud along the line of sight (Elvis et al. 1993). VLBI and X-ray observations reveal a distorted jet-like structure and relativistically beamed emission region in PKS 0528+134. The detection of excess absorption over a wide redshift range favors the intrinsic absorption model. The absorbing gas may also shape the radio morphology and spectrum of a GPS source.

1. Introduction

Giga-Hertz Peaked Spectrum (GPS) sources are distinguished from other classes of compact radio sources by their radio spectra, which display sharp peaks between about 1 and 10 GHz. GPS sources usually have unresolved or simple double structure over milliarcsecond (mas) scales and lack extended emission over scales $\gtrsim 0''.05$. The variability and polarization of GPS sources are generally significantly lower than those of core-jet sources, and moderate X-ray emission has been measured by the Einstein Observatory (Worrall & Wilkes 1990). The spatial distribution of GPS sources favors high redshifts (O'Dea et al. 1991).

Two possible scenarios for the distinctive features displayed by GPS sources are that these objects represent a stage in the normal evolution of extragalactic radio sources ("evolutionary models") or that they are examples of radio sources suffering from extreme environmental conditions ("environmental models"). The evolutionary models suggest that the interstellar medium surrounding a GPS source is not dense enough to restrict the outflow(s) from the central engine. Therefore, what we see in a GPS source is the "naked" emitter itself. The compactness of GPS sources and lack of extended structure is explained by their supposed relative youth ($\lesssim 10^4$ years). The peaked spectra are caused by synchrotron self-absorption, although the reason for the ~ 1–10 rather than ~ 10–1000 GHz spectral down-turn of most compact sources is not understood in this model. According to the environmental models, a GPS source is embedded deep within a dense surrounding medium, which smothers the nuclear emission, confining the source to a very compact region by temporarily cutting off the energy supply to more extended regions. The medium can also cause the 1–10 GHz spectral peak via thermal free-free absorption.

---

[1] Also at Harvard-Smithsonian Center for Astrophysics

X-ray observations could be helpful toward understanding the nature of GPS sources. Here we report the results of our ROSAT PSPC observations of the GPS sources OQ 208 and PKS 0528+134. Strong confinement of the emission region in OQ 208 has been invoked in an effort to interpret the radio observations (de Bruyn 1990), while the quasar PKS 0528+134 is the brightest known extragalactic γ-ray emitter (Hunter et al. 1993). We examine the parameters observed at both radio and X-ray frequencies with respect to the synchrotron self-Compton (SSC) model in an attempt to determine the X-ray radiation mechanism in the emitting region.

2. Source Characteristics and ROSAT X-ray Observations

Table 1 lists the basic features of the two observed sources. The radio spectrum is peaked at ν_m with flux density S_m and follows a power-law ($S_\nu \propto \nu^{-\alpha}$) for $\nu > \nu_m$. The Galactic neutral hydrogen column density N_H is measured from the 21 cm line emission (Stark et al. 1984). Table 2 summarizes the results of our ROSAT PSPC observations.

Table 1: Characteristics of PKS 0528+134 and OQ 208

Source	ID	z	α	ν_m (GHz)	S_m (Jy)	pc-scale structure	N_H (10^{20} cm^{-2})
0528+134	QSO	2.07	0.6	7	1.1	bent jet	25
1404+286	RG	0.077	1.3	4.8	3.0	point-like	1.4

The spectral analysis is carried out with a single component power-law spectral model; uncertainties represent the 68% confidence level. The last column in Table 2 lists the ratio, R_H, of equivalent X-ray absorbing column density to the Galactic column density. In both sources, X-ray absorption in excess of the Galactic value is measured.

Table 2: ROSAT PSPC Observations of PKS 0528+134 and OQ 208

Source	Obs. Date	Net Counts	S(1keV) (μJy)	α_x	N_{H_x} (10^{20} cm^{-2})	R_H
0528+134	1991 Apr 16	269	$1.59^{+0.24}_{-0.23}$	$2.2^{+0.38}_{-0.32}$	84.4^{+7}_{-5}	3.4
1404+286	1991 Jul 12	128	0.032 ± 0.016	1.97 ± 0.73	16.7 ± 7.9	12

3. Synchrotron Self-Compton Model

The SSC model incorporates synchrotron self-absorption as the cause of the spectral peak and first-order inverse Compton scattering as the X-ray emission mechanism (e.g., Marscher 1987). For a given spectral peak, optically thin spectral index, and angular size, the SSC model can predict the X-ray flux density from self-Compton scattering. This analysis should be particularly accurate for GPS sources, since the peaked spectra allow accurate measurements of the parameters S_m and ν_m.

For OQ 208, VLBI observations at about the turnover frequency give a source size of 0.94 milliarcseconds with no bulk relativistic motion detected (Charlot 1990; de Bruyn 1990). The SSC model then gives a magnetic field $B \approx 0.008$ gauss and a self-Compton flux density of ~ 0.08 μJy at 1 keV. This is only a factor of about 3 greater than the X-ray flux density measured by our ROSAT PSPC observations.

For PKS 0528+134, our 1990–91 VLBI observations measure an angular size of 0.36 mas. The SSC model then gives $B \approx 0.0014\delta$ gauss, where δ is the Doppler factor of bulk relativistic motion. The predicted self-Compton flux density is $\sim 5000\delta^{-5.2}\mu$Jy at 1 keV. Given the measured X-ray flux density of 1.6 μJy, we find that bulk relativistic motion with a Doppler factor $\delta \gtrsim 4.7$ is necessary. This is consistent with the beaming requirements imposed by the γ-ray observations (Hunter et al. 1993).

4. Discussion

Detection of excess X-ray absorption in a low-redshift GPS source, OQ 208, suggests that the absorber is intrinsic to the source itself or the immediate surroundings. Possible candidates include a cooling flow or interstellar/intra-cluster clouds. The parsec-scale bent jet observed in PKS 0528+134 and strong confinement of OQ 208 seen at radio frequencies both suggest that the interstellar medium is dynamically important in these sources. This is consistent with the excess X-ray absorption detected by our ROSAT observations. The common appearance of GPS sources among the excess X-ray absorbers might suggest a common origin of both phenomena. Our results are also consistent with the high detection rate of excess absorption among high-redshift quasars reported by Elvis et al. (1993). The detection of excess X-ray absorption in the low-redshift object OQ 208 clearly favors the intrinsic absorption model.

The predicted SSC X-ray flux density of OQ 208 agrees with the value measured by ROSAT to within the accuracy of the calculation; hence, no relativistic beaming is required. The overprediction of the SSC X-ray flux density of PKS 0528+134 relative to the ROSAT value suggests relativistic beaming with $\delta \gtrsim 5$, which is consistent with the γ-ray properties of this source. This can be tested by further VLBI observations to check for apparent superluminal motion. The role of relativistic beaming in GPS sources and the relationship of PKS 0528+134 to other GPS sources remain unclear.

ACKNOWLEDGMENTS: The authors thank Drs. Martin Elvis and Fabrizio Fiore for discussions regarding OQ 208. This material is based on research supported by NSF grant AST-9116525 and by NASA grants NAG5-1943 and NAG5-1637.

References

de Bruyn, A.G. 1990, *Compact Steep-Spectrum and GHz-Peaked Radio Sources*, ed. by Fanti, C., Fanti, R, O'Dea, C.P., & Schilizzi, R.T., 206

Charlot, P. 1990, *Astron. Astrophys.*, **229**, 51

Hunter, S.D., et al. 1993, *Astrophys. J.*, **409**, 134

Elvis, M., Fiore, F., Wilkes, B., McDowell, J., & Bechtold, J. 1993, *Astrophys. J.*, in press

Marscher, A.P. 1987, in *Superluminal Radio Sources*, ed. Zensus, J.A. & Pearson, T.J. (Cambridge: Cambridge University Press), 280

O'Dea, C.P., Baum, S.A., & Stanghellini, C. 1991, *Astrophys. J.*, **380**, 66

Stark, A.A., Heiles, C., Bally, J., & Linke, R. 1984, Bell Labs, available on-line at U.S. ROSAT Science Data Center

Worrall, D.M., & Wilkes, B.J. 1990, *Astrophys. J.*, **360**, 396

Zhang, Y.F., et al. 1994, *Astrophys. J.*, submitted

X-RAY EMISSION IN POWERFUL RADIO GALAXIES AND QUASARS

D.M. Worrall
Harvard-Smithsonian Center for Astrophysics, Cambridge, MA 02138-1596
and
C.R. Lawrence, T.J. Pearson, A.C.S. Readhead
OVRO 105-24, California Institute of Technology, Pasadena, CA 91125

Email ID
dmw@cfa.harvard.edu

ABSTRACT

ROSAT is the first mission to have detected X-ray emission in radio galaxies which are both powerful ($l_{178 \text{ MHz}} > 10^{27}$ W Hz^{-1} sr^{-1}) and distant ($z > 0.4$), enabling tests of "unified schemes" through a comparison of the X-ray and radio properties of powerful quasars and radio galaxies. These radio galaxies are faint in X-rays, but, nevertheless, ROSAT is capable of resolving any associated X-ray-emitting gas of cluster dimension.

We present ROSAT PSPC observations of two such radio galaxies and suggest that there is a component of unresolved X-ray emission in powerful, high-redshift radio galaxies which may be related to the radio core; this will be tested by ROSAT observations of other powerful radio galaxies.

1. ROSAT PSPC Observations of two Radio Galaxies

We observed 3C 220.3 and 3C 280 with the ROSAT PSPC during the AO1 pointed phase of the mission (Table 1). 3C 220.3 was undetected. 3C 280 gives 71 ± 12 net counts ($0.2 - 1.9$ keV). For absorption only by gas in our Galaxy, the PSPC spectrum fits a power law with $0.5 < \alpha < 2.0$ and a 1 keV flux density of 1.7 ± 0.9 nJy. Some intrinsic absorption and a steeper power-law slope cannot be excluded, although there are sufficient counts in the low-energy channels to suggest that such absorption has a column density $\lesssim 4 \times 10^{21}$ atoms cm^{-2}. A Raymond-Smith thermal spectral model also agrees with the data; for Galactic absorption only, any temperature $\gtrsim 0.4$ keV is acceptable.

Table 1

Radio Galaxy	z	$\log N_H^a$	ROR[b]	Date	Exposure Time (s)
3C 220.3	0.685	20.515	700072	1991 Feb 27	8,791
3C 280	0.998	20.086	700073	1991 Jun 2-3	48,051

a. Galactic values from Stark et al. (1992).
b. ROSAT Observation Request number

2. X-ray Radial Profile of 3C 280

A point source alone gives a poor fit to the X-ray radial profile of 3C 280. A thermal β model (e.g., Sarazin 1986) gives a good fit, but the fit is improved still further if the source is modeled with a combination of a point source and β model (Fig. 1). The evidence for unresolved emission is suggestive, rather than compelling; moreover, the X-ray data do not distinguish between a thermal or a non-thermal origin for the possible unresolved component. However, the evidence for non-thermal X-rays from the nuclei of low-redshift radio galaxies (e.g., Fabbiano et al. 1984; Worrall & Birkinshaw 1994) and radio-loud quasars (e.g., Worrall et al. 1987) leads us to consider seriously the possibility that 3C 280 also emits non-thermal X-radiation from its radio core.

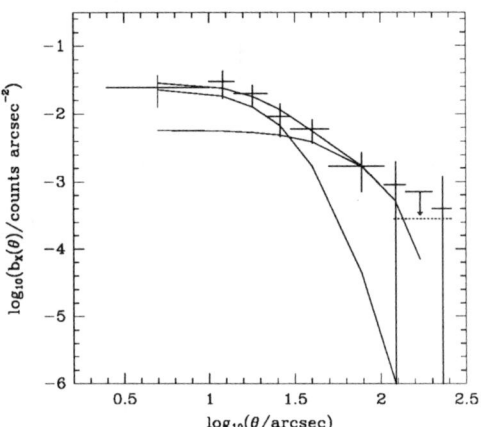

Figure 1: Background-subtracted PSPC radial profile for 3C 280. The best-fit model (upper curve), shown convolved with the PRF and background subtracted for comparison with the data, is a combination of an unresolved component (narrow curve) and a β model of core radius 65" for $\beta = 2/3$ (broad curve). The dotted line shows the contribution of the model to the background annulus.

3. Radio Core-related X-rays in Quasars and Galaxies

In Figure 2 we compare the unresolved X-ray and core radio emissions for the two radio galaxies with those of radio-loud quasars which both have similar total (isotropic) power and were observed with the *Einstein* IPC (Wilkes et al. 1994). The luminosity-luminosity and flux-flux plots are similar because the sources lie in a relatively narrow band of redshift. The fraction, R, of 5 GHz flux density in the core component of the quasars has a bimodal distribution, in which core-dominated quasars with $R > 0.5$ (filled squares) can be clearly distinguished from lobe-dominated quasars (open squares). From Figure 2 we find:

(a.) A correlation between the core X-ray and radio emission in core-dominated quasars.

(b.) 3C 280 lies on an extrapolation of the correlation.

(c.) Lobe-dominated quasars have X-ray emission in excess of the correlation.

The inference from point (a), that the X-rays from core-dominated quasars are beamed, is supported by earlier work. Point (b) is most simply explained if the same core X-ray to radio relationship holds for radio galaxies as for core-

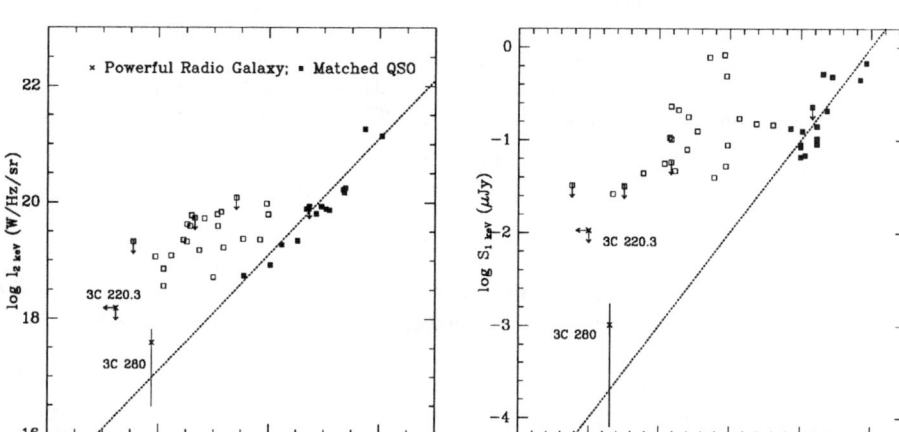

Figure 2: Core X-ray and radio flux densities for quasars (squares) matched in isotropic radio power with the radio galaxies 3C 220.3 and 3C 280. A line of slope unity (dotted) connects 3C 280 with the core-dominated quasars (filled squares), but not the lobe dominated quasars (open squares). Luminosity-luminosity (left) and flux-flux (right) plots are both presented to show that common-distance spreading has only a minor influence on the luminosity-luminosity plot and is not responsible for the correlation.

dominated quasars; this will be tested by ROSAT observations of other powerful radio galaxies. Point (c) suggests that lobe-dominated quasars contain an additional source of X-ray emission, possibly related to the nuclear X-ray emission in radio-quiet quasars. We infer that this compact X-ray component is obscured in radio galaxies, and dominated by beamed emission from the jet in core-dominated quasars. Further details of this work can be found in Worrall et al. (1994).

This work was primarily funded by NASA grants NAG5-1706 and NAG5-1882.

References

Fabbiano, G., Miller, L., Trinchieri, G., Longair, M. & Elvis, M. 1984, ApJ. 277. 115
Sarazin, C.L. 1986, Rev.Mod.Phys., 58, 1
Stark, A.A., Gammie, C.F., Wilson, R.W., Bally, J., Linke, R.A., Heiles. C. & Hurwitz, M. 1992, ApJS, 79, 77
Wilkes, B.J., Tananbaum, H., Worrall, D.M., Avni, Y., Oey, M.S. & Flanagan. J. 1993, ApJS, in press
Worrall, D.M. & Birkinshaw, M. 1994, ApJ, in press
Worrall, D.M., Giommi, P., Tananbaum, H. & Zamorani, G. 1987, ApJ. 313. 596
Worrall, D.M., Lawrence, C.R., Pearson, T.J. & Readhead, A.C.S. 1994. ApJ (Letters), in press

OPTICAL/UV/SOFT X-RAY QUASAR SPECTRA: MODELS vs. OBSERVATIONS.

Aneta Siemiginowska, Fabrizio Fiore, Martin Elvis,
Belinda J. Wilkes, Jonathan C. McDowell, Smita Mathur
Harvard-Smithsonian Center for Astrophysics, 60 Garden St., Cambridge, MA 02138

Email ID
siemiginowska@cfa.harvard.edu

ABSTRACT

We compare the optical to soft X-ray spectral energy distribution of six bright low-redshift (0.048<z<0.155) radio-quiet quasars with models of quasars where the soft excess is interpreted in terms of (a) optically thick thermal emission from the innermost region of an accretion disk in Schwarzschild and Kerr geometries, (b) reprocessing from ionized gas. Pure disk models even in a Kerr geometry cannot easily reproduce the observed optical to soft X-ray energy distribution. The range of parameters which give reasonable predictions is narrow: high inclinations, high accretion rates and central black hole mass $\sim 10^8 M_\odot$. We consider modifications of the disk models: 1) an underlying power law component extending from the infrared ($3\mu m$) to the X-ray; 2) reprocessing of hard X-rays. The modified models can explain the optical-UV and soft X-ray observations. Free-free models are discussed by Fiore in these Proceedings.

1. Data

The X-ray data are taken from Fiore et al. (1994a), where the ROSAT observations were fitted by two power laws and thermal bremsstrahlung plus power law models. IR, optical and UV observations are from Elvis et al. (1994) and Fiore et al. (1994b).

2. Accretion Disk Models

We consider a geometrically thin accretion disk around a supermassive ($10^6 - 10^8 M_\odot$) black hole, with accretion rates: 0.01 - 1.0 $\dot{M}_{Edd}$ and the efficiency η: 0.08 for non-rotating and 0.324 for rotating black hole.

(1) Accretion disks in Schwarzschild geometry, unless supplemented by a hot corona, systematically underpredict the soft X-ray emission.

(2) Accretion disk models in a Kerr geometry cannot reproduce the optical-UV slope for the range of parameters for which they can match the observed soft X-ray color. They are too steep (Fig.1).

(3) Only high inclination, high accretion rate, $M_{bh} = 10^8 M_\odot$ models can reproduce simultaneously the observed soft X-ray color and UV luminosity. Models with lower mass can reproduce the soft X-ray color but underpredict the UV luminosity (Fig.2). Models that could account for the soft X-ray color tend to overpredict the soft X-ray luminosity.

3. An IR-to-X-ray Underlying Power Law

When a power law (Elvis et al. 1986, Carleton et al. 1987) extending between

IR ($3\mu m$) and the X-rays with $\alpha_{IRX} = 1.25$ ' was added to the accretion disk spectra:

(1) the model can reproduce both the soft X-ray and the optical-UV colors (Fig.3);

(2) the $3\mu m$ and 0.4keV luminosities predicted by the models appeared in the observed range.

4. Comments on Disk Irradiation

Recent models of an accretion disk illuminated by an external X-ray source (Ross & Fabian, 1992; Matt et al. 1993) shows that the X-ray spectrum flattens with respect to accretion disk models. Reflection could well be an important component in the 0.2-2 keV region. For low accretion rates (corresponding to $\frac{L}{L_{Edd}} = 0.15$), and low ionization parameters, many lines are present in the soft X-ray band (O VIII Kα, Fe XVII&XVIII Lα lines). For higher accretion rates ($\frac{L}{L_{Edd}} = 0.30$) the line strength is much reduced. The limits on the emission line strength found in Fiore et al. (1994a) argue for high accretion rates ($\frac{L}{L_{Edd}} \sim 0.3$).

Also the outer regions of the disk which contribute to the optical-UV part of the spectrum can be modified by irradiation. Irradiation may cause the flattening of the spectrum as well as the wide range of observed optical-UV slopes.

4. Conclusions

We compared the observed optical-UV to soft X-ray spectral energy distribution of a sample of six radio-quiet, low-redshift quasars with the predictions of Big Blue Bump and soft X-ray component models.

(1) Assuming a physical link between the Big Blue Bump and the soft X-ray component, free-free (in general emission from ionized plasma) models can explain the observed soft X-ray color and the mean optical-UV color, but not the soft X-ray color and the spread in optical-UV color (Fiore these Proceedings).

(2) Pure disk models even in a Kerr geometry do not seem to have the required flexibility to account for the observed spread in optical-UV and soft X-ray slopes and luminosities. The flat soft X-ray component slope found by the PSPC requires high inclinations and high accretion rates, which overestimate the soft X-ray luminosity, when producing the correct UV luminosity.

(3) The model with an underlying power law component extending from the infrared ($3\mu m$) to the X-ray added to accretion disk emission, can explain *both* the optical-UV and soft X-ray slopes and luminosities *and* the observed $3\mu m$ luminosity.

ACKNOWLEDGEMENTS: This work was supported by NASA grants NAGW-2201 (LTSARP), NAG5-1872, NAG5-1883 and NAG5-1536 (ROSAT), and NASA contracts NAS5-30934 (RSDC), NAS5-30751 (HEAO-2) and NAS8-39073 (ASC).

References

Carleton, N.P., Elvis, M., Fabbiano, G., Willner, S.P., Lawrence, A., & Ward, M. 1987, ApJ, 318, 595.
Elvis, M. et al. 1986, ApJ, 310, 291
Elvis M. et al. 1994, ApJ Supp., submitted.
Fiore, F., Elvis, E., Siemiginowska, A., Wilkes, B.J., & McDowell, J.C. 1994a, ApJ,

submitted.

Fiore, F., Elvis, E., Siemiginowska, A., Wilkes, B.J., & McDowell, J.C., Mathur, S. 1994b, ApJ, submitted.

Matt, G. Fabian, A.C., & Ross, R.R. 1993, MNRAS, 264, 839.

Ross R.R. & Fabian A.C. 1993, MNRAS, 261, 74.

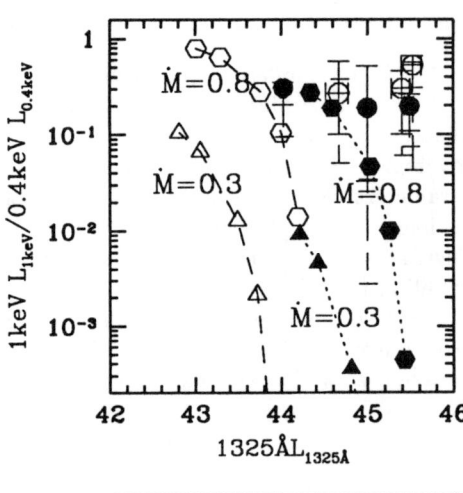

Figure 1. The Soft X-ray color plotted against the optical-UV color. Filled and open circles with error bars identify the quasars, the dashed errors bars indicate the error resulting from a power law plus free-free fit. Dashed lines identify accretion disk models in a Kerr geometry. Filled hexagons and triangles indicate disk inclinations, $\mu = \cos\theta = 1, 0.75, 0.5, 0.2, 0.1$ (soft X-ray color increases with inclination) with $M = 10^8 M_\odot$, accretion rates $0.3\ L_{Edd}$ and $0.8\ L_{Edd}$ respectively. Open hexagons and triangles identify disk models with $M = 10^7 M_\odot$ for the same inclinations and accretion rates.

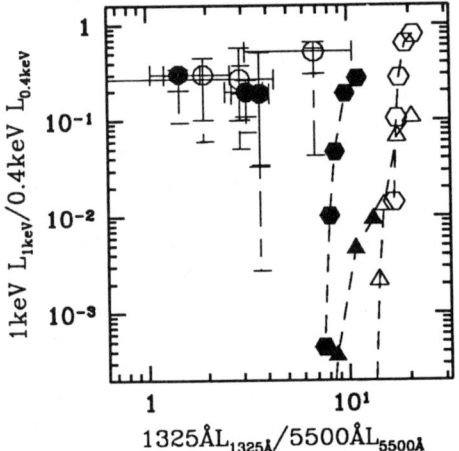

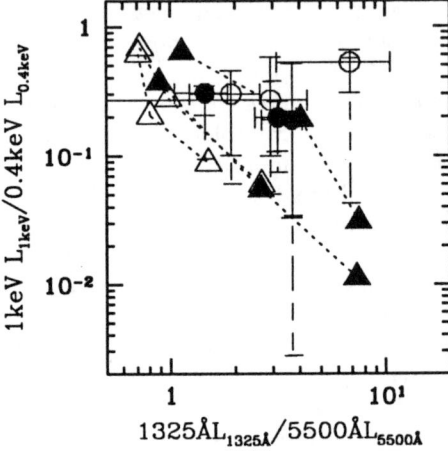

Figure 2. The soft X-ray color vs. the 1325Å luminosity. Filled exagons and triangles identify accretion disk models (Kerr) with $M_{bh} = 10^8 M_\odot$ and open exagon and triangles $M_{bh} = 10^7 M_\odot$ with accretion rates: $\dot{M} = 0.3\dot{M}_{crit}$, $\dot{M} = 0.8\dot{M}_{crit}$ and $\mu=1; 0.75; 0.5; 0.2; 0.1$.

Figure 3. The soft X-ray vs. the optical-UV color. Filled triangles identify accretion disk plus power law models with $M_{bh} = 10^8 M_\odot$ and $\dot{M} = 0.3\dot{M}_{crit}$, $\mu=0.5$ and four values of the power law normalization. The power law slope is fixed to 1.25. Open triangles the same for $M_{bh} = 10^7 M_\odot$.

PSPC Observations of the 1 Jy BL Lacs.

Rita M. Sambruna [1] and C. Megan Urry [1]
John Stocke [2] and Eric Perlman [2]
Ron Kollgaard [3] and Eric Feigelson [3]
Diana Worrall [4]
Paolo Padovani [5]
Laura Maraschi [6]
Aldo Treves [7]

Email ID
sambruna@stsci.edu

ABSTRACT

Preliminary results of a spectral analysis of ROSAT PSPC pointed observations for a complete sample of radio-selected BL Lac objects (1 Jy sample) are presented. At present, data are available for 24 of the 34 sources, 20 of which have enough statistical significance to allow spectral analysis. The average spectral index for the whole group is 2.39 ± 0.02 (weighted mean). Study with more complex models, as well as spatial and temporal analyses, are underway.

1. Dataset and Analysis

The 1 Jy sample is the first homogeneous, flux limited ($S_\nu \geq 1$ Jy at 5 GHz) sample of radio selected BL Lacs [ref. 1]. It contains 34 objects which were identified on the basis of the radio spectrum and the properties of the optical counterparts. Data from PSPC pointed observations are available for 24 out of the 34 BL Lacs of the complete sample. The data were analyzed using the IRAF PROS package. The source photons were extracted from a source-centered circle of typical radius 2 arcmin. The background was measured in an annulus of inner radius ~ 3 arcmin and outer radius ~ 10 arcmin centered on the source; background data contaminated with distinguishable X-ray sources were excluded. Time intervals corresponding to anomalous, variable background were rejected.

The background-subtracted source counts are listed in Table 1 for each object, together with the date of observation and the exposure time (corrected for bad time intervals). For 20 out of 24 objects there are sufficient net counts ($\gtrsim 100$) to allow spectral analysis. For the bright sources 1652+398 (Mrk 501) and 2005-489 a more detailed study is in progress, as well as a temporal and spectral analysis for the whole sample (Sambruna et al. 1994, in preparation).

The source spectra were analyzed individually with XSPEC. In order to accom-

---

[1] Space Telescope Science Institute, 3700 San Martin Dr., Baltimore, MD 21218
[2] University of Colorado, CASA, Campus Box 389, Boulder, CO 80309
[3] Penn. State Univ., Dept. of Astron. & Astrophys., 525 Darvey Lab, University Park, PA 16802
[4] Harv.-Smith. Center for Astrophysics, 60 Garden St., Cambridge, MA 02138
[5] II Universita' di Roma, via E. Carnevale, I-00173 Roma, Italy
[6] Dipartimento di Fisica, via Dodecaneso 33, 16146 Genova, Italy
[7] SISSA, Strada Costiera 11, 34014 Trieste, Italy

modate somewhat the assumption of Gaussian errors, the spectra of the weaker sources were rebinned in such a way that the new bins contain at least 10 counts. To take into account the instrumental gain shift, the appropriate matrix was used in the fits (92 Mar 11 for observations taken before Oct 14, 1991; 93 Jan 12 for those after). Our fitting was performed over channels 4 – 33 only, in order to exclude the earth contamination at the low energies and the the effects of the poorly known response matrix at the high ones.

As a first, simple approximation we fit the data to a single power-law model, either with the column density of absorbing material free to vary or fixed at the Galactic values, reported in Table 1 and taken from [refs. 2,3].

2. Results and Discussion

A single power law with fixed absorption is an acceptable representation of the spectra of all sources, except for Mrk 501 and 2005-489, whose data are of very high signal to noise. More complex models are under study for the latter two objects, which are not discussed further here. For the remaining sources, the total reduced χ^2 is 1.12 for 376 degrees of freedom, with a corresponding $P_{\chi^2} = 5.6\%$. No correlations or trends are apparent when the best-fit power-law photon indices are plotted against the redshift (Figure 1). When the absorption is left free to vary, acceptable fits are also found (total reduced $\chi^2 = 1.10/357$, $P_{\chi^2} = 8.3\%$). The fitted absorptions are in general in good agreement with the Galactic values within 90% confidence errors. Two notable exceptions are 0954+658 and 2007+777, for which the fitted N_H is higher than the Galactic value. Both objects have properties consistent with FRII radio sources [ref. 4].

The distribution of the spectral indices from the fits with N_H fixed is shown in Figure 2a. A simple average yields $\langle \Gamma \rangle = 2.10$ and standard deviation 0.43. A weighted average gives $\langle \Gamma \rangle = 2.39 \pm 0.02$.

The average value of the photon index derived from the ROSAT PSPC compares well with the average photon index derived for an inhomogeneous sample of radio selected BL Lacs observed with the *Einstein* IPC [ref. 5] and EXOSAT LE + ME [ref. 6]. Radio-selected BL Lacs appear to have flatter slopes in X-rays than X-ray selected BL Lacs [refs. 5,6,7]. although the large spread in the distribution prevents any firm statement. In the frame of accelerating jet models [e.g. ref. 8], X-rays are emitted nearly isotropically at the base of the jet via synchrotron mechanism, with lower frequency radiation coming from the outer, increasingly beamed regions. Depending on the viewing angle ϕ to the jet, different energy distributions are expected in X-rays. Radio-selected objects, corresponding to smaller angles ($\phi \lesssim 15°$) [ref. 9], would be characterized by flatter indices than X-ray selected ones ($15° < \phi < 30°$), because of a Compton component becoming important in the thicker, outer regions. Our finding that radio-selected BL Lacs have, on average, harder spectra than X-ray selected objects, supports this scenario.

References: 1: Stickel, M., Padovani, P., Urry, C.M., Fried, J.W., & Kühr, H. 1991, ApJ 374, 431 **2:** Stark, A.A. et al. 1992, ApJ Suppl. 79, 77 **3:** Elvis, M., Lockman, F.J., & Wilkes, B.J. 1989, AJ 97, 777 **4:** Kollgaard, R.I., Wardle,J.F.C, Roberts, D.H., & Gabuzda, D.C. 1992, AJ 104, 1687 **5:** Worrall, D.M. & Wilkes, B.J. 1990, ApJ 360, 396 **6:** Sambruna, R.M., Barr, P., Giommi,P., Maraschi, L., Tagliaferri, G., & Treves, A. 1993, ApJ, in press **7:** Perlman, E. & Stocke, J. 1993, BAAS, Vol. 25, No. 2, p.791 **8:** Ghisellini, G. & Maraschi, L. 1989, ApJ 340, 181 **9:** Ghisellini, G., Padovani, P., Celotti, A., & Maraschi, L. 1993, ApJ 407, 65

Table 1: ROSAT PSPC observations of the 1 Jy BL Lacs

Object	Redshift	Galactic N_H ($\times 10^{20}$ cm^{-2})	Date of Obs.	Exposure (s)	Net Counts
0048-097	...	3.50	4 Jul 93	8363	3156 ± 60
0118-272	> 0.557	1.53	10 Jul 93	2635	343 ± 20
0426-380	> 1.030	1.94	2 Aug 93	4048	171 ± 15
0537-441	0.896	4.00	10 Apr 91	2598	891 ± 31
0735+178	> 0.424	4.73	28 Oct 92	6684	533 ± 27
0814+425	0.258?	4.92	6 Oct 92	6120	76 ± 11
0820+225	0.951	4.24	29 Oct 92	3728	60 ± 9
0828+493	0.548	3.94	3 Oct 92	4821	65 ± 10
0851+202	0.306	2.75	10 Nov 91	6785	3806 ± 63
			16 Apr 91	3202	803 ± 30
0954+658	0.367	4.28	17 Apr 91	5672	327 ± 19
1144-379	1.048	8.67	7 Jul 93	7747	840 ± 30
1147+245	...	2.03	27 May 93	10942	180 ± 19
1308+326	0.997	1.08	8 Jun 91	7824	633 ± 30
1514-241	0.049	8.80	17 Aug 93	2915	7 ± 5
1519-273	...	8.90	17 Aug 93	2552	256 ± 17
1538+149	0.605	3.23	28 Jan 93	6915	242 ± 18
1652+398	0.033	1.73	25 Feb 91	7281	42589 ± 209
1749+096	0.320	9.61	17 Mar 91	9507	726 ± 29
1749+701	0.770	4.01	9 Nov 92	3702	314 ± 20
1803+784	0.684	4.02	25 Jul 92	5602	570 ± 25
1823+568	0.664	4.20	19 Jun 92	5900	784 ± 30
2005-489	0.071	4.60	27 Apr 92	9787	29524 ± 174
2007+777	0.342	8.90	11 Dec 91	4116	131 ± 13
2200+420	0.069	20.15	22 Dec 92	6746	328 ± 18

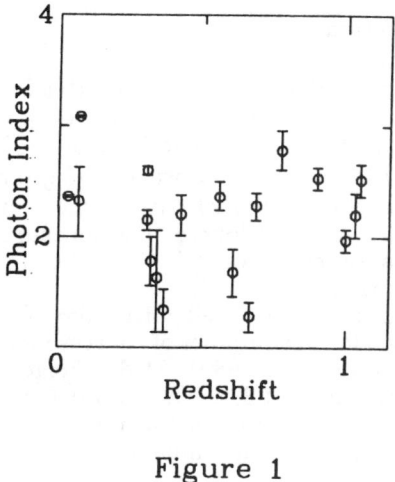

Figure 1

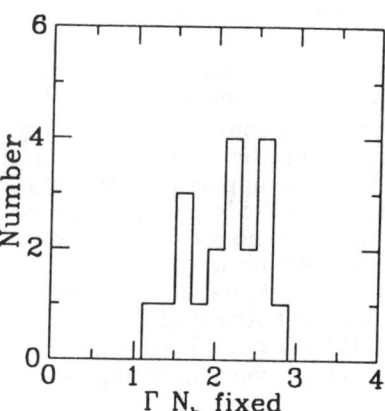

Figure 2

X-RAYS FROM THE LOBES OF FORNAX A

S. A. Laurent-Muehleisen, E. D. Feigelson, R. I. Kollgaard
Department of Astronomy & Astrophysics, Pennsylvania State University

Email ID
salm@astro.psu.edu, edf@astro.psu.edu, rik@astro.psu.edu

and E. B. Fomalont
National Radio Astronomy Observatory, Charlottesville

Email ID
efomalon@suru.cv.nrao.edu

ABSTRACT

We present a deep *ROSAT* PSPC image of the radio lobes of the nearby galaxy Fornax A (= NGC 1316) and find, after image processing, X-ray emission closely mimicking the radio emission. We argue that this is the long-sought inverse Compton radiation produced when cosmic microwave background photons are upscattered off synchrotron emitting electrons in the radio lobes. The derived magnetic field in the lobes is compared with equipartition values and we also discuss the possibility of thermal Bremsstrahlung emission as a source of lobe X-rays. This observation constitutes the best case for detection of inverse Compton X-rays in radio lobes to date.

1. Introduction

When a low energy photon encounters a high energy particle with a Lorentz factor of γ, its energy is increased by $\sim\gamma^2$ via the inverse Compton (IC) process. The synchrotron emitting lobes of radio galaxies are a plentiful source of high energy electrons with $\gamma \approx 10^2 - 10^4$ and the cosmic microwave background is an ubiquitous source of low energy photons ($\nu \sim 10^{11}$Hz). IC X-ray emission therefore must be associated with all radio-emitting lobes. The measurement provides a rare opportunity to test the oft-adopted hypothesis that lobe particles and magnetic fields are in energy equipartition.

However, no convincing case of IC emission associated with radio lobes has been observed, although it has been searched for in a number of sources including Centaurus A (Marshall & Clark 1981; Morini *et al.* 1989), M 87 (Feigelson *et al.* 1987), and Abell 1367 (Gavazzi & Trinchieri 1983). We chose Fornax A (NGC 1316) as the best target for IC X-ray emission as the radio lobes are strong, have a convenient angular size, and are not contaminated by cluster gas.

2. Observations of Fornax A (NGC 1316)

We present a deep image of the radio galaxy Fornax A obtained with the *ROSAT* Position Sensitive Proportional Counter (PSPC) consisting of 16 separate orbital intervals taken between January 13 and January 20, 1992 for a total of 25.48 ksec. The color figure shows a 1.5 GHz image of Fornax A (Fomalont *et al.* 1989) with X-ray contours of the 0.9-2.0 keV flux smoothed to $5'$.

Scattered solar X-rays and the particle induced background have been modeled and subtracted from the data (Snowden et al. 1994). In addition, afterpulses and point sources have been removed from the X-ray image.

3. Are the X-rays Inverse Compton?

If a large fraction of the observed X-ray flux is produced via the thermal Bremsstrahlung mechanism, a significant Faraday depolarization ought to be present. We have used the homogeneous spherical model of Cioffi & Jones (1980) and the equipartition magnetic field to predict the thermal electron density. The resulting thermal flux is a factor of 10-200 below that observed, depending on exact values of the electron density and temperature. However, fits to a Raymond-Smith plasma are consistent with the hard portion of the X-ray lobe spectra, although with electron densities a factor of 5-10 larger than the depolarization observations suggest. We conclude the observed X-ray emission is probably **not** thermal in nature.

4. Is the Lobe Magnetic Field in Equipartition?

Burbidge (1956) first showed the minimum energy condition for radio lobes corresponds to roughly an equipartition between the magnetic field and particle energy densities. This equipartition magnetic field has often been assumed to be the value of the true field, although there is little observational evidence for this. Because the ratio of synchrotron emission to IC emission is a function of the magnetic field strength, the detection of IC X-ray lobe emission provides a direct measure of the true magnetic field. For the east lobe we find $B_{ic}=2.5\mu G$ and $B_{eq}=2.5\mu G$. For the west lobe we find $B_{ic}=1.8\mu G$ and $B_{eq}=2.9\mu G$. These calculations assume the radio spectral index is -0.8 ($S_\nu \propto \nu^\alpha$), the B-field is perpendicular to the line-of-sight, the ratio of the energy in electrons to protons is unity, the upper and lower cutoffs of the synchrotron spectrum are 0.01 and 100 GHz, respectively, and the filling factor is unity. Decreasing the filling factor by a factor of 10, as suggested by the observed radio filamentation, increases the equipartition field by only a factor of ~2.

5. Conclusions

We present the first clear case for inverse Compton X-ray emission from radio lobes. The derived magnetic field is shown to be very close to the equipartition value. The possibility that the X-ray emission is thermal is discussed and found to be unlikely.

References

Burbidge, G., 1956, ApJ, 124, 416.
Cioffi, D. F. & Jones, T. W., 1980, AJ, 85, 368.
Feigelson, E., Wood, P., Schreier, E., Harris, D. & Reid, J., 1987, ApJ, 312, 101.
Fomalont, E. B., Ebneter, K., van Breugel, W., Ekers, R., ApJLett, 346, L17.
Gavazzi, G. & Trinchieri, G., 1983, ApJ, 270, 410.
Marshall, F. J. & Clark, G. W., 1981, ApJ, 245, 840.
Morini, J., Anselmo, F., & Molteni, D., 1989, ApJ, 347, 750.
Snowden, S., McCammon, D., Burrows, D. & Mendenhall, J., 1994, ApJ, in press.

X-RAYS AND RELATIVISTIC BEAMING IN RADIO-SELECTED BL LACERTAE OBJECTS

R. I. Kollgaard, E. D. Feigelson
Dept. of Astronomy & Astrophysics, Penn State University, University Park, PA 16802
Email ID: rik@astro.psu.edu, edf@astro.psu.edu

D. C. Gabuzda
Dept. of Physics & Astronomy, University of Calgary, Calgary, AB T2N 1N4, Canada
Email ID: denise@bear.ras.ucalgary.ca

R. M. Sambruna
SISSA/ISAS, Strada Costiera 11, Trieste, Italy &
Space Telescope Science Institute, 3700 San Martin Drive, Baltimore, MD 21218
Email ID: sambruna@stsci.edu

C. M. Urry
Space Telescope Science Institute, 3700 San Martin Drive, Baltimore, MD 21218
Email ID: cmu@stsci.edu

ABSTRACT

We have used X-ray and VLBI observations of twenty three BL Lac objects from the 1 Jy sample of radio bright sources to test models for the origin of X-ray emission in BL Lacs. We find a correlation between the X-ray luminosity and the parsec scale radio emission, consistent with the idea that the X-rays are relativistically beamed. A simple approximation for the synchrotron self-Compton process has been used to estimate the minimum Doppler factor for bulk relativistic motion. Relativistic motion is implied for $\sim$ 75% of the sources, although there are no clear trends between the minimum Doppler factor and other traits. These results suggest that the soft X-ray spectra do not contain a significant inverse-Compton contribution, but this needs to be tested with more sophisticated models.

1. Introduction

Combined X-ray and radio observations are a powerful tool for understanding the physics and dynamics of nonthermal regions in Active Galactic Nuclei. The high energy electrons in AGN jets require that some of the synchrotron photons be boosted to X-ray energies via inverse-Compton scattering. However, the actual contribution of this synchrotron self-Compton (SSC) mechanism to the observed X-ray emission is unclear even in BL Lacertae objects where nonthermal processes dominate. Direct synchrotron emission may be important.

2. Observations

The '1 Jy' sample is a well-defined, flux limited sample of the 34 radio brightest BL Lac objects (Stickel et al. 1991). BL Lac objects are generally modeled as having a relativistic jet oriented close to the line of sight (e.g., Kollgaard 1994). Their radio-X-ray continuum spectral index is steeper than BL Lac objects selected from X-ray surveys, suggesting that the X-rays are subject to significantly less beaming than the radio emission (e.g., Padovani & Urry 1990; Ghisellini et al. 1993). Possible explanations include an isotropic X-ray component or an "accelerating" jet, where the X-rays originate in the inner jet with a lower bulk Lorentz factor (Ghisellini et al. 1985).

We have made pointed PSPC observations of 10 sources from the '1 Jy' sample of BL Lac objects. With the additional of spectral information from additional sources (Sambruna et al., these proceedings) X-ray data for 23 of the 1 Jy BL Lacs are available for analysis.

3. Testing Beaming Models

Making the reasonable assumption that the core radio emission L_R is highly beamed, we expect the X-ray luminosity L_X to be correlated if it is also dominated by a beamed component. Radio data on the parsec-scale cores has been taken from various sources in the literature, in particular an ongoing VLBI monitoring program of the entire 1 Jy sample (Gabuzda et al. 1993; Gabuzda et al., in preparation). By definition BL Lac objects are variable so in order to determine a 'typical' radio flux we have averaged the available radio data. Observations made at 2.3 and 8.4 GHz have been extrapolated to 5 GHz by assuming the core has an optically thick spectrum with $\alpha = 0.5$ ($S \sim \nu^{+\alpha}$). X-rays have been detected from each of the 23 BL Lacs, and we have excluded only those sources without VLBI data or known redshifts.

We find that, except for the source with the weakest radio emission there is a strong correlation with $\log L_X \propto (1.0 \pm 0.2) \log L_R$. This is consistent with the logarithmic slope 0.93 ± 0.09 found for quasars with strong radio cores (Kembhavi 1993). If the radio emission in radio bright BL Lacs is beamed (which is very likely) we conclude the same is true for the X-rays. The outlying radio-weak source is Mkn 501 which has properties similiar to X-ray selected BL Lac objects (Laurent-Muehleisen et al. 1993).

4. Simple SSC Models

In principle, VLBI data can be used to directly test whether the SSC mechanism significantly contributes to the observed X-ray emission, though in practice there are many difficulties (Marscher 1987). We have collected VLBI data on both the parsec-core radio flux and angular size for those sources with X-ray data. Using the simple approximation that the core is dominated by a homogeneous component with a turnover frequency at the frequency of observations, the minimum Doppler factor $D_{\min}$ for bulk relativistic motion can be estimated (Ghisellini et al. 1993). Although an estimate of the expected X-ray flux is extremely sensitive to the VLBI derived paramters, this is less true for $D_{\min}$. For $\sim 75\%$ of the objects $D_{\min} > 1$, requiring relativistic bulk motion. The values of $D_{\min}$ for the sample range from 0.2 for 0828+493 where there is limited VLBI data, to 1803+784 where values of $D_{\min} > 7$ are typical. More sophisticated testing of SSC models is underway.

The estimated $D_{\min}$ values are compared to other properties of the objects, including the radio core and extended power, X-ray power, photon index, and core polarization. No trends are found, though some are expected from simple SSC models. For example, one might expect a tendency for $D_{\min}$ to be larger for sources with flatter X-ray spectra if the SSC mechanism dominates. We tentatively conclude that beamed

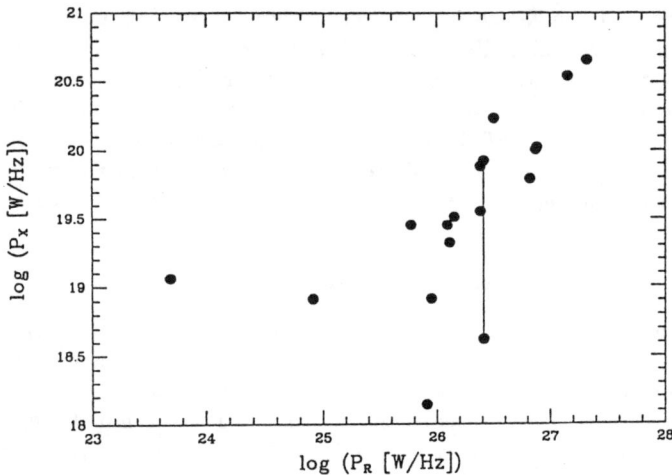

Figure 1: The 1 keV X-ray power P_X observed by ROSAT plotted against the parsec-scale 5 GHz radio power P_R determined from VLBI observations. The two connected points correspond to OJ 287 which for which very different X-ray flux densities were observed with ROSAT. The correlation is insensitive to which X-ray value is used.

synchrotron, rather than inverse Compton, dominates the soft X-ray emission of radio-selected BL Lac objects.

While most BL Lac objects have the radio morphology of low luminosity radio galaxies (FR I), a few of the radio brightest have the edge-brightened structure of higher luminosity radio galaxies (Kollgaard et al. 1992). We did note a tendency for those BL Lacs with known FR II morphologies to have a larger D_{min} than ones with known FR I morphologies. This could be due to inverse-Compton emission being more important for the soft X-ray spectra of intrinsically more powerful sources. Further study is in progress to see if this trend is real.

Acknowledgements: This work has been partially supported by NASA under grant NAGW-2120.

References

Gabuzda et al. 1992, ApJ, 388, 40
Ghisellini, G., Maraschi, L. & Treves, A. 1985, AA, 146, 204
Ghisellini, G., Padovani, P., Celotti, A. & Maraschi, L. 1993, ApJ, 407, 65
Kembhavi, A. 1993, MNRAS, 264, 683
Kollgaard, R. I., Wardle, J. F. C., Roberts, D. H. & Gabuzda, D. C. 1992, AJ, 104, 1687
Kollgaard, R. I., 1994, Vistas in Astronomy, to appear
Laurent-Muehleisen et al. 1993, AJ, 106, 875
Marscher, A. P. 1987, in Superluminal Radio Sources, p. 280
Padovani, P. & Urry, C. M. 1990, ApJ, 356, 75
Stickel, M., Padovani, P., Urry, C. M., Fried, J. W. & Kühr, H. 1991, ApJ, 374, 431

MULTIFREQUENCY STUDIES OF OPTICALLY QUIET QUASARS

R. I. Kollgaard, S. A. Laurent-Muehleisen, E. D. Feigelson
Dept. of Astronomy & Astrophysics, Penn State University, University Park, PA 16802
Email ID: rik@astro.psu.edu, salm@astro.psu.edu, edf@astro.psu.edu

H. Spinrad
Dept. of Astronomy, University of California, Berkeley, CA 94720
Email ID: spinrad@bigz.berkeley.edu

W. Brinkmann
Max-Planck Institut für Extraterrestrische Physik,
Giessenbachstrasse, 85740 Garching, Germany
Email ID: wpb@ibma.ipp-garching.mpg.de

ABSTRACT

We have made multiband observations of six candidate Optically Quiet Quasars, and confirm the inverted Spectral Energy Distribution in four. OQQs could be ordinary radio-loud AGN whose SED is significantly modified by absorption either intrinsic to the source or along the line of sight, or the X-rays could be dominated by a strong inverse-Compton contribution. Our four OQQs exhibit a range of properties and more than one explanation is possible.

1. Introduction

"Optically Quiet Quasars" (OQQs) are X-ray and radio-loud sources with very weak optical emission (Ledden & O'Dell 1983; Bregman et al. 1985). OQQs have an inverted Spectral Energy Distribution (SED) with the X-rays lying significantly above a simple extrapolation of the IR-optical continuum. The most probable explanations for the inverted SED are: (1) significant extinction due to absorbing material somewhere along the line of sight; or (2) an unusually strong X-ray component due to a large contribution from inverse-Compton scattering within the radio synchrotron core.

Multiband observations are needed to distinguish between these (and other) possibilities. For instance, one might expect a source with a high intrinsic column density to exhibit other signs of a dense environment such as a compact radio structure, depolarization, and other traits of GHz-peaked radio sources. This scenario has recently been found for the nearby GHz-peaked source OQ 208, where an N_H column density 12× greater than the galactic value was detected in the soft X-ray spectrum (Zhang & Marscher, these proceedings). A strong inverse-Compton contribution would likely be associated with rapid variability, high polarization, and other properties of blazars. This model is consistent with the X-ray emission observed from the powerful γ-ray blazar PKS 0528+134 (Zhang & Marscher, these proceedings). However, the inverted SED can also appear if there is excess absorption due to an unrelated system which intercepts the line of sight. In this case, the only requirement is that the AGN be a strong radio–X-ray source behind the intervening absorption system. This scenario has been suggested as a possibility for the absorption seen in three high redshift quasars (Elvis et al. 1993; Elvis et al. these proceedings), though as these authors point out, two of these three quasars are GHz-peaked sources. Other evidence for an intervening system would be the presence of other absorption features, such as MgII lines or the $\lambda 2175$ dust feature at an intermediate redshift.

2. Observations

We have observed 6 OQQ candidates with *ROSAT*, the Lick Observatory 3-m Shane reflector, and the Very Large Array. Three sources were previously classified as OQQs and were targets of pointed PSPC observations. Three additional OQQ candidates were selected from a comparison of *ROSAT* all-sky survey images of the North Ecliptic Pole (NEP) with unbiased radio (Kollgaard et al. 1993) and optical surveys.

0026+346. Originally classified as an OQQ from *Einstein* observations (Bregman et al. 1985), but our PSPC observations did not detect it. The original *Einstein* report is probably spurious, and the object is not an OQQ.

0406+121 ($z = 1.02?$). Ledden & O'Dell (1983) first identified this as an OQQ, which our observations confirm. Optical spectroscopy reveals only a single narrow emission line at $\lambda 7530$Å ($W_\lambda < 5$Å). Our radio observations find a compact structure with a flat spectrum, while VLBI shows a compact parsec-scale structure, possibly with a small jet (Akujor & Porcas 1992). Our PSPC spectrum can be fit with a powerlaw but has insufficient counts to unambiguously determine N_H and α_E simultaneously. The fits are consistent with both a very hard X-ray spectrum ($\alpha_E = 0.2 \pm 0.4$) and galactic N_H, as well as for excess absorption ($N_H \sim 10^{22}$cm^{-2} at $z = 1$) for a spectral index fixed at a typical quasar value of $\alpha_E = 1.2$.

0500+019 ($z \sim 1$?). Ledden & O'Dell first identified this source as an OQQ, which our data confirm. Hodges et al. (1984) found it to be a GHz-peaked radio source with a very compact parsec-scale structure. We do not have additional radio observations, and our optical spectrum (taken during bad seeing conditions) is inconclusive, although its continuum is very red. As with 0406+121 the PSPC spectrum is consistent with hard X-rays ($\alpha_E = 0.4 \pm 0.4$) and galactic absorption, and with a large excess absorption ($N_H \sim 10^{22}$cm^{-2}) and a typical quasar X-ray spectral index.

1746+642 ($z=1.23$). We identified this OQQ candidate from the NEP region. Additional multiband observations show a modestly inverted SED. Optical spectroscopy reveals several narrow lines. The radio structure consists of large radio lobes, and possibly a compact jet component $\sim 0.15''$ from the core. The lobes exhibit asymmetric polarization, with one being significantly more polarized at 1.5 GHz than the other.

1747+653 ($z=1.51$). New observations of this OQQ candidate from the NEP show a modestly inverted spectrum. Optical spectroscopy finds several emission lines. The radio structure is very compact with an unresolved steep spectrum feature $\sim 1''$ from the core.

1815+680 ($z=0.2338$). New observations of this OQQ candidate from the NEP do not show an inverted spectrum. However, optical spectroscopy shows a number of narrow emission lines and a strong E galaxy-type absorption component, implying that the AGN itself might have an inverted SED. Radio imaging shows a large double lobed structure with asymmetric polarization.

3. Discussion

Although four of these sources share an inverted SED, they otherwise have a wide range of properties which do not all fit simply into one of the proposed scenarios. While the GHz-peaked spectrum of 0500+019 is consistent with significant absorption intrinsic to the quasar, the properties of 0406+121 indicate it is a BL Lac object, which are generally thought to have a highly relativistic jet oriented close to the line of sight. Such a jet could produce significant inverse-Compton emission. The highly asymmetric radio lobe polarization observed in 1746+642 also suggests a jet oriented near the line of sight, if the more polarized lobe is closer to us (and thus seen through less depolarizing gas around the galaxy) than the less polarized lobe (e.g., Laing 1988). We note both 0406+121 and 0500+019 have very red optical continua while 1746+642 and 1747+653 do not.

The presence of inverse-Compton emission can be tested for the two sources with VLBI observations, by combining the radio and X-ray data to estimate the minimum Doppler factor for any bulk relativistic motion (e.g., Ghisellini et al. 1993). However, the results are inconclusive, with no bulk motion *required* to produce the observed X-ray emission. This test can be improved, though, with higher resolution VLBI data.

A wide range of OQQ properties is consistent with absorption by an intervening object which just happens to intercept the line of sight. As the probability of intercepting a Lyman-α system, for example, is very small at low redshifts this is also consistent with all four confirmed OQQs having $z > 1$ (Elvis et al. 1994). The column densities suggested by our PSPC data of 0406+121 and 0500+019 imply sufficient quantities of dust (assuming galactic gas/dust ratios) to reduce the optical emission from an ordinary radio-loud quasar and BL Lac object and produce an inverted SED. However, more evidence for an absorbing system is needed to test this scenario for our sources.

We suspect that the class of OQQs will require more than one explanation. For instance, some may have intrinsic absorption or intervening clouds, while others may be inverse-Compton dominated objects.

Acknowledgements: This work has been partially supported by NASA under grant NAGW-2120.

References

Akujor, C. E. & Porcas, R. W. 1992, in Extragalactic Radio Sources – From Beams to Jets, ed. J. Roland, H. Sol & G. Pelletier, p. 134
Bregman, J. N., Glassgold, A. E., Huggins, P. J. & Kinney, A. L. 1985, ApJ, 291, 505
Elvis, M., Fiore, F., Mathur, S. & Wilkes, B. J. 1994, ApJ, to appear
Ghisellini, G., Padovani, P., Celotti, A. & Maraschi, L. 1993, ApJ, 407, 65
Hodges, M. W., Mutel, R. L. & Phillips, R. B. 1984, AJ, 89, 1327
Laing, R. A. 1988, Nature, 331, 149
Kollgaard, R. I., Brinkmann, W., Chester, M. M., Feigelson, E. D., Hertz, P., Reich, P. & Wielebinski, R. 1994, ApJSupp, to appear
Ledden, J. E. & O'Dell, S. L. 1983, ApJ, 270, 434

ACTIVE AND PASSIVE GALAXIES IN DEEP ROSAT SURVEYS

Richard E. Griffiths, Roberto Della Ceca
Johns Hopkins University, Baltimore, MD 21218, USA.

Brian J. Boyle
Institute of Astronomy, Madingley Rd, Cambridge, CB3 0HA, UK.

I. Georgantopoulus, G.C. Stewart
Department of Physics and Astronomy, Leicester, LE1 7RH, UK.

and T. Shanks
Department of Physics, Durham, DH1 3LE, UK.

Email ID
griffith@mds.pha.jhu.edu

ABSTRACT

ROSAT deep survey X-ray sources have been identified with active galaxies (Active Galactic Nuclei and Narrow Emission-Line galaxies) and also with early-type, "passive galaxies". In this paper we discuss preliminary results of the galaxy content, both early-type and emission-line galaxies, of 5 ROSAT fields.

1. Introduction

We have performed $\sim 30-60$ ks surveys of 5 fields (QSF1, QSF3, SGP2, SGP3 and SGP4) which had previously been surveyed for UV-excess quasars by Boyle et al. (1990). Three hundred and eighty two X-ray sources have been found in these fields and 215 of them belong to a complete X-ray sample. We have secured optical identifications for $\sim 75\%$ of the sources in the complete sample, using multi-object spectrographs at the AAT. Amongst the sources so far identified, there are 110 AGN, 10 early-type galaxies, 16 Narrow-Line (starburst ?) galaxies and 23 stars. The AGN content of the first two of these fields has been described by Boyle et al (1993), who derive the AGN luminosity function and the constraints on its evolution. AGN account for $40-70\%$ of the X-ray background between 1 and 2 keV, i.e. they probably do *not* account for all the observed soft X-ray background. In this summary, we focus on the non-AGN counterparts to the deep survey sources, and describe the galaxy content of the identifications, both early-type and emission-line galaxies. The galaxy content of the first two fields (QSF1, QSF3) is described in Griffiths et al (1993). A Hubble constant of 50 km s^{-1} and a Friedmann universe with a deceleration parameter $q_o = 0$ are assumed.

2. Early-type and Narrow-Line Galaxies

The X-ray to optical flux ratios for the extragalactic X-ray sources in the complete sample are shown in figure 1. As can be seen in this figure the Narrow-Line (NL) galaxies and the early-type (non-emission line, GAL) galaxies tend to have a lower $f_x/f(V)$ than the AGN. Moreover the number of NL galaxies seems to increase at the lower x-ray fluxes. If this trend is confirmed from the identification of the X-ray sources found in deeper soft X-ray surveys (e.g. the Lockman hole deep X-ray survey) then these objects could be an important population at X-ray fluxes of $\sim 10^{-15}\ erg\ cm^{-2}\ s^{-1}$.

In figures 2 and 3 we show the distribution of the NL and early-type galaxies in the $L_x - z$ and $L_x - L_o$ plane, respectively. We have included in these figures all the sources (in the complete and non-complete sample) so far identified. The area inside the dashed line in figure 3 represents the locus populated by the "Normal Galaxies" (E+S0) observed with the *Einstein* Observatory (adapted from Fabbiano et al., 1992).

The early-type (E to Sab) galaxies have X-ray luminosities $10^{42} - 10^{43.5}$ $ergs\ s^{-1}$, but show no optical signatures indicative of such large X-ray luminosities. The hypotheses for the origin of this emission are as follows: a) the X-ray emission may arise from coronal bremstrahlung from hot gas constrained by dark matter surrounding the ellipticals; b) the galaxies could be the brightest members of small groups of galaxies and the X-ray sources could be the centroid of extended emission from the group. The EMSS sources MS0116.3-0116 (Maccagni et al., 1987) and MS1111.8-3754 (Maccagni et al., 1988) could be a prototype for this class of X-ray selected objects. These objects show an X-ray luminosity $\sim 10^{43}$ and $\sim 3.6 \times 10^{44}\ ergs\ s^{-1}$ and are found in a sparse group of galaxies and in a poor cluster, respectively; and c) the X-ray emission may arise in the active galactic nuclei of otherwise passive early-type galaxies. The EMSS contains interesting examples of such objects, such as MS0002.5-4205 and MS0007.1-0231, with about twenty other similar objects (Stocke et al., 1991).

The emission-line (NL) galaxies have optical spectra typical of "starburst" galaxies, although we cannot rule out the presence of Seyfert 2 nuclei in several cases. Several of these galaxies have a luminosity in excess of $10^{42.5}\ erg\ s^{-1}$. For these objects it is unlikely that supernova remnants can account for the observed emission, which may instead be attributed to massive X-ray binaries in the star-forming regions of these galaxies. Using the 16 NL galaxies in the complete sample we have been able to derive, for the first time, a preliminary luminosity function, starting from an X-ray selected sample. In figure 4 we have compared this luminosity function with the local ($z < 0.1$) X-ray luminosity function of the AGN in the EMSS (Maccacaro et al., 1991). As can be seen in this figure, for $L_x < 5 \times 10^{42}\ erg\ s^{-1}$, the spatial density of NL galaxies could be higher than the spatial density of typical broad-line objects, expecially when we consider that a fraction of the EMSS AGN could be NL objects. Using the derived luminosity function of the NL galaxies, we have estimated directly their contribution to the cosmic X-ray background. On the assumptions of no evolution we find that their contribution to the 1–2 keV X-ray background is $\sim 10\%$. This value could increase up to $\sim 20\%$ if these galaxies undergo a cosmic evolution similar to that of AGN (Boyle et al., 1993).

3. Conclusions

AGN contribute about a half of the soft XRB, and possibly as much as 70%; X-ray luminous galaxies are the next most important contributors, and may make up the bulk of the remainder. The galaxies are of various classes, the most important of which are the "passive" ellipticals and the Narrow-Line (starburst ?) galaxies. Understanding the origin of their X-ray emission is a very important task and further work is in progress (Griffiths et al. 1993).

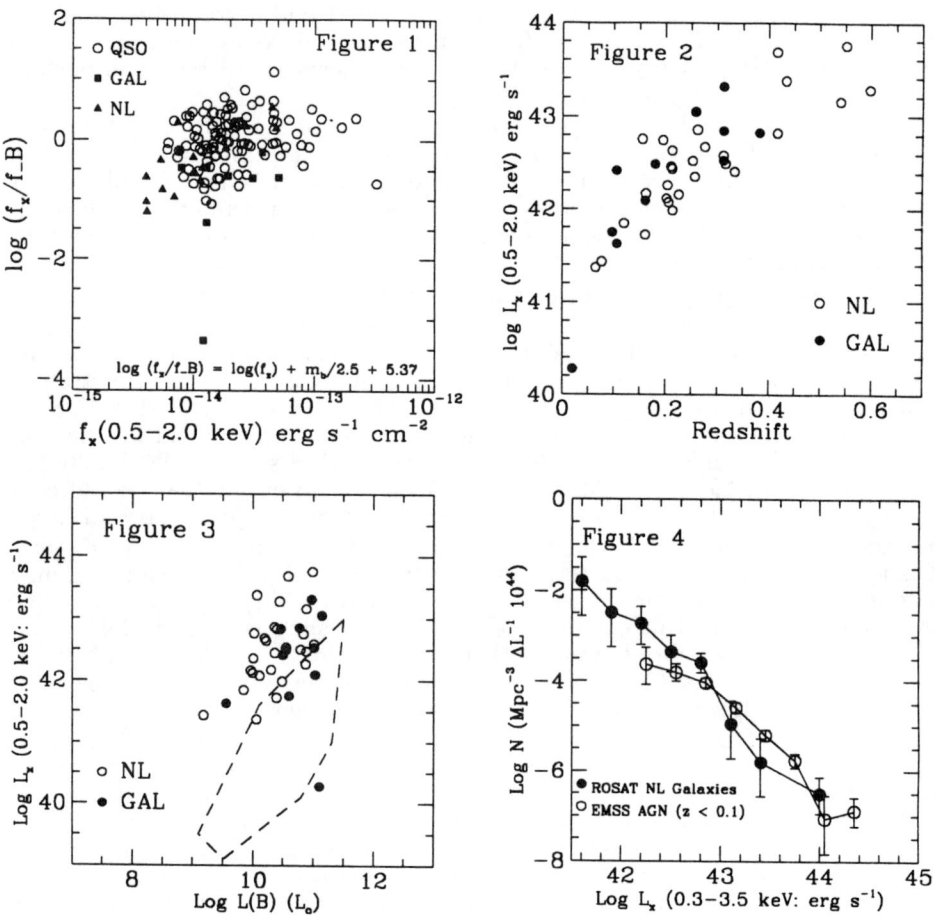

References

Boyle, B.J., et al., 1990, MNRAS, 243, 1.
Boyle, B.J., et al., 1993, MNRAS, 260, 49.
Fabbiano, G., et al., 1992, Ap.J.Suppl., 80, 531
Griffiths, R. E., Georgantopoulos, I., Boyle, B. J.,
   Stewart, G. C., and Shanks, T., 1993, in preparation.
Maccacaro, T., et al., 1991, Ap. J., 374, 117.
Maccagni, D., et al., 1987, Ap.J., 316, 132.
Maccagni, D., et al., 1988, Ap.J.Lett., 334, L1.
Stocke, J.T., et al., 1991, Ap.J.Suppl.Ser., 76, 813.

THE ROSAT PSPC SPECTRUM OF PKS2155-304

Diane M. Gilmore and C. Megan Urry
Space Telescope Science Institute, Baltimore, MD 21218

ABSTRACT

The x-ray bright BL Lac object PKS2155-304 was observed with the PSPC on two separate occasions: May 6,1991 and November 17, 1992. There was a 45% drop in flux between the observations. The later observations were well fit with a simple power law; however, the earlier spectrum appears more complex, and requires an additional feature. We modelled this as an absorption notch or edge. Similar absorption by hot gas, apparently intrinsic to the source, has been detected with other instruments in this and other BL Lac objects. However, we believe the source of the feature is instrumental and is probably the result of a small temporal gain shift ($\sim$1-2%) in the response matrix.

1. Introduction

Previous observations of the x-ray bright BL Lac object PKS2155-304 have detected an x-ray absorption feature at $\sim$0.6 to 0.7 keV in spectra taken with the Objective Grating Spectrometer (OGS) aboard the *Einstein Observatory* (Canizares and Kruper, 1984), as well as in spectra from the Broad Band X-ray Telescope (BBXRT) in December 1990 at roughly the same energy (Madejski et al., 1992). The feature is thought to be due to Lyα absorption of O_{VIII} (E_{rest} $\sim$654 eV) at the redshift of the source, z $\sim$0.117 (Bowyer at al., 1984). We present results from our Rosat PSPC-B observations of the source, in which we have attempted to detect and characterize the x-ray absorption feature.

2. Spectral Analysis

Photons were collected from a 2 arcminute circular source region. Background regions were annuli centered on the source, from 3.5 to 10 arcminutes in radius. Periods of high-background count rate were clipped prior to spectral analysis. Using the standard MPE SASS binning, 34-channel spectra were extracted, and analyzed using XSPEC version 8.2 (Shafer et al., 1991). The final spectra had more than 18 cts/bin in fit channels 4 to 30 enabling χ^2 statistics (Cash, 1979) to be used.

The count rate measured in the energy range from 0.14 to 2.02 keV (channels 4 to 30 of the MPE SASS processing) was $\sim$35 cts/sec for the May 1991 data, and $\sim$19 cts/sec for the Nov 92 data. Since the source is bright relative to the background ($\sim$0.5 cts/s), vignetting would have had a minimal affect on the spectral analysis.

To compensate for the residual gain shift in spectra taken after the gain switch which occured on Oct 14, 1991, we used two different response matrices to analyze spectra from the two epochs; for the May 1991 data we used pspcb_92mar11.rsp, the Nov 1992 data were analyzed with pspcb_93jan12.rsp. To avoid overinterpretation of spectral features, we added (in quadrature) a 2% systematic error to all spectra (1993, Turner, J. and George, I.). The resultant spectral fits are shown in Table I.

A power law cutoff at low energies with free local column density produced acceptable fits for the summed spectra of both epochs. When fit individually however,

Table I. Spectral fits using pi bins 4 through 30 (0.14 - 2.02 keV). Fits with parameter N_H frozen are fixed at the galactic value (Lockman and Savage, in prep.).

A) Power law fits. (The first two entries are for November 1992 summed spectra).

	χ^2	N_H (exp20)	Γ_1	Normalization (exp-2)	Flux[f] (exp-10)
sum'[b]	0.73	1.48(1.30-1.67)	2.57(2.49-2.66)	3.21(3.12-3.31)	1.27
sum'[a]	0.77	1.36 (frozen)	2.52(2.50-2.54)	3.20(3.13-3.28)	1.25
sum[b]	0.64	1.31(1.19-1.43)	2.38(2.33-2.44)	6.46(6.34-6.59)	2.37
sum[a]	0.65	1.36 (frozen)	2.41(2.39-2.42)	6.47(6.38-6.56)	2.38
obi2[b]	1.34	1.34(1.19-1.50)	2.38(2.31-2.45)	6.46(6.30-6.62)	2.35
obi2[a]	1.28	1.36 (frozen)	2.39(2.37-2.40)	6.46(6.34-6.58)	2.36

B) Power law plus added notch fits.

	χ^2	N_H (exp20)	Γ_1	Normalization (exp-2)	E_{line} keV	Flux[f] (exp-10)
sum[d]	0.35	1.46(1.24-1.70)	2.45(2.36-2.55)	6.55(6.37-6.75)	0.45(0.00-0.65)	2.29
sum[c]	0.40	1.36 (frozen)	2.41(2.39-2.43)	6.51(6.37-6.66)	0.44(0.26-0.63)	2.29
obi2[d]	0.69	1.67(1.38-2.05)	2.54(2.39-2.71)	6.64(6.40-6.92)	0.38(0.36-0.54)	2.05
obi2[c]	1.00	1.36 (frozen)	2.39(2.36-2.43)	6.53(6.33-6.73)	0.43(0.32-0.57)	2.19

C) Power law plus added edge fits.

	χ^2	N_H (exp20)	Γ_1	Normalization (exp-2)	E_{edge} keV	Flux[f] (exp-10)
sum[d]	0.41	1.62[u]	2.53(2.35-2.72)	6.78[u]	0.28[u]	2.20
sum[c]	0.54	1.36 (frozen)	2.42(2.38-2.47)	6.55(6.37-6.73)	0.27[u]	2.32
obi2[d]	0.78	2.04(1.47-2.70)	2.71(2.45-2.99)	7.16(6.57-7.87)	0.27(0.26-0.28)	2.07
obi2[c]	1.24	1.36 (frozen)	2.41(2.36-2.49)	6.54(6.32-6.78)	0.27[u]	2.30

Key: u = unconstrained
a = 90% error for 1 interesting parameter
b = 90% error for 2 interesting parameters
c = 90% error for 3 interesting parameters
d = 90% error for 4 interesting parameters
f = Absorbed flux in $ergs/cm^2/sec$

the second exposure of the first epoch appeared more complex (χ^2_{DOF}=1.34 for 24 dof), and required an added component which we modelled as an absorption feature (a notch or an edge). Both models produced acceptable fits however, a notch improved the fits most significantly (at a significance of $\gg$99%).

The November 1992 data yielded good fits (χ^2_{DOF} <1.0) with a simple cut-off power law of photon index Γ ~2.6 and neutral column density N_H ~1.48×10^{20} atoms/cm^2. Adding a notch did not improve the fits significantly.

The column density and photon index (for simple power law fits) increased between epochs. The spectral differences are significant, as demonstrated by the disjoint χ^2 contours plotted in Figure 1. The spectrum steepened (softened) by $\delta\Gamma$ ~0.2 (and N_H rose by ~10%), while the flux dropped by ~45%. This correlation between spectral hardness and source intensity has been seen before in this and other BL Lac objects.

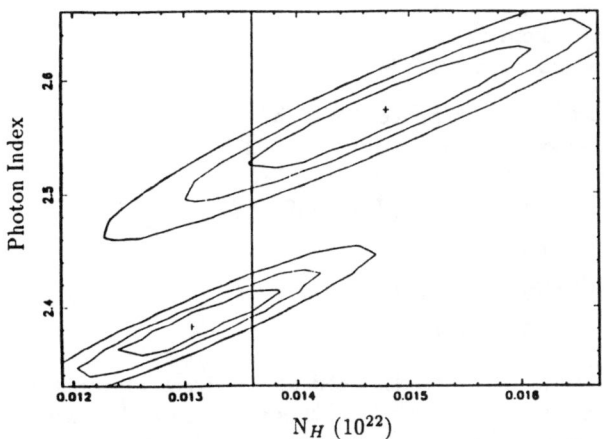

Figure 1. χ^2 contour levels of 68%, 90%, and 99% (for two interesting parameters) for the best power law fits to the summed spectra of the two epochs. The contours at the lower left are from the May 1991 observation, when the source was ~45% brighter. Galactic N_H is indicated by the verticle line.

3. Discussion

The residuals of a power law fit to our earlier data set for PKS2155 show a deficiency of measured counts from 0.4 to 0.5 keV, and an excess of counts from 0.7 to 1.0 keV. Residual systematic errors of this type occur in fits of newer data (taken after the gain switch of October 14, 1991) using the old matrix, pspcb_92mar11.rsp. The deficiency of counts at 0.4 keV in our spectra (~14%) is comparable in size to that seen in the 1992 calibration spectrum of MRK421 (~11%) (Turner J. and George, I., 1993).

Brinkmann (1993, in prep) has found that a change in gain as small as ~1% can cause noticeable changes in the spectral fit at or near the carbon edge for steep ($\Gamma \sim 2.7$) spectral sources like PKS2155. Given that the edge appears at an energy of 0.28 keV - the location of the carbon edge, and that the shape and size of the feature is similar to the feature found in the residuals of the MRK421 data, it is likely that we have detected a response matrix error in our early (A0) data. We find no other spectral features in our PSPC data for either epoch. The presence of a higher energy line cannot be ruled out, however, given the problems with temporal gain variations in the PSPC, detection of an x-ray feature such as the one found in the OGS and BBXRT data is problematic.

4. References

Bevington, P. 1969, "Data Reduction and Error Analysis for the Physical Sciences".
Brinkmann, 1993, in preparation. Bowyer, S. et al. 1984, ApJ Letters, 278, L103.
Canizares, C., and Kruper, J. 1984, ApJ Letters 278, L99.
Cash, 1979, Ap.J. 228, 939.
Hasinger, et al., 1992, Legacy No. 2.
Lampton, M., Margon, B., Bowyer, S., 1976, Ap.J. 208, 177.
Lockman and Savage, in prep.
Madejski, G., et al. 1991, in "Frontiers in Xray Astronomy", eds. Y. Tanaka and
    K. Koyama, Univ. Academy Press, Tokyo, p.583
Seward, F., and Wilkes, B. 1991, ROSAT-PROS Users Guide, V. 1.1
Shafer, R., Haberl, F., Arnaud, K., and Tennant, A. 1991, Xspec User's Guide,
    version 2, ESA TM-09.
Turner, T.J. and George, I. M., ed. 1993, Rosat PSPC Calibration Guide.

ROSAT OBSERVATIONS OF 3C390.3

Matthew Fremont[1,3], J.H. Beall[1,2,3],
B.N. Dorland[1], and W.A. Snyder[1]

[1]E.O. Hulburt Center for Space Research
Naval Research Laboratory, Washington, DC 20375.
[2]CSI, George Mason University, Fairfax, VA.
[3]St. John's College, Annapolis, MD 21404.

Email ID
beall@ssd0.nrl.navy.mil

ABSTRACT

ROSAT observed 3C 390.3 at various intervals between 18 March 91 through 30 March 91 using the PSPC instrument, providing a total exposure time of 11.5 ksec. We have constructed a background-subtracted light curve of the source for this interval. The light curve shows no significant variability on the time scales of 200 seconds to 10 days. We also present a spectral fit to these data and discuss a search for angular structure in 3C 390.3

Text

We present herein a preliminary analysis of the ROSAT observations of 3C390.3 taken at various intervals between 18 March 91 and 30 March 91. The ROSAT PSPC map of 3C 390.3 has been used to determine the source and background regions for obtaining a background-subtracted light curve. The background region is the region within a large ellipse centered on the source and within the central support ring of the mirror, but excluding the regions determined to be observed or suspected sources not associated with the AGN. The large background region thus obtained was used to accumulate good counting statistics in order to show the quasi-periodic nature of the background due to scattering of X-rays off of the earth's atmosphere near earth-occultation of the source (see Snowden and Freyberg 1992).

The positions of SIMBAD-listed sources within the field of view were also determined for this image. It should be noted that the SIMBAD position of 3C 390.3 differs from that given in the the NASA/IPAC Extragalactic Database (NED). As a consequence, the labeled position of 3C390.3 appears to differ from those of EQ 1945+794 and IRAS 1845+794E, both of which we understand to be the same object (EQ 1945+794 is a reference from a radio catalog (see Walter, 1989), and NED identifies the infrared and radio sources as the same object). Further, the quasar QSO 1845+797E is almost certainly the same object as 3C 390.3. NED lists no other objects in the immediate vicinity of 3C 390.3, and the the limited precision of the listed position of QSO 1845+797E is likely the cause of its apparent offset from 3C 390.3. However, this bears further investigation, as we shall show in the discussion on angular structure in the central source (see below).

The background obtained as outlined above shows significant variations during each orbit. These will be discussed in detail in a subsequent paper, but are due to scattered solar X-rays (Snowden and Freyberg 1992). We have therefore excluded these high count rate intervals from the data used to produce the background-subtracted light curves and the X-ray image. With this exclusion of less than 8 % of the data, 10,638 seconds of useful exposure remain. In this data set the background exhibits small orbit-to-orbit variations (< 10 %) and is generally either constant or can be modeled by a linear increase or decrease, typically less than 10 %, over the course of an observation interval.

The background-subtracted light curve thus obtained for 3C390.3 shows no statistically significant variability on time scales from a few hundred to $< 10^4$ s. From these data, a background-subtracted spectrum of 3C 390.3 has been obtained. The data were fit by an absorbed power-law with spectral index of 0.76 and absorption $NH = 10^{20.76}$ (fit as a free parameter). The reduced chi-squared for these parameters is approximately 1.

We now turn to a consideration of the time-integrated image of 3C 390.3 for the low-background portions of the observation. A perusal of the initial SAO images reveals an apparent small lobe extending northwest from the central point source. We investigate this "lobe" in some detail, since it is aligned with the well-known radio jet associated with this source.

We have selected the central portion of the IRAF image (pixels 240 through 280 in both the x and y axes) and displayed these using the ROSAT IDL AUL's Gouraud shading algorithm. The image shows some enhancement of the background region in the vicinity of the "lobe." A contour plot in σ-space with coutours drawn at $2.0, 2.5, 3.0, 4.0, 5.0, 6.0, 10.0, and 20\,\sigma$ also shows this apparent structure (see Figure 1).

Since these data represent the σ values of each pixel, it is likely that the summed significance will be $> 4 - 5\sigma$. It is therefore reasonable to fit two sources in this region of the image to determine a limit to the confidence level with which we can say that the "lobe" is an actual detection.

At the suggestion of W. Voges of MPI, we have investigated the arrival time of the photons associated with the "lobe," and have found no preferred arrival time. Presumably, a preferred arrival time for the photons in the "jet" would be consistent with the hypothesis that there is a one-time error in the pointing axis solution for the spacecraft. Additionally, the distance of the "lobe" from 3C390. 3 is approximately 3 times the FWHM of the detector response function. Finally, the energy bands we have used for this analysis are above 0.2 keV. The lobe is, therefore, unlikely to be due to "ghosting" of the low-energy ROSAT PSPC data (see Nousek and Lesser 1993).

We intend to continue our study of 3C390. 3 along several lines, including work to determine whether or not this artifact is in truth an linear X-ray structure (i.e., "jet"). First, we intend to model the angular distribution of pie-shaped segments of the data using the algorithm developed by G. Reichert (private communication), which is shortly to be incorporated in the IDL AUL. These investigations will be presented in a subsequent paper.

ACKNOWLEDGEMENTS

This research has made use of the SIMBAD Database, operated at CDS, Strasbourg, France, and the NASA/IPAC Extragalactic Database (NED) which is oper-

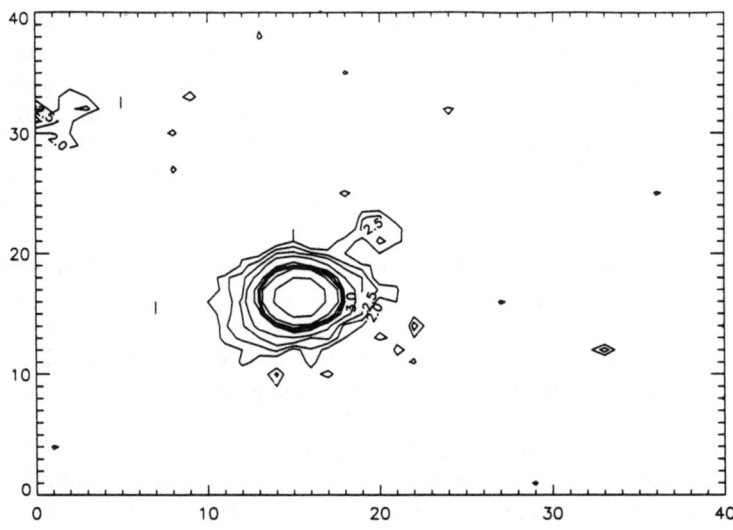

Figure 1: σ-space contour plot of central image of ROSAT 3C 390.3 observation. The contours show a marginally significant extension of the central source in the direction of the radio jet (see text).

ated by the Jet Propulsion Laboratory, CALTECH, under contract with the National Aeronautics and Space Administration, and the IDL Astronomical Users Library (W. Landsman).

The authors would also like to extend a special thanks to M. P. Kowalski of the E. O. Hulbert Center for Space Research, W. Voges of MPE/Garching, D. Worrall of CfA, and A. Wilson of STScI for many helpful discussions.

The present work was supported in part by the Office of Naval Research.

REFERENCES

Snowden, S.L. and Freyberg, M. J., The Scattered Solar X-ray Background of the *ROSAT* PSPC, 1992 Ap.J., 404:403-411.

Walter, H. G., A Celestial Reference Frame Based on Extragalactic Radio Sources, Astron. Astrophys. Suppl 1989, 79, 283-289.

Nousek, J.A., and Lesser, A., Electronic Ghost Images Around Soft ROSAT Point Sources, ROSAT Newsletter No.8, June 1993

List of Participants

THE 1st ANNUAL
ROSAT SCIENCE SYMPOSIUM
and DATA ANALYSIS WORKSHOP
University of Maryland
Conference Center
University College
College Park, Maryland

8 - 10 November 1993

List of Participants

Appleton, Philip N.	Iowa State University
Audley, Damian	NASA/GSFC
Barrett, Paul	USRA-NASA/GSFC
Beall, Jim	Naval Research Laboratory
Bechtold, Jill	Steward Observatory
Birkinshaw, Mark	Harvard Smithsonian/CFA
Blair, William P.	The Johns Hopkins University
Boldt, Elihu	NASA/GSFC
Borne, Kirk	STScI
Breen, Jeffrey Oliver	University of Virginia
Bregman, Joel N.	University of Michigan
Brown, Alexander	JILA
Brunner, Hermann	University of Tübingen
Buls, Fred	University of Georgia
Buote, David	MIT
Burns, Jack O.	New Mexico State University
Burrows, David	Penn State University
Caillault, J. P.	University of Georgia
Ceca, Robert Della	The Johns Hopkins University

List of Participants

Cecil, George	University of North Carolina
Chester, Margaret M.	Penn State University
Cheung, Cynthia	NASA/GSFC
Chu, You-Hua	University of Illinois
Conroy, Maureen	SAO
Corcoran, Michael	USRA-NASA/GSFC
Cox, Caroline	University of Michigan
Craig, Bill	LLNL
Cruddace, Ray	Naval Reseach Laboratory
Dahlem, Michael	STScI
Daines, Stuart	Harvard/Smithsonian-CFA
Daly, Ruth A.	Princeton University
Damiani, Francesco	Osservatorio Astronomico Palermo
Daniels, Julian	Johns Hopkins University/MSGC
Danner, Rudolf	California Institute of Technology
David, Larry	SAO
Davis, David	UMD-NASA/GSFC
DePonte, Janet	SAO
Dickel, John	University of Illinois
Donahue, Megan	STScI
Dow, Kimberly	SAO
Drake, Stephen A.	USRA-NASA/GSFC
Elvis, Martin	Harvard/Smithsonian CFA
Finley, David S.	Eureka Scientific, Inc.
Finley, John P.	Purdue University
Fiore, Fabrizio	Harvard/Smithsonian CFA
Fleming, Tom	University of Arizona

List of Participants 439

Forman, William	Harvard University
Freeman, Peter E.	University of Chicago
Gagné, Marc	University of Georgia
Geier, Sven	University of Maryland
George, Ian M.	USRA-NASA/GSFC
Gilmore, Diane M.	STScI
Gorenstein, Paul	Harvard/Smithsonian CFA
Green, Richard	Kitt Peak National Obs.
Griffiths, Richard E.	The Johns Hopkins University
Guo, Zhiyu	Penn State University
Gursky, Herbert	Naval Research Laboratory
Halpern, Jules	Columbia University
Harnden, Frank R.	Harvard/Smithsonian CFA
Hearty III, Thomas J.	University of Georgia
Heckman, Tim	The Johns Hopkins University
Henry, Dick	The Johns Hopkins University
Henry, Patrick	University of Hawaii
Hertz, Paul	Naval Research Laboratory
Holdridge, David	USRA
Holt, Steve	NASA/GSFC
Hughes, Jack P.	Harvard/Smithsonian CFA
Hwang, Una	MIT
Jones, Christine	Harvard/Smithsonian CFA
Kashyap, Vinay	University of Chicago
Kassim, Namir	Naval Research Laboratory
Kerr, Frank J.	University of Maryland/USRA
Kollgaard, Ron	Penn State University

List of Participants

Kowalski, Michael P.	NRL
Kronberg, Philipp	University of Toronto
Laor, Ari	Institute for Advanced Studies
Laurent-Muehlesein, Sally	Penn State University
Leighly, Karen	NASA/GSFC
Leiter, Darryl	NASA/GSFC
Levenson, Nancy	University of California-Berkeley
Long, Knox	STScI
Loewenstein, Michael	USRA-NASA/GSFC
Madejski, Grzegorz	USRA-NASA/GSFC
Maggio, Antonio	Osservatorio Astronomico Palermo
Marshall, Frank	NASA/GSFC
Martin, Crystal	Steward Observatory
Mattox, John	NASA/GSFC
Mendenhall, Jeffrey A.	Penn State University
Miyaji, Takamitsu	NASA/GSFC
Morse, Jon A.	STScI
Murphy, Rosa	University of Illinois at Urbana
Mushotzky, Richard	NASA/GSFC
Nguyen, Minhthuyen	Hughes STX
Nousek, John	Penn State University
Ogelman, Hakki	University of Wisconsin
Ormes, Jonathan F.	NASA/GSFC
Pesce, Joseph E.	STScI
Petre, Robert	NASA/GSFC
Pisarski, Rich	NASA/GSFC
Plucinsky, Paul P.	Harvard-Smithsonian/CFA

Reichert, Gail	USRA-NASA/GSFC
Rho, Jeonghee	NASA/GSFC
Rhode, Katherine	SAO
Riegler, Guenter	NASA Headquarters
Rohrbach, Gail	NASA/GSFC
Rosati, Piero	The Johns Hopkins University
Rose, Jim	University of North Carolina
Saba, Jack	Hughes STX
Saken, Jon M.	STScI
Sambruna, Rita M.	STScI
Sanders, Wilton	University of Wisconsin
Schlegel, Eric	USRA-NASA/GSFC
Schmitt, Jürgen	MPE
Schulman, Eric	University of Michigan
Schweizer, Francois	Carnegie
Sembay, Steven	University of Leicester
Serlemitsos, Peter	NASA/GSFC
Shaya, Edward	The University of Maryland
Siddiqui, Hassan I.	University of Leicester
Siemiginowska, Ameta	SAO
Slavin, Jonathan	NASA/GSFC
Smale, Karen	Hughes STX
Snowden, Steve	USRA-NASA/GSFC
Soong, Yang	USRA-NASA/GSFC
Sparks, William B.	STScI
Stahle, Caroline	NASA/GSFC
Stocke, John	CASA/University of Colorado

List of Participants

Swank, Jean	NASA/GSFC
Szkody, Paula	University of Washington
Takamitsu, Miyaji	NASA/GSFC
Treves, Aldo	SISSA
Trimble, Virginia	UMD/University of California
Trümper, Joachim	MPE
Turner, Tracey Jane	USRA-NASA/GSFC
Vallance, Robert J.	The University of Birmingham
Verter, Fran	USRA-NASA/GSFC
Voges, Wolfgang	MPE
Vrtilek, Saeqa Dil	University of Maryland/SAO
Wachter, Stefanie	University of Washington
Waldron, Wayne L.	Applied Research Corporation
Walter, Fred	SUNY
Walterbos, René	New Mexico State University
Wan, Lin	Princeton University
Wang, Q. Daniel	JILA/CASA
Watson, Michael	University of Leicester
White, Nick	NASA/GSFC
White III, Raymond E.	University of Alabama
White, Stephen	University of Maryland
Wilson, Andrew S.	STScI/University of Maryland
Winkler, Frank	Middlebury College
Worrall, Diana	SAO/CFA
Yusaf, Remana	NASA/GSFC
Zepka, Alex	Cornell University
Zhang, Weiping	USRA-NASA/GSFC

Zhang, Yun Fei	Harvard/Smithsonian-CFA
Zycki, Piotr	Copernicus Astronomical Center

Author Index

Author Index

Author Index

A

Andersen, V., 352
Angelini, L., 272
Appleton, P. N., 349
Arnaud, K. A., 375
Aschenbach, B., 3
Ayres, T. R., 36

B

Beall, J. H., 432
Bechtold, J., 105
Begelman, M., 59
Birkinshaw, M., 110, 378, 397
Blair, W. P., 315, 339
Bocchino, F., 309
Boldt, E., 216
Boyle, B. J., 426
Braun, R., 173
Breen, J. O., 366
Bregman, J. N., 164, 345
Brinkmann, W., 423
Brown, A., 36
Brunner, H., 211, 372
Burg, R., 260
Burns, J., 183
Burrows, D. N., 16, 223, 249, 299
Butcher, J., 383

C

Caillault, J.-P., 7
Caldwell, N., 380
Cecil, G., 358
Charles, P. A., 339
Chu, Y.-H., 154, 328, 342
Ciapi, A., 400
Cominsky, L. R., 80
Conroy, M. A., 233, 236
Corcoran, M. F., 159
Cordes, J. M., 51
Cutri, R., 105

D

Dahlem, M., 333
Daines, S., 369
Daly, R. A., 386, 392
Damiani, F., 269
David, L. P., 178
Day, C., 159
DeChristopher, B. M., 315
Della Ceca, R., 426
DePonte, J., 236
De Pree, C., 358
Dickel, J. R., 328
Doe, S., 183
Done, C., 59
Donahue, M., 75, 375
Dorland, B. N., 432
Drake, S. A., 272

E

Edgar, R. J., 3
Edge, A., 383
Egger, R., 3
Elvis, M., 105, 121, 412

F

Fabbiano, G., 349
Feigelson, E. D., 415, 418, 420, 423
Filippenko, A. V., 85
Finley, J. P., 41
Fiore, F., 59, 95, 105, 121, 412
Fomalont, E. B., 418
Forman, W., 129, 178, 369
Fossati, G., 400
Freeman, P. E., 294
Fremont, M., 432

G

Gabuzda, D. C., 420

Geier, S., 275
Georgantopoulus, I., 426
Giacconi, R., 260
Gilmore, D. M., 429
Gordon, S. M., 339
Gorenstein, P., 205
Goscha, D. L., 328
Graham, J. R., 322
Griffiths, R. E., 426
Guo, Z., 299

H

Harnden, Jr., F. R., 239, 269
Heckman, T. M., 139
Hertz, P., 80, 325
Hester, J. J., 318, 322
Hughes, J. P., 144, 205, 378

J

Jedrzejewski, R. I., 389
Johnstone, R., 383
Jones, C., 129, 178, 369
Jones, L. V., 392

K

Kashyap, V., 239
Kassim, N. E., 325
Kollgaard, R. I., 415, 418, 420, 423
Kowalski, M. P., 363

L

Lamb, D. Q., 294
Laor, A., 121
Laurent-Muehleisen, S. A., 418, 423
Lawrence, C. R., 409
Leiter, D., 216
Leonardi, A., 380
Levenson, N. A., 322
Lilly, S. J., 386, 392
Linsky, J. L., 36

Loewenstein, M., 355
Loken, C., 183
Long, K. S., 312, 315, 339
Lundgren, S. C., 51
Ly, Y., 80

M

Macchetto, F., 389
Madejski, G., 59
Maggio, A., 244, 309
Maraschi, L., 400, 415
Marcum, P. M., 349
Marscher, A. P., 403, 406
Mathur, S., 412
McCammon, D., 3, 223
McDowell, J. C., 105, 121, 412
Mendenhall, J. A., 16, 223, 249
Metcalf, T. H., 294
Micela, G., 239, 269
Morris, S. L., 375
Morse, J. A., 252, 258
Murphy, R., 328, 342
Mushotzky, R. F., 59, 85

N

Norman, C. A., 173

O

Owen, F., 183

P

Padovani, P., 415
Pearson, T. J., 409
Perlman, E. S., 75, 415
Petre, R., 159, 318, 320, 322, 355
Pinknew, J., 183
Plucinsky, P. P., 3
Pravdo, S. H., 272

R

Rawley, G., 159
Raychaudhury, S., 366
Raymond, J. C., 322
Readhead, A. C. S., 409
Reichert, G. A., 85
Rho, J.-H., 318, 320
Rieke, M., 105
Roettiger, K., 183
Rosati, P., 260
Rose, J. A., 380
Rosner, R., 239

S

Saken, J. M., 315
Sambruna, R. M., 415, 420
Sanders, W. T., 299
Schachter, J., 75
Schlegel, E. M., 195, 336
Schmitt, J. H. M. M., 24, 159
Schulman, E., 345
Sciortino, S., 239, 244, 269, 309
Serlemitsos, P., 59
Shanks, T., 426
Siddiqui, H., 383
Siemiginowska, A., 105, 412
Sikora, M., 59
Smale, A. P., 272
Snowden, S. L., 3, 223
Snyder, W. A., 432
Sokoloski, J. L., 386
Spinrad, H., 423
Staubert, R., 372
Steakley, M. F., 173
Stewart, G. C., 383, 426
Stocke, J. T., 75, 375, 415
Swank, J., 159
Szkody, P., 291

T

Tagliaferri, G., 400
Treves, A., 400, 415
Tucker, W. H., 205

Turner, T. J., 59
Tyson, A., 369

U

Urry, C. M., 415, 420, 429

V

Van Dyk, S. D., 325
Voges, W., 183

W

Waldron, W. L., 279
Walter, F. M., 263, 282, 287
Walterbos, R. A. M., 173
Wan, L., 392
Wang, Q. D., 75, 173, 301
Wasserman, I. M., 51
Watson, M. G., 64
Weiler, K. W., 325
Weimer, S., 372
Wendker, H. J., 275
Westphal, H., 372
White, N. E., 272
White, R., 183
White III, R. E., 352
Wilkes, B. J., 105, 121, 412
Williamson, C., 352
Wilson, A. S., 115, 358
Winkler, P. F., 312, 315
Winrich, C., 349
Wood, K. S., 80
Worrall, D. M., 110, 397, 409, 415

Z

Zepka, A. F., 51
Zhang, Y. F., 403, 406
Zoonematkermani, S., 287

Source Index

Index

Specific Sources Index

Page numbers refer to the FIRST page of the article.

Each source is listed under a category which should describe its most significant property.

AGN - listed individually under: BL Lacs, Galaxies-Radio, GigaHertz-Peaked Sources, LINERS, Quasars, Seyfert I galaxies, Seyfert II galaxies

Black hole candidates
- MS0317.7-6647   75

BL Lac objects (Blazars)
- 0048-097   415
- 0118-272   415
- 0426-380   415
- 0537-441   415
- 0735+178   415
- 0814+425   415
- 0820+225   415
- 0828+493   415
- 0851+202   415
- 0954+658   415
- 1144-379   415
- 1147+245   415
- 1308+326   415
- 1514-241   415
- 1519-273   415
- 1538+149   415
- 1652+398   415
- 1749+096   415
- 1749+701   415
- 1803+784   415
- 1823+568   415
- 2005-489   415
- 2007+777   415
- 2200+420   415
- Mrk 421   429
- Mrk 501   420
- OJ 287   420
- PKS 1034-293   400
- PKS 1335-127   400
- PKS 2155-304   429

Cataclysmic Variables
- 0132-65   64
- 0153-59   64
- 0203+29   64
- 0453-42   64
- 0501-03   64
- 0515+0104   287
- 0531-46   64
- 1002-19   64
- 1007-20   64
- 1015+09   64
- 1149+28   64
- 1307+53   64
- 1313-32   64
- 1844-74   64
- 1938-46   64
- 2107-05   64
- 2316-05   64
- H0542-407   294
- PG 1346+082   294
- RE0751+14   64
- RX J1940.2-1025   59
- S193   291
- TT Boo   291

453

BY Cam 64
YY Dra 291
AH Eri 291
UZ For 64
AM Her 64
Nova Her 1991 = V838 Her 291
EX Hya 64
V426 Oph 294
SS UMi 291
HV Vir 291
Clusters and groups (galaxy and super)
0016+1609 378, 392
0024+1654 392
0147.8-3941 392
0302+1658 392
0302.5+1717 392
0303+17 392
0418.3-3844 392
1130+34 386, 392
1245+34 386, 392
1333.3+1725 392
1512.4+3647 392
1621.5+2640 392
2053.7-0449 392
3C19 392
3C215 392
3C244.1 392
3C268.3 392
3C295 392
3C330 392
53W076 386, 392
53W080 386, 392
A85 178
A168 183
A262 178
A370 369, 392
A478 178
A514 183
A539 178
A665 378
A895 392
A908 392
A913 392
A963 369
A1367 418
A1569 183
A1589 183
A1689 369
A1795 178, 372
A2029 178
A2142 383
A2147 183
A2199 383
A2589 178
A3634 183
A3558 366
A3571 366
Centaurus 389
Coma 380
DC2048-52 380
Eridanus 352
HC 62 178
MS0451-03 375
MS0735+74 375
MS2137 369
N79-299A 183
NGC 3607 355
NGC 5044 group 129, 178
Perseus 363
PKS 0116+08 392
S49-132 183
Shapley 366
Clusters (stellar)

Cep OB3   154
Chamaeleon I   7
Hyades   7
IC 2391   7
IC 2602   7
L1495E (Tau)   7
ρ Oph   7
NGC 2264   7
NGC 6475   7
Orion   7
α Per   7
Pleiades   7, 239
Praesepe   7
Sco-Cen OB   301
Tau-Aur   7, 269
Trapezium   7, 275
Diffuse emission
  η Car   159
  B361   16
  Cepheus   16
  Chamaeleon I   16
  Coalsack   16
  Draco enhancement   249
  Eridanus   299
  L1495   16
  L1551   16
  Lockman Hole   260
  MBM 12   16
  Mon OB1   16
  Mon OB2   16
  Monogem Ring   3, 16, 41
  R CrA   16
Galaxies, Dwarf
  IC 2608   349
  UGC 5927   349
Galaxies, Elliptical
  M87   129, 418

NGC 1399   129
NGC 1400   352
NGC 1407   352
NGC 3607   355
NGC 3608   355
NGC 4472   129
NGC 4636   178
NGC 4696   389
NGC 4697   355
NGC 6251   397
Galaxies, Interacting
  Arp 270 = NGC 3395/3396   349
  NGC 1792   333
  NGC 1808   333
  NGC 3395/3396 = Arp 270   349
  NGC 4038/4039   349
Galaxies, Spiral
  IC 342   164
  M31   129
  M33   154, 164, 339, 345
  M51   129
  M82   164
  M101   154, 164, 342
  NGC 253   164
  NGC 604   154
  NGC 891   164
  NGC 1313   75
  NGC 2146   164
  NGC 3079   164
  NGC 3628   164
  NGC 4244   164
  NGC 4565   164
  NGC 4631   164, 173
  NGC 5194   129
  NGC 5529   164

NGC 5907   164
NGC 6946   164, 336
Galaxies, Miscellaneous
  LMC   129, 154, 164
Galaxies, Starburst
  D94   380
  M82   139
  NGC 253   139
  NGC 1569   139
GigaHertz-Peaked Sources   403, 406
HII regions
  NGC 2359   154
  NGC 3199   154
  NGC 5447   342
  NGC 5455   342
  NGC 5461   342
  NGC 5462   342
  NGC 5471   342
  NGC 6164/65   154
  NGC 6888   154
  NGC 7635   154
  RCW 58   154
  S155   154
  S308   154
LINERS
  M81   85
  MCG-6-30-15   85
  NGC 2639   85
  NGC 3079   85
  NGC 3998   85
  NGC 4579   85
  NGC 5506   85
  Pictor A   85
Molecular clouds (see also Diffuse Emission)
  R CrA   16, 301

Nebulae
  0540-693   144, 154
  30 Dor   154, 342
  η Car   159
  Carina nebula   154
  Cas A   144
  Coalsack   16
  E0102.2-72   144
  G292.0+1.8   144
  Honeycomb nebula   154
  N44   342
  N132   144
  Orion   275
  R136   154
  R140   154
  Rosette Nebula   154
Pulsars
  0437-47   51
  0531+21 (Crab)   41
  0540-69   41
  0630+18 (Geminga)   41, 51
  0631+10   51
  0656+14   41
  0833-45   41
  1055-52   41
  1259-63   51
  1509-58   41
  1800-21   41
  1706-44.0   41
  1823-13   41
  1843+20   51
  1905+04   51
  1908+04   51
  1929+10   51
  1929+53   41
  1951+32   41
  1957+20   51

Source Index 457

2224+65 51
2259+586 320
Geminga 41, 51
Vela pulsar 41
Radio galaxies
  1130+34 386
  1245+34 386
  3C 220.3 409
  3C 280 409
  3C 465 183
  4C 35.03 110
  53W076 386
  Cen A 418
  Cyg A 392
  Fornax A 418
  NGC 315 110
  NGC 326 110
  NGC 1316 418
  NGC 2484 110
  NGC 4065 183
  NGC 4061 183
  NGC 4261 110
  NGC 6251 110
  NGC 7503 183
Quasars
  0000-263 105
  0026+346 423
  0207-389 105
  0406+121 423
  0420-388 105
  0438-436 105
  0500+019 423
  1114+445 95, 121
  1115+407 121
  1211+143 95
  1216+069 121
  1226+023 121
  1244+026 95
  1309+355 121
  1322+659 121
  1352+183 121
  1415+451 121
  1512+370 121
  1543+489 121
  1746+642 423
  1747+653 423
  1815+680 423
  1845+797E 432
  1946+768 105
  2126-158 105
  3C273 95
  3C390.3 432
  4C39.25 406
  53W080 386
  EQ 1945+794 432
  Mkn 110 95
  Mkn 205 95
  MS0002.5-4205 426
  MS0007.1-0231 426
  MS0116.3-0116 426
  MS1111.8-3754 426
  NAB0205+025 95
  OQ 208 403
  PKS 0528+134 403
  QSO 1845+794E 432
  Ton 1542 95
Seyfert galaxies
  Mkn 3 115
  NGC 1068 115
  NGC 2110 115
  NGC 3516 115, 252, 268
  NGC 4151 115
  NGC 4258 (M106) 115, 358
  NGC 6814 59

Seyfert I   252, 268
Shadows
　B361   16
　Cepheus   16
　Chamaeleon I   16
　Coalsack   16
　L1495   16
　L1551   16
　MBM 12   16
　Mon OB1   16
　Mon OB2   16
　R CrA   16
Stars
　α Aqr   36
　β Aqr   36
　γ Aql   36
　Algol (β Per)   24, 272
　η Car   159
　RZ Cas   272
　T Cen   80
　U Cep   272
　UV Cet   24
　α CrB   24
　γ Dra   36
　TW Dra   272
　λ Eri   24
　α Gem   24
　YY Gem   24
　θ Her   36
　α Hya   36
　β Ind   36
　AR Lac   24
　θ^1C Ori   7
　ζ Ori   24
　DY Peg   80
　β Per (Algol)   24, 272
　ζ Pup   24

RR Sco   80
DH Tau   269
DI Tau   269
GH Tau   269
GI Tau   269
GK Tau   269
HL Tau   269
XZ Tau   269
V807 Tau   269
V826 Tau   269
V827 Tau   269
V830 Tau   269
α TrA   24, 36
CF Tuc   24
HD 93129   159
HD 93162   159
HD 93204   159
HD 93205   159
HD 93403   159
HZ 43   24, 252
LHS 3003   24
Star-forming regions
　Orion   154, 282
　Orion OB 1   299
　Trapezium   275
Supernovae
　SN 1978K   195
　SN 1986J   195
　SN 1987A   205
　SN 1993J   195
Supernova Remnants
　0540-69   41, 144
　3C400.2   315
　Cas A   144
　CTB 80   41
　CTB 109   320
　Cygnus Loop   322

E0102.2-72   144
G8.7-0.1   41
G18.9-1.1   325
G109.1-1.0   320
G292.0+1.8   144
G326.3-1.8   325
IC 443   318
Monogem Ring   41
N132D   144
SN 1006   312
W30   41
W49B   328
Vela SNR   41, 309
X-ray binaries, high-mass
   MS0317.7-6647   75
X-ray binaries, low-mass
   EXO 0748-676   80

Subject Index

Index

Index to ROSAT Science Symposium Proceedings
Numbers refer to the FIRST page of the article.

A

α_{ox} 105
Absorption
    circumstellar 275
    galactic 397
    intrinsic 105, 275, 279, 397
    shadows 16
Active Galactic Nuclei - see individual entries for the following sources:
    BL Lacs
    GigaHertz Peaked Sources
    Radio galaxies
    Seyfert I
    Seyfert II
    Starburst Galaxies
    contributors to X-ray background 216
    variability 432
Accretion disk sources 64, 121, 211
Algol binaries 272
Aspect solutions and issues 252

B

Background (X-ray) - see X-ray Background
Beamed emission 403, 406, 420
Binary stars 272, 294
Black holes 75, 216

C

BL Lacs (blazars) 110, 400, 415, 420, 429
    radio-selected 415

Cataclysmic Variables 59, 64, 287, 291, 294
    period distribution of 64
    polars 64
Circumstellar matter 272, 336
Clusters, galaxy
    cooling flows 363, 389
    cosmological implications 178
    distant 375, 386, 392
    evolution or merging 375, 378, 380, 392
    lensing 369
    profiles (temperature, mass) 363, 372, 383
    supercluster 366
Clusters, stellar (and associations) 7, 282
Compton scattering 418, 420, 423
Cooling Flow 129, 372, 375, 383, 389
Correlations 121
Cosmology 178

D

Dark matter 178, 352, 372, 383
Data analysis
    exposure maps 223, 249
    extended sources 144, 223, 249, 258
    methods 244, 263, 294
    particle background 223, 249

scattered solar X-rays 223, 249
spectral fitting 244
source detection 239, 260
time filters 236
Diffuse emission
   analysis of 144, 249
   extragalactic 173, 339
   in Milky Way 3, 16, 159, 275, 299
Distance estimators 325

E
Elliptical galaxies - see Galaxies, elliptical
Emission bands
   1/4 keV 16
   3/4 keV 16
Entrainment (AGN) 358
Evolution
   cluster 380, 386, 420
   orbital (binaries) 80
   stellar 7
Excess emission 64, 85
Extended emission 397

F
Fe II lines 121

G
Galactic fountains 164
Galaxies, elliptical 129, 352, 397
Galaxies, S0 355
Galaxies, spiral 164, 173, 336, 339, 342, 345
Galaxy groups - see Clusters, galaxy
Galaxy interactions 333, 349
Galaxy ISM 115, 336
Gas
   hot ionized 333
GigaHertz Peaked Sources 403, 406, 423

H
H I 397
H II regions (all sizes) 154, 342
Haloes, galaxy 173
High-mass X-ray Binary - see X-ray binaries, high mass
HRI (ROSAT) 252
   point spread function 252, 258

I
Intracluster medium 183, 375, 386, 409, 420
Intragroup gas 355
ISM, 301
   bubbles 16, 275
   of galaxies 139, 345, 355
   H II complexes 154
   stars 205

J
Jets 358, 403, 406, 420, 432

L
Lensing 369
LINERS 85
Low-mass X-ray binary - see X-ray binaries, low mass

M
Magnetic fields 64, 418
Mergers (Cluster) 389
Models
   non-equilibrium ionization (SNR) 309, 318, 320

radioactive decay 195
reverse shock 195
Synchrotron Self-Compton 403, 406, 420
Molecular clouds 301

N
Nebulae - see Diffuse Emission
Neutron star(s) 75
Non-equilibrium ionization model (SNR) 309

O
Outflows 173

P
Polarization of
   quasars 400
Profiles
   mass 372
   temperature 372
PSPC signature in data
   exposure maps 223, 249
   spectra 244
   wobble 80
Pulsars
   X-ray 51, 75
   isolated, radio 41

Q
Quasars 105, 121, 211
   high-redshift 105
   low-redshift 95, 412
   optically quiet 423
   polarization 400
   radio-loud 105, 409
   radio-quiet 105, 412
   survey 211, 426

R
Radio galaxies 110, 183, 397, 409, 418, 420, 423
Relativistic beaming 110, 409
RS CVn stars 272

S
Sedov phase (SNR) 325
Seyfert Galaxies 59, 115, 252, 258
   Seyfert I 252, 258
Shadows 16, 75, 301
Shocks, AGN 358
Shocks, SNR 144, 154, 216, 309, 312, 322, 328
Software
   IDL 223, 239, 263
   PROS 233, 236, 239
   XSPEC 244
Source counts 239
Spectral features
   absorption 423, 429
   emission 121
   Fe II 121
   Hβ 121
   [O III] 121
   soft excess 64, 95, 412
Spiral Galaxies - see Galaxies, spiral
Starburst galaxies 139, 164, 173
Stars
   activity 7, 269
   chromospheres 36
   coronae 7, 36, 244, 272
   early type 7, 24, 275
   late type 7, 24
   luminosity functions 7
   neutron 7
   O type 159, 279, 349

Pre-main sequence 7, 269, 275, 282
rotation 7, 269
RS CVn type 272
T Tau type (weak, classical) 7, 269, 275
winds 154, 299
WR type 349
X-ray emission 7
Star formation
galaxy interactions 333
regions of 7, 282
Statistics 80, 239, 294
Substructure (cluster) 183, 363, 380
Sunyaev-Zel'dovich effect 378
Superbubbles 3, 154, 164
Superluminal motion 406
Supernovae 129, 195, 205
Supernovae Remnants 3, 144, 205, 309, 312, 315, 318, 320, 322, 325, 328, 336, 339
Superwinds 164
Synchrotron radiation 418

T
Timing 59, 80
Bayesian analysis 294
Transition region (stellar) 36

V
Variability
X-ray 7, 59, 121, 403, 406, 432

W
White dwarfs 24
Wide-Area Tail sources 183
Winds, wind-blown bubbles 154, 299

X
X-ray background 216, 301, 426
X-ray binaries
low mass 80
X-ray emission
non-thermal 110, 397, 409
X-ray excess, soft 64, 95, 406, 412

AIP Conference Proceedings

		L.C. Number	ISBN
No. 43	Particles and Fields – 1977 (APS/DPF, Argonne, IL)	78-55683	0-88318-142-8
No. 44	Future Trends in Superconductive Electronics (Charlottesville, 1978)	77-9240	0-88318-143-6
No. 45	New Results in High Energy Physics – 1978 (Vanderbilt Conference)	78-67196	0-88318-144-4
No. 46	Topics in Nonlinear Dynamics (La Jolla Institute)	78-57870	0-88318-145-2
No. 47	Clustering Aspects of Nuclear Structure and Nuclear Reactions (Winnipeg, 1978)	78-64942	0-88318-146-0
No. 48	Current Trends in the Theory of Fields (Tallahassee, FL, 1978)	78-72948	0-88318-147-9
No. 49	Cosmic Rays and Particle Physics – 1978 (Bartol Conference)	79-50489	0-88318-148-7
No. 50	Laser-Solid Interactions and Laser Processing – 1978 (Boston, MA)	79-51564	0-88318-149-5
No. 51	High Energy Physics with Polarized Beams and Polarized Targets (Argonne, IL, 1978)	79-64565	0-88318-150-9
No. 52	Long-Distance Neutrino Detection – 1978 (C. L. Cowan Memorial Symposium)	79-52078	0-88318-151-7
No. 53	Modulated Structures – 1979 (Kailua Kona, Hawaii)	79-53846	0-88318-152-5
No. 54	Meson-Nuclear Physics – 1979 (Houston, TX)	79-53978	0-88318-153-3
No. 55	Quantum Chromodynamics (La Jolla, CA, 1978)	79-54969	0-88318-154-1
No. 56	Particle Acceleration Mechanisms in Astrophysics (La Jolla, CA, 1979)	79-55844	0-88318-155-X
No. 57	Nonlinear Dynamics and the Beam-Beam Interaction (Brookhaven, NY, 1979)	79-57341	0-88318-156-8
No. 58	Inhomogeneous Superconductors – 1979 (Berkeley Springs, WV)	79-57620	0-88318-157-6
No. 59	Particles and Fields – 1979 (APS/DPF Montreal)	80-66631	0-88318-158-4
No. 60	History of the ZGS (Argonne, IL, 1979)	80-67694	0-88318-159-2
No. 61	Aspects of the Kinetics and Dynamics of Surface Reactions (La Jolla Institute, 1979)	80-68004	0-88318-160-6
No. 62	High Energy e^+e^- Interactions (Vanderbilt, 1980)	80-53377	0-88318-161-4
No. 63	Supernovae Spectra (La Jolla, CA, 1980)	80-70019	0-88318-162-2

No. 64	Laboratory EXAFS Facilities – 1980 (Univ. of Washington)	80-70579	0-88318-163-0
No. 65	Optics in Four Dimensions – 1980 (ICO, Ensenada)	80-70771	0-88318-164-9
No. 66	Physics in the Automotive Industry – 1980 (APS/AAPT Topical Conference)	80-70987	0-88318-165-7
No. 67	Experimental Meson Spectroscopy – 1980 (Sixth International Conference, Brookhaven, NY)	80-71123	0-88318-166-5
No. 68	High Energy Physics – 1980 (XX International Conference, Madison, WI)	81-65032	0-88318-167-3
No. 69	Polarization Phenomena in Nuclear Physics – 1980 (Fifth International Symposium, Santa Fe, NM)	81-65107	0-88318-168-1
No. 70	Chemistry and Physics of Coal Utilization – 1980 (APS, Morgantown)	81-65106	0-88318-169-X
No. 71	Group Theory and its Applications in Physics – 1980 (Latin American School of Physics, Mexico City)	81-66132	0-88318-170-3
No. 72	Weak Interactions as a Probe of Unification (Virginia Polytechnic Institute – 1980)	81-67184	0-88318-171-1
No. 73	Tetrahedrally Bonded Amorphous Semiconductors (Carefree, AZ, 1981)	81-67419	0-88318-172-X
No. 74	Perturbative Quantum Chromodynamics (Tallahassee, FL, 1981)	81-70372	0-88318-173-8
No. 75	Low Energy X-Ray Diagnostics – 1981 (Monterey, CA)	81-69841	0-88318-174-6
No. 76	Nonlinear Properties of Internal Waves (La Jolla Institute, 1981)	81-71062	0-88318-175-4
No. 77	Gamma Ray Transients and Related Astrophysical Phenomena (La Jolla Institute, 1981)	81-71543	0-88318-176-2
No. 78	Shock Waves in Condensed Matter – 1981 (Menlo Park, NJ)	82-70014	0-88318-177-0
No. 79	Pion Production and Absorption in Nuclei – 1981 (Indiana University Cyclotron Facility)	82-70678	0-88318-178-9
No. 80	Polarized Proton Ion Sources (Ann Arbor, MI, 1981)	82-71025	0-88318-179-7
No. 81	Particles and Fields – 1981: Testing the Standard Model (APS/DPF, Santa Cruz, CA)	82-71156	0-88318-180-0
No. 82	Interpretation of Climate and Photochemical Models, Ozone and Temperature Measurements (La Jolla Institute, 1981)	82-71345	0-88318-181-9
No. 83	The Galactic Center (Cal. Inst. of Tech., 1982)	82-71635	0-88318-182-7
No. 84	Physics in the Steel Industry (APS/AISI, Lehigh University, 1981)	82-72033	0-88318-183-5

No. 85	Proton-Antiproton Collider Physics – 1981 (Madison, WI)	82-72141	0-88318-184-3
No. 86	Momentum Wave Functions – 1982 (Adelaide, Australia)	82-72375	0-88318-185-1
No. 87	Physics of High Energy Particle Accelerators (Fermilab Summer School, 1981)	82-72421	0-88318-186-X
No. 88	Mathematical Methods in Hydrodynamics and Integrability in Dynamical Systems (La Jolla Institute, 1981)	82-72462	0-88318-187-8
No. 89	Neutron Scattering – 1981 (Argonne National Laboratory)	82-73094	0-88318-188-6
No. 90	Laser Techniques for Extreme Ultraviolet Spectroscopy (Boulder, CO, 1982)	82-73205	0-88318-189-4
No. 91	Laser Acceleration of Particles (Los Alamos, NM, 1982)	82-73361	0-88318-190-8
No. 92	The State of Particle Accelerators and High Energy Physics (Fermilab, 1981)	82-73861	0-88318-191-6
No. 93	Novel Results in Particle Physics (Vanderbilt, 1982)	82-73954	0-88318-192-4
No. 94	X-Ray and Atomic Inner-Shell Physics – 1982 (International Conference, U. of Oregon)	82-74075	0-88318-193-2
No. 95	High Energy Spin Physics – 1982 (Brookhaven National Laboratory)	83-70154	0-88318-194-0
No. 96	Science Underground (Los Alamos, NM, 1982)	83-70377	0-88318-195-9
No. 97	The Interaction Between Medium Energy Nucleons in Nuclei – 1982 (Indiana University)	83-70649	0-88318-196-7
No. 98	Particles and Fields – 1982 (APS/DPF University of Maryland)	83-70807	0-88318-197-5
No. 99	Neutrino Mass and Gauge Structure of Weak Interactions (Telemark, 1982)	83-71072	0-88318-198-3
No. 100	Excimer Lasers – 1983 (OSA, Lake Tahoe, NV)	83-71437	0-88318-199-1
No. 101	Positron-Electron Pairs in Astrophysics (Goddard Space Flight Center, 1983)	83-71926	0-88318-200-9
No. 102	Intense Medium Energy Sources of Strangeness (UC-Santa Cruz, CA, 1983)	83-72261	0-88318-201-7
No. 103	Quantum Fluids and Solids – 1983 (Sanibel Island, FL)	83-72440	0-88318-202-5
No. 104	Physics, Technology and the Nuclear Arms Race (APS, Baltimore, MD, 1983)	83-72533	0-88318-203-3
No. 105	Physics of High Energy Particle Accelerators (SLAC Summer School, 1982)	83-72986	0-88318-304-8

No. 106	Predictability of Fluid Motions (La Jolla Institute, 1983)	83-73641	0-88318-305-6
No. 107	Physics and Chemistry of Porous Media (Schlumberger-Doll Research, 1983)	83-73640	0-88318-306-4
No. 108	The Time Projection Chamber (TRIUMF, Vancouver, 1983)	83-83445	0-88318-307-2
No. 109	Random Walks and Their Applications in the Physical and Biological Sciences (NBS/La Jolla Institute, 1982)	84-70208	0-88318-308-0
No. 110	Hadron Substructure in Nuclear Physics (Indiana University, 1983)	84-70165	0-88318-309-9
No. 111	Production and Neutralization of Negative Ions and Beams (3rd Int'l Symposium) (Brookhaven, NY, 1983)	84-70379	0-88318-310-2
No. 112	Particles and Fields – 1983 (APS/DPF, Blacksburg, VA)	84-70378	0-88318-311-0
No. 113	Experimental Meson Spectroscopy – 1983 (7th International Conference, Brookhaven, NY)	84-70910	0-88318-312-9
No. 114	Low Energy Tests of Conservation Laws in Particle Physics (Blacksburg, VA, 1983)	84-71157	0-88318-313-7
No. 115	High Energy Transients in Astrophysics (Santa Cruz, CA, 1983)	84-71205	0-88318-314-5
No. 116	Problems in Unification and Supergravity (La Jolla Institute, 1983)	84-71246	0-88318-315-3
No. 117	Polarized Proton Ion Sources (TRIUMF, Vancouver, 1983)	84-71235	0-88318-316-1
No. 118	Free Electron Generation of Extreme Ultraviolet Coherent Radiation (Brookhaven/OSA, 1983)	84-71539	0-88318-317-X
No. 119	Laser Techniques in the Extreme Ultraviolet (OSA, Boulder, CO, 1984)	84-72128	0-88318-318-8
No. 120	Optical Effects in Amorphous Semiconductors (Snowbird, UT, 1984)	84-72419	0-88318-319-6
No. 121	High Energy e^+e^- Interactions (Vanderbilt, 1984)	84-72632	0-88318-320-X
No. 122	The Physics of VLSI (Xerox, Palo Alto, CA, 1984)	84-72729	0-88318-321-8
No. 123	Intersections Between Particle and Nuclear Physics (Steamboat Springs, CO, 1984)	84-72790	0-88318-322-6
No. 124	Neutron-Nucleus Collisions: A Probe of Nuclear Structure (Burr Oak State Park, 1984)	84-73216	0-88318-323-4
No. 125	Capture Gamma-Ray Spectroscopy and Related Topics – 1984 (Int'l Symposium, Knoxville, TN)	84-73303	0-88318-324-2

No. 126	Solar Neutrinos and Neutrino Astronomy (Homestake, 1984)	84-63143	0-88318-325-0
No. 127	Physics of High Energy Particle Accelerators (BNL/SUNY Summer School, 1983)	85-70057	0-88318-326-9
No. 128	Nuclear Physics with Stored, Cooled Beams (McCormick's Creek State Park, IN, 1984)	85-71167	0-88318-327-7
No. 129	Radiofrequency Plasma Heating (Sixth Topical Conference) (Callaway Gardens, GA, 1985)	85-48027	0-88318-328-5
No. 130	Laser Acceleration of Particles (Malibu, CA, 1985)	85-48028	0-88318-329-3
No. 131	Workshop on Polarized ^{3}He Beams and Targets (Princeton, NJ, 1984)	85-48026	0-88318-330-7
No. 132	Hadron Spectroscopy – 1985 (International Conference, Univ. of Maryland)	85-72537	0-88318-331-5
No. 133	Hadronic Probes and Nuclear Interactions (Arizona State University, 1985)	85-72638	0-88318-332-3
No. 134	The State of High Energy Physics (BNL/SUNY Summer School, 1983)	85-73170	0-88318-333-1
No. 135	Energy Sources: Conservation and Renewables (APS, Washington, DC, 1985)	85-73019	0-88318-334-X
No. 136	Atomic Theory Workshop on Relativistic and QED Effects in Heavy Atoms (Gaithersburg, MD, 1985)	85-73790	0-88318-335-8
No. 137	Polymer-Flow Interaction (La Jolla Institute, 1985)	85-73915	0-88318-336-6
No. 138	Frontiers in Electronic Materials and Processing (Houston, TX, 1985)	86-70108	0-88318-337-4
No. 139	High-Current, High-Brightness, and High-Duty Factor Ion Injectors (La Jolla Institute, 1985)	86-70245	0-88318-338-2
No. 140	Boron-Rich Solids (Albuquerque, NM, 1985)	86-70246	0-88318-339-0
No. 141	Gamma-Ray Bursts (Stanford, CA, 1984)	86-70761	0-88318-340-4
No. 142	Nuclear Structure at High Spin, Excitation, and Momentum Transfer (Indiana University, 1985)	86-70837	0-88318-341-2
No. 143	Mexican School of Particles and Fields (Oaxtepec, México, 1984)	86-81187	0-88318-342-0
No. 144	Magnetospheric Phenomena in Astrophysics (Los Alamos, NM, 1984)	86-71149	0-88318-343-9
No. 145	Polarized Beams at SSC & Polarized Antiprotons (Ann Arbor, MI & Bodega Bay, CA, 1985)	86-71343	0-88318-344-7
No. 146	Advances in Laser Science—I (Dallas, TX, 1985)	86-71536	0-88318-345-5

No. 147	Short Wavelength Coherent Radiation: Generation and Applications (Monterey, CA, 1986)	86-71674	0-88318-346-3
No. 148	Space Colonization: Technology and The Liberal Arts (Geneva, NY, 1985)	86-71675	0-88318-347-1
No. 149	Physics and Chemistry of Protective Coatings (Universal City, CA, 1985)	86-72019	0-88318-348-X
No. 150	Intersections Between Particle and Nuclear Physics (Lake Louise, Canada, 1986)	86-72018	0-88318-349-8
No. 151	Neural Networks for Computing (Snowbird, UT, 1986)	86-72481	0-88318-351-X
No. 152	Heavy Ion Inertial Fusion (Washington, DC, 1986)	86-73185	0-88318-352-8
No. 153	Physics of Particle Accelerators (SLAC Summer School, 1985) (Fermilab Summer School, 1984)	87-70103	0-88318-353-6
No. 154	Physics and Chemistry of Porous Media—II (Ridgefield, CT, 1986)	83-73640	0-88318-354-4
No. 155	The Galactic Center: Proceedings of the Symposium Honoring C. H. Townes (Berkeley, CA, 1986)	86-73186	0-88318-355-2
No. 156	Advanced Accelerator Concepts (Madison, WI, 1986)	87-70635	0-88318-358-0
No. 157	Stability of Amorphous Silicon Alloy Materials and Devices (Palo Alto, CA, 1987)	87-70990	0-88318-359-9
No. 158	Production and Neutralization of Negative Ions and Beams (Brookhaven, NY, 1986)	87-71695	0-88318-358-7
No. 159	Applications of Radio-Frequency Power to Plasma: Seventh Topical Conference (Kissimmee, FL, 1987)	87-71812	0-88318-359-5
No. 160	Advances in Laser Science—II (Seattle, WA, 1986)	87-71962	0-88318-360-9
No. 161	Electron Scattering in Nuclear and Particle Science: In Commemoration of the 35th Anniversary of the Lyman-Hanson-Scott Experiment (Urbana, IL, 1986)	87-72403	0-88318-361-7
No. 162	Few-Body Systems and Multiparticle Dynamics (Crystal City, VA, 1987)	87-72594	0-88318-362-5
No. 163	Pion–Nucleus Physics: Future Directions and New Facilities at LAMPF (Los Alamos, NM, 1987)	87-72961	0-88318-363-3
No. 164	Nuclei Far from Stability: Fifth International Conference (Rosseau Lake, ON, 1987)	87-73214	0-88318-364-1

No. 165	Thin Film Processing and Characterization of High-Temperature Superconductors (Anaheim, CA, 1987)	87-73420	0-88318-365-X
No. 166	Photovoltaic Safety (Denver, CO, 1988)	88-42854	0-88318-366-8
No. 167	Deposition and Growth: Limits for Microelectronics (Anaheim, CA, 1987)	88-71432	0-88318-367-6
No. 168	Atomic Processes in Plasmas (Santa Fe, NM, 1987)	88-71273	0-88318-368-4
No. 169	Modern Physics in America: A Michelson-Morley Centennial Symposium (Cleveland, OH, 1987)	88-71348	0-88318-369-2
No. 170	Nuclear Spectroscopy of Astrophysical Sources (Washington, DC, 1987)	88-71625	0-88318-370-6
No. 171	Vacuum Design of Advanced and Compact Synchrotron Light Sources (Upton, NY, 1988)	88-71824	0-88318-371-4
No. 172	Advances in Laser Science—III: Proceedings of the International Laser Science Conference (Atlantic City, NJ, 1987)	88-71879	0-88318-372-2
No. 173	Cooperative Networks in Physics Education (Oaxtepec, Mexico, 1987)	88-72091	0-88318-373-0
No. 174	Radio Wave Scattering in the Interstellar Medium (San Diego, CA, 1988)	88-72092	0-88318-374-9
No. 175	Non-neutral Plasma Physics (Washington, DC, 1988)	88-72275	0-88318-375-7
No. 176	Intersections Between Particle and Nuclear Physics (Third International Conference) (Rockport, ME, 1988)	88-62535	0-88318-376-5
No. 177	Linear Accelerator and Beam Optics Codes (La Jolla, CA, 1988)	88-46074	0-88318-377-3
No. 178	Nuclear Arms Technologies in the 1990s (Washington, DC, 1988)	88-83262	0-88318-378-1
No. 179	The Michelson Era in American Science: 1870–1930 (Cleveland, OH, 1987)	88-83369	0-88318-379-X
No. 180	Frontiers in Science: International Symposium (Urbana, IL, 1987)	88-83526	0-88318-380-3
No. 181	Muon-Catalyzed Fusion (Sanibel Island, FL, 1988)	88-83636	0-88318-381-1
No. 182	High T_c Superconducting Thin Films, Devices, and Applications (Atlanta, GA, 1988)	88-03947	0-88318-382-X
No. 183	Cosmic Abundances of Matter (Minneapolis, MN, 1988)	89-80147	0-88318-383-8
No. 184	Physics of Particle Accelerators (Ithaca, NY, 1988)	89-83575	0-88318-384-6

No. 185	Glueballs, Hybrids, and Exotic Hadrons (Upton, NY, 1988)	89-83513	0-88318-385-4
No. 186	High-Energy Radiation Background in Space (Sanibel Island, FL, 1987)	89-83833	0-88318-386-2
No. 187	High-Energy Spin Physics (Minneapolis, MN, 1988)	89-83948	0-88318-387-0
No. 188	International Symposium on Electron Beam Ion Sources and their Applications (Upton, NY, 1988)	89-84343	0-88318-388-9
No. 189	Relativistic, Quantum Electrodynamic, and Weak Interaction Effects in Atoms (Santa Barbara, CA, 1988)	89-84431	0-88318-389-7
No. 190	Radio-frequency Power in Plasmas (Irvine, CA, 1989)	89-45805	0-88318-397-8
No. 191	Advances in Laser Science—IV (Atlanta, GA, 1988)	89-85595	0-88318-391-9
No. 192	Vacuum Mechatronics (First International Workshop) (Santa Barbara, CA, 1989)	89-45905	0-88318-394-3
No. 193	Advanced Accelerator Concepts (Lake Arrowhead, CA, 1989)	89-45914	0-88318-393-5
No. 194	Quantum Fluids and Solids—1989 (Gainesville, FL, 1989)	89-81079	0-88318-395-1
No. 195	Dense Z-Pinches (Laguna Beach, CA, 1989)	89-46212	0-88318-396-X
No. 196	Heavy Quark Physics (Ithaca, NY, 1989)	89-81583	0-88318-644-6
No. 197	Drops and Bubbles (Monterey, CA, 1988)	89-46360	0-88318-392-7
No. 198	Astrophysics in Antarctica (Newark, DE, 1989)	89-46421	0-88318-398-6
No. 199	Surface Conditioning of Vacuum Systems (Los Angeles, CA, 1989)	89-82542	0-88318-756-6
No. 200	High T_c Superconducting Thin Films: Processing, Characterization, and Applications (Boston, MA, 1989)	90-80006	0-88318-759-0
No. 201	QED Structure Functions (Ann Arbor, MI, 1989)	90-80229	0-88318-671-3
No. 202	NASA Workshop on Physics From a Lunar Base (Stanford, CA, 1989)	90-55073	0-88318-646-2
No. 203	Particle Astrophysics: The NASA Cosmic Ray Program for the 1990s and Beyond (Greenbelt, MD, 1989)	90-55077	0-88318-763-9
No. 204	Aspects of Electron-Molecule Scattering and Photoionization (New Haven, CT, 1989)	90-55175	0-88318-764-7

No. 205	The Physics of Electronic and Atomic Collisions (XVI International Conference) (New York, NY, 1989)	90-53183	0-88318-390-0
No. 206	Atomic Processes in Plasmas (Gaithersburg, MD, 1989)	90-55265	0-88318-769-8
No. 207	Astrophysics from the Moon (Annapolis, MD, 1990)	90-55582	0-88318-770-1
No. 208	Current Topics in Shock Waves (Bethlehem, PA, 1989)	90-55617	0-88318-776-0
No. 209	Computing for High Luminosity and High Intensity Facilities (Santa Fe, NM, 1990)	90-55634	0-88318-786-8
No. 210	Production and Neutralization of Negative Ions and Beams (Brookhaven, NY, 1990)	90-55316	0-88318-786-8
No. 211	High-Energy Astrophysics in the 21st Century (Taos, NM, 1989)	90-55644	0-88318-803-1
No. 212	Accelerator Instrumentation (Brookhaven, NY, 1989)	90-55838	0-88318-645-4
No. 213	Frontiers in Condensed Matter Theory (New York, NY, 1989)	90-6421	0-88318-771-X 0-88318-772-8 (pbk.)
No. 214	Beam Dynamics Issues of High-Luminosity Asymmetric Collider Rings (Berkeley, CA, 1990)	90-55857	0-88318-767-1
No. 215	X-Ray and Inner-Shell Processes (Knoxville, TN, 1990)	90-84700	0-88318-790-6
No. 216	Spectral Line Shapes, Vol. 6 (Austin, TX, 1990)	90-06278	0-88318-791-4
No. 217	Space Nuclear Power Systems (Albuquerque, NM, 1991)	90-56220	0-88318-838-4
No. 218	Positron Beams for Solids and Surfaces (London, Canada, 1990)	90-56407	0-88318-842-2
No. 219	Superconductivity and Its Applications (Buffalo, NY, 1990)	91-55020	0-88318-835-X
No. 220	High Energy Gamma-Ray Astronomy (Ann Arbor, MI, 1990)	91-70876	0-88318-812-0
No. 221	Particle Production Near Threshold (Nashville, IN, 1990)	91-55134	0-88318-829-5
No. 222	After the First Three Minutes (College Park, MD, 1990)	91-55214	0-88318-828-7
No. 223	Polarized Collider Workshop (University Park, PA, 1990)	91-71303	0-88318-826-0
No. 224	LAMPF Workshop on (π, K) Physics (Los Alamos, NM, 1990)	91-71304	0-88318-825-2

No. 225	Half Collision Resonance Phenomena in Molecules (Caracas, Venezuela, 1990)	91-55210	0-88318-840-6
No. 226	The Living Cell in Four Dimensions (Gif sur Yvette, France, 1990)	91-55209	0-88318-794-9
No. 227	Advanced Processing and Characterization Technologies (Clearwater, FL, 1991)	91-55194	0-88318-910-0
No. 228	Anomalous Nuclear Effects in Deuterium/Solid Systems (Provo, UT, 1990)	91-55245	0-88318-833-3
No. 229	Accelerator Instrumentation (Batavia, IL, 1990)	91-55347	0-88318-832-1
No. 230	Nonlinear Dynamics and Particle Acceleration (Tsukuba, Japan, 1990)	91-55348	0-88318-824-4
No. 231	Boron-Rich Solids (Albuquerque, NM, 1990)	91-53024	0-88318-793-4
No. 232	Gamma-Ray Line Astrophysics (Paris-Saclay, France, 1990)	91-55492	0-88318-875-9
No. 233	Atomic Physics 12 (Ann Arbor, MI, 1990)	91-55595	088318-811-2
No. 234	Amorphous Silicon Materials and Solar Cells (Denver, CO, 1991)	91-55575	088318-831-7
No. 235	Physics and Chemistry of MCT and Novel IR Detector Materials (San Francisco, CA, 1990)	91-55493	0-88318-931-3
No. 236	Vacuum Design of Synchrotron Light Sources (Argonne, IL, 1990)	91-55527	0-88318-873-2
No. 237	Kent M. Terwilliger Memorial Symposium (Ann Arbor, MI, 1989)	91-55576	0-88318-788-4
No. 238	Capture Gamma-Ray Spectroscopy (Pacific Grove, CA, 1990)	91-57923	0-88318-830-9
No. 239	Advances in Biomolecular Simulations (Obernai, France, 1991)	91-58106	0-88318-940-2
No. 240	Joint Soviet-American Workshop on the Physics of Semiconductor Lasers (Leningrad, USSR, 1991)	91-58537	0-88318-936-4
No. 241	Scanned Probe Microscopy (Santa Barbara, CA, 1991)	91-76758	0-88318-816-3
No. 242	Strong, Weak, and Electromagnetic Interactions in Nuclei, Atoms, and Astrophysics: A Workshop in Honor of Stewart D. Bloom's Retirement (Livermore, CA, 1991)	91-76876	0-88318-943-7
No. 243	Intersections Between Particle and Nuclear Physics (Tucson, AZ, 1991)	91-77580	0-88318-950-X

No. 244	Radio Frequency Power in Plasmas (Charleston, SC, 1991)	91-77853	0-88318-937-2
No. 245	Basic Space Science (Bangalore, India, 1991)	91-78379	0-88318-951-8
No. 246	Space Nuclear Power Systems (Albuquerque, NM, 1992)	91-58793	1-56396-027-3 1-56396-026-5 (pbk.)
No. 247	Global Warming: Physics and Facts (Washington, DC, 1991)	91-78423	0-88318-932-1
No. 248	Computer-Aided Statistical Physics (Taipei, Taiwan, 1991)	91-78378	0-88318-942-9
No. 249	The Physics of Particle Accelerators (Upton, NY, 1989, 1990)	92-52843	0-88318-789-2
No. 250	Towards a Unified Picture of Nuclear Dynamics (Nikko, Japan, 1991)	92-70143	0-88318-951-8
No. 251	Superconductivity and its Applications (Buffalo, NY, 1991)	92-52726	1-56396-016-8
No. 252	Accelerator Instrumentation (Newport News, VA, 1991)	92-70356	0-88318-934-8
No. 253	High-Brightness Beams for Advanced Accelerator Applications (College Park, MD, 1991)	92-52705	0-88318-947-X
No. 254	Testing the AGN Paradigm (College Park, MD, 1991)	92-52780	1-56396-009-5
No. 255	Advanced Beam Dynamics Workshop on Effects of Errors in Accelerators, Their Diagnosis and Corrections (Corpus Christi, TX, 1991)	92-52842	1-56396-006-0
No. 256	Slow Dynamics in Condensed Matter (Fukuoka, Japan, 1991)	92-53120	0-88318-938-0
No. 257	Atomic Processes in Plasmas (Portland, ME, 1991)	91-08105	0-88318-939-9
No. 258	Synchrotron Radiation and Dynamic Phenomena (Grenoble, France, 1991)	92-53790	1-56396-008-7
No. 259	Future Directions in Nuclear Physics with 4π Gamma Detection Systems of the New Generation (Strasbourg, France, 1991)	92-53222	0-88318-952-6
No. 260	Computational Quantum Physics (Nashville, TN, 1991)	92-71777	0-88318-933-X
No. 261	Rare and Exclusive B&K Decays and Novel Flavor Factories (Santa Monica, CA, 1991)	92-71873	1-56396-055-9

No. 262	Molecular Electronics—Science and Technology (St. Thomas, Virgin Islands, 1991)	92-72210	1-56396-041-9
No. 263	Stress-Induced Phenomena in Metallization: First International Workshop (Ithaca, NY, 1991)	92-72292	1-56396-082-6
No. 264	Particle Acceleration in Cosmic Plasmas (Newark, DE, 1991)	92-73316	0-88318-948-8
No. 265	Gamma-Ray Bursts (Huntsville, AL, 1991)	92-73456	1-56396-018-4
No. 266	Group Theory in Physics (Cocoyoc, Morelos, Mexico, 1991)	92-73457	1-56396-101-6
No. 267	Electromechanical Coupling of the Solar Atmosphere (Capri, Italy, 1991)	92-82717	1-56396-110-5
No. 268	Photovoltaic Advanced Research & Development Project (Denver, CO, 1992)	92-74159	1-56396-056-7
No. 269	CEBAF 1992 Summer Workshop (Newport News, VA, 1992)	92-75403	1-56396-067-2
No. 270	Time Reversal—The Arthur Rich Memorial Symposium (Ann Arbor, MI, 1991)	92-83852	1-56396-105-9
No. 271	Tenth Symposium Space Nuclear Power and Propulsion (Vols. I–III) (Albuquerque, NM, 1993)	92-75162	1-56396-137-7 (set)
No. 272	Proceedings of the XXVI International Conference on High Energy Physics (Vols. I and II) (Dallas, TX, 1992)	93-70412	1-56396-127-X (set)
No. 273	Superconductivity and Its Applications (Buffalo, NY, 1992)	93-70502	1-56396-189-X
No. 274	VIth International Conference on the Physics of Highly Charged Ions (Manhattan, KS, 1992)	93-70577	1-56396-102-4
No. 275	Atomic Physics 13 (Munich, Germany, 1992)	93-70826	1-56396-057-5
No. 276	Very High Energy Cosmic-Ray Interactions: VIIth International Symposium (Ann Arbor, MI, 1992)	93-71342	1-56396-038-9
No. 277	The World at Risk: Natural Hazards and Climate Change (Cambridge, MA, 1992)	93-71333	1-56396-066-4

No. 278	Back to the Galaxy (College Park, MD, 1992)	93-71543	1-56396-227-6
No. 279	Advanced Accelerator Concepts (Port Jefferson, NY, 1992)	93-71773	1-56396-191-1
No. 280	Compton Gamma-Ray Observatory (St. Louis, MO, 1992)	93-71830	1-56396-104-0
No. 281	Accelerator Instrumentation Fourth Annual Workshop (Berkeley, CA, 1992)	93-072110	1-56396-190-3
No. 282	Quantum 1/f Noise & Other Low Frequency Fluctuations in Electronic Devices (St. Louis, MO, 1992)	93-072366	1-56396-252-7
No. 283	Earth and Space Science Information Systems (Pasadena, CA, 1992)	93-072360	1-56396-094-X
No. 284	US-Japan Workshop on Ion Temperature Gradient-Driven Turbulent Transport (Austin, TX, 1993)	93-72460	1-56396-221-7
No. 285	Noise in Physical Systems and 1/f Fluctuations (St. Louis, MO, 1993)	93-72575	1-56396-270-5
No. 286	Ordering Disorder: Prospect and Retrospect in Condensed Matter Physics: Proceedings of the Indo-U.S. Workshop (Hyderabad, India, 1993)	93-072549	1-56396-255-1
No. 287	Production and Neutralization of Negative Ions and Beams: Sixth International Symposium (Upton, NY, 1992)	93-72821	1-56396-103-2
No. 288	Laser Ablation: Mechanismas and Applications-II: Second International Conference (Knoxville, TN, 1993)	93-73040	1-56396-226-8
No. 289	Radio Frequency Power in Plasmas: Tenth Topical Conference (Boston, MA, 1993)	93-72964	1-56396-264-0
No. 290	Laser Spectroscopy: XIth International Conference (Hot Springs, VA, 1993)	93-73050	1-56396-262-4
No. 291	Prairie View Summer Science Academy (Prairie View, TX, 1992)	93-73081	1-56396-133-4
No. 292	Stability of Particle Motion in Storage Rings (Upton, NY, 1992)	93-73534	1-56396-225-X
No. 293	Polarized Ion Sources and Polarized Gas Targets (Madison, WI, 1993)	93-74102	1-56396-220-9
No. 294	High-Energy Solar Phenomena A New Era of Spacecraft Measurements (Waterville Valley, NH, 1993)	93-74147	1-56396-291-8

No. 295	The Physics of Electronic and Atomic Collisions: XVIII International Conference (Aarhus, Denmark, 1993)	93-74103	1-56396-290-X
No. 296	The Chaos Paradigm: Developments an Applications in Engineering and Science (Mystic, CT, 1993)	93-74146	1-56396-254-3
No. 297	Computational Accelerator Physics (Los Alamos, NM, 1993)	93-74205	1-56396-222-5
No. 298	Ultrafast Reaction Dynamics and Solvent Effects (Royaumont, France, 1993)	93-074354	1-56396-280-2
No. 299	Dense Z-Pinches: Third International Conference (London, 1993)	93-074569	1-56396-297-7
No. 300	Discovery of Weak Neutral Currents: The Weak Interaction Before and After (Santa Monica, CA, 1993)	94-70515	1-56396-306-X
No. 301	Eleventh Symposium Space Nuclear Power and Propulsion (3 Vols.) (Albuquerque, NM, 1994)	92-75162	1-56396-305-1 (Set) 156396-301-9 (pbk. set)
No. 302	Lepton and Photon Interactions/ XVI International Symposium (Ithaca, NY, 1993)	94-70079	1-56396-106-7
No. 303	Slow Positron Beam Techniques for Solids and Surfaces Fifth International Workshop (Jackson Hole, WY 1992)	94-71036	1-56396-267-5
No. 304	The Second Compton Symposium (College Park, MD, 1993)	94-70742	1-56396-261-6
No. 305	Stress-Induced Phenomena in Metallization Second International Workshop (Austin, TX, 1993)	94-70650	1-56396-251-9
No. 306	12th NREL Photovoltaic Program Review (Denver, CO, 1993)	94-70748	1-56396-315-9
No. 307	Gamma-Ray Bursts Second Workshop (Huntsville, AL 1993)	94-71317	1-56396-336-1
No. 308	The Evolution of X-Ray Binaries (College Park, MD 1993)	94-76853	1-56396-329-9
No. 309	High-Pressure Science and Technology—1993 (Colorado Springs, CO 1993)	93-72821	1-56396-219-5 (Set)
No. 310	Analysis of Interplanetary Dust (Houston, TX 1993)	94-71292	1-56396-341-8